Introduction to the Finite Element Method

To my children Kristian, Signe, Tobias and
to my beloved Jannie, who makes life a gift

Niels

Introduction to the Finite Element Method

Niels Saabye Ottosen and Hans Petersson
University of Lund, Sweden

Harlow, England • London • New York • Boston • San Francisco • Toronto • Sydney • Dubai • Singapore • Hong Kong
Tokyo • Seoul • Taipei • New Delhi • Cape Town • São Paulo • Mexico City • Madrid • Amsterdam • Munich • Paris • Milan

Pearson Education Limited
Edinburgh Gate
Harlow
Essex CM20 2JE
England

and Associated Companies throughout the world

Visit us on the World Wide Web at:
http://www.pearsoned.co.uk

First published 1992 by
Prentice Hall

© Prentice Hall Europe 1992

Typeset in 10/12pt Times
by P & R Typesetters Ltd., Salisbury, Wilts

Library of Congress Cataloging-in-Publication Data

Ottosen, Niels Saabye.
    Introduction to the finite element method/Niels Saabye Ottosen
and Hans Petersson.
        p.   cm.
    Includes bibliographical references and index.
    ISBN 0-13-473877-2
    1. Finite element method.   I. Petersson, Hans.   II. Title.
TA347.F5088   1992
620'.001'51535—dc20   91-41612
                                                                    CIP

British Library Cataloguing in Publication Data

Ottosen, Niels Saabye
    Introduction to the finite element method.
    I. Title II. Petersson, Hans
    515

ISBN-13: 978-0-13-473877-2

Contents

Preface

This textbook is intended to be used as an introductory text for students in various fields of engineering. It presents a joint effort by the Divisions of Structural Mechanics and Solid Mechanics at the University of Lund to combine their experiences in the teaching of the finite element method.

The text deals with the formulation of the finite element method and the intention is that the reader should be able to formulate the finite element approach for arbitrary differential equations. To facilitate this approach the weak formulation of the differential equations in question is used in combination with the Galerkin method.

To achieve this objective, and in order to be able to evaluate the physical importance of the different terms appearing in the equations, a profound understanding of the physical significance of the various differential equations discussed is essential. Therefore, emphasis is given to a detailed derivation of the differential equations encountered. In this manner, the book serves partly as an introduction to the various differential equations typically occurring within engineering mechanics and partly as an introduction to the finite element formulation of these equations. By this two-fold approach it is hoped that the reader will be left with a firm grasp of many of those problems that he or she may encounter in his or her later professional career.

The background required by the student is modest, as only an introductory knowledge of mechanics, matrix algebra, and differential and integral calculus is required. All differential equations of the physical problems dealt with are derived, and the exposition is self-contained with detailed derivations so that the book may also be used for individual studies.

Most textbooks within this field are concerned either with the derivation of the equations of mechanics or the formulation of the finite element method itself. We believe that the present text represents a modern trend in teaching engineering disciplines, since all the differential equations dealt with are derived before the corresponding finite element formulation is established. While this approach may be inconvenient for students of mathematics, it is likely to be very helpful for engineering students. In particular, a clear perception of the physical assumptions behind the different theories is favoured by this approach.

It is also hoped that the mixture of problems covered will enable the student both to obtain an overview of some important parts of engineering mechanics and to appreciate the wide field of applications for the finite element method.

The equations are numbered according to the chapter in which they appear, for instance (3.27). The same system applies for references to figures and tables. References to the literature are collected in alphabetical order at the end of the book. When a definition is made or a concept is explained for the first time, *italic* lettering will be used.

Suggestions to the Instructor

Different possible course contents are outlined below:

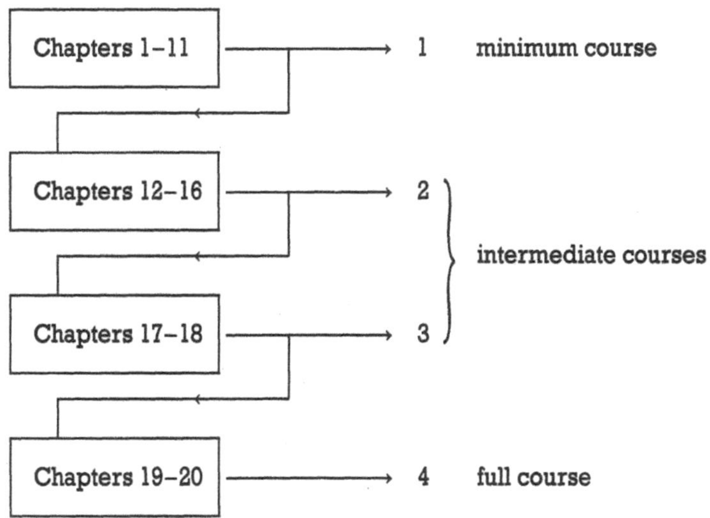

The minimum course proposed for an introduction to the finite element method is indicated by route 1 above. It only considers problems where the unknown function is a scalar, for instance the temperature in heat flow problems.

In addition, topics in solid mechanics are treated in the intermediate course outlines 2 and 3. Here, Chapter 14 on Saint-Venant torsion problems may be omitted without any adverse effect on the following chapters. However, Saint-Venant torsion provides an interesting example of a solid mechanics problem that is analogous to the two-dimensional heat flow problem. Course route 3 includes the finite element formulations of beams and plates that are of great practical importance.

Finally, the full course indicated by route 4 also treats isoparametric finite elements and numerical integration, and it should be noted that Chapters 19 and 20 may be combined directly with the minimum course.

As an indication of the amount of time required, course proposals 3 and 4 may require 3–6 effective working weeks depending on the extent of exercises and projects set by the instructor.

Acknowledgements

This textbook has grown out of years of experience of teaching the finite element method. Many people have therefore influenced the course material in a direct or indirect manner. In particular, we would like to express our sincere gratitude to K.-G. Olsson, who read the entire manuscript and provided a number of valuable suggestions and improvements. The comments given by Dr B. Bodelind and P.-E. Austrell are also appreciated. Moreover, we are sincerely grateful to P. Nilsson and B. Zadig for their accurate and expedient typing and drawing, respectively, which were executed within very tight time limits.

1

Introduction

1.1 Basic description

All the physical phenomena encountered in engineering mechanics are modelled by differential equations, and usually the problem addressed is too complicated to be solved by classical analytical methods. The *finite element method* is a *numerical* approach by which general differential equations can be solved in an *approximate* manner (cf. Figure 1.1).

The differential equation or equations, which describe the physical problem considered, are assumed to hold over a certain region. This region may be one-, two- or three-dimensional. It is a characteristic feature of the finite element method that instead of seeking approximations that hold directly over the entire region, the region is divided into smaller parts, so-called *finite elements*, and the approximation is then carried out over each element. For instance, even though the variable varies in a highly non-linear manner over the entire region, it may be a fair approximation to assume that the variable varies in a linear fashion over each element. The collection of all elements is called a *finite element mesh* (cf. Figure 1.2).

When the type of approximation which is to be applied over each element has been selected, the corresponding behaviour of each element can then be determined. This can be performed because the approximation made over each element is fairly simple. Having determined the behaviour of all elements, these elements are then patched together, using some specific rules, to form the entire region, which eventually enables us to obtain an approximate solution for the behaviour of the entire body.

The finite element (FE) method can be applied to obtain approximate solutions for arbitrary differential equations. In this textbook we shall only be concerned with *boundary value problems*, i.e. differential equations where certain information is known *a priori* for the unknowns at the boundary. For simplicity, only linear boundary problems will be considered. However, even *initial value problems* can be solved by the FE method. Initial value problems are typical for transient phenomena like wave propagation and transient heat conduction.

As the FE method is a numerical means of solving general differential equations, it can be applied to various physical phenomena. In order to emphasize this aspect,

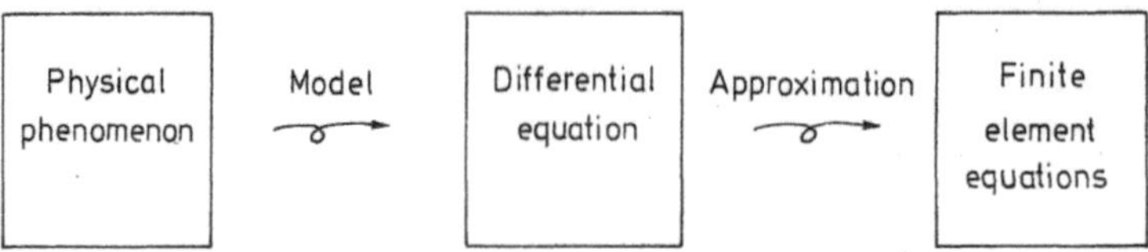

Figure 1.1 Steps in engineering mechanics analysis

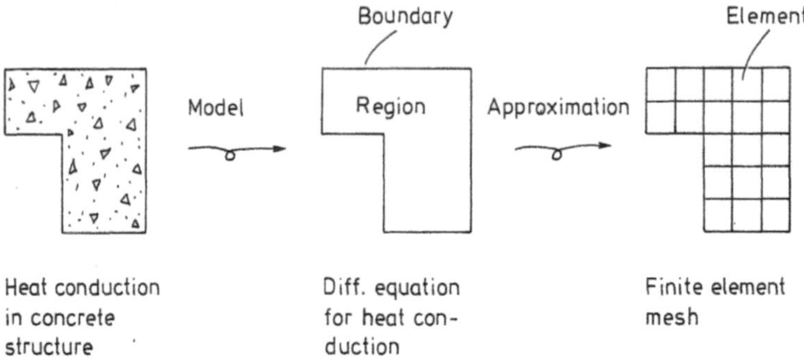

Figure 1.2 Illustration of modelling steps

we shall be concerned here with the FE formulation of such diverse problems as heat conduction, torsion of elastic shafts, diffusion, groundwater flow, and the elastic behaviour of one-, two- and three-dimensional bodies, including beam and plate analysis.

As previously mentioned, it is a characteristic feature of the FE method that the region, i.e. the body, is divided into smaller parts, i.e. the elements, for which a rather simple approximation is adopted. This approximation is usually a polynomial. The approximation over each element means that an approximation is adopted for how the variable changes over the element. This approximation is, in fact, some kind of interpolation over the element, where it is assumed that the variable is known at certain points in the element. These points are called *nodal points* and they are often located at the boundary of each element. The precise manner in which the variable changes between its values at the nodal points is expressed by the specific approximation, which may be linear, quadratic, cubic, etc.

In order to be more specific about this approximation technique, consider the true temperature distribution $T(x)$ along the one-dimensional fin shown in Figure 1.3. Assume that the temperature is known at the five positions shown in Figure 1.4. These positions correspond to the nodal points and need not be equally spaced.

Suppose now that the fin is divided into four elements over which the temperature is assumed to vary linearly between the nodal points. This situation is shown in Figure 1.5(a) and the resulting approximate temperature distribution along the fin is illustrated in Figure 1.5(b).

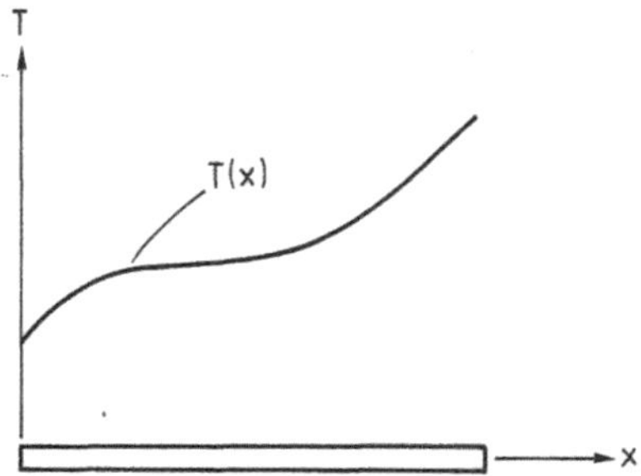

Figure 1.3   Temperature distribution along one-dimensional fin

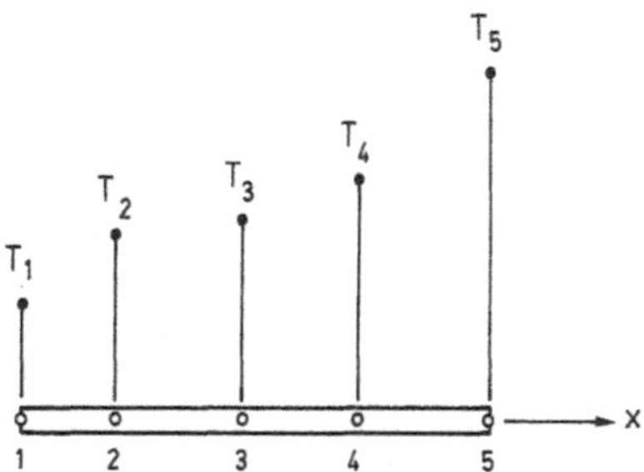

Figure 1.4   Nodal points and temperature values at the nodal points

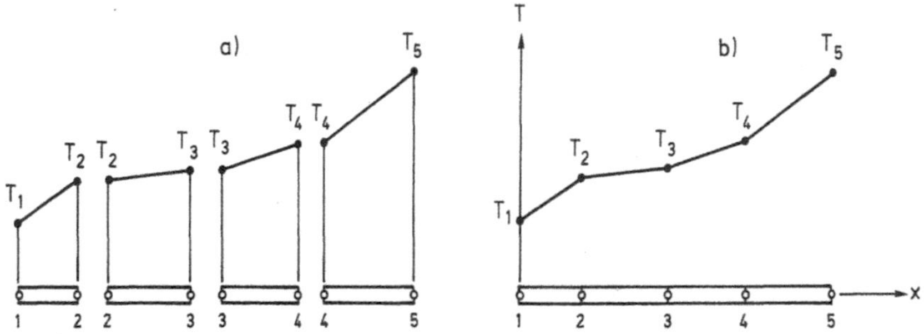

Figure 1.5   (a) Four elements with linear temperature variation within each element; (b) resulting approximate temperature distribution along the fin

Now, assume instead that the fin is divided into two elements and that within each element a quadratic variation is adopted. This situation is shown in Figure 1.6(a) and the resulting approximate temperature distribution along the fin then takes the form illustrated in Figure 1.6(b).

We observe that irrespective of whether the linear or quadratic approximation is used, the approximate temperature distribution along the fin is known once the

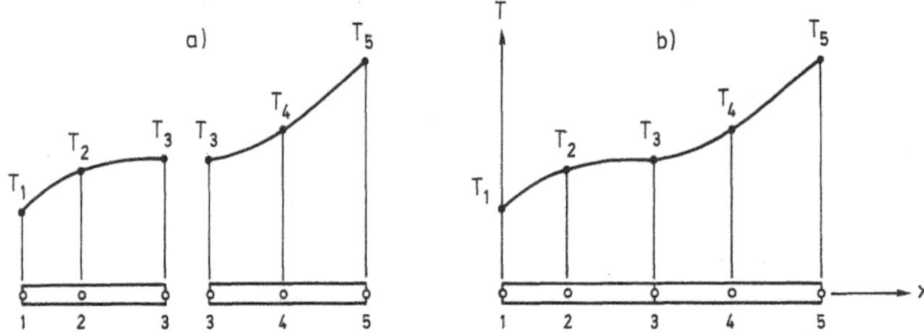

Figure 1.6 (a) Two elements with quadratic temperature variation within each element; (b) resulting approximate temperature distribution along the fin

temperatures at the nodal points are known, i.e. these temperatures values are now the unknowns of the problem. In this way the original problem with, in principle, infinitely many unknowns, i.e. *degrees of freedom* (d.o.f.), has been replaced by a problem with a finite number of unknowns. In the present situation, the number of unknowns is five. In general, it is obvious that the more unknowns, the more accurate the approximate solution.

A system with a finite number of unknowns is called a *discrete system* in contrast to the original *continuous system* with an infinite number of unknowns. It will turn out later that the determination of the values of the variable at the nodal points follows from the solution of a certain system of equations. In the case considered in Figures 1.3–1.6 this system of equations would consist of five equations with five unknowns, but in general the system often involves thousands of unknowns. Obviously such systems cannot be solved by hand and, therefore, the FE method relies entirely on the availability of efficient computers.

As already touched upon, the FE method can be applied to arbitrary differential equations. Moreover, arbitrary geometries of bodies consisting of arbitrary materials can be analyzed. It is therefore no surprise that, in general, the FE method today presents the most powerful approach for solving differential equations that occur in engineering, physics and mathematics.

In its most basic form the FE method shares many common features with simple matrix structural analysis. In this analysis, structures consisting of trusses and beams are treated in a systematic manner with certain displacements as the unknowns. We shall present such an analysis in Chapter 3. However, today's FE approach emphasizes that arbitrary differential equations can be solved, and this much more general viewpoint is the one we shall adopt.

The emergence of the FE method took place in the early 1960s and since then its use has spread to virtually all fields of engineering. Some of the prominent names associated with the development of the FE method are Argyris, Clough and Zienkiewicz, and for a historical account of different major contributions, we may

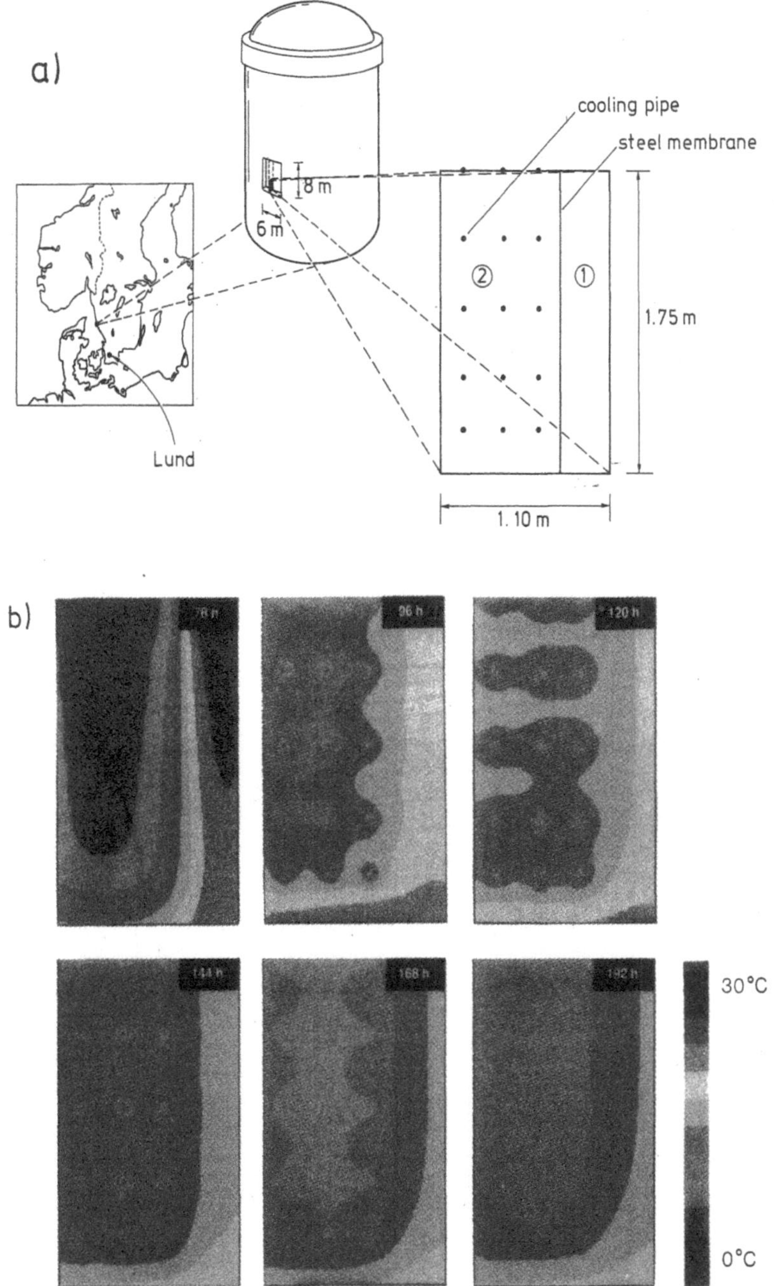

Figure 1.7 (a) Casting of concrete in containment vessel after replacement of nuclear equipment; casting of region 2 occurs 72 hours after casting of region 1; (b) resulting temperature development

refer, for instance, to Segerlind (1976), Stasa (1985), Zienkiewicz (1970, 1983) and Zienkiewicz and Taylor (1989).

In practice, so-called *general-purpose* FE programs are often used to analyze different problems. Even though very efficient computer programs are available today, we want to emphasize that in order to obtain meaningful results, a profound understanding of the underlying theories and their limitations is required by the user. Otherwise, the FE program may produce results which may look neat and convincing when presented in beautiful colour graphics, but which may be irrelevant. We also want to underline that even though a large number of FE programs are available today, research within different FE approaches is more active than ever. This research is directed, for instance, towards more efficient and reliable formulations as well as towards various non-linear problems. To illustrate this research activity we may refer to Cook *et al.* (1989), who state that 10 papers on the FE method were published in 1961, 134 in 1966, and 844 in 1971. By 1976, the cumulative total of FE publications exceeded 7000, and by 1986, the total was about 20 000 papers.

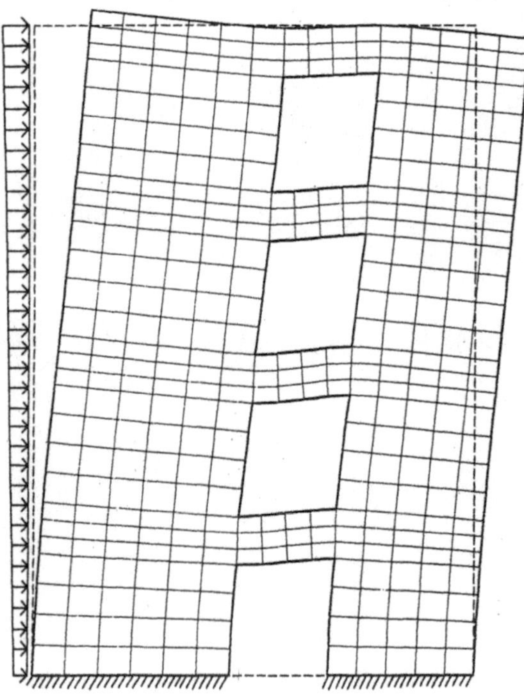

Figure 1.8   Finite element mesh and deflections of a laterally loaded structural component

1.2 Some specific applications

As already stressed, the FE method is applicable in widely different fields and Figures 1.7–1.14, which show some results obtained at the University of Lund, illustrate this important fact.

A nuclear containment vessel is shown in Figure 1.7. In order to replace some of the nuclear equipment, a hole was made in the concrete containment vessel and, after the replacement, concrete was cast to restore the containment shell. The concrete was cast in two steps: first in region 1 within the steel membrane and 3 days later in region 2 outside this membrane. Owing to the hydration of cement during hardening of the concrete, heat is developed and the resulting heat expansion may result in cracking of the concrete. To avoid such cracking, the fresh concrete in region 2 is cold, about 3°C, and, in addition, heat is removed during the hardening process by means of cooling pipes. The casting of region 2 occurred 72 hours after the casting of region 1 and Figure 1.7 shows the resulting temperature development.

Figure 1.8 shows the FE mesh and the deflections of an ordinary structural component due to lateral loading.

In Figures 1.9 and 1.10, the passenger cabin of a car is shown. The problem is to investigate the noise level, i.e. the air pressure variation in the cabin caused by engine vibrations. This problem constitutes a so-called coupled problem in which the

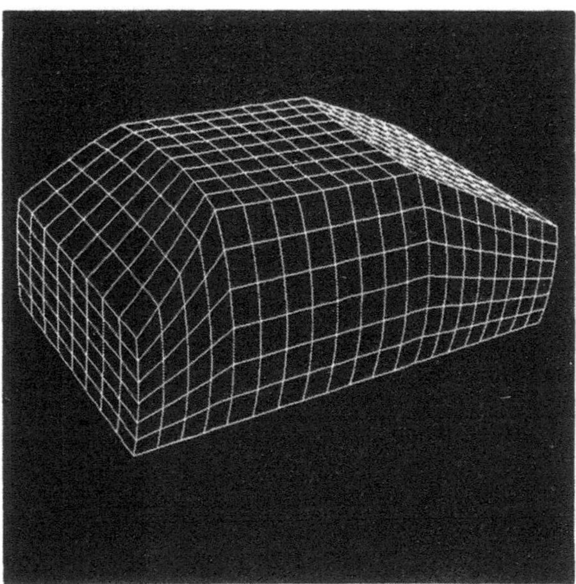

Figure 1.9 Finite element mesh of the structural part, i.e. the passenger cabin itself. The air inside the cabin is also divided into finite elements. This cabin is exposed to engine vibrations

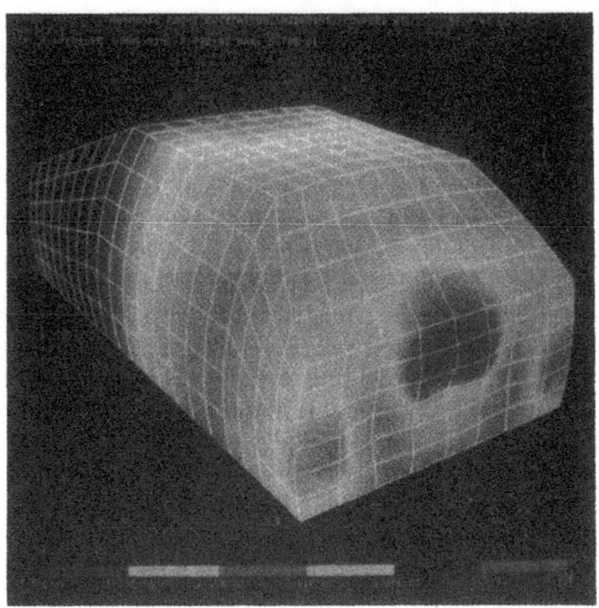

Figure 1.10   Result in terms of air pressure along the inside of the passenger cabin and structural deformation of the cabin

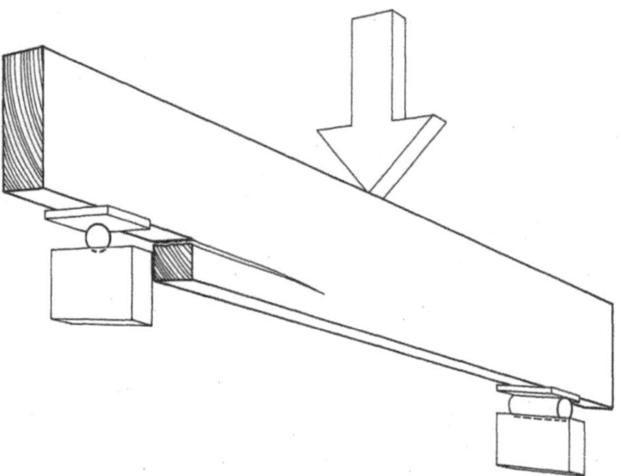

Figure 1.11   Wooden beam tested in laboratory

structural vibrations of the passenger cabin influence the motion of the air within the cabin and vice versa.

Figures 1.11 and 1.12 show the behaviour of a wooden beam in which a horizontal crack develops due to the loading.

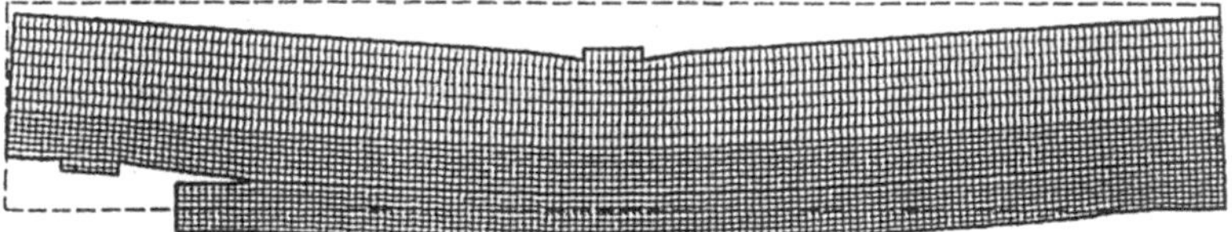

Figure 1.12   Displacement and finite element mesh

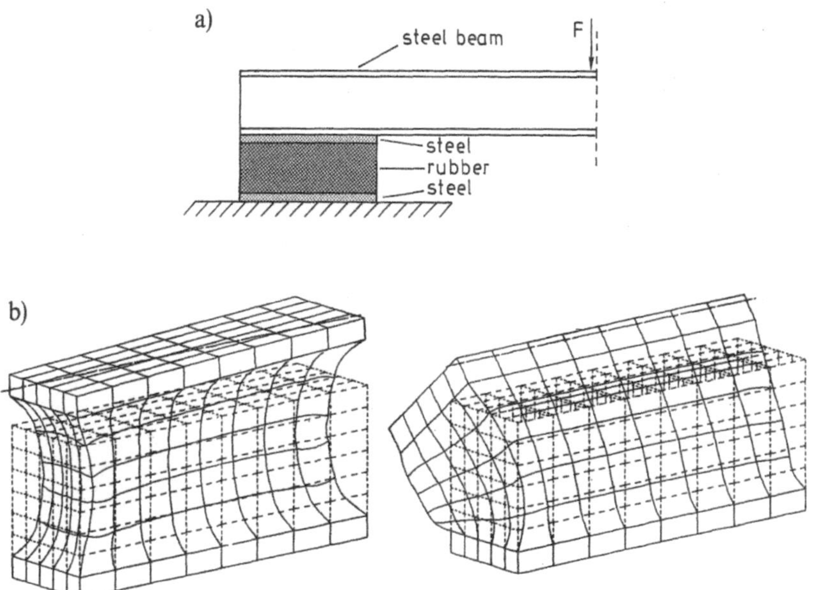

Figure 1.13   (a) Steel beam supported by rubber components; (b) deformation modes of
rubber component at some resonance frequencies

A steel beam supported by rubber components is shown in Figure 1.13(a). To
investigate its dynamic response in detail, the deformation modes at some resonance
frequencies of the rubber component itself were determined (cf. Figure 1.13(b)).

Finally, a water-filled pipe is shown in Figure 1.14(a). The deflection modes of
the pipe at some eigenfrequencies, i.e. at resonance, are shown in Figure 1.14(b)
together with the corresponding water pressures in Figure 1.14(c). Just as in
Figures 1.9 and 1.10, this problem constitutes a coupled problem in which the pipe
deflection influences the water pressure and vice versa.

The problems illustrated above show that widely different and complicated
physical phenomena can be analyzed by the FE method. Indeed, the main objective

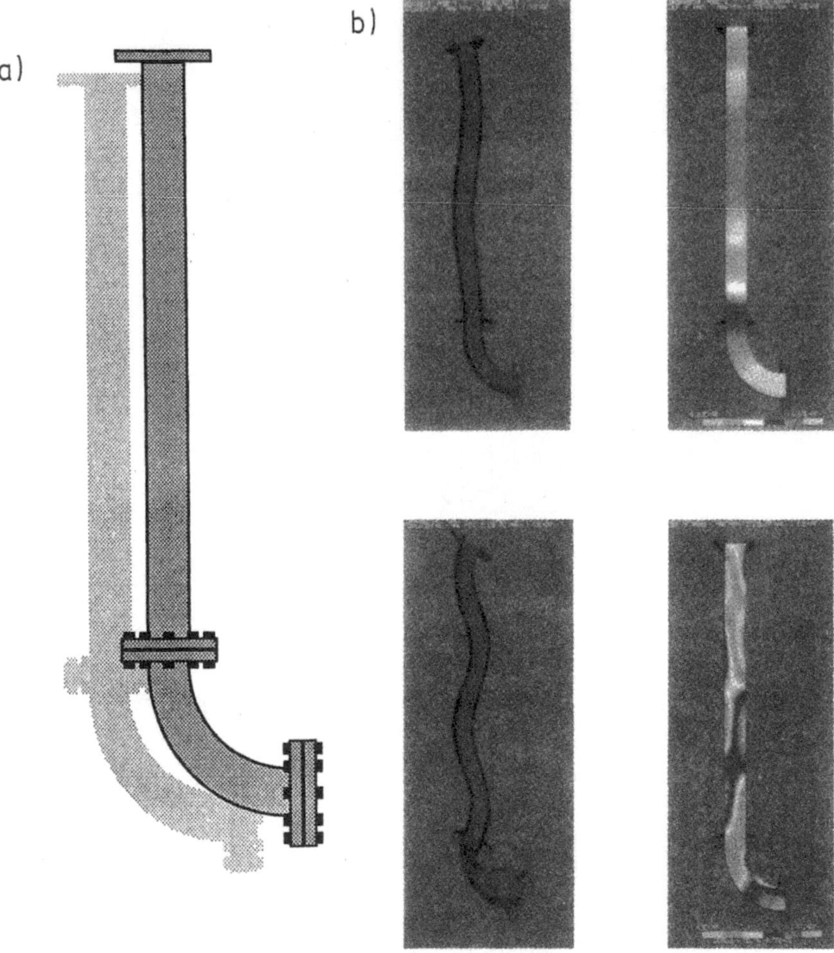

Figure 1.14   (a) Water-filled copper tube; (b) deflection modes of tube at some eigenfrequencies; (c) corresponding water pressures at resonance

of this textbook is to enable the reader to derive the FE formulation of various problems and to appreciate the unified approach offered by this method. Before this can be achieved, some preliminary tools are necessary and, first, we shall provide a review of basic matrix algebra.

2

Matrix algebra

The finite element method is a numerical approach which results in the establishment of systems of equations often involving thousands of unknowns. To enable one to deal with such expressions in a compact fashion which emphasizes the physical content, use of matrix algebra turns out to be convenient. In this chapter we shall present a short review of some basic matrix algebra. No proofs will be given, as the intention is simply to recall elementary results. For a more detailed treatment the reader is referred, for instance, to Bathe (1982), Hildebrand (1965) and Strang (1980). Apart from providing a review of matrix algebra, the present chapter also serves as an introduction to the notation adopted in this textbook.

2.1 Definitions

In general, a *matrix* consists of a collection of certain quantities which are termed the *components* of the matrix. The components are ordered in *rows* and *columns* and if the number of rows or columns is equal to one, the matrix is *one dimensional*, otherwise it is *two dimensional*.

As an example of a one-dimensional matrix, consider a *column matrix*, where the number of columns is equal to one. A column matrix is denoted usually by a lower-case letter in bold type; for instance

$$\mathbf{a} = \begin{bmatrix} a_1 \\ a_2 \\ a_3 \end{bmatrix}; \quad \mathbf{b} = \begin{bmatrix} b_1 \\ b_2 \\ b_3 \end{bmatrix}; \quad \mathbf{c} = \begin{bmatrix} c_1 \\ c_2 \\ c_3 \\ c_4 \end{bmatrix} \tag{2.1}$$

where c_1, c_2, c_3 and c_4 are the components of the matrix $\mathbf{c}$. Apart from being one or two dimensional, the specific *dimension of a matrix* is given by the number of rows and columns, i.e. column matrix $\mathbf{c}$ of (2.1) has the dimension 4×1. The *transpose*

$\mathbf{a}^T$ of $\mathbf{a}$ is given by the *row matrix*

$$\mathbf{a}^T = [a_1 \quad a_2 \quad a_3] \tag{2.2}$$

which has the dimension 1×3.

A two-dimensional matrix is denoted usually by an upper-case letter in bold type; for instance

$$\mathbf{B} = \begin{bmatrix} B_{11} & B_{12} & B_{13} \\ B_{21} & B_{22} & B_{23} \\ B_{31} & B_{32} & B_{33} \end{bmatrix}; \quad \mathbf{C} = \begin{bmatrix} C_{11} & C_{12} \\ C_{21} & C_{22} \\ C_{31} & C_{32} \\ C_{41} & C_{42} \end{bmatrix} \tag{2.3}$$

where $\mathbf{B}$ is termed a *square matrix* since the number of rows and columns are equal. The dimensions of $\mathbf{B}$ and $\mathbf{C}$ are 3×3 and 4×2, respectively. The transpose $\mathbf{B}^T$ of $\mathbf{B}$ is obtained by interchanging rows and columns in $\mathbf{B}$, i.e.

$$\mathbf{B}^T = \begin{bmatrix} B_{11} & B_{21} & B_{31} \\ B_{12} & B_{22} & B_{32} \\ B_{13} & B_{23} & B_{33} \end{bmatrix} \tag{2.4}$$

and the matrix $\mathbf{B}$ is *symmetric* if $\mathbf{B} = \mathbf{B}^T$. In what follows we shall often encounter symmetric matrices. If only the diagonal components in a matrix $\mathbf{A}$ are different from zero, $\mathbf{A}$ is termed a *diagonal matrix*. As an example, $\mathbf{A}$ given by

$$\mathbf{A} = \begin{bmatrix} 2 & 0 & 0 & 0 \\ 0 & 1 & 0 & 0 \\ 0 & 0 & -1 & 0 \\ 0 & 0 & 0 & 3 \end{bmatrix} \tag{2.5}$$

is a diagonal matrix. Obviously, a diagonal matrix is a square matrix. If all the diagonal components of a diagonal matrix are equal to unity, the matrix is a *unit matrix*. A unit matrix is written as $\mathbf{I}$ and an example is

$$\mathbf{I} = \begin{bmatrix} 1 & 0 & 0 \\ 0 & 1 & 0 \\ 0 & 0 & 1 \end{bmatrix} \tag{2.6}$$

A *zero matrix* $\mathbf{0}$ is defined as a matrix where all components are equal to zero. Examples are

$$\mathbf{0} = \begin{bmatrix} 0 & 0 \\ 0 & 0 \end{bmatrix}; \quad \mathbf{0} = \begin{bmatrix} 0 & 0 & 0 \\ 0 & 0 & 0 \end{bmatrix} \tag{2.7}$$

2.2 Addition and subtraction

If two matrices have the same dimensions they can be added and subtracted, with addition and subtraction being carried out for each component of the matrices. For instance, we have

$$\mathbf{c} = \mathbf{a} \pm \mathbf{b} = \begin{bmatrix} a_1 \\ a_2 \\ a_3 \end{bmatrix} \pm \begin{bmatrix} b_1 \\ b_2 \\ b_3 \end{bmatrix} = \begin{bmatrix} a_1 \pm b_1 \\ a_2 \pm b_2 \\ a_3 \pm b_3 \end{bmatrix} \tag{2.8}$$

Likewise we have

$$\mathbf{c}^{\mathrm{T}} = \mathbf{a}^{\mathrm{T}} \pm \mathbf{b}^{\mathrm{T}} = \begin{bmatrix} a_1 & a_2 & a_3 \end{bmatrix} \pm \begin{bmatrix} b_1 & b_2 & b_3 \end{bmatrix}$$
$$= \begin{bmatrix} a_1 \pm b_1 & a_2 \pm b_2 & a_3 \pm b_3 \end{bmatrix} \tag{2.9}$$

and for two-dimensional matrices we have

$$\mathbf{C} = \mathbf{A} \pm \mathbf{B} = \begin{bmatrix} A_{11} & A_{12} \\ A_{21} & A_{22} \\ A_{31} & A_{32} \end{bmatrix} \pm \begin{bmatrix} B_{11} & B_{12} \\ B_{21} & B_{22} \\ B_{31} & B_{32} \end{bmatrix} = \begin{bmatrix} A_{11} \pm B_{11} & A_{12} \pm B_{12} \\ A_{21} \pm B_{21} & A_{22} \pm B_{22} \\ A_{31} \pm B_{31} & A_{32} \pm B_{32} \end{bmatrix} \tag{2.10}$$

It follows that

$$\mathbf{a} \pm \mathbf{b} = \pm \mathbf{b} + \mathbf{a}$$
$$\mathbf{a}^{\mathrm{T}} \pm \mathbf{b}^{\mathrm{T}} = \pm \mathbf{b}^{\mathrm{T}} + \mathbf{a}^{\mathrm{T}} \tag{2.11}$$
$$\mathbf{A} \pm \mathbf{B} = \pm \mathbf{B} + \mathbf{A}$$

Moreover, it appears directly that

$$(\mathbf{a} \pm \mathbf{b})^{\mathrm{T}} = \mathbf{a}^{\mathrm{T}} \pm \mathbf{b}^{\mathrm{T}}$$
$$(\mathbf{a}^{\mathrm{T}} \pm \mathbf{b}^{\mathrm{T}})^{\mathrm{T}} = \mathbf{a} \pm \mathbf{b} \tag{2.12}$$
$$(\mathbf{A} \pm \mathbf{B})^{\mathrm{T}} = \mathbf{A}^{\mathrm{T}} \pm \mathbf{B}^{\mathrm{T}}$$

2.3 Multiplication

Any matrix can be multiplied by a number c and this implies that each component is multiplied by the number c. Examples are

$$c\mathbf{a} = c\begin{bmatrix} a_1 \\ a_2 \\ a_3 \end{bmatrix} = \begin{bmatrix} ca_1 \\ ca_2 \\ ca_3 \end{bmatrix} \tag{2.13}$$

$$cA = c \begin{bmatrix} A_{11} & A_{12} \\ A_{21} & A_{22} \\ A_{31} & A_{32} \end{bmatrix} = \begin{bmatrix} cA_{11} & cA_{12} \\ cA_{21} & cA_{22} \\ cA_{31} & cA_{32} \end{bmatrix} \tag{2.14}$$

The *scalar product* of two column matrices $\mathbf{a}$ and $\mathbf{b}$ having the same dimension is defined according to

$$\mathbf{a}^T\mathbf{b} = \sum_{i=1}^{n} a_i b_i \tag{2.15}$$

where n is the number of rows. It follows that

$$\mathbf{a}^T\mathbf{b} = \mathbf{b}^T\mathbf{a} \tag{2.16}$$

With $\mathbf{a}$ and $\mathbf{b}$ given by (2.1) we obtain

$$\mathbf{a}^T\mathbf{b} = [a_1 \quad a_2 \quad a_3] \begin{bmatrix} b_1 \\ b_2 \\ b_3 \end{bmatrix} = a_1 b_1 + a_2 b_2 + a_3 b_3 \tag{2.17}$$

The *length* of $\mathbf{a}$ or $\mathbf{a}^T$ is denoted by $|\mathbf{a}|$ and defined by

$$|\mathbf{a}| = (a_1^2 + a_2^2 + \cdots + a_n^2)^{1/2} \tag{2.18}$$

where n is the number of rows in $\mathbf{a}$. From (2.15) and (2.18) we have that

$$|\mathbf{a}| = (\mathbf{a}^T\mathbf{a})^{1/2} \tag{2.19}$$

Let us now generalize the multiplication rule (2.15) to more complicated situations. Let $\mathbf{A}$ have the dimension $m \times n$ and $\mathbf{x}$ the dimension $n \times 1$; then the matrix product $\mathbf{Ax}$ defines the row matrix $\mathbf{c}$ with the dimension $m \times 1$ according to

$$\mathbf{c} = \mathbf{Ax} \quad \text{where} \quad c_i = \sum_{j=1}^{n} A_{ij} x_j \tag{2.20}$$

In symbolic form we can write

$$\underset{m \times 1}{\mathbf{c}} = \underset{(m \times n)}{\mathbf{A}} \; \underset{(n \times 1)}{\mathbf{x}} \tag{2.21}$$

As an example we obtain

$$\mathbf{c} = \mathbf{Ax} = \begin{bmatrix} A_{11} & A_{12} \\ A_{21} & A_{22} \\ A_{31} & A_{32} \end{bmatrix} \begin{bmatrix} x_1 \\ x_2 \end{bmatrix} = \begin{bmatrix} A_{11}x_1 + A_{12}x_2 \\ A_{21}x_1 + A_{22}x_2 \\ A_{31}x_1 + A_{32}x_2 \end{bmatrix}$$

Likewise, let now $\mathbf{x}$ have the dimension $m \times 1$ and $\mathbf{A}$ the dimension $m \times n$; then the matrix product $\mathbf{x}^{\mathrm{T}}\mathbf{A}$ defines the row matrix $\mathbf{c}^{\mathrm{T}}$ with the dimension $1 \times n$ according to

$$\mathbf{c}^{\mathrm{T}} = \mathbf{x}^{\mathrm{T}}\mathbf{A} \quad \text{where} \quad c_j = \sum_{i=1}^{n} x_i A_{ij} \tag{2.22}$$

In symbolic form we have

$$\underset{1 \times n}{\mathbf{c}^{\mathrm{T}}} = \underset{(1 \times m)}{\mathbf{x}^{\mathrm{T}}} \underset{(m \times n)}{\mathbf{A}} \tag{2.23}$$

An example is

$$\mathbf{c}^{\mathrm{T}} = \mathbf{x}^{\mathrm{T}}\mathbf{A} = \begin{bmatrix} x_1 & x_2 \end{bmatrix} \begin{bmatrix} A_{11} & A_{12} & A_{13} \\ A_{21} & A_{22} & A_{23} \end{bmatrix}$$

$$= \begin{bmatrix} x_1 A_{11} + x_2 A_{21} & x_1 A_{12} + x_2 A_{22} & x_1 A_{13} + x_2 A_{23} \end{bmatrix}$$

Finally, matrix multiplication of two-dimensional matrices can be carried out. If $\mathbf{A}$ has the dimension $m \times n$ and $\mathbf{B}$ the dimension $n \times p$, the matrix product $\mathbf{AB}$ defines the matrix $\mathbf{C}$ with the dimension $m \times p$ according to

$$\boxed{\mathbf{C} = \mathbf{AB} \quad \text{where} \quad C_{ij} = \sum_{k=1}^{n} A_{ik} B_{kj}} \tag{2.24}$$

In symbolic form we have

$$\underset{m \times p}{\mathbf{C}} = \underset{(m \times n)}{\mathbf{A}} \underset{(n \times p)}{\mathbf{B}} \tag{2.25}$$

As an example we have

$$\mathbf{C} = \mathbf{AB} = \begin{bmatrix} A_{11} & A_{12} \\ A_{21} & A_{22} \\ A_{31} & A_{32} \end{bmatrix} \begin{bmatrix} B_{11} & B_{12} \\ B_{21} & B_{22} \end{bmatrix}$$

$$= \begin{bmatrix} A_{11}B_{11} + A_{12}B_{21} & A_{11}B_{12} + A_{12}B_{22} \\ A_{21}B_{11} + A_{22}B_{21} & A_{21}B_{12} + A_{22}B_{22} \\ A_{31}B_{11} + A_{32}B_{21} & A_{31}B_{12} + A_{32}B_{22} \end{bmatrix}$$

We emphasize that matrix multiplication is only defined if the matrices possess the correct dimensions. Moreover, if $\mathbf{A}$ has the dimension $m \times n$ and $\mathbf{B}$ has the dimension $n \times m$ both the multiplications $\mathbf{AB}$ and $\mathbf{BA}$ are defined, but in general we have

$$\mathbf{AB} \neq \mathbf{BA} \tag{2.26}$$

This follows from the fact that $\mathbf{AB}$ defines a matrix with the dimension $m \times m$, whereas $\mathbf{BA}$ defines a matrix with the dimension $n \times n$. Moreover, even if $m = n$ (2.26) will in general hold true. As an example consider

$$\mathbf{AB} = \begin{bmatrix} A_{11} & A_{12} \\ A_{21} & A_{22} \end{bmatrix} \begin{bmatrix} B_{11} & B_{12} \\ B_{21} & B_{22} \end{bmatrix} = \begin{bmatrix} A_{11}B_{11} + A_{12}B_{21} & A_{11}B_{12} + A_{12}B_{22} \\ A_{21}B_{11} + A_{22}B_{21} & A_{21}B_{12} + A_{22}B_{22} \end{bmatrix} \tag{2.27}$$

and

$$\mathbf{BA} = \begin{bmatrix} B_{11} & B_{12} \\ B_{21} & B_{22} \end{bmatrix} \begin{bmatrix} A_{11} & A_{12} \\ A_{21} & A_{22} \end{bmatrix} = \begin{bmatrix} B_{11}A_{11} + B_{12}A_{21} & B_{11}A_{12} + B_{12}A_{22} \\ B_{21}A_{11} + B_{22}A_{21} & B_{21}A_{12} + B_{22}A_{22} \end{bmatrix} \tag{2.28}$$

A comparison of (2.27) and (2.28) shows that (2.26) will hold unless the components of $\mathbf{A}$ and $\mathbf{B}$ possess some specific properties.

From definition (2.24) and the definition of the unit matrix $\mathbf{I}$ it follows that

$$\mathbf{AI} = \mathbf{A} \tag{2.29}$$

We also recall that

$$\boxed{(\mathbf{AB})^{\mathrm{T}} = \mathbf{B}^{\mathrm{T}}\mathbf{A}^{\mathrm{T}}} \tag{2.30}$$

It appears that (2.30) implies

$$(\mathbf{ABC})^{\mathrm{T}} = ((\mathbf{AB})\mathbf{C})^{\mathrm{T}} = \mathbf{C}^{\mathrm{T}}(\mathbf{AB})^{\mathrm{T}} = \mathbf{C}^{\mathrm{T}}\mathbf{B}^{\mathrm{T}}\mathbf{A}^{\mathrm{T}} \tag{2.31}$$

and we also have

$$(\mathbf{Ax})^{\mathrm{T}} = \mathbf{x}^{\mathrm{T}}\mathbf{A}^{\mathrm{T}} \tag{2.32}$$

Moreover, it follows trivially that if c denotes a number then

$$c\mathbf{AB} = \mathbf{A}(c\mathbf{B}) \tag{2.33}$$

Finally, we recall that the distribution law from ordinary algebra also holds for matrices; for example

$$\begin{aligned} (\mathbf{A} + \mathbf{B})\mathbf{x} &= \mathbf{Ax} + \mathbf{Bx} \\ \mathbf{x}^{\mathrm{T}}(\mathbf{A} + \mathbf{B}) &= \mathbf{x}^{\mathrm{T}}\mathbf{A} + \mathbf{x}^{\mathrm{T}}\mathbf{B} \\ (\mathbf{A} + \mathbf{B})\mathbf{C} &= \mathbf{AC} + \mathbf{BC} \\ \mathbf{C}(\mathbf{A} + \mathbf{B}) &= \mathbf{CA} + \mathbf{CB} \end{aligned} \tag{2.34}$$

2.4 Determinant

For the square matrix $\mathbf{A}$, it is possible to calculate the *determinant* det $\mathbf{A}$ of $\mathbf{A}$. Let $\mathbf{A}$ be of dimension $n \times n$. If $n = 1$ then, by definition, det $\mathbf{A} = A_{11}$. Before it is possible to determine det $\mathbf{A}$ for $n \geq 2$ some definitions have to be made.

If row number i and column number k are deleted, a new square matrix with dimension $(n-1) \times (n-1)$ emerges. The determinant of this matrix is called a *minor* of $\mathbf{A}$ and is denoted by $\det M_{ik}$. The *cofactor* of $\mathbf{A}$ is denoted by A_{ik}^{c} and is defined by

$$A_{ik}^{c} = (-1)^{i+k} \det M_{ik} \tag{2.35}$$

According to the *expansion formula* of Laplace we then have

$$\det \mathbf{A} = \sum_{k=1}^{n} A_{ik} A_{ik}^{c} \tag{2.36}$$

where i indicates any row number in the range $1 \leq i \leq n$.

As an example consider $\mathbf{A}$ given by

$$\mathbf{A} = \begin{bmatrix} A_{11} & A_{12} \\ A_{21} & A_{22} \end{bmatrix} \tag{2.37}$$

It follows that

$$\det M_{11} = A_{22}; \quad \det M_{12} = A_{21}; \quad \det M_{21} = A_{12}; \quad \det M_{22} = A_{11} \tag{2.38}$$

Suppose that in (2.36) we choose $i = 1$; then with (2.35) and (2.38) we obtain the familiar result

$$\begin{aligned} \det \mathbf{A} &= \sum_{k=1}^{2} A_{1k} A_{1k}^{c} = A_{11} A_{11}^{c} + A_{12} A_{12}^{c} \\ &= A_{11}(-1)^{1+1} \det M_{11} + A_{12}(-1)^{1+2} \det M_{12} \\ &= A_{11} A_{22} - A_{12} A_{21} \end{aligned} \tag{2.39}$$

The reader may verify that the same result is obtained if instead we choose $i = 2$ in (2.36).

It appears that (2.36) expresses the determinant as a certain linear combination of all components A_{ik} in the arbitrary row i. It turns out that the expansion formula of Laplace can also be written as a linear combination of all components in an arbitrary column. In fact, we have

$$\det \mathbf{A} = \sum_{k=1}^{n} A_{kj} A_{kj}^{c} \tag{2.40}$$

where j indicates any column number in the range $1 \leq j \leq n$. As an example consider $\mathbf{A}$ as given by (2.37) and choose $j = 1$. Then with (2.40), (2.35) and (2.38) we obtain

$$\det \mathbf{A} = \sum_{k=1}^{2} A_{k1} A_{k1}^{c} = A_{11} A_{11}^{c} + A_{21} A_{21}^{c}$$

$$= A_{11}(-1)^{1+1} \det M_{11} + A_{21}(-1)^{2+1} \det M_{21}$$

$$= A_{11} A_{22} - A_{21} A_{12}$$

in accordance with (2.39).

In practice, the direct use of the expansion formulae of Laplace in numerical calculations is unsuitable since many operations are necessary. For theoretical considerations, however, these expansion formulae are very important as they enable one to establish a number of properties for determinants. In particular, it can be shown that if:

---

1. a row (or column) consists of zeros, the determinant is zero;
2. if two rows (or columns) are proportional, the determinant is zero;
3. if a row (or column) is multiplied by a factor k, the determinant is also multiplied by the factor k.

Moreover, for later purposes it is convenient to recall that a *row* (or *column*) *operation* is an operation where all components of one row (or column) are multiplied by a certain factor and then added to another row (or column). We then have:

---

4. row (or column) operations do not change the determinant;
5. if two rows (or columns) are interchanged, the determinant changes its sign.

Finally, we note that

$$\det \mathbf{A}^{T} = \det \mathbf{A} \tag{2.41}$$

$$\det(\mathbf{AB}) = \det \mathbf{A} \det \mathbf{B} \tag{2.42}$$

whereas

$$\det(\mathbf{A} + \mathbf{B}) \neq \det \mathbf{A} + \det \mathbf{B} \tag{2.43}$$

2.5 Inverse matrix

The *inverse* $\mathbf{A}^{-1}$ of a square matrix $\mathbf{A}$ is defined by

$$\mathbf{A}^{-1}\mathbf{A} = \mathbf{A}\mathbf{A}^{-1} = \mathbf{I} \tag{2.44}$$

By means of the cofactor A_{ik}^{c} of $\mathbf{A}$ as defined by (2.35), it is possible to present an explicit expression for the inverse matrix $\mathbf{A}^{-1}$.

For all values of i and k, we are able to construct a square matrix of dimension $n \times n$ by means of (2.35). If this matrix is transposed, we obtain the *adjoint matrix* of $\mathbf{A}$ given by

$$\text{adj } \mathbf{A} = \begin{bmatrix} A_{11}^{c} & A_{12}^{c} & \cdots & A_{1n}^{c} \\ A_{21}^{c} & A_{22}^{c} & \cdots & A_{2n}^{c} \\ \vdots & \vdots & & \vdots \\ A_{n1}^{c} & A_{n2}^{c} & \cdots & A_{nn}^{c} \end{bmatrix}^{\mathrm{T}} \tag{2.45}$$

It turns out that the inverse matrix is given by

$$\boxed{\mathbf{A}^{-1} = \text{adj } \mathbf{A}/\det \mathbf{A}} \tag{2.46}$$

It is important to note that $\mathbf{A}^{-1}$ only exists if $\det \mathbf{A} \neq 0$. A square matrix $\mathbf{A}$ for which $\det \mathbf{A} = 0$ is called a *singular matrix*. We also recall that

$$(\mathbf{A}^{-1})^{\mathrm{T}} = (\mathbf{A}^{\mathrm{T}})^{-1} \tag{2.47}$$

$$(\mathbf{AB})^{-1} = \mathbf{B}^{-1}\mathbf{A}^{-1} \tag{2.48}$$

and the latter expression implies

$$(\mathbf{ABC})^{-1} = \mathbf{C}^{-1}(\mathbf{AB})^{-1} = \mathbf{C}^{-1}\mathbf{B}^{-1}\mathbf{A}^{-1} \tag{2.49}$$

We note that if

$$\mathbf{A}^{-1} = \mathbf{A}^{\mathrm{T}} \tag{2.50}$$

then the matrix is called an *orthogonal matrix*. If $\mathbf{A}$ is orthogonal, we have $\mathbf{A}^{\mathrm{T}}\mathbf{A} = \mathbf{A}\mathbf{A}^{\mathrm{T}} = \mathbf{I}$.

2.6 Linear equations: number of equations is equal to number of unknowns

The solution of systems of linear equations often involving thousands of unknowns is characteristic of numerical approaches like the finite element method. It is therefore of importance to have a firm understanding of the fundamental properties of different types of linear equation systems.

Consider first the system of equations

$$\boxed{\mathbf{Ax} = \mathbf{b}} \tag{2.51}$$

where $\mathbf{A}$ is a square matrix of dimension $n \times n$ and $\mathbf{x}$ and $\mathbf{b}$ both have the dimension

$n \times 1$. If $\mathbf{b} = \mathbf{0}$ then (2.51) is called a *homogeneous system of equations*, otherwise it is termed an *inhomogeneous system of equations*.

Assume that

$$\det \mathbf{A} \neq 0 \tag{2.52}$$

then by premultiplying (2.51) by $\mathbf{A}^{-1}$ we obtain

$$\mathbf{A}^{-1}\mathbf{A}\mathbf{x} = \mathbf{A}^{-1}\mathbf{b}$$

which, due to (2.44), results in the solution

$$\boxed{\mathbf{x} = \mathbf{A}^{-1}\mathbf{b}} \tag{2.53}$$

It appears that the homogeneous system of equations

$$\mathbf{A}\mathbf{x} = \mathbf{0} \tag{2.54}$$

for which $\det \mathbf{A} \neq 0$ only has the trivial solution $\mathbf{x} = \mathbf{0}$. Therefore the non-trivial solution of the homogeneous system of equations (2.54) requires that $\det \mathbf{A} = 0$. Homogeneous systems of equations are of importance in many applications, for instance in buckling and vibration problems.

We conclude that for homogeneous systems of equations, we have

$$\tag{2.55}$$
> $\mathbf{A}\mathbf{x} = \mathbf{0}$
>
> - If $\det \mathbf{A} = 0$, a non-trivial solution exists
> - If $\det \mathbf{A} \neq 0$, no non-trivial solution exists

For inhomogeneous systems of equations, we recall that

$$\tag{2.56}$$
> $\mathbf{A}\mathbf{x} = \mathbf{b}; \quad \mathbf{b} \neq \mathbf{0}$
>
> - If $\det \mathbf{A} \neq 0$, one unique solution given by (2.53) exists
> - If $\det \mathbf{A} = 0$, no unique solution exists. Depending on the specific $\mathbf{b}$-matrix we may have no solution or an infinity of solutions

With these important statements in mind let us now investigate in more detail the solution of (2.56) when $\det \mathbf{A} \neq 0$ and where matrix $\mathbf{A}$ is termed the *coefficient matrix*. In principle, this solution is given by (2.53). In practical numerical calculations, however, the direct establishment of the inverse $\mathbf{A}^{-1}$ is all too cumbersome and other approaches are used. A very common and efficient approach is *Gauss elimination*.

In Gauss elimination, the equations are combined in such a manner that the lower left part of the new coefficient matrix consists of zeros. To obtain this objective, a multiple of the first equation is added to the second equation such that the first component of the second row in the new coefficient matrix becomes zero. A multiple of the first equation is then added to the third equation such that the first component of the third row in the new coefficient matrix becomes zero. This process is continued until all the components of the first column – except the first component – are zero. This process is, in fact, an elimination of the x_1-variable from all equations except the first. From this new system of equations, and using the second equation, the x_2-variable is eliminated from the equations below the second equation. We continue this process until the lower left part of the new coefficient matrix consists of zeros.

Therefore, by this elimination technique, (2.56) is transformed into the form

$$\mathbf{A}'\mathbf{x} = \mathbf{b}' \tag{2.57}$$

where we write $\mathbf{A}'$ and $\mathbf{b}'$ instead of $\mathbf{A}$ and $\mathbf{b}$, respectively, to emphasize that the new coefficient matrix and right-hand side have changed as a result of the elimination process. In accordance with what was described above $\mathbf{A}'$ has the form

$$\mathbf{A}' = \begin{bmatrix} A'_{11} & A'_{12} & A'_{13} & \cdots & A'_{1n} \\ 0 & A'_{22} & A'_{23} & \cdots & A'_{2n} \\ 0 & 0 & A'_{33} & \cdots & A'_{3n} \\ \vdots & \vdots & \vdots & & \vdots \\ 0 & 0 & 0 & \cdots & A'_{nn} \end{bmatrix} \tag{2.58}$$

where the lower left part below the diagonal consists of zeros. This form is also expressed by saying that the system of equations has been *triangularized*. The diagonal elements $A'_{11}, A'_{22}, \ldots, A'_{nn}$ are termed *pivot elements*.

The system of equations (2.57) with the coefficient matrix (2.58) is indeed very easy to solve. From the last equation, i.e. equation number n, we obtain $x_n = b'_n / A'_{nn}$. This solution is substituted into equation number $n - 1$ to provide x_{n-1} and we continue in this manner until all x-components have been determined. This process is called *back-substitution*.

To illustrate this Gauss elimination technique, consider the system of equations

$$\begin{bmatrix} 200 & -100 & 0 \\ -100 & 200 & -100 \\ 0 & -100 & 100 \end{bmatrix} \begin{bmatrix} x_1 \\ x_2 \\ x_3 \end{bmatrix} = \begin{bmatrix} 8 \\ 8 \\ 2 \end{bmatrix} \tag{2.59}$$

Multiply the first equation by $1/2$ and add to the second. This results in

$$\begin{bmatrix} 200 & -100 & 0 \\ 0 & 150 & -100 \\ 0 & -100 & 100 \end{bmatrix} \begin{bmatrix} x_1 \\ x_2 \\ x_3 \end{bmatrix} = \begin{bmatrix} 8 \\ 12 \\ 2 \end{bmatrix}$$

The first component in the third row is already zero. Therefore, multiply the second equation by 2/3 and add to the third equation to obtain

$$\begin{bmatrix} 200 & -100 & 0 \\ 0 & 150 & -100 \\ 0 & 0 & \frac{100}{3} \end{bmatrix} \begin{bmatrix} x_1 \\ x_2 \\ x_3 \end{bmatrix} = \begin{bmatrix} 8 \\ 12 \\ 10 \end{bmatrix} \tag{2.60}$$

The system of equations has now been triangularized in accordance with (2.58). Now comes the process of back-substitution. From the third equation we obtain $x_3 = 30/100$; the second equation then provides $x_2 = 28/100$ and the first equation results in $x_1 = 18/100$, i.e.

$$\begin{bmatrix} x_1 \\ x_2 \\ x_3 \end{bmatrix} = \frac{1}{100} \begin{bmatrix} 18 \\ 28 \\ 30 \end{bmatrix}$$

Having illustrated the Gauss elimination technique, we return to the general triangularized form of (2.57) and (2.58). Let us calculate the determinant det $\mathbf{A}'$ of the coefficient matrix (2.58). To do this it is convenient to use the expansion formula of Laplace given by (2.40) where the expansion takes place along some column. In the present case we choose the first column, i.e. $j = 1$. As the only non-zero component in this column is A'_{11} we obtain from (2.40)

$$\det A' = \sum_{k=1}^{n} A'_{kj} A'^{c}_{kj} = \sum_{k=1}^{n} A'_{k1} A'^{c}_{k1} = A'_{11} A'^{c}_{11} \tag{2.61}$$

From (2.35) the cofactor A'^{c}_{11} is

$$A'^{c}_{11} = \det M_{11} \tag{2.62}$$

To determine the minor det M_{11} we delete the first row and first column in the $\mathbf{A}'$-matrix and calculate the determinant of the remaining matrix. It appears readily that steps identical with (2.61) and (2.62) will result. Therefore, the reader may easily verify that we end up with the simple result

$$\det \mathbf{A}' = A'_{11} A'_{22} \ldots A'_{nn} \tag{2.63}$$

i.e. the determinant is the product of all the pivot elements.

As the determinant of $\mathbf{A}'$ turns out to be very easy to identify, it is of interest to investigate how this determinant relates to the determinant of the original coefficient matrix $\mathbf{A}$ of (2.56). For this purpose we observe that $\mathbf{A}'$ is obtained from $\mathbf{A}$ by making suitable row operations. Referring to statement (4) on page 18 we immediately conclude that

$$\boxed{\det \mathbf{A}' = \det \mathbf{A}} \tag{2.64}$$

and the reader may check this conclusion by using (2.59) and (2.60). Often the establishment of the determinant is most easily based on (2.63) and (2.64).

2.7 Linear equations: number of equations is different from number of unknowns

Having treated the important case of linear equations where the number of equations and unknowns is the same, we shall for later purposes investigate the properties of homogeneous systems of equations where we have more unknowns than equations. That is, we consider the linear homogeneous system of equations $\mathbf{Ax} = \mathbf{0}$ where $\mathbf{A}$ has the dimension $m \times n$, whereas $\mathbf{x}$ has the dimension $n \times 1$ and $\mathbf{0}$ the dimension $m \times 1$; moreover, $n > m$. The following result is obtained:

$$
\underset{(m \times n)\,(n \times 1)\qquad m \times 1}{\mathbf{A} \quad \mathbf{x} \;=\; \mathbf{0}}
$$

- For $n > m$, i.e. more unknowns than equations, we have at least $n - m$ non-trivial solutions

(2.65)

2.8 Quadratic forms and positive definiteness

For a square matrix $\mathbf{A}$ consider the quantity $\mathbf{x}^T\mathbf{Ax}$, which is a number. This quantity is called a *quadratic form*. If

$$
\mathbf{x}^T\mathbf{Ax} > 0 \quad \text{for all } \mathbf{x} \neq \mathbf{0}
$$

(2.66)

then the matrix $\mathbf{A}$ is said to be *positive definite*. In what follows, we shall often encounter positive definite matrices.

It follows directly from (2.66) that the determinant of a positive definite matrix is different from zero. To show this assume that (2.66) holds and that $\det \mathbf{A} = 0$. Since $\det \mathbf{A} = 0$, it is possible to determine a non-trivial solution to the homogeneous system of equations $\mathbf{Ax} = \mathbf{0}$, which means that (2.66) is violated. Therefore $\det \mathbf{A} \neq 0$ must hold for positive definite matrices, i.e.

$$
\text{If } \mathbf{A} \text{ is positive definite then } \det \mathbf{A} \neq 0
$$

(2.67)

Note that the inverse argument does not hold, i.e. if $\det \mathbf{A} \neq 0$ we cannot conclude that $\mathbf{A}$ is positive definite.

Finally, if a matrix is positive definite it is required that all the diagonal components of the matrix are positive. To show this, choose $\mathbf{x}$ as

$$
\mathbf{x}^T = [0 \ \dots \ 0 \ x_k \ 0 \ \dots \ 0]
$$

then it follows from (2.66) that

$$\mathbf{x}^T\mathbf{A}\mathbf{x} = A_{kk}x_k^2 > 0$$

which proves the statement above.

We also mention that the square matrix $\mathbf{A}$ is called *positive semi-definite* if

$$\boxed{\mathbf{x}^T\mathbf{A}\mathbf{x} \geq 0 \quad \text{for all } \mathbf{x} \neq \mathbf{0}} \tag{2.68}$$

2.9 Partitioning

To facilitate matrix manipulations, it may be useful to *partition* a matrix into *submatrices*. A submatrix is a matrix that is obtained from the original matrix by including only the components of certain rows and columns. The idea is demonstrated below, where the dashed lines are the lines of partitioning:

$$\mathbf{A} = \begin{bmatrix} A_{11} & A_{12} & A_{13} \\ A_{21} & A_{22} & A_{23} \\ A_{31} & A_{32} & A_{33} \end{bmatrix} \tag{2.69}$$

Using this partitioning the matrix $\mathbf{A}$ may be written as

$$\mathbf{A} = \begin{bmatrix} \mathbf{B} & \mathbf{C} \\ \mathbf{D} & \mathbf{E} \end{bmatrix} \tag{2.70}$$

where

$$\mathbf{B} = \begin{bmatrix} A_{11} & A_{12} \\ A_{21} & A_{22} \end{bmatrix}; \quad \mathbf{C} = \begin{bmatrix} A_{13} \\ A_{23} \end{bmatrix}$$
$$\mathbf{D} = [A_{31} \quad A_{32}]; \quad \mathbf{E} = [A_{33}] \tag{2.71}$$

As an example of the use of partitioning, consider the system of equations

$$\mathbf{A}\mathbf{x} = \mathbf{f} \tag{2.72}$$

where $\mathbf{A}$ is given by (2.69) and $\mathbf{x}$ and $\mathbf{f}$ are given by

$$\mathbf{x} = \begin{bmatrix} x_1 \\ x_2 \\ x_3 \end{bmatrix}; \quad \mathbf{f} = \begin{bmatrix} f_1 \\ f_2 \\ f_3 \end{bmatrix} \tag{2.73}$$

Define the matrices **y**, **z**, **g** and **h** by

$$\mathbf{y} = \begin{bmatrix} x_1 \\ x_2 \end{bmatrix}; \quad \mathbf{z} = [x_3]; \quad \mathbf{g} = \begin{bmatrix} f_1 \\ f_2 \end{bmatrix}; \quad \mathbf{h} = [f_3] \tag{2.74}$$

Then the system of equations (2.72) may be written as

$$\begin{bmatrix} \mathbf{B} & \mathbf{C} \\ \mathbf{D} & \mathbf{E} \end{bmatrix} \begin{bmatrix} \mathbf{y} \\ \mathbf{z} \end{bmatrix} = \begin{bmatrix} \mathbf{g} \\ \mathbf{h} \end{bmatrix} \tag{2.75}$$

Multiplication of the matrices implies

$$\begin{aligned} \mathbf{By} + \mathbf{Cz} &= \mathbf{g} \\ \mathbf{Dy} + \mathbf{Ez} &= \mathbf{h} \end{aligned} \tag{2.76}$$

By inspection, the reader may check that (2.72) and (2.76) are, in fact, identical. The important point to observe is that the partitioning must be made in such a manner that the multiplications in (2.76) are meaningful, i.e. that the submatrices possess the correct dimensions. By partitioning matrices, it follows that not only multiplications, but also additions and subtractions can be carried out.

2.10 Differentiation and integration

Consider the matrix **A** given by

$$\mathbf{A} = \begin{bmatrix} A_{11} & A_{12} & A_{13} \\ A_{21} & A_{22} & A_{23} \end{bmatrix} \tag{2.77}$$

If the components of this matrix depend on a variable x, we define *matrix differentiation* as

$$\frac{\mathrm{d}\mathbf{A}}{\mathrm{d}x} = \begin{bmatrix} \dfrac{\mathrm{d}A_{11}}{\mathrm{d}x} & \dfrac{\mathrm{d}A_{12}}{\mathrm{d}x} & \dfrac{\mathrm{d}A_{13}}{\mathrm{d}x} \\ \dfrac{\mathrm{d}A_{21}}{\mathrm{d}x} & \dfrac{\mathrm{d}A_{22}}{\mathrm{d}x} & \dfrac{\mathrm{d}A_{23}}{\mathrm{d}x} \end{bmatrix} \tag{2.78}$$

i.e. all components are differentiated with respect to x. As an example, consider the system of equations

$$\mathbf{A}(x)\mathbf{b} = \mathbf{f}(x) \tag{2.79}$$

where **A** and **f** depend on x, whereas **b** is constant. Differentiation of (2.79) yields

$$\frac{\mathrm{d}\mathbf{A}}{\mathrm{d}x}\mathbf{b} = \frac{\mathrm{d}\mathbf{f}}{\mathrm{d}x} \tag{2.80}$$

If the matrix multiplication in (2.79) is carried out and if differentiation is then performed, it is easily shown that the result is identical with (2.80). We conclude that the usual rules for differentiation hold even for matrix differentiations.

In a similar manner, *integration of the matrix* $\mathbf{A}$ given by (2.77) is defined as an integration of each component, i.e.

$$\int \mathbf{A}\,dx = \begin{bmatrix} \displaystyle\int A_{11}\,dx & \displaystyle\int A_{12}\,dx & \displaystyle\int A_{13}\,dx \\[2ex] \displaystyle\int A_{21}\,dx & \displaystyle\int A_{22}\,dx & \displaystyle\int A_{23}\,dx \end{bmatrix} \tag{2.81}$$

3

Direct approach – truss analysis

As already touched upon in the introduction, the finite element method is based on the concept that the body in question is divided into regions, so-called *elements*, for which the behaviour can be described in a simpler manner than the behaviour of the entire body. When the behaviour of the elements has been formulated, these elements are patched together according to some rules, thereby enabling one to predict the behaviour of the entire body.

We shall discuss later how the behaviour of the elements themselves can be obtained for general situations using the finite element method, but in some simple cases this behaviour can be obtained directly. The present chapter is devoted to a study of such a simple situation, namely the response of elastic springs which may form a conventional truss structure. We emphasize that apart from the manner in which the element characteristics are established, the remaining steps in the calculation are the same as for the general finite element method. Therefore, this chapter serves as a prelude to many of the features that we will encounter in the following chapters.

3.1 Simple structure consisting of springs

We shall present a systematic solution to the response of a structure consisting of interacting elastic springs. For this purpose, consider the structure shown in Figure 3.1, where two elastic springs with *stiffnesses* k_1 and k_2 are connected in series along the x-axis and where the loading consists of three *external forces* F_1, F_2 and F_3. These forces are directed along the springs. The positive direction of forces is chosen to be in the direction of the x-axis.

In order to analyze the entire structure in a systematic manner, which can easily be applied to more complicated spring systems, we split the structure into smaller parts, i.e. elements, for which we directly know the response. The ends of each element are called nodal points. In the present situation it is natural to split the structure into two elements corresponding to three nodal points as shown in Figure 3.2.

The structure in Figure 3.1 can now be illustrated as shown in Figure 3.3.

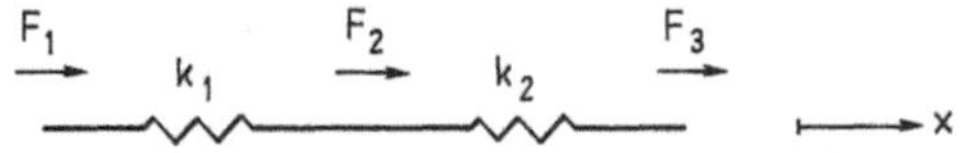

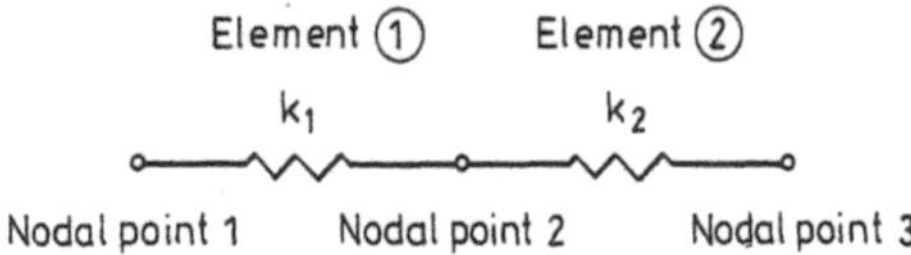

Figure 3.1 Two springs loaded by external forces

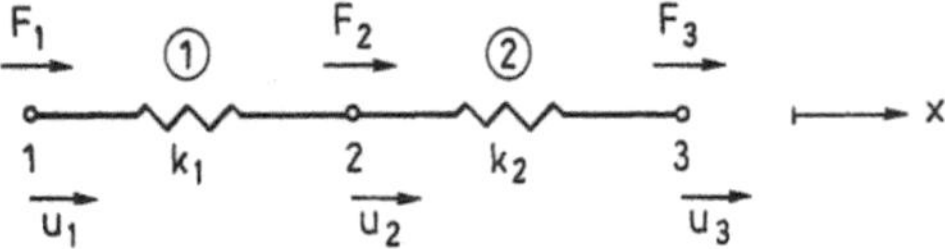

Figure 3.2 Division of structure into elements

Figure 3.3 Structure characterized by forces, displacements, elements and nodal points

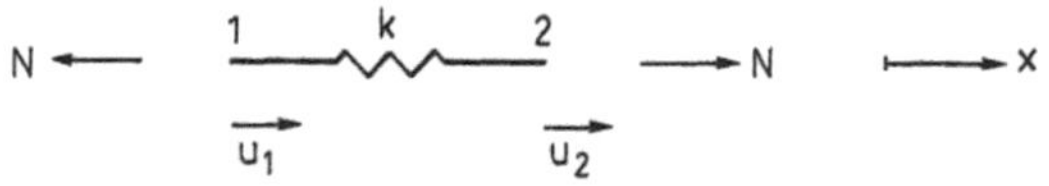

Figure 3.4 Force N within an element

In this figure the *displacements* u_1, u_2 and u_3 of the nodal points are also indicated. In order to be systematic, the positive direction of the displacements is the same as the positive direction of the forces, i.e. in the direction of the x-axis. The external forces F_1, F_2 and F_3 are assumed to be applied at the nodal points. It appears that, by means of Figure 3.3, we have obtained a systematic description of the structure considered.

Let us now evaluate the response of an element, i.e. one spring, and begin by recalling some elementary facts. For this purpose, consider first Figure 3.4, where an element in equilibrium is loaded by two oppositely directed forces. The force N within the element is given by

$$N = k(u_2 - u_1) \tag{3.1}$$

where u_1 and u_2 are the displacements of the end points of the element. These displacements are positive when they are directed in the direction of the x-axis. Moreover, it appears that when the spring is exposed to a tensile force, N is positive.

Expression (3.1) is an example of a *constitutive* law, i.e. a relation which describes how the material behaves. In the present case we have a linear elastic response in accordance with *Hooke's law*.

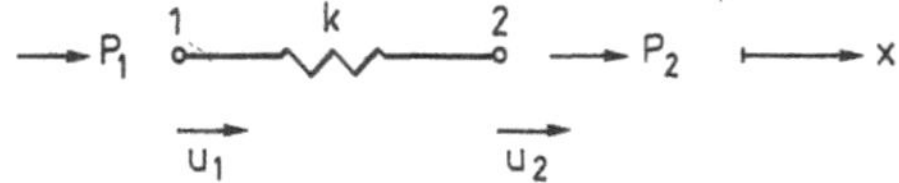

Figure 3.5   Characterization of one element

Instead of the description of the element shown by Figure 3.4 we shall adopt the one indicated in Figure 3.5, where the end points as well as the forces P_1 and P_2 and the displacements u_1 and u_2 at the end points are identified. We emphasize that P_1 and P_2 are the *element forces*, i.e. forces acting on the element, and these forces may be different from the external forces acting on the nodal points of Figure 3.3. Likewise, one should note that u_1 and u_2 of Figure 3.5 are *element displacements*, i.e. displacements at the end points of the element. By comparison of Figures 3.4 and 3.5, we have $P_2 = N$, i.e. we obtain from (3.1) that

$$P_2 = k(u_2 - u_1) \qquad . \tag{3.2}$$

Moreover, it follows that $P_1 = -N$, i.e.

$$P_1 = k(u_1 - u_2) \tag{3.3}$$

We then observe that

$$P_1 + P_2 = 0 \tag{3.4}$$

This equation is an example of a *balance equation*, which in the present situation is an *equilibrium* equation expressing the fact that the sum of all forces acting on the element is equal to zero.

We can combine (3.2) and (3.3) into

$$\begin{bmatrix} k & -k \\ -k & k \end{bmatrix} \begin{bmatrix} u_1 \\ u_2 \end{bmatrix} = \begin{bmatrix} P_1 \\ P_2 \end{bmatrix}$$

or

$$\boxed{\mathbf{K}^e \mathbf{a}^e = \mathbf{f}^e} \tag{3.5}$$

where

$$\mathbf{K}^e = \begin{bmatrix} k & -k \\ -k & k \end{bmatrix}; \quad \mathbf{a}^e = \begin{bmatrix} u_1 \\ u_2 \end{bmatrix}; \quad \mathbf{f}^e = \begin{bmatrix} P_1 \\ P_2 \end{bmatrix} \tag{3.6}$$

$\mathbf{K}^e$ is termed the *element stiffness matrix*, $\mathbf{a}^e$ the *nodal displacement vector* for the element and $\mathbf{f}^e$ the *element force vector*. We observe that $\mathbf{K}^e$ is symmetric. Equation (3.5) is the *element stiffness relation* and it describes the relation between element displacements and element forces at the end points.

Let us now split the structure shown in Figure 3.3 so that the response of each element becomes apparent. This is shown in Figure 3.6 where $u_1^{\text{①}}$ refers to the element

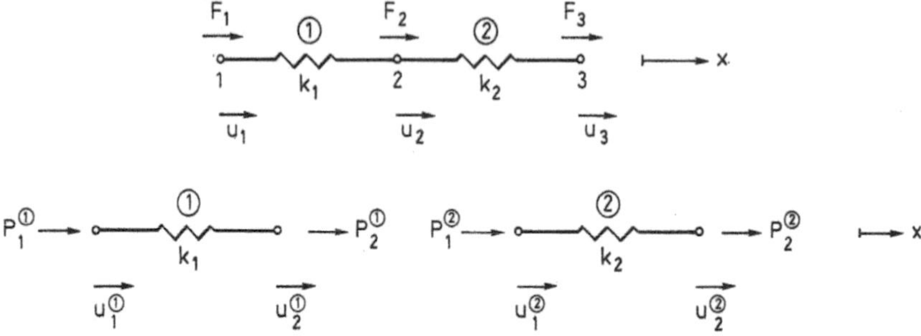

Figure 3.6   Identification of structure and splitting into elements

displacement at end point 1 of element 1 whereas, for instance, $P_1^{\circled{2}}$ refers to the element force at end point 1 of element 2.

With this notation and referring to (3.5), we obtain for element 1

$$\begin{bmatrix} k_1 & -k_1 \\ -k_1 & k_1 \end{bmatrix} \begin{bmatrix} u_1^{\circled{1}} \\ u_2^{\circled{1}} \end{bmatrix} = \begin{bmatrix} P_1^{\circled{1}} \\ P_2^{\circled{1}} \end{bmatrix} \tag{3.7}$$

and, likewise, we obtain for element 2

$$\begin{bmatrix} k_2 & -k_2 \\ -k_2 & k_2 \end{bmatrix} \begin{bmatrix} u_1^{\circled{2}} \\ u_2^{\circled{2}} \end{bmatrix} = \begin{bmatrix} P_1^{\circled{2}} \\ P_2^{\circled{2}} \end{bmatrix} \tag{3.8}$$

Equations (3.7) and (3.8) describe the behaviour of two elements independently of each other without recognizing that the structure comprises these elements. In reality the displacements u_1, u_2, u_3 of the structure are related to the element displacements, $u_1^{\circled{1}}, u_2^{\circled{1}}$ and $u_1^{\circled{2}}, u_2^{\circled{2}}$. From Figure 3.6 appears that

$$u_1^{\circled{1}} = u_1; \quad u_2^{\circled{2}} = u_3 \tag{3.9}$$

Moreover, the two springs are connected at nodal point 2, i.e.

$$u_2^{\circled{1}} = u_1^{\circled{2}} = u_2 \tag{3.10}$$

With (3.9) and (3.10), equations (3.7) and (3.8) become

$$\begin{bmatrix} k_1 & -k_1 \\ -k_1 & k_1 \end{bmatrix} \begin{bmatrix} u_1 \\ u_2 \end{bmatrix} = \begin{bmatrix} P_1^{\circled{1}} \\ P_2^{\circled{1}} \end{bmatrix} \tag{3.11}$$

and

$$\begin{bmatrix} k_2 & -k_2 \\ -k_2 & k_2 \end{bmatrix} \begin{bmatrix} u_2 \\ u_3 \end{bmatrix} = \begin{bmatrix} P_1^{\circled{2}} \\ P_2^{\circled{2}} \end{bmatrix} \tag{3.12}$$

The conditions (3.9) and (3.10) express the fact that the elements are *connected* or *compatible*, i.e. (3.9) and (3.10) express conditions for *displacement compatibility*.

As we have three unknowns in terms of the displacements u_1, u_2 and u_3 of the structure, it is convenient to rewrite (3.11) and (3.12) to obtain the forms

$$
\begin{bmatrix} k_1 & -k_1 & 0 \\ -k_1 & k_1 & 0 \\ 0 & 0 & 0 \end{bmatrix}
\begin{bmatrix} u_1 \\ u_2 \\ u_3 \end{bmatrix} =
\begin{bmatrix} P_1^{\textcircled{1}} \\ P_2^{\textcircled{1}} \\ 0 \end{bmatrix}
\tag{3.13}
$$

and

$$
\begin{bmatrix} 0 & 0 & 0 \\ 0 & k_2 & -k_2 \\ 0 & -k_2 & k_2 \end{bmatrix}
\begin{bmatrix} u_1 \\ u_2 \\ u_3 \end{bmatrix} =
\begin{bmatrix} 0 \\ P_1^{\textcircled{2}} \\ P_2^{\textcircled{2}} \end{bmatrix}
\tag{3.14}
$$

The relations (3.13) and (3.14) are termed the *expanded element relations* since they comprise the element relations written in an expanded form – so that they involve the unknowns u_1, u_2 and u_3 of the entire structure. We may write (3.13) applicable for element 1 as

$$
\boxed{\mathbf{K}_1^{\mathrm{ee}}\mathbf{a} = \mathbf{f}_1^{\mathrm{ee}}}
\tag{3.15}
$$

where

$$
\mathbf{K}_1^{\mathrm{ee}} = \begin{bmatrix} k_1 & -k_1 & 0 \\ -k_1 & k_1 & 0 \\ 0 & 0 & 0 \end{bmatrix};\quad
\mathbf{a} = \begin{bmatrix} u_1 \\ u_2 \\ u_3 \end{bmatrix};\quad
\mathbf{f}_1^{\mathrm{ee}} = \begin{bmatrix} P_1^{\textcircled{1}} \\ P_2^{\textcircled{1}} \\ 0 \end{bmatrix}
\tag{3.16}
$$

$\mathbf{K}_1^{\mathrm{ee}}$ is termed the *expanded element stiffness matrix* for element 1, $\mathbf{a}$ the *nodal displacement vector* for the entire structure and $\mathbf{f}_1^{\mathrm{ee}}$ the *expanded element force vector* for element 1. We observe that $\mathbf{K}_1^{\mathrm{ee}}$ is symmetric. Likewise, for element 2, we obtain from (3.14)

$$
\boxed{\mathbf{K}_2^{\mathrm{ee}}\mathbf{a} = \mathbf{f}_2^{\mathrm{ee}}}
\tag{3.17}
$$

where

$$
\mathbf{K}_2^{\mathrm{ee}} = \begin{bmatrix} 0 & 0 & 0 \\ 0 & k_2 & -k_2 \\ 0 & -k_2 & k_2 \end{bmatrix};\quad
\mathbf{f}_2^{\mathrm{ee}} = \begin{bmatrix} 0 \\ P_1^{\textcircled{2}} \\ P_2^{\textcircled{2}} \end{bmatrix}
\tag{3.18}
$$

In the establishment of the element stiffness relation (3.5) we have used the equilibrium condition for each element (cf. (3.4)). The element forces P_1 and P_2 acting on the element and shown in Figure 3.5 enter this element stiffness relation. For the structure in question, all element forces are shown in Figure 3.6. It is obvious, however, that these element forces are related in one way or another to the external forces F_1, F_2, F_3 acting on the structure. This relation can be derived by expressing the fact that each nodal point of the entire structure is in equilibrium. The external forces

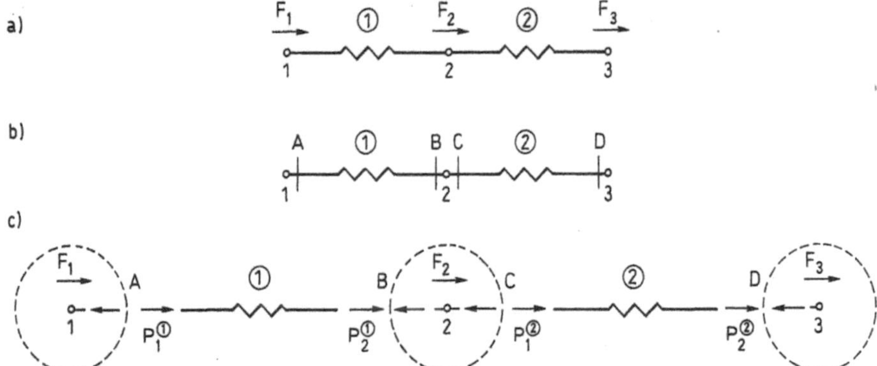

Figure 3.7   Forces acting on each nodal point

F_1, F_2, F_3 on the structure are shown again in Figure 3.7(a). Let us now cut the structure at points A, B, C and D as shown in Figure 3.7(b) so that the structure is divided into free bodies. Using the principle of *action and reaction*, all the forces on all the nodal points are illustrated in Figure 3.7(c).

Equilibrium for nodal point 1 requires that $F_1 - P_1^{(1)} = 0$, i.e.

$$F_1 = P_1^{(1)} \tag{3.19}$$

Likewise equilibrium for nodal points 2 and 3 requires that

$$F_2 = P_2^{(1)} + P_1^{(2)}; \quad F_3 = P_2^{(2)} \tag{3.20}$$

respectively. Equations (3.19) and (3.20) can be combined into

$$\begin{bmatrix} P_1^{(1)} \\ P_2^{(1)} \\ 0 \end{bmatrix} + \begin{bmatrix} 0 \\ P_1^{(2)} \\ P_2^{(2)} \end{bmatrix} = \begin{bmatrix} F_1 \\ F_2 \\ F_3 \end{bmatrix} \tag{3.21}$$

The convention that the positive directions for element forces and external forces are the same is responsible for this simple result. Let us define the *force vector* $\mathbf{f}$ by

$$\mathbf{f} = \begin{bmatrix} F_1 \\ F_2 \\ F_3 \end{bmatrix} \tag{3.22}$$

Use of (3.16), (3.18) and (3.22) enables us to write (3.21) as

$$\boxed{\mathbf{f}_1^{ee} + \mathbf{f}_2^{ee} = \mathbf{f}} \quad \text{or} \quad \mathbf{f} = \sum_{\alpha=1}^{2} \mathbf{f}_\alpha^{ee} \tag{3.23}$$

We are now in a position to establish the system of equations that determines the response of the entire structure. From (3.15), (3.17) and (3.23) it follows that

$$\mathbf{K}_1^{ee}\mathbf{a} + \mathbf{K}_2^{ee}\mathbf{a} = \mathbf{f} \tag{3.24}$$

i.e.

$$\boxed{\mathbf{Ka} = \mathbf{f}} \tag{3.25}$$

where

$$\boxed{\mathbf{K} = \mathbf{K}_1^{ee} + \mathbf{K}_2^{ee}} \quad \text{or} \quad \mathbf{K} = \sum_{\alpha=1}^{2} \mathbf{K}_\alpha^{ee} \tag{3.26}$$

Written explicitly (3.25) takes the form

$$\begin{bmatrix} k_1 & -k_1 & 0 \\ -k_1 & k_1 + k_2 & -k_2 \\ 0 & -k_2 & k_2 \end{bmatrix} \begin{bmatrix} u_1 \\ u_2 \\ u_3 \end{bmatrix} = \begin{bmatrix} F_1 \\ F_2 \\ F_3 \end{bmatrix} \tag{3.27}$$

In (3.25) $\mathbf{K}$ is the *total* or *global stiffness matrix* of the structure, the column matrix $\mathbf{a}$ contains the displacements of the nodal points of the structure and the column matrix $\mathbf{f}$ contains the external forces applied at the nodal points of the structure. We observe from (3.27) that $\mathbf{K}$ is *symmetric*. A system of equations like (3.25) is typical of finite element calculations and the symmetry of $\mathbf{K}$ is also typical. The process by which the system of equations (3.25) is constructed from the response of each element for the entire structure (cf. (3.23) and (3.26)) is called *assembling*. In this assembling process we have used conditions for compatibility, namely that the elements are connected, as well as equilibrium conditions for the nodal points. We shall find that the relations (3.23) and (3.26) used in the assembling process are characteristic of general finite element formulations.

We emphasize that the element stiffness relation was based on a constitutive relation – Hooke's law, equation (3.1) – and the condition of equilibrium for the element (cf. (3.4)). The assembling process, by which the system of equations (3.25) for the entire structure is constructed, is based on compatibility (i.e. the elements are connected, cf. (3.9) and (3.10)) as well as on requiring equilibrium for each nodal point (cf. (3.21) or (3.23)). Equation (3.23) is remarkable as it states that the external force at a nodal point is equal to the sum of all element forces related to that nodal point. This simple result follows from equilibrium – the principle of action and reaction – and from the fact that we have chosen the same positive direction for external forces and element forces. From (3.23) it follows that we can just add the expanded element stiffness matrices given by (3.16) and (3.18) to obtain the total stiffness matrix $\mathbf{K}$ (cf. (3.26)). Once these points are fully understood by the reader, it is in fact very easy to construct the equation system (3.25) directly without going through all the detailed steps given above.

If all the displacements $\mathbf{a}$ are known, the forces $\mathbf{f}$ which must be applied in order to obtain these displacements can be found directly from (3.25). In practice, the objective is most often to determine the displacements **a** when (3.25) is given. However, to solve (3.25) we know from Chapter 2 that the determinant of **K** is essential. From (3.25) and (3.27) it is easily shown that

$$\boxed{\det \mathbf{K} = 0} \tag{3.28}$$

i.e. the system of equations (3.25) possesses no unique solution. It is important to understand the physical evaluation of this mathematical statement. In establishing (3.25) we have, among other things, used the fact that the structure is in equilibrium. Indeed, by adding the three equations of (3.27) it follows that $F_1 + F_2 + F_3 = 0$, which is the equilibrium condition for the entire structure. However, if no displacements are prescribed *a priori*, the structure is in equilibrium for infinitely many positions; it may even move with a constant velocity. Therefore, in order to obtain a unique solution for the displacements we must specify that *rigid-body motions* are not allowed. This can be achieved by specifying at least one nodal point displacement.

Let us, for example, specify that $u_1 = 0$. Then (3.27) can be written as

$$-k_1 u_2 = F_1 \tag{3.29}$$

and

$$\begin{bmatrix} k_1 + k_2 & -k_2 \\ -k_2 & k_2 \end{bmatrix} \begin{bmatrix} u_2 \\ u_3 \end{bmatrix} = \begin{bmatrix} F_2 \\ F_3 \end{bmatrix} \tag{3.30}$$

The determinant of this coefficient matrix is $k_1 k_2$, which is different from zero, i.e. (3.30) possesses a unique solution given by

$$u_2 = \frac{1}{k_1}(F_2 + F_3); \quad u_3 = \frac{F_2}{k_1} + \left(\frac{1}{k_1} + \frac{1}{k_2} \right) F_3 \tag{3.31}$$

Assume that the external forces F_2 and F_3 are given; then the displacements u_2 and u_3 are determined from (3.31). Knowing the u_2-value, the external force F_1 is determined from (3.29). We observe that if F_2 and F_3 are positive then F_1 is negative, i.e. the external force F_1 is directed in the negative direction of the x-axis. In fact, F_1 is the *reaction force* at nodal point 1 which must be applied in order to keep the structure in equilibrium. From (3.29) and (3.30) it is easy to convince oneself that $F_1 + F_2 + F_3 = 0$ holds, as required for equilibrium. When all the displacements u_1, u_2 and u_3 are known, the force N in each element can be determined according to (3.1).

The specification that $u_1 = 0$ is an example of a *boundary condition*, and we emphasize that the enforcement of at least one such boundary condition is necessary in order to solve the system of equations, otherwise the determinant is equal to zero (cf. (3.28)). Moreover, when u_1 is specified then the reaction force F_1 is initially unknown. In contrast, when F_2 and F_3 are specified, the displacements u_2 and u_3 are initially unknown. More generally, for a given structure it is not possible to prescribe the force and the displacement at the same position.

We have already mentioned that due to the enforcement of the boundary conditions the determinant of the stiffness matrix appearing in (3.30) is different from zero. It is of interest to investigate whether this symmetric stiffness matrix is positive definite. Let $\mathbf{x}$ be an arbitrary column matrix; then we obtain

$$[x_1 \quad x_2]\begin{bmatrix} k_1 + k_2 & -k_2 \\ -k_2 & k_2 \end{bmatrix}\begin{bmatrix} x_1 \\ x_2 \end{bmatrix} = k_1 x_1^2 + k_2(x_1 - x_2)^2 > 0 \qquad (3.32)$$

i.e. this stiffness matrix is certainly positive definite (cf. (2.66)). The fact that the stiffness matrix of (3.30) is symmetric, positive definite and has a determinant different from zero is characteristic of finite element applications.

The systematic approach described above may seem somewhat cumbersome, but it lends itself directly to a computerized solution, where the stiffness relation for each element and the specification of how the elements are connected are the only information required in order to establish the system of equations governing the response of the entire structure. The enforcement of the boundary conditions then enables one to solve this system of equations.

The approach adopted for the spring system is an example of *matrix structural analysis*. Moreover, it is a so-called *displacement method* since the displacements are considered as the unknowns. However, apart from the establishment of the element stiffness relation, which could be constructed directly here, the remaining steps in the analysis turn out to be identical with those of the general finite element (FE) method. We can therefore summarize the following basic steps:

Steps in the FE method

1. Establishment of stiffness relations for each element. Material properties and equilibrium conditions for each element are used in this establishment.
2. Enforcement of compatibility, i.e. the elements are connected.
3. Enforcement of equilibrium conditions for the whole structure, in the present case for the nodal points.
4. By means of 2. and 3. the system of equations is constructed for the whole structure. This step is called assembling.
5. In order to solve the system of equations for the whole structure, the boundary conditions are enforced.
6. Solution of the system of equations.

Here, a very detailed derivation of the steps has been adopted, but once these steps have been understood, the reader may verify that it is rather straightforward to establish the system of equations for the whole structure.

Let us emphasize the following points:

1. All elements are located along the x-axis. For each element, and referring to Figure 3.5, end point 1 is always located to the left of end point 2, i.e. no ambiguity exists as to how the element is oriented. We also note that these end points are often called *local* or *element nodal points* as distinguished from the (*global*) *nodal points* of the entire structure. Since the elements only can be directed in one way, it follows that for a given global nodal point the external force applied at that nodal point is always equal to the sum of all the element forces related to that nodal point. This observation is a consequence of equilibrium as well as of the systematic orientation of forces. An example is given by (3.21).

2. This important point, in combination with the fact that the elements are joined together to form the structure, i.e. the compatibility requirement, imply the essential assembling procedure which establishes the global stiffness matrix $\mathbf{K}$ as the sum of the expanded element stiffnesses and the right-hand side $\mathbf{f}$ of (3.25) as the external forces at the nodal points.

To illustrate that once the steps above are understood one may directly establish the system of equations which controls the behaviour of the entire structure, let us consider Figure 3.8. In this structure, five nodal points exist where the external forces $F_1 \ldots F_5$ may be applied.

The displacements of the nodal points are denoted by $u_1 \ldots u_5$. Moreover, the stiffnesses of the five springs are given by $k_1 \ldots k_5$. The reader may verify that the following system of equations can be written down directly:

$$
\begin{array}{ccccc}
1 & 2 & 3 & 4 & 5
\end{array}
$$

$$
\begin{array}{c}
1 \\ 2 \\ 3 \\ 4 \\ 5
\end{array}
\begin{bmatrix}
k_1 & -k_1 & 0 & 0 & 0 \\
-k_1 & k_1 + k_2 + k_4 & -k_2 & -k_4 & 0 \\
0 & -k_2 & k_2 + k_3 & -k_3 & 0 \\
0 & -k_4 & -k_3 & k_3 + k_4 + k_5 & -k_5 \\
0 & 0 & 0 & -k_5 & k_5
\end{bmatrix}
\begin{bmatrix}
u_1 \\ u_2 \\ u_3 \\ u_4 \\ u_5
\end{bmatrix}
=
\begin{bmatrix}
F_1 \\ F_2 \\ F_3 \\ F_4 \\ F_5
\end{bmatrix}
$$

$$(3.33)$$

We recall that before this system of equations can be solved, the boundary conditions must be enforced. We also observe that the stiffness matrices of (3.33) and (3.27) are

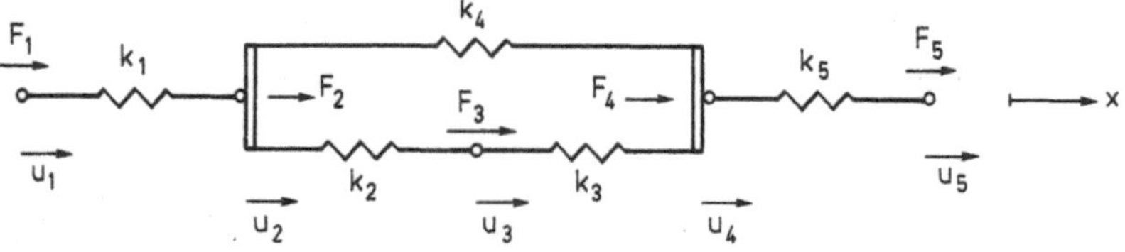

Figure 3.8 Structure consisting of five springs

banded, i.e. non-zero terms exist only about the diagonal. This is a common feature of FE equations.

3.2 Truss analysis – local and global coordinates

The spring element can even be applied to the important engineering problem of structures consisting of *trusses*, i.e. bars joined by frictionless hinges which imply that only tension or compression exists in the bars. A typical truss structure is shown in Figure 3.9 and apart from its engineering importance the analysis of truss structures provides us with the opportunity to discuss the concept of *local* and *global coordinate systems.*

With reference to the spring element, Figure 3.4, let us first establish the stiffness k for a bar loaded by the forces N as shown in Figure 3.10. The bar has length L and its cross-section is given by A.

By definition, the force N is related to the normal stress σ by

$$N = A\sigma \tag{3.34}$$

The bar is assumed to behave as linearly elastic according to Hooke's law, i.e.

$$\sigma = E\varepsilon \tag{3.35}$$

where E is Young's modulus and ε is the normal strain. By definition, ε is determined as

$$\varepsilon = \frac{u_2 - u_1}{L} \tag{3.36}$$

where u_1 and u_2 are the displacements of end points 1 and 2, respectively. The positive direction of these displacements is in the direction of the x-axis (cf. Figure 3.10).

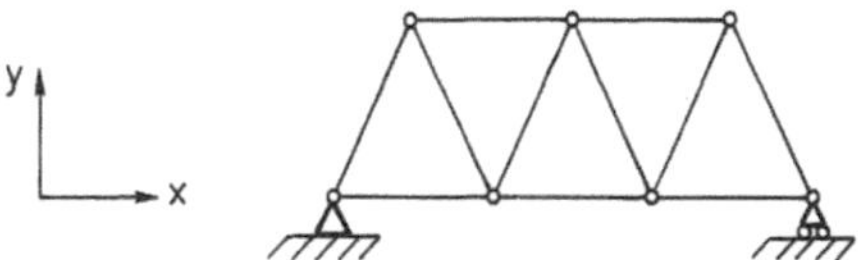

Figure 3.9 Two-dimensional truss structure consisting of bars joined by frictionless hinges

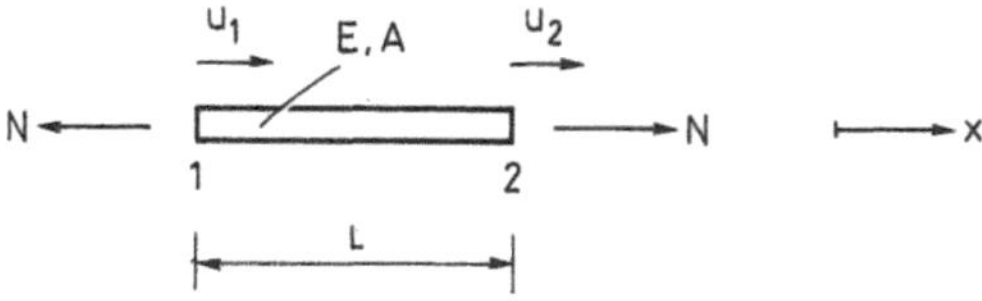

Figure 3.10 Loading of bar

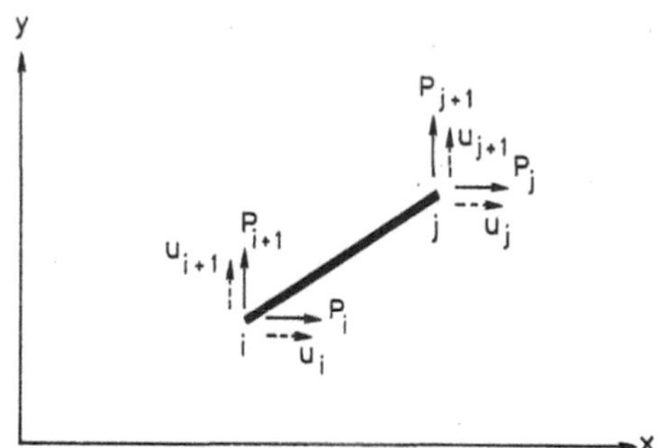

Figure 3.11   Displacement and force components of bar element in global xy-coordinate system

Combining (3.34)–(3.36) we obtain

$$N = k(u_2 - u_1) \tag{3.37}$$

where the stiffness k is given by

$$\boxed{k = AE/L} \tag{3.38}$$

It appears that when k is determined according to (3.38), the spring element can be applied to a bar oriented along the x-axis.

Now the important point is that the structure shown in Figure 3.9 is two dimensional, whereas our previous spring analysis was confined to springs joined along the x-axis. To enable us to analyze a structure like that shown in Figure 3.9, we consider a bar element located in the xy-coordinate system as shown in Figure 3.11. This element has the global nodal points i and j as shown.

At each global nodal point we may have element forces and element displacements with components in the x- and y-directions as shown in Figure 3.11. At nodal point i, we have the element forces P_i and P_{i+1} as well as the element displacements u_i and u_{i+1}. Likewise at nodal point j, we have the element forces P_j and P_{j+1} as well as the element displacements u_j and u_{j+1}. The positive directions of these forces and displacements are in the direction of the coordinate axes as shown in the figure. Our objective is to determine the relation between the forces P_i, P_{i+1}, P_j, P_{j+1} and the displacements u_i, u_{i+1}, u_j, u_{j+1}.

For this purpose, and before any loading is applied, we introduce a local $x'y'$-coordinate system with origin at end point 1 and with the x'-axis directed along the bar axis from end point 1 towards end point 2. Moreover, the y'-axis is chosen such that the $x'y'$-system can be obtained by a rotation of the xy-system. This rotation is given by the angle ϕ. We then arrive at the situation shown in Figure 3.12.

In this figure we also have element forces and element displacements at each end point, but now these forces and displacements are given in accordance with their components in the x'- and y'-directions. At end point 1 we have the element forces P'_1 and P'_2 as well as the element displacements u'_1 and u'_2. Likewise at end point 2,

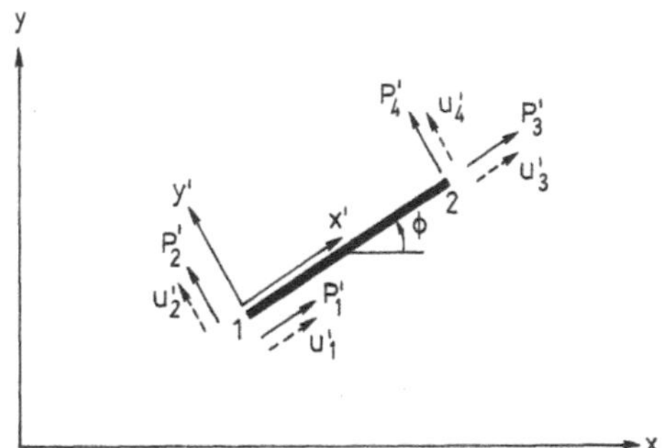

Figure 3.12 Displacement and force components of bar element in local $x'y'$-coordinate system

we have the element forces P'_3 and P'_4 as well as the element displacements u'_3 and u'_4. The positive directions of these forces and displacements are in the direction of the local coordinate axes, as shown in Figure 3.12.

From Figure 3.5 and equations (3.5) and (3.6) we obtain

$$k \begin{bmatrix} 1 & -1 \\ -1 & 1 \end{bmatrix} \begin{bmatrix} u'_1 \\ u'_3 \end{bmatrix} = \begin{bmatrix} P'_1 \\ P'_3 \end{bmatrix} \tag{3.39}$$

where the stiffness k is determined from (3.38). Equation (3.39) can be rewritten in the form

$$\boxed{\mathbf{K}^{e'} \mathbf{a}^{e'} = \mathbf{f}^{e'}} \tag{3.40}$$

where

$$\mathbf{a}^{e'} = \begin{bmatrix} u'_1 \\ u'_2 \\ u'_3 \\ u'_4 \end{bmatrix}; \quad \mathbf{f}^{e'} = \begin{bmatrix} P'_1 \\ P'_2 \\ P'_3 \\ P'_4 \end{bmatrix} \tag{3.41}$$

and

$$\mathbf{K}^{e'} = k \begin{bmatrix} 1 & 0 & -1 & 0 \\ 0 & 0 & 0 & 0 \\ -1 & 0 & 1 & 0 \\ 0 & 0 & 0 & 0 \end{bmatrix} \tag{3.42}$$

We observe that displacements along the y'-axis, i.e. u'_2 and u'_4, create no forces in the bar. This is a consequence of our assumption that the displacements are small. If large displacements were allowed, the components u'_2 and u'_4 might result in an elongation of the bar and thus in the development of forces. Moreover, the so-called *local element stiffness matrix* $\mathbf{K}^{e'}$ turns out to be symmetric.

Let us now establish a relation between the displacements u_i, u_{i+1}, u_j, u_{j+1} and $u'_1 \ldots u'_4$. As shown in Figure 3.12, the x'-axis makes an angle ϕ with the global x-axis. By geometrical arguments, it follows directly from Figures 3.11 and 3.12 that

$$u_i = u'_1 \cos \phi - u'_2 \sin \phi; \quad u_{i+1} = u'_1 \sin \phi + u'_2 \cos \phi \tag{3.43}$$

Likewise, we obtain

$$u_j = u'_3 \cos \phi - u'_4 \sin \phi; \quad u_{j+1} = u'_3 \sin \phi + u'_4 \cos \phi \tag{3.44}$$

Equations (3.43) and (3.44) can be combined into

$$\boxed{\mathbf{a}^e = \mathbf{L}^T \mathbf{a}^{e'}} \tag{3.45}$$

where

$$\mathbf{a}^e = \begin{bmatrix} u_i \\ u_{i+1} \\ u_j \\ u_{j+1} \end{bmatrix}; \quad \mathbf{a}^{e'} = \begin{bmatrix} u'_1 \\ u'_2 \\ u'_3 \\ u'_4 \end{bmatrix} \tag{3.46}$$

and

$$\mathbf{L}^T = \begin{bmatrix} \cos \phi & -\sin \phi & 0 & 0 \\ \sin \phi & \cos \phi & 0 & 0 \\ 0 & 0 & \cos \phi & -\sin \phi \\ 0 & 0 & \sin \phi & \cos \phi \end{bmatrix} \tag{3.47}$$

where $\mathbf{L}$ denotes the so-called *transformation matrix*. This matrix possesses a remarkable property. With $\mathbf{L}^T$ given by (3.47) it can be confirmed that the following relation holds:

$$\boxed{\mathbf{L}^T \mathbf{L} = \mathbf{L} \mathbf{L}^T = \mathbf{I}} \tag{3.48}$$

The transformation matrix $\mathbf{L}$ is therefore an orthogonal matrix (cf. (2.44) and (2.50)) and we have that

$$\mathbf{L}^{-1} = \mathbf{L}^T \tag{3.49}$$

Premultiplying (3.45) by $\mathbf{L}$ and using relation (3.48), we then conclude that

$$\boxed{\mathbf{a}^{e'} = \mathbf{L} \mathbf{a}^e} \tag{3.50}$$

As both forces and displacements are vector quantities, the element forces P_i, P_{i+1}, P_j, P_{j+1} and $P'_1 \ldots P'_4$ are related in exactly the same manner as the element displacements. Similarly with (3.45) and (3.46) we obtain that

$$\boxed{\mathbf{f}^e = \mathbf{L}^T \mathbf{f}^{e'}} \tag{3.51}$$

where

$$\mathbf{f}^e = \begin{bmatrix} P_i \\ P_{i+1} \\ P_j \\ P_{j+1} \end{bmatrix}; \quad \mathbf{f}^{e'} = \begin{bmatrix} P'_1 \\ P'_2 \\ P'_3 \\ P'_4 \end{bmatrix} \tag{3.52}$$

With these preliminary remarks, we are now in a position to determine the relation between the element forces $\mathbf{f}^e$ and the element displacements $\mathbf{a}^e$. Insertion of (3.50) into (3.40) yields

$$\mathbf{K}^{e'} \mathbf{L} \mathbf{a}^e = \mathbf{f}^{e'}$$

and premultiplication by $\mathbf{L}^T$ and use of (3.51) result in

$$\boxed{\mathbf{K}^e \mathbf{a}^e = \mathbf{f}^e} \tag{3.53}$$

where the element stiffness matrix $\mathbf{K}^e$ is given by

$$\boxed{\mathbf{K}^e = \mathbf{L}^T \mathbf{K}^{e'} \mathbf{L}} \tag{3.54}$$

We observe that once the local element stiffness relation (3.42) is known the *global element stiffness relation* (3.53) can be obtained by means of the transformation matrix $\mathbf{L}$.

It may be of interest to obtain an explicit expression for the global element stiffness matrix $\mathbf{K}^e$ from (3.54) with $\mathbf{K}^{e'}$ given by (3.42) and $\mathbf{L}^T$ given by (3.47). Simple calculations show that

$$\mathbf{K}^e = k \begin{bmatrix} \cos^2 \phi & \cos \phi \sin \phi & -\cos^2 \phi & -\cos \phi \sin \phi \\ \cos \phi \sin \phi & \sin^2 \phi & -\cos \phi \sin \phi & -\sin^2 \phi \\ -\cos^2 \phi & -\cos \phi \sin \phi & \cos^2 \phi & \cos \phi \sin \phi \\ -\cos \phi \sin \phi & -\sin^2 \phi & \cos \phi \sin \phi & \sin^2 \phi \end{bmatrix} \tag{3.55}$$

and it appears that the global element stiffness matrix $\mathbf{K}^e$ is symmetric.

One question may be raised: namely, how do we choose the local end point 1 and local end point 2? To illustrate this problem, consider again the bar element shown in Figure 3.11. When choosing the local coordinate system we then have the two possibilities shown in Figure 3.13.

In Figure 3.13(a) we have chosen the same possibility as shown in Figure 3.12,

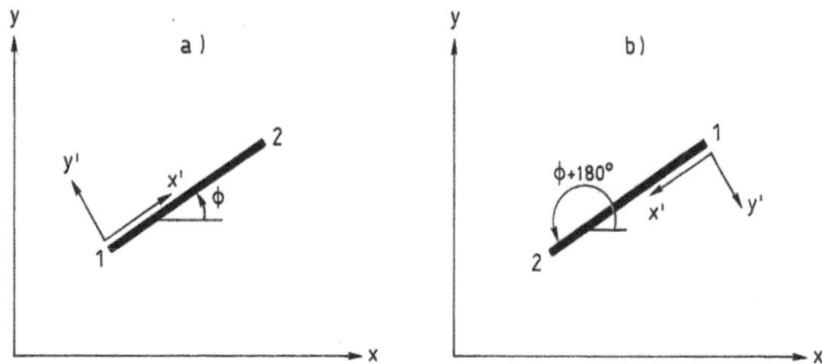

Figure 3.13   Two possibilities for choosing the local coordinate system

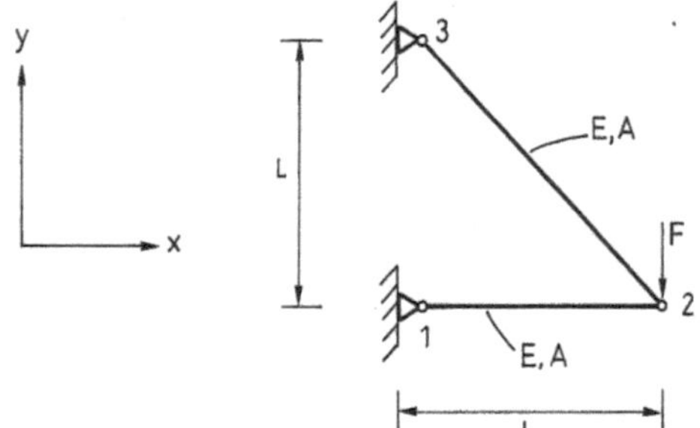

Figure 3.14   Simple truss structure

but the alternative choice shown in Figure 3.13(b) might also be taken. In Figure 3.13(a) the local coordinate system has been obtained from the global coordinate system by a rotation angle ϕ, whereas Figure 3.13(b) requires the rotation angle $\phi + 180°$.

With the choice shown in Figure 3.13(a) we derived the result given by (3.53), where the global element stiffness matrix $\mathbf{K}^e$ is given by (3.55). It appears that the only information related to the choice of the local $x'y'$-coordinate system is the angle ϕ and it also appears directly from (3.55) that if ϕ is replaced by $\phi + 180°$, the global element stiffness matrix $\mathbf{K}^e$ remains the same. This means that the two possible choices of the local coordinate system shown in Figure 3.13 result in the same global stiffness matrix.

As an example of the use of the element stiffness matrix, consider the simple truss structure shown in Figure 3.14.

The bars have the same E-modulus and cross-sectional area A. For clarity let us split the structure into two elements and also identify all displacements and forces acting on each element. This is shown in Figure 3.15, where we have already made use of the fact that the elements are joined to form the structure, i.e. the displacements shown in Figure 3.15 are the displacements of the global nodal points.

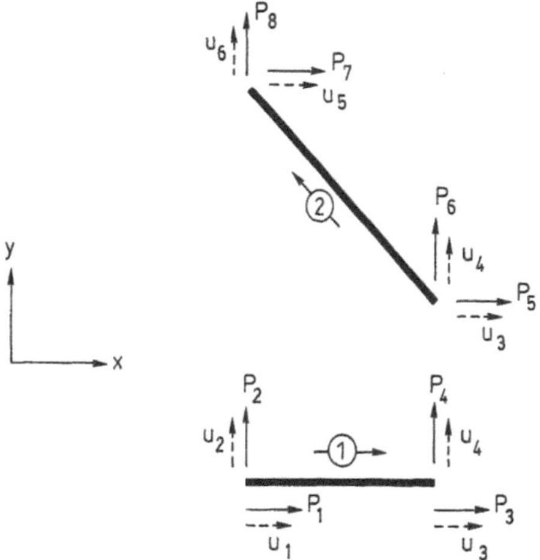

Figure 3.15 Forces and displacements acting on the elements

With reference to (3.38) and (3.55) and using $\phi = 0$ for element 1 we obtain for this element

$$\frac{AE}{L}\begin{bmatrix} 1 & 0 & -1 & 0 & 0 & 0 \\ 0 & 0 & 0 & 0 & 0 & 0 \\ -1 & 0 & 1 & 0 & 0 & 0 \\ 0 & 0 & 0 & 0 & 0 & 0 \\ 0 & 0 & 0 & 0 & 0 & 0 \\ 0 & 0 & 0 & 0 & 0 & 0 \end{bmatrix}\begin{bmatrix} u_1 \\ u_2 \\ u_3 \\ u_4 \\ u_5 \\ u_6 \end{bmatrix} = \begin{bmatrix} P_1 \\ P_2 \\ P_3 \\ P_4 \\ 0 \\ 0 \end{bmatrix} \quad \text{or} \quad \mathbf{K}_1^{ee}\mathbf{a} = \mathbf{f}_1^{ee} \qquad (3.56)$$

For element 2 the length is $L\sqrt{2}$ and $\phi = 135°$. As $\cos 135° = -1/\sqrt{2}$ and $\sin 135° = 1/\sqrt{2}$, we obtain

$$\frac{AE}{L\sqrt{2}}\begin{bmatrix} 0 & 0 & 0 & 0 & 0 & 0 \\ 0 & 0 & 0 & 0 & 0 & 0 \\ 0 & 0 & \tfrac{1}{2} & -\tfrac{1}{2} & -\tfrac{1}{2} & \tfrac{1}{2} \\ 0 & 0 & -\tfrac{1}{2} & \tfrac{1}{2} & \tfrac{1}{2} & -\tfrac{1}{2} \\ 0 & 0 & -\tfrac{1}{2} & \tfrac{1}{2} & \tfrac{1}{2} & -\tfrac{1}{2} \\ 0 & 0 & \tfrac{1}{2} & -\tfrac{1}{2} & -\tfrac{1}{2} & \tfrac{1}{2} \end{bmatrix}\begin{bmatrix} u_1 \\ u_2 \\ u_3 \\ u_4 \\ u_5 \\ u_6 \end{bmatrix} = \begin{bmatrix} 0 \\ 0 \\ P_5 \\ P_6 \\ P_7 \\ P_8 \end{bmatrix} \quad \text{or} \quad \mathbf{K}_2^{ee}\mathbf{a} = \mathbf{f}_2^{ee} \qquad (3.57)$$

As already indicated, expressions (3.56) and (3.57) are in fact the expanded element relations since each element is described by the nodal displacements for the entire structure. Let us now consider all the external forces that may be applied to the structure. These external forces are shown in Figure 3.16, and we note that they are considered as positive quantities when they are in the direction of the coordinate axes, just like the element forces.

We emphasize that these external forces act on the nodal points, whereas the forces shown in Figure 3.15 act on the elements. Using the principle of action and reaction and considering the equilibrium for each nodal point, we obtain from Figures 3.15 and 3.16 that

$$
\begin{aligned}
F_1 &= P_1; & F_2 &= P_2 \\
F_3 &= P_3 + P_5; & F_4 &= P_4 + P_6 \\
F_5 &= P_7; & F_6 &= P_8
\end{aligned}
\tag{3.58}
$$

These expressions can be combined into

$$
\begin{bmatrix} P_1 \\ P_2 \\ P_3 \\ P_4 \\ 0 \\ 0 \end{bmatrix}
+
\begin{bmatrix} 0 \\ 0 \\ P_5 \\ P_6 \\ P_7 \\ P_8 \end{bmatrix}
=
\begin{bmatrix} F_1 \\ F_2 \\ F_3 \\ F_4 \\ F_5 \\ F_6 \end{bmatrix}
\quad \text{or} \quad \mathbf{f}_1^{ee} + \mathbf{f}_2^{ee} = \mathbf{f}
\tag{3.59}
$$

It appears that these equilibrium conditions are similar to those obtained from Figure 3.7 and equation (3.21). Adding (3.56) and (3.57) and using (3.59) we obtain

$$
\frac{AE}{L}
\begin{bmatrix}
1 & 0 & -1 & 0 & 0 & 0 \\
0 & 0 & 0 & 0 & 0 & 0 \\
-1 & 0 & 1+\dfrac{1}{2\sqrt{2}} & -\dfrac{1}{2\sqrt{2}} & -\dfrac{1}{2\sqrt{2}} & \dfrac{1}{2\sqrt{2}} \\
0 & 0 & -\dfrac{1}{2\sqrt{2}} & \dfrac{1}{2\sqrt{2}} & \dfrac{1}{2\sqrt{2}} & -\dfrac{1}{2\sqrt{2}} \\
0 & 0 & -\dfrac{1}{2\sqrt{2}} & \dfrac{1}{2\sqrt{2}} & \dfrac{1}{2\sqrt{2}} & -\dfrac{1}{2\sqrt{2}} \\
0 & 0 & \dfrac{1}{2\sqrt{2}} & -\dfrac{1}{2\sqrt{2}} & -\dfrac{1}{2\sqrt{2}} & \dfrac{1}{2\sqrt{2}}
\end{bmatrix}
\begin{bmatrix} u_1 \\ u_2 \\ u_3 \\ u_4 \\ u_5 \\ u_6 \end{bmatrix}
=
\begin{bmatrix} F_1 \\ F_2 \\ F_3 \\ F_4 \\ F_5 \\ F_6 \end{bmatrix}
\tag{3.60}
$$

which is the system of equations which controls the behaviour of the structure. In an obvious notation we write

$$
\boxed{\mathbf{Ka} = \mathbf{f}} \quad \text{where} \quad \mathbf{K} = \mathbf{K}_1^{ee} + \mathbf{K}_2^{ee}
\tag{3.61}
$$

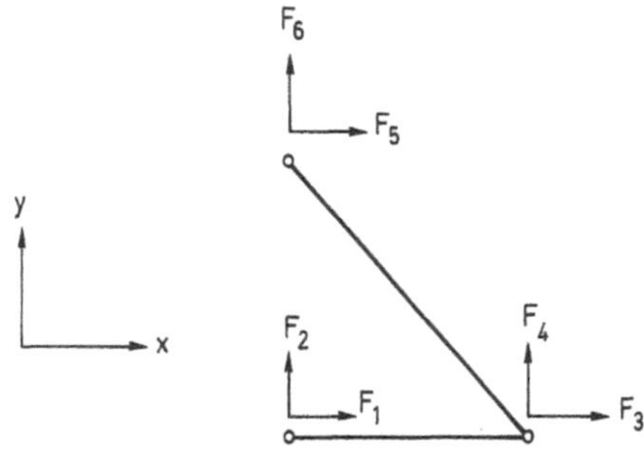

Figure 3.16   External forces applied to the structure

We observe that, due to the systematic choice of direction of forces, any external force applied to a nodal point will always be equal to the sum of those element forces joining that nodal point. Examples are given by (3.21) and (3.59). Once this important point is understood, it suffices to consider only the expanded element stiffness matrices and then to add all these matrices to obtain the global stiffness matrix $\mathbf{K}$ (cf. also (3.26)). The right-hand side $\mathbf{f}$ of the total system of equations is always equal to the external forces and we need not show this for every application, because we have now proved that this is a general statement.

In (3.60) we have not enforced the actual boundary conditions for the structure. Referring to Figures 3.14 and 3.15, we know that the displacements at the supports are zero, i.e.

$$u_1 = u_2 = 0; \quad u_5 = u_6 = 0 \tag{3.62}$$

At nodal point 2 we do not know the displacements u_3 and u_4, but instead we know the external forces. Referring to Figures 3.14 and 3.16 we have

$$F_3 = 0; \quad F_4 = -F \tag{3.63}$$

Using (3.62) and (3.63) in (3.60) we obtain

$$-\frac{AE}{L}u_3 = F_1 \tag{3.64}$$

$$0 = F_2 \tag{3.65}$$

and

$$\frac{AE}{L}\begin{bmatrix} 1 + \dfrac{1}{2\sqrt{2}} & -\dfrac{1}{2\sqrt{2}} \\[3mm] -\dfrac{1}{2\sqrt{2}} & \dfrac{1}{2\sqrt{2}} \end{bmatrix}\begin{bmatrix} u_3 \\[2mm] u_4 \end{bmatrix} = \begin{bmatrix} 0 \\[2mm] -F \end{bmatrix} \tag{3.66}$$

as well as

$$\frac{AE}{L2\sqrt{2}}\begin{bmatrix} -1 & 1 \\ 1 & -1 \end{bmatrix}\begin{bmatrix} u_3 \\ u_4 \end{bmatrix} = \begin{bmatrix} F_5 \\ F_6 \end{bmatrix} \tag{3.67}$$

The solution of (3.66) yields

$$u_3 = -\frac{FL}{AE}; \quad u_4 = -(1 + 2\sqrt{2})\frac{FL}{AE} \tag{3.68}$$

and from this solution the external forces F_1, F_2, F_5 and F_6, which are in fact the reaction forces from the supports, can be determined. We obtain

$$F_1 = F; \quad F_2 = 0; \quad F_5 = -F; \quad F_6 = F \tag{3.69}$$

Finally, let us evaluate the physical meaning of the stiffness matrix **K**. In general, we obtain an equation in the form of (3.61). Assume that the nodal displacements **a** are given by

$$\mathbf{a} = \begin{bmatrix} u_1 \\ u_2 \\ \vdots \\ u_j \\ \vdots \\ u_n \end{bmatrix} = \begin{bmatrix} 0 \\ 0 \\ \vdots \\ 1 \\ \vdots \\ 0 \end{bmatrix} \tag{3.70}$$

where n is the number of degrees of freedom (d.o.f.) of the structure. In (3.70) all nodal displacements are equal to zero except $u_j = 1$. Using this **a**-matrix in (3.61) results in

$$\begin{bmatrix} K_{1j} \\ K_{2j} \\ \vdots \\ K_{jj} \\ \vdots \\ K_{nj} \end{bmatrix} = \begin{bmatrix} f_1 \\ f_2 \\ \vdots \\ f_j \\ \vdots \\ f_n \end{bmatrix} \tag{3.71}$$

Therefore, when the loading of the structure is such that all displacements are kept at zero except $u_j = 1$, this loading implies that the external forces **f** are given by (3.71) That is, the column j of **K** represents the external nodal forces associated with the activation of the unit displacement $u_j = 1$. However, we note that activation of one d.o.f. creates nodal forces in only that element or elements which contain the d.o.f. in question. Other elements are not strained and produce no nodal forces. As component K_{ij} of **K** is the nodal force in node i created by a unit displacement of u_j, we conclude that K_{ij} is zero unless the degrees of freedom given by i and j are both present in at least one element. The reader is invited to check these general observations with the stiffness matrix given by (3.60).

This discussion completes the description of the discrete approach. We emphasize that apart from the element stiffness relation, which could be established here directly, the remaining steps are identical to those of the general FE method (cf. the steps on page 35). The critical point in the discrete approach was the assembling process, which was treated here in great detail. In the general FE method, however, this assembling process is taken care of automatically, but it is important that the reader is able to understand the physics behind this assembling scheme. Indeed, this is one of the major objectives of this chapter. For further information on the topics dealt with, reference is made to Cook *et al.* (1989), Przemieniecki (1968), Stasa (1985), Thelandersson (1984) and Zienkiewicz and Taylor (1989).

4

Strong and weak formulations – one-dimensional heat equation

For the problems considered in the previous chapter, the finite element (FE) equations could be formulated directly. In general, however, the FE method is a numerical method to solve arbitrary differential equations. To achieve this objective, it is a characteristic feature of the FE approach that the differential equations in question are first reformulated into an equivalent form, the so-called *weak formulation*. In this chapter we shall therefore discuss the formulation in terms of so-called *strong* and *weak forms*. To facilitate this discussion we shall consider a simple differential equation, which turns out to govern one-dimensional heat flow as well as other important physical phenomena like elastic bars, flexible strings, etc. To obtain a firm background, it is convenient first to establish this differential equation.

4.1 One-dimensional heat equation – strong form

Consider a fin with an inhomogeneous temperature distribution $T(x)$, as shown in Figure 4.1(a). It is assumed that heat flows only along the fin, implying that we are considering a one-dimensional problem. The cross-sectional area of the fin is given by $A(x)$. Heat Q may be transferred to the fin across its outer surface or created internally, and Q denotes the heat input per unit time and per unit length of the fin, i.e. Q has the dimension $[J/s\,m]$. Q is measured as positive if heat is supplied to the fin.

The heat supply Q may, for instance, be created by heating of an electric wire embodied within the fin. However, irrespective of how the heat Q is supplied, this heat supply can in reality only take place if a temperature variation normal to the fin axis is present. That is, we do not have a one-dimensional heat problem, but we imagine that the thickness of the fin is small in relation to its length so that temperature variations along it are much larger than temperature variations normal to the fin axis. It is with this model approximation in mind that we consider the heat flow as one dimensional. Another example of one-dimensional heat flow is the case of an infinitely large wall with constant thickness as shown in Figure 4.1(b).

Let us now establish the differential equation which controls the heat flow in the

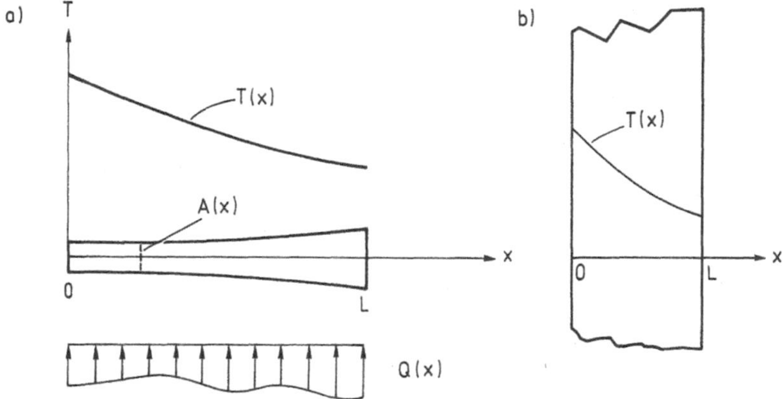

Figure 4.1 (a) Heat conduction in one-dimensional fin; (b) one-dimensional heat flow in infinitely large wall with constant thickness

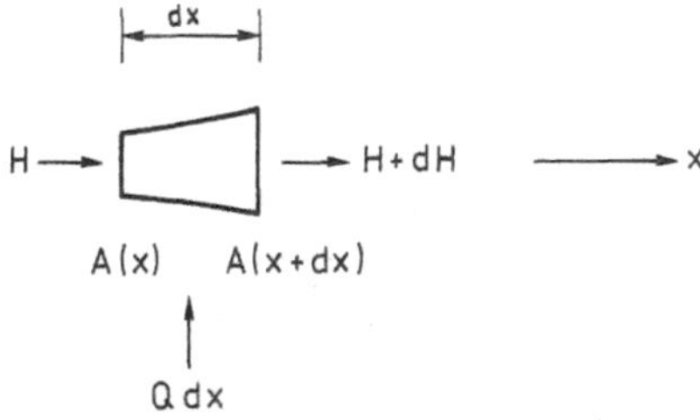

Figure 4.2 Infinitely small part of fin

fin. For this purpose, consider an infinitely small part dx of the fin, as shown in Figure 4.2.

In this figure, H denotes the heat inflow per unit time at position x and $H + dH$ denotes the heat outflow per unit time at position $x + dx$. Both H and $H + dH$ are considered positive when directed along the x-axis. It appears that we also have a heat supply per unit time given by $Q\,dx$.

Let us first establish a balance or *conservation equation*. As we only consider *stationary* conditions, i.e. the problem is assumed to be time independent, we can express the fact that the total heat inflow per unit time equals the total heat outflow per unit time. With reference to Figure 4.2, we have

$$H + Q\,dx = H + dH$$

i.e.

$$\frac{dH}{dx} = Q \tag{4.1}$$

By definition, we have

$$H(x) = A(x)q(x)$$

where q is the so-called *flux*. It appears that q is the energy which passes through a unit area per unit time, i.e. q has the dimension $[J/m^2\, s]$. Moreover, the flux q is considered positive when heat flows in the x-direction. With this expression for H, (4.1) takes the form

$$\boxed{\frac{\mathrm{d}}{\mathrm{d}x}(Aq) = Q} \tag{4.2}$$

We next invoke a *constitutive relation*, i.e. a relation which describes how heat flows within the material. With T being the temperature and k the *thermal conductivity* of the material, *Fourier's law* of heat conduction from 1822 then states that the flux q is given by

$$\boxed{q = -k\frac{\mathrm{d}T}{\mathrm{d}x}} \tag{4.3}$$

It appears that the thermal conductivity k has the dimension $[J/^\circ C\, m\, s]$ and it is convenient to take this material parameter as a positive quantity. The flux q is measured as positive when heat flows in the x-direction. As an example consider Figure 4.1 where T decreases with x, i.e. $\mathrm{d}T/\mathrm{d}x < 0$. With (4.3) this implies that the flux $q > 0$ in accordance with the physical fact that heat flows from hotter to cooler regions. This example illustrates why a minus sign appears in the constitutive equation (4.3). Insertion of (4.3) into (4.2) yields the following differential equation:

$$\frac{\mathrm{d}}{\mathrm{d}x}\left(Ak\frac{\mathrm{d}T}{\mathrm{d}x}\right) + Q = 0; \quad 0 \le x \le L \tag{4.4}$$

and if the factor Ak is constant, we obtain

$$Ak\frac{\mathrm{d}^2 T}{\mathrm{d}x^2} + Q = 0; \quad 0 \le x \le L \tag{4.5}$$

The *one-dimensional heat equation* established applies for the region considered, i.e. $0 \le x \le L$ (cf. Figure 4.1).

In order to solve this differential equation, we require *boundary conditions* at the ends of the fin. As we have a second-order differential equation we know that two boundary conditions are required corresponding to the two ends of the fin. At the ends we may assume either that the temperature T is given or that the flux q is given. If, for example, one end of the fin is completely insulated, the flux q is zero. As another example of boundary conditions, assume that at $x = 0$ the flux q is given, whereas at $x = L$ the temperature T is given, i.e.

$$q(x = 0) = -\left(k\frac{dT}{dx}\right)_{x=0} = h \tag{4.6}$$

$$T(x = L) = g \tag{4.7}$$

where h and g are known quantities.

The so-called *strong formulation* of our problem is then given by the differential equation (4.4) together with the boundary conditions (4.6) and (4.7):

Strong form of one-dimensional heat flow

$$\frac{d}{dx}\left(Ak\frac{dT}{dx}\right) + Q = 0; \quad 0 \leq x \leq L \tag{4.8}$$

$$q(x = 0) = -\left(k\frac{dT}{dx}\right)_{x=0} = h \tag{4.9}$$

$$T(x = L) = g \tag{4.10}$$

It is quite obvious that this heat conduction formulation also applies, for instance, to *diffusion* problems where ions move due to differences in ion concentration. Let the concentration be denoted by c, where c has the dimension [ions/m^3]. The ion flux q is then given by the so-called *Fick's law* from 1855

$$q = -D\frac{dc}{dx} \tag{4.11}$$

where q is now the number of ions passing through a unit area per unit time, i.e. q has the dimension [ions/m^2 s]. The *diffusion coefficient D* then takes the dimension [m^2/s]. It appears that Fick's law (4.11) is similar to Fourier's law (4.3). Moreover, for diffusion problems the balance equation for ions is similar to (4.2), i.e. both the differential equation and the boundary conditions for diffusion problems become analogous with those for heat conduction.

Next consider *electric currents* for which we have the equivalent to *Ohm's law* from 1827, i.e.

$$q = -\gamma\frac{dV}{dx} \tag{4.12}$$

where the flux q is the electric charge passing through a unit area per unit time. We recall that electric charge is measured in coulombs and the ampere is defined as coulombs per unit time. That is, the flux q has the dimension [Coulomb/m^2 s] or [Amp/m^2]. Moreover, in (4.12) V is the *voltage* and γ is the *electric conductivity*, which has the dimension [Amp/Volt m], i.e. [1/Ohm m]. The balance equation for electric charge is similar to (4.2), i.e. the heat equation also controls electric currents.

4.2 Axially loaded elastic bar

In order to illustrate that the heat equation is also encountered in other engineering applications, consider an elastic bar loaded in its axial direction by body forces b per unit axial length, i.e. b has the dimension [N/m] (cf. Figure 4.3). The body force is measured as positive if it acts in the direction of the x-axis. The cross-sectional area of the bar is $A(x)$.

To establish the differential equation for this problem we again invoke a balance principle. In the present case this balance equation takes the form of an equilibrium condition. For this purpose consider the infinitely small part $\mathrm{d}x$ of the bar, as shown in Figure 4.4, where N denotes the tensile force at the left end and $N + \mathrm{d}N$ denotes the tensile force at the right end. It appears that $b\,\mathrm{d}x$ is the total body force acting on the small body.

Equilibrium requires that the sum of all forces is equal to zero, i.e.

$$-N + b\,\mathrm{d}x + N + \mathrm{d}N = 0$$

or

$$\frac{\mathrm{d}N}{\mathrm{d}x} + b = 0 \tag{4.13}$$

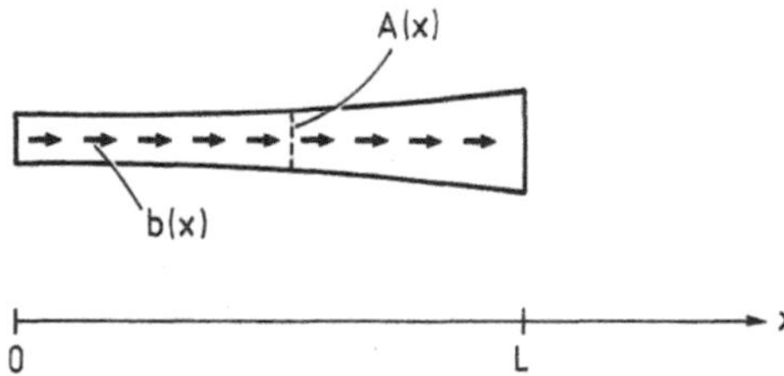

Figure 4.3   Axially loaded elastic bar

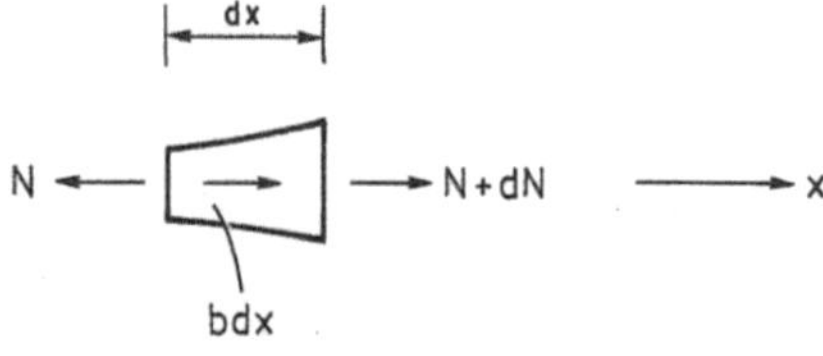

Figure 4.4   Infinitely small part of bar

The force N is given by

$$N = A(x)\sigma(x) \tag{4.14}$$

where A is the cross-sectional area and σ is the normal stress. With (4.14), (4.13) takes the form

$$\boxed{\frac{\mathrm{d}}{\mathrm{d}x}(A\sigma) + b = 0} \tag{4.15}$$

Let us now make use of the constitutive relation, which describes the manner in which the material deforms due to loading. In the present case, we assume *linear elasticity*, i.e. *Hooke's law* from 1676 states that

$$\boxed{\sigma = E\varepsilon} \tag{4.16}$$

where E is Young's modulus and ε is the normal strain, i.e. relative elongation. From elementary mechanics we have the *kinematic relation*

$$\varepsilon = \frac{\mathrm{d}u}{\mathrm{d}x} \tag{4.17}$$

where u is the displacement of the bar. This displacement is measured as positive when it is in the direction of the x-axis. Using (4.16) and (4.17) in (4.15) we obtain

$$\boxed{\frac{\mathrm{d}}{\mathrm{d}x}\left(AE\frac{\mathrm{d}u}{\mathrm{d}x}\right) + b = 0; \quad 0 \le x \le L} \tag{4.18}$$

and if the term AE is constant, this expression reduces to

$$AE\frac{\mathrm{d}^2u}{\mathrm{d}x^2} + b = 0; \quad 0 \le x \le L \tag{4.19}$$

The similarity with the heat equation (4.8) is obvious.

In order to solve (4.18), *boundary conditions* are required. These may be given in terms of either prescribed forces or displacements at the ends. As an example, assume that at $x = 0$ the force N is given, whereas at $x = L$ the displacement u is given, i.e.

$$\boxed{\begin{aligned} N(x = 0) = (A\sigma)_{x=0} = \left(AE\frac{\mathrm{d}u}{\mathrm{d}x}\right)_{x=0} = h \\ u(x = L) = g \end{aligned}} \tag{4.20}$$
$$\tag{4.21}$$

where h and g are known quantities. The similarity of (4.20) and (4.21) with (4.9) and (4.10) is obvious; we have demonstrated that the behaviour of an axially loaded elastic bar is also controlled by the same equations as for one-dimensional heat flow.

4.3 Flexible string

A further example, which is also similar to the equation for one-dimensional heat flow, is the deflection of a transversely loaded flexible string. Such a string possesses stiffness in its axial direction, but no stiffness towards bending.

The string is shown in Figure 4.5, where $w(x)$ is the lateral deflection in the z-direction and p is the lateral loading per unit axial length, i.e. p has the dimension [N/m]. The lateral loading p is measured as positive in the z-direction. Moreover, S denotes the tensile force in the string.

In order to establish the pertinent differential equation we again make use of a balance principle, which in the present case is obviously that of equilibrium. Consider therefore an infinitely small part dx of the string, as shown in Figure 4.6.

At position x the tensile force in the string is S, the deflection is w and θ denotes the angle that the string makes with the x-axis. The horizontal and vertical components of S are given by H and V, respectively. At position $x + dx$, the string force is $S + dS$, $H + dH$ (the horizontal component) and $V + dV$ (the vertical component). Moreover, it appears that $p\,dx$ is the total external force acting in the z-direction on the part dx.

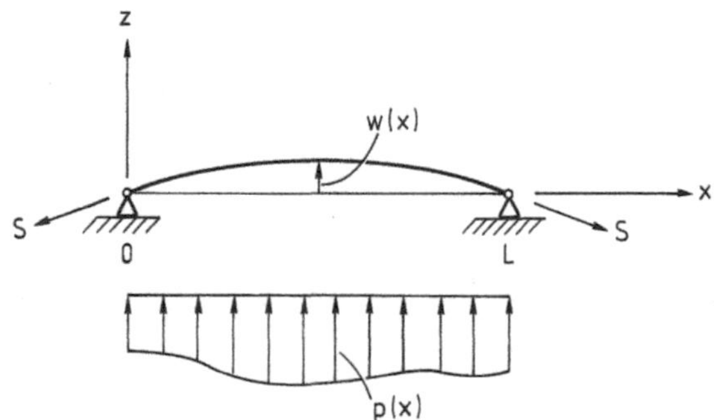

Figure 4.5   Laterally loaded flexible string

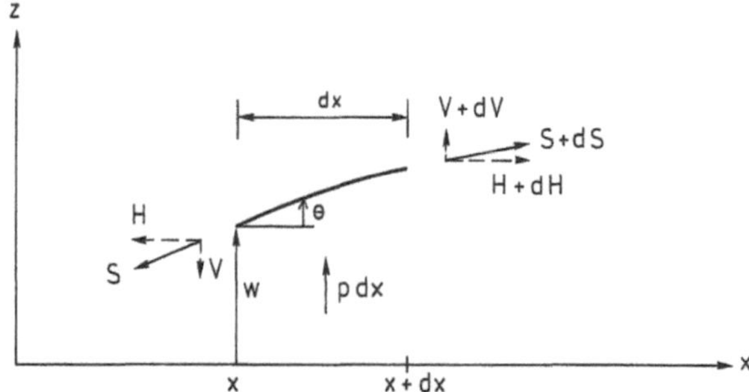

Figure 4.6   Infinitely small part of string

Horizontal equilibrium requires

$$-H + H + \mathrm{d}H = 0$$

i.e.

$$H = \text{constant} \tag{4.22}$$

The condition for vertical equilibrium is

$$-V + p\,\mathrm{d}x + V + \mathrm{d}V = 0$$

i.e.

$$\frac{\mathrm{d}V}{\mathrm{d}x} + p = 0 \tag{4.23}$$

As S is tangential to the flexible string we have

$$\mathrm{tg}\,\theta = \frac{V}{H} \tag{4.24}$$

Moreover

$$\mathrm{tg}\,\theta = \frac{\mathrm{d}w}{\mathrm{d}x}; \quad \text{i.e.} \quad V = H\frac{\mathrm{d}w}{\mathrm{d}x} \tag{4.25}$$

Inserting (4.25) in (4.23) and noting that H is constant, we get

$$H\frac{\mathrm{d}^2 w}{\mathrm{d}x^2} + p = 0 \tag{4.26}$$

in analogy with (4.8). Moreover $H = S\cos\theta$ and if we restrict ourselves to situations where the slope θ is small, we have $\cos\theta \simeq 1$, i.e.

$$\boxed{H = S = \text{constant}} \tag{4.27}$$

for small deflections. This means that the tensile force in the string is independent of the loading. Note that this is a consequence of our assumption of small deflections. With (4.27), (4.26) becomes

$$\boxed{S\frac{\mathrm{d}^2 w}{\mathrm{d}x^2} + p = 0; \quad 0 \leq x \leq L} \tag{4.28}$$

Referring to Figure 4.5, the boundary conditions for the string are of the form $w(x = 0) = 0$ and $w(x = L) = 0$. This type of boundary condition may also occur for heat flow. Note that no constitutive relation was involved in the derivation of (4.28). This is a consequence of our assumption of small deflections which implies that $S = \text{constant}$, and as no change occurs in the tensile force S, the constitutive relation is not involved.

4.4 Weak form of one-dimensional heat flow

Having established the differential equation and boundary conditions (i.e. the so-called strong form) of one-dimensional heat flow (cf. (4.8)–(4.10)), and having illustrated that this formulation also covers other physical phenomena, we shall now introduce the important concept of a *weak formulation*. The weak formulation is in fact a reformulation of the strong form and it is from this weak form that the FE approach is established.

To establish the weak form of the strong form given by (4.8)–(4.10), multiply (4.8) by an arbitrary function – the so-called *weight function* – $v(x)$ to obtain

$$v\left[\frac{\mathrm{d}}{\mathrm{d}x}\left(Ak\frac{\mathrm{d}T}{\mathrm{d}x}\right) + Q\right] = 0$$

We may even integrate this equation over the pertinent region, i.e. $0 \le x \le L$, to obtain

$$\int_0^L v\left[\frac{\mathrm{d}}{\mathrm{d}x}\left(Ak\frac{\mathrm{d}T}{\mathrm{d}x}\right) + Q\right]\mathrm{d}x = 0 \tag{4.29}$$

In order to evaluate this expression further, we recall from elementary analysis the rule for *integration by parts*. By definition we have

$$\int_a^b \frac{\mathrm{d}y}{\mathrm{d}x}\,\mathrm{d}x = \int_a^b \mathrm{d}y = [y(x)]_a^b \tag{4.30}$$

Put

$$y(x) = \phi(x)\psi(x)$$

i.e.

$$\frac{\mathrm{d}y}{\mathrm{d}x} = \frac{\mathrm{d}\phi}{\mathrm{d}x}\psi + \phi\frac{\mathrm{d}\psi}{\mathrm{d}x} \tag{4.31}$$

With (4.31), (4.30) implies that

$$\int_a^b \left(\frac{\mathrm{d}\phi}{\mathrm{d}x}\psi + \phi\frac{\mathrm{d}\psi}{\mathrm{d}x}\right)\mathrm{d}x = [\phi(x)\psi(x)]_a^b$$

i.e.

$$\boxed{\int_a^b \phi\frac{\mathrm{d}\psi}{\mathrm{d}x}\,\mathrm{d}x = [\phi\psi]_a^b - \int_a^b \frac{\mathrm{d}\phi}{\mathrm{d}x}\psi\,\mathrm{d}x} \tag{4.32}$$

With this result we conclude that

$$\int_0^L v\frac{\mathrm{d}}{\mathrm{d}x}\left(Ak\frac{\mathrm{d}T}{\mathrm{d}x}\right)\mathrm{d}x = \left[vAk\frac{\mathrm{d}T}{\mathrm{d}x}\right]_0^L - \int_0^L \frac{\mathrm{d}v}{\mathrm{d}x}Ak\frac{\mathrm{d}T}{\mathrm{d}x}\,\mathrm{d}x \tag{4.33}$$

Use of this expression in (4.29) implies that

$$\int_0^L \frac{\mathrm{d}v}{\mathrm{d}x} Ak \frac{\mathrm{d}T}{\mathrm{d}x}\,\mathrm{d}x = \left[vAk\frac{\mathrm{d}T}{\mathrm{d}x} \right]_0^L + \int_0^L vQ\,\mathrm{d}x \tag{4.34}$$

We claimed previously that the weight function $v(x)$ is arbitrary. In fact, this is not completely true, since we must require that the derivations above are meaningful. We shall not probe into this issue, but just note that the mathematical restrictions that we must place on the function $v(x)$ are clearly quite modest. We refer to Becker *et al.* (1981), Hughes (1987) and Strang and Fix (1973) for a detailed mathematical treatment.

It is important that, due to the integration by parts, (4.34) contains terms related to the boundary, i.e. the ends of the fin. These terms are given by the first expression on the right-hand side of (4.34). Now, using $q = -k\,\mathrm{d}T/\mathrm{d}x$ as well as the boundary condition (4.9) we get

$$\left[vAk\frac{\mathrm{d}T}{\mathrm{d}x} \right]_0^L = \left(vAk\frac{\mathrm{d}T}{\mathrm{d}x} \right)_{x=L} - \left(vAk\frac{\mathrm{d}T}{\mathrm{d}x} \right)_{x=0} = -(vAq)_{x=L} + (vA)_{x=0}h \tag{4.35}$$

where the flux $q(x = L)$ is unknown, whereas the flux $q(x = 0) = h$ is known. Using (4.35) in (4.34) we obtain the *weak form* of one-dimensional heat flow:

Weak form of one-dimensional heat flow

$$\int_0^L \frac{\mathrm{d}v}{\mathrm{d}x} Ak \frac{\mathrm{d}T}{\mathrm{d}x}\,\mathrm{d}x = -(vAq)_{x=L} + (vA)_{x=0}h + \int_0^L vQ\,\mathrm{d}x \tag{4.36}$$

$$T(x = L) = g \tag{4.37}$$

---

where (4.37) is just a repetition of (4.10) and where the weight function v is an arbitrary function.

The weak form was obtained from the strong form by suitable manipulations, i.e. the strong form implies the weak form. It seems natural to investigate whether the converse statement is true, i.e. whether the weak form implies the strong form. The answer is in fact affirmative, as will be shown next. In order to prove that the weak form implies the strong form, the key point is that (4.36) is required to hold not only for one function $v(x)$ but also for *any* choice of $v(x)$.

To show that the weak form implies the strong form, we accept (4.36) and (4.37). We can eliminate the term on the left-hand side of (4.36) by using the rule of integration by parts given in (4.33). Expression (4.36) then takes the form

$$\left[vAk\frac{\mathrm{d}T}{\mathrm{d}x} \right]_0^L - \int_0^L v\frac{\mathrm{d}}{\mathrm{d}x}\left(Ak\frac{\mathrm{d}T}{\mathrm{d}x} \right)\mathrm{d}x = -(vAq)_{x=L} + (vA)_{x=0}h + \int_0^L vQ\,\mathrm{d}x$$

i.e.

$$\int_0^L v\left[\frac{\mathrm{d}}{\mathrm{d}x}\left(Ak\frac{\mathrm{d}T}{\mathrm{d}x}\right) + Q\right]\mathrm{d}x$$

$$-(vA)_{x=L}\left(k\frac{\mathrm{d}T}{\mathrm{d}x} + q\right)_{x=L} + (vA)_{x=0}\left(k\frac{\mathrm{d}T}{\mathrm{d}x} + h\right)_{x=0} = 0 \tag{4.38}$$

We recall that the weight function v is an arbitrary function. We can therefore put

$$v = \phi\left[\frac{\mathrm{d}}{\mathrm{d}x}\left(Ak\frac{\mathrm{d}T}{\mathrm{d}x}\right) + Q\right] \quad \text{where} \quad \phi\begin{cases} = 0, & x = 0 \\ > 0, & 0 < x < L \\ = 0, & x = L \end{cases} \tag{4.39}$$

This implies that $v(x = 0) = v(x = L) = 0$. Using (4.39) in (4.38) we therefore obtain

$$\int_0^L \phi\left[\frac{\mathrm{d}}{\mathrm{d}x}\left(Ak\frac{\mathrm{d}T}{\mathrm{d}x}\right) + Q\right]^2 \mathrm{d}x = 0$$

from which we conclude that

$$\frac{\mathrm{d}}{\mathrm{d}x}\left(Ak\frac{\mathrm{d}T}{\mathrm{d}x}\right) + Q = 0; \quad 0 \leq x \leq L \tag{4.40}$$

We immediately observe that this is the differential equation in the strong form (cf. (4.8)). Moreover, using (4.40) in (4.38) we are left with

$$-(vA)_{x=L}\left(k\frac{\mathrm{d}T}{\mathrm{d}x} + q\right)_{x=L} + (vA)_{x=0}\left(k\frac{\mathrm{d}T}{\mathrm{d}x} + h\right)_{x=0} = 0 \tag{4.41}$$

As the weight function v is arbitrary, we now choose a v-function for which $v(x = 0) = 0$ and $v(x = L) \neq 0$. From (4.41) we then conclude that

$$-\left(k\frac{\mathrm{d}T}{\mathrm{d}x}\right)_{x=L} = q(x = L)$$

Likewise, choosing a v-function for which $v(x = 0) \neq 0$ and $v(x = L) = 0$, we obtain from (4.41) that

$$-\left(k\frac{\mathrm{d}T}{\mathrm{d}x}\right)_{x=0} = h$$

which is exactly the boundary condition given by (4.9).

Therefore, we have shown that the weak form given by (4.36) and (4.37) implies the strong form given by (4.8)–(4.10). Moreover, we already know that the strong form implies the weak form, i.e. the weak and strong formulations of our problem are identical.

Let us return to the boundary conditions (4.9) and (4.10). We have seen that the flux q (or h) emerges by itself in the boundary terms of the weak formulation.

As the flux may be prescribed at the boundary, it is therefore called a *natural boundary condition*. Occasionally, the term *Neumann boundary condition* is used. On the other hand, the boundary condition given by (4.10) and (4.37), which prescribes the value of the variable itself, is called an *essential boundary condition*. Occasionally the term *Dirichlet boundary condition* is used. Boundary conditions like (4.9) and (4.10), which contain both natural and essential conditions, are also called *mixed boundary conditions*.

We emphasize that (4.36) of the weak formulation is identical with the fulfilment of both (4.8) and (4.9) and vice versa: whereas (4.8) expresses the balance principle within the body, (4.9) expresses the balance principle at the boundary.

4.4.1 *Advantages of the weak formulation compared with the strong form*

We have proved that the weak and strong formulations are identical, so one may ask why we have taken the trouble to obtain the weak form. The answer is fundamental for the establishment of the FE method: it is on the weak form that the FE formulation is based.

In order to understand this point consider the differential equation in the strong form, i.e. (4.8). It appears that the unknown function T is differentiated twice. The FE approach is an approximate method, so in one way or another we have to replace the true T-function by an approximate one. If this approximation is made directly in (4.8), we need to deal with an approximating function, which is at least twice differentiable. In the weak form, however, only the first derivative of the temperature T enters (cf. (4.36)). That is, if we choose the weak form as the basis for the approximation, we may deal with approximating functions which only need to be differentiable once. This aspect clearly favours the weak form compared with the strong form, and it also suggests the terminology of weak and strong forms.

Another point, closely related to the matter discussed above, is that the weak form provides in fact a more general formulation than the strong form. This statement may seem contradictory, as we have just proved that the two forms are equivalent. However, in doing so we implicitly assumed that the variable T could be differentiated as many times as was necessary, i.e. twice. What is of importance is that the weak formulation (4.36) and (4.37) holds in an unchanged form even when *discontinuities* occur, whereas the presence of discontinuities requires a modification of the strong form.

To illustrate this issue, consider again the fin of Figure 4.1, but assume now that the thermal conductivity k experiences a discontinuity at point $x = a$, as shown in Figure 4.7, where $k = k_1$ holds in region 1 to the left of point $x = a$ and $k = k_2$ holds in region 2 to the right of point $x = a$.

Despite the discontinuity of the thermal conductivity at point $x = a$, the amount of heat leaving region 1 at point $x = a$ must equal the amount of heat entering region

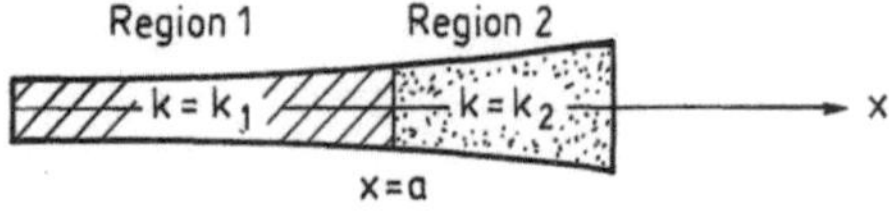

Figure 4.7 Discontinuity of thermal conductivity

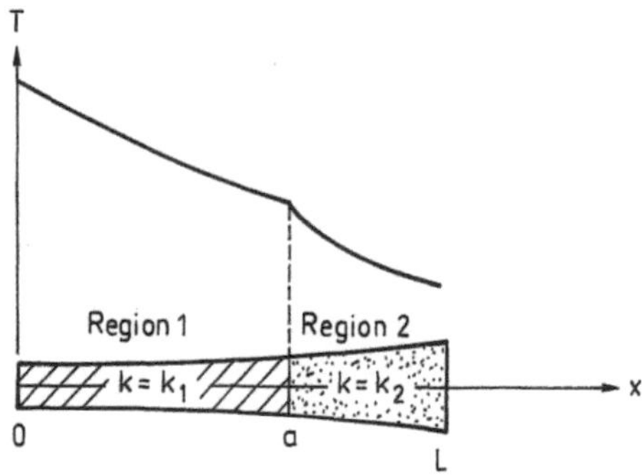

Figure 4.8 Non-smooth temperature distribution at $x = a$

2 at point $x = a$, i.e.

$$-\left(Ak_1 \frac{dT}{dx} \right)_{x=a} = -\left(Ak_2 \frac{dT}{dx} \right)_{x=a} \tag{4.42}$$

As $k_1 \neq k_2$ the derivative dT/dx takes different values to the left and right of point $x = a$. As the temperature T must be a continuous function, we obtain the temperature distribution shown in principle in Figure 4.8. It appears that the temperature T is continuous, but non-smooth at point $x = a$.

It follows directly that in regions 1 and 2, the differential equation (4.8) still holds, i.e.

$$\frac{d}{dx}\left(Ak_1 \frac{dT}{dx} \right) + Q = 0; \quad 0 \leq x \leq a \tag{4.43}$$

$$\frac{d}{dx}\left(Ak_2 \frac{dT}{dx} \right) + Q = 0; \quad a \leq x \leq L \tag{4.44}$$

Moreover, we again have the boundary conditions (4.9) and (4.10), i.e.

$$q(x = 0) = -\left(k_1 \frac{dT}{dx} \right)_{x=0} = h \tag{4.45}$$

$$T(x = L) = g \tag{4.46}$$

Equations (4.42)–(4.46) now constitute the strong form of our problem. Next let us establish the weak formulation.

For this purpose, multiply (4.43) by an arbitrary function v – the weight function – and integrate over the pertinent region, i.e.

$$\int_0^a v\left[\frac{d}{dx}\left(Ak_1\frac{dT}{dx}\right) + Q\right] dx = 0 \tag{4.47}$$

Likewise for region 2 we obtain from (4.44)

$$\int_a^L v\left[\frac{d}{dx}\left(Ak_2\frac{dT}{dx}\right) + Q\right] dx = 0 \tag{4.48}$$

Integrating (4.47) by parts similarly to (4.33) we derive

$$\int_0^a \frac{dv}{dx}Ak_1\frac{dT}{dx}\,dx = \left[vAk_1\frac{dT}{dx}\right]_0^a + \int_0^a vQ\,dx \tag{4.49}$$

In a like manner, we obtain from (4.48)

$$\int_a^L \frac{dv}{dx}Ak_2\frac{dT}{dx}\,dx = \left[vAk_2\frac{dT}{dx}\right]_a^L + \int_a^L vQ\,dx \tag{4.50}$$

In (4.49) and (4.50), the derivative dT/dx appears in the integrands. As T is non-smooth at point $x = a$ (cf. Figure 4.8) the derivative dT/dx experiences a discontinuity at this point. We then obtain the distribution of dT/dx as shown in Figure 4.9.

The important point is that even though dT/dx experiences a discontinuity, an integration of dT/dx over the region is still well defined. With this observation in mind, one may add (4.49) and (4.50) to obtain

$$\int_0^L \frac{dv}{dx}Ak\frac{dT}{dx}\,dx = \left(vAk_2\frac{dT}{dx}\right)_{x=L} - \left(vAk_1\frac{dT}{dx}\right)_{x=0} + \int_0^L vQ\,dx \tag{4.51}$$

where use is made of (4.42) and where $k = k_1$ for $0 \leq x \leq a$ and $k = k_2$ for $a < x \leq L$.

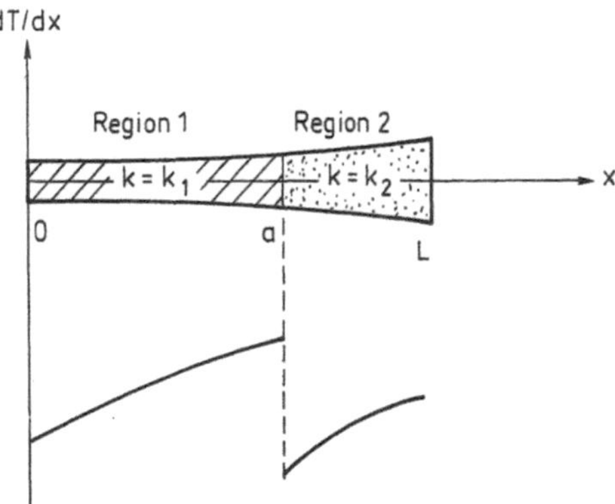

Figure 4.9   Discontinuity of dT/dx at $x = a$

Using the boundary condition (4.45) as well as $q = -k_2\,dT/dx$, we get

$$\int_0^L \frac{dv}{dx}Ak\frac{dT}{dx}\,dx = -(vAq)_{x=L} + (vA)_{x=0}h + \int_0^L vQ\,dx \tag{4.52}$$

Moreover, we also have the essential boundary condition (4.46), namely

$$T(x = L) = g \tag{4.53}$$

We observe that the weak formulation (4.52) and (4.53) is exactly the same as the one given by (4.36) and (4.37) even though we are now considering a discontinuous problem. By comparison, the strong forms of the continuous problem (4.8)–(4.10) and of the discontinuous problem (4.42)–(4.46) differ.

In this sense, we have shown that the weak form is more general than the strong form, as the weak form applies irrespective of possible discontinuities. For a more detailed discussion of these matters the reader may consult Becker *et al.* (1981), and for a general in-depth discussion of the weak formulation reference may be made to Zienkiewicz and Taylor (1989).

4.4.2 *Alternative derivation of heat equation*

Previously we established the differential equation for one-dimensional heat flow by considering an infinitely small part dx of the body (cf. Figure 4.2). For later purposes it is of interest to establish this differential equation directly by using simple rules of integration.

The fin is shown again in Figure 4.10, where for simplicity we disregard any possible discontinuities. Let us consider the arbitrary region $a < x < b$.

The balance or conservation principle for this region states that the heat per unit time entering the region equals the heat per unit time leaving the body, i.e.

$$(Aq)_{x=a} + \int_a^b Q\,dx = (Aq)_{x=b}$$

which can be written as

$$\int_a^b Q\,dx = [Aq]_a^b \quad \text{or} \quad \int_a^b Q\,dx = -\left[Ak\frac{dT}{dx}\right]_a^b \tag{4.54}$$

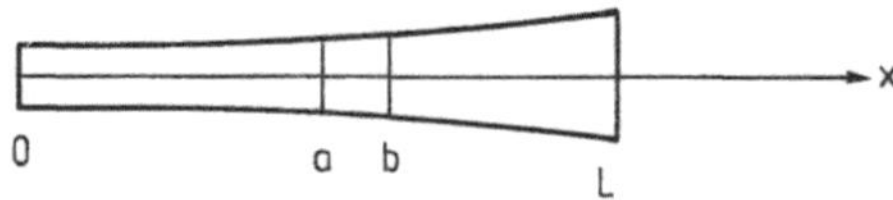

Figure 4.10   Fin with arbitrary region $a < x < b$

Table 4.1 Examples of second-order differential equations

Differential equation	Physical problem	Quantities	Constitutive law
$\dfrac{d}{dx}\left(Ak\dfrac{dT}{dx}\right)+Q=0$	One-dimensional heat flow	$T=$ temperature $A=$ area $k=$ thermal conductivity $Q=$ heat supply	Fourier $q=-k\,dT/dx$ $q=$ heat flux
$\dfrac{d}{dx}\left(AE\dfrac{du}{dx}\right)+b=0$	Axially loaded elastic bar	$u=$ displacement $A=$ area $E=$ Young's modulus $b=$ axial loading	Hooke $\sigma=E\,du/dx$ $\sigma=$ stress
$S\dfrac{d^2w}{dx^2}+p=0$	Transversely loaded flexible string	$w=$ deflection $S=$ string force $p=$ lateral loading	
$\dfrac{d}{dx}\left(AD\dfrac{dc}{dx}\right)+Q=0$	One-dimensional diffusion	$c=$ iron concentration $A=$ area $D=$ diffusion coefficient $Q=$ ion supply	Fick $q=-D\,dc/dx$ $q=$ ion flux
$\dfrac{d}{dx}\left(A\gamma\dfrac{dV}{dx}\right)+Q=0$	One-dimensional electric current	$V=$ voltage $A=$ area $\gamma=$ electric conductivity $Q=$ electric charge supply	Ohm $q=-\gamma\,dV/dx$ $q=$ electric charge flux
$\dfrac{d}{dx}\left(A\dfrac{D^2}{32\mu}\dfrac{dp}{dx}\right)+Q=0$	Laminar flow in pipe (Poiseuille flow)	$p=$ pressure $A=$ area $D=$ diameter $\mu=$ viscosity $Q=$ fluid supply	$q=-(D^2/32\mu)\,dp/dx$ $q=$ volume flux $q=$ mean velocity

By definition we have

$$\left[Ak\frac{dT}{dx}\right]_a^b = \int_a^b \frac{d}{dx}\left(Ak\frac{dT}{dx}\right)dx$$

and use of this expression in (4.54) implies that

$$\int_a^b \left[\frac{d}{dx}\left(Ak\frac{dT}{dx}\right)+Q\right]dx = 0$$

As this integral holds for an arbitrary region, i.e. arbitrary a- and b-values, we conclude that

$$\frac{d}{dx}\left(Ak\frac{dT}{dx}\right)+Q=0; \quad 0 \leq x \leq L$$

which is exactly the differential equation established previously in (4.8).

In the present one-dimensional case this alternative derivation does not provide any significant advantages over the approach adopted in Figure 4.2. For two- and three-dimensional problems, however, we shall see later that an analogous formulation to that illustrated above results in great simplifications when establishing differential equations for various physical problems.

4.5 Concluding remarks

We have seen that various physical problems lead to the same type of differential equation and boundary conditions and some examples are given in Table 4.1. The differential equation and boundary conditions together were referred to as the strong form of the problem. We showed that this strong form could be reformulated into an equivalent weak form.

It was emphasized that the weak form is the one on which the FE approach is based. The reason is that the order of differentiation of the unknown is lower in the weak form than in the strong form, thus facilitating the approximation process. Moreover, the weak form applies without changes to continuous as well as discontinuous problems, implying the important conclusion that the FE method also holds for continuous as well as discontinuous problems.

We discussed these aspects and their various ramifications for a specific differential equation since the same type of arguments can be put forward for more complicated situations like those encountered in two- and three-dimensional problems. Therefore, the conclusions of the present chapter will carry over to the more advanced problems that will be dealt with in later chapters.

5

Gradient, Gauss' divergence theorem and the Green–Gauss theorem

For a one-dimensional situation, we showed in Chapter 4 that in order to obtain the weak form from the strong form, integration by parts was necessary (cf. (4.33)). In two or three dimensions, a similar type of integration needs to be performed, and for this purpose we shall make use of the *Green–Gauss theorem*, which is based on *Gauss' divergence theorem*. These theorems are of fundamental importance for many applications. Before presenting a heuristic proof for Gauss' divergence theorem and the closely related Green–Gauss theorem, we shall discuss the concept of the *gradient* of a function.

5.1 Gradient

Consider the two-dimensional quantity ϕ which depends on the coordinates x and y, i.e. $\phi = \phi(x, y)$. As ϕ varies with position and as ϕ only comprises one quantity, we say that ϕ constitutes a *scalar field* over the region in which x and y are defined. As an example, ϕ may be taken to be the temperature distribution over a two-dimensional body and Figure 5.1 shows a principle sketch of the ϕ-distribution.

To investigate how $\phi = \phi(x, y)$ varies with position, we use the chain rule to obtain

$$d\phi = \frac{\partial \phi}{\partial x} dx + \frac{\partial \phi}{\partial y} dy \tag{5.1}$$

What (5.1) expresses is illustrated in Figure 5.2, where point P_1 has coordinates (x, y) and at that point we have $\phi(x, y)$. At the adjacent point P_2 with coordinates $(x + dx, y + dy)$ the function takes the value $\phi + d\phi$, where $d\phi$ is given by (5.1).

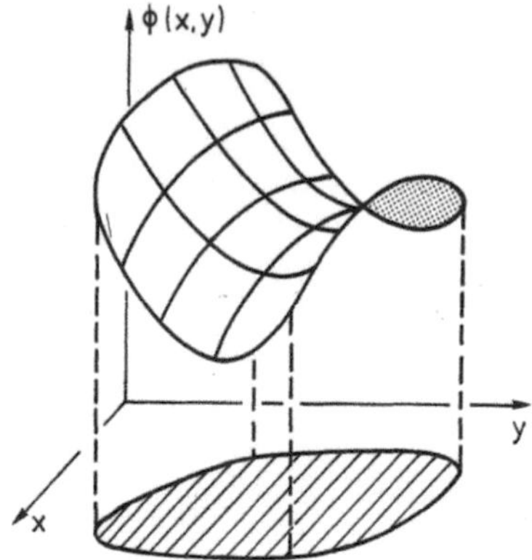

Figure 5.1   Distribution of ϕ over region in xy-plane

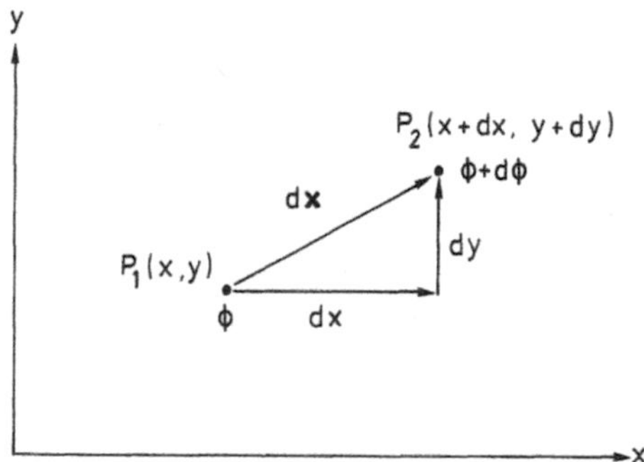

Figure 5.2   Change of ϕ with position

Let us introduce column matrices $\nabla\phi$ and d**x** defined by

$$\nabla\phi = \begin{bmatrix} \dfrac{\partial\phi}{\partial x} \\ \dfrac{\partial\phi}{\partial y} \end{bmatrix} \qquad \mathbf{dx} = \begin{bmatrix} dx \\ dy \end{bmatrix} \tag{5.2}$$

From Figure 5.2 it appears that the column matrix d**x** is in fact a vector with components dx and dy in the x- and y-directions, respectively. Likewise, we can consider the column matrix $\nabla\phi$ as a vector with components $\partial\phi/\partial x$ and $\partial\phi/\partial y$ in the x- and y-directions, respectively. The vector $\nabla\phi$ is called the *gradient* of ϕ. With

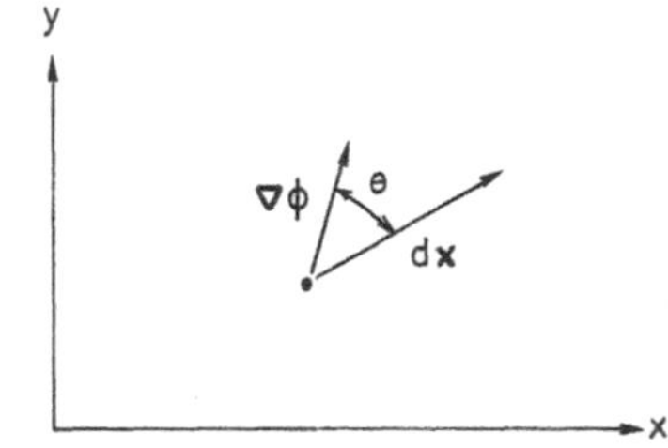

Figure 5.3 Angle θ between vectors $\nabla\phi$ and dx

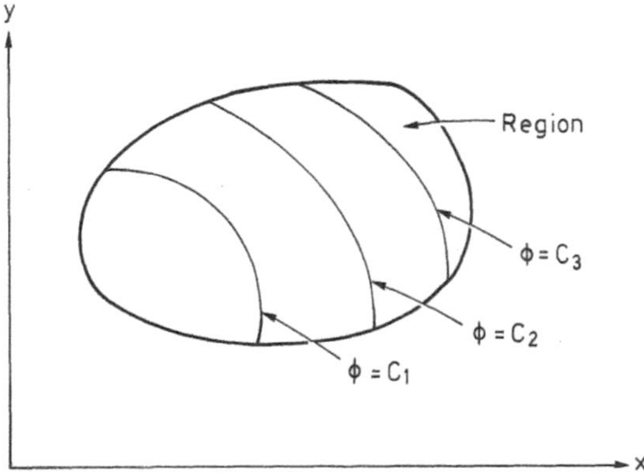

Figure 5.4 Contour curves in the xy-plane; C_1, C_2 and C_3 are constants

(5.2), (5.1) can be rewritten according to

$$\mathrm{d}\phi = \left[\frac{\partial\phi}{\partial x}\ \frac{\partial\phi}{\partial y}\right]\left[\begin{matrix}\mathrm{d}x\\\mathrm{d}y\end{matrix}\right] = (\nabla\phi)^{\mathrm{T}}\,\mathrm{d}x \tag{5.3}$$

i.e. $\mathrm{d}\phi$ is the scalar product of $\nabla\phi$ and dx. We recall from elementary vector algebra that this scalar product is given by

$$\mathrm{d}\phi = (\nabla\phi)^{\mathrm{T}}\,\mathrm{d}x = |\nabla\phi|\,|\mathrm{d}x|\cos\theta \tag{5.4}$$

where $|\nabla\phi|$ and $|\mathrm{d}x|$ are the lengths of the corresponding vectors (cf. (2.18)) and θ is the angle between the two vectors; see Figure 5.3.

As $\phi = \phi(x, y)$, it is obvious that we can identify so-called *contour curves* along which ϕ remains constant. This is achieved by putting $\phi(x, y) = C$, where C is a constant, and it follows that this expression provides a relation between x and y, i.e. it describes a curve – a contour curve – in the xy-plane. Such contour curves are illustrated in Figure 5.4.

Let us now assume that we move along a contour curve for which ϕ is constant. For such a change in position the vector dx is directed tangentially to the contour curve (cf. Figure 5.5).

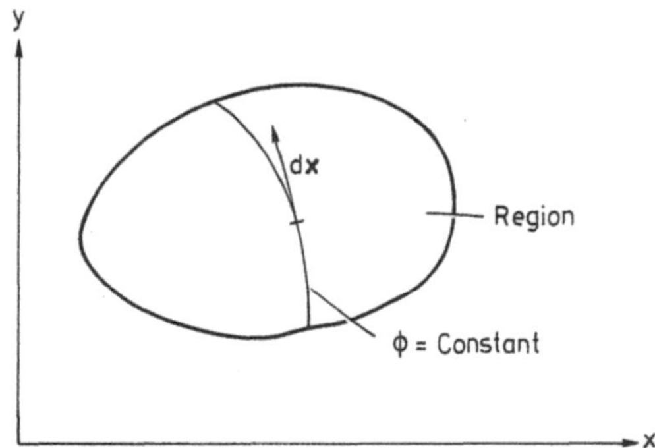

Figure 5.5   Change of position along contour curve

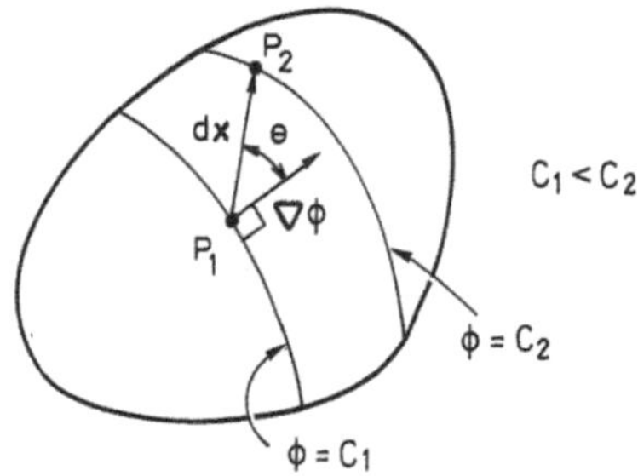

Figure 5.6   Two infinitely close contour curves

Moreover, as ϕ is constant along the contour curve we have from (5.4) that

$$d\phi = (\nabla\phi)^\mathrm{T}\, d\mathbf{x} = |\nabla\phi||d\mathbf{x}|\cos\theta = 0$$

which can be fulfilled only if $\cos\theta = 0$, i.e. the two vectors are orthogonal. As $d\mathbf{x}$ is tangential to the contour curve, we conclude that the gradient $\nabla\phi$ is orthogonal to the contour curve. However, given a certain contour curve this leaves us with two possible and opposite directions of the vector $\nabla\phi$.

To determine the correct direction of these two possibilities, consider Figure 5.6 where two infinitely close contour curves are shown and where $C_1 < C_2$.

Assume that we are at point P_1 and move to the adjacent point P_2 as indicated by the vector $d\mathbf{x}$. Since ϕ is larger at point P_2 than at point P_1, we have from (5.4) that

$$d\phi = (\nabla\phi)^\mathrm{T}\, d\mathbf{x} = |\nabla\phi||d\mathbf{x}|\cos\theta > 0 \tag{5.5}$$

This expression can only be true if $\cos\theta > 0$, i.e. the gradient $\nabla\phi$ is directed as shown in Figure 5.6. We conclude the following:

> The gradient $\nabla\phi$ is orthogonal to the contour curve and the direction of $\nabla\phi$ is in the direction of increasing ϕ-values.

Let us now assume that we are at the position given by point P_1 of Figure 5.6, i.e. the gradient $\nabla\phi$ is given and fixed. For any given length $|\mathrm{dx}|$ of dx we conclude from (5.4) that the largest possible increase of ϕ occurs when dx is in the direction of the gradient $\nabla\phi$.

The change of ϕ with position is given by (5.3). However, for later purposes it is convenient to rewrite this expression slightly. The change of position is given by the vector dx. Define the length of dx by $\mathrm{d}m$, i.e.

$$\mathrm{d}m = |\mathrm{dx}| = (\mathrm{d}x^2 + \mathrm{d}y^2)^{1/2} \tag{5.6}$$

It is then possible to define a unit vector **m** in the direction of dx. With m_x and m_y being the components of the **m**-vector, this vector is given by

$$\mathbf{m} = \begin{bmatrix} m_x \\ m_y \end{bmatrix} = \frac{\mathrm{dx}}{\mathrm{d}m} \quad \text{and} \quad |\mathbf{m}| = 1 \tag{5.7}$$

Dividing (5.3) by $\mathrm{d}m$ and using (5.7), it follows that

$$\boxed{\frac{\mathrm{d}\phi}{\mathrm{d}m} = (\nabla\phi)^{\mathrm{T}}\mathbf{m}} \tag{5.8}$$

We observe that $\mathrm{d}\phi/\mathrm{d}m$ is the slope of ϕ in the direction of the unit vector **m**. This slope is illustrated in Figure 5.7.

The results above can be generalized directly to three dimensions where $\phi = \phi(x, y, z)$. From (5.3) we have

$$\mathrm{d}\phi = (\nabla\phi)^{\mathrm{T}}\,\mathrm{dx} \tag{5.9}$$

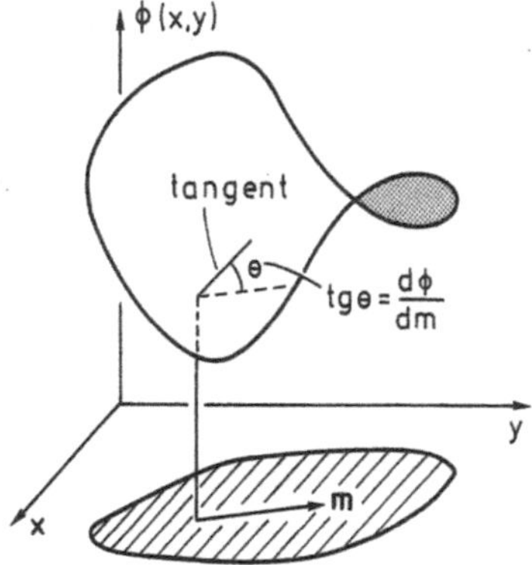

Figure 5.7　Illustration of slope $\mathrm{d}\phi/\mathrm{d}m$

where the gradient $\nabla\phi$ and dx are given by

$$\nabla\phi = \begin{bmatrix} \dfrac{\partial\phi}{\partial x} \\[2mm] \dfrac{\partial\phi}{\partial y} \\[2mm] \dfrac{\partial\phi}{\partial z} \end{bmatrix} \qquad dx = \begin{bmatrix} dx \\[2mm] dy \\[2mm] dz \end{bmatrix} \qquad\qquad (5.10)$$

From (5.9) we conclude that

$$\frac{d\phi}{dm} = (\nabla\phi)^{\mathrm{T}}\mathbf{m} \qquad\qquad (5.11)$$

where the unit vector $\mathbf{m}$ is given by

$$\mathbf{m} = \begin{bmatrix} m_x \\ m_y \\ m_z \end{bmatrix} = \frac{dx}{dm} \quad \text{and} \quad dm = |dx|; \quad \text{i.e.} \ |\mathbf{m}| = 1 \qquad\qquad (5.12)$$

5.2 Gauss' divergence theorem and the Green–Gauss theorem

The fundamental divergence theorem of Gauss and the closely related Green–Gauss theorem will now be derived in a heuristic manner. More mathematical proofs may be found in Kaplan (1981), Kreysig (1979) and Sokolnikoff and Redheffer (1958).

For the function $\phi = \phi(x, y)$ defined over the area – i.e. region – A, consider the *area integral*

$$\int_A \frac{\partial\phi}{\partial x}\, dA$$

The region is shown in Figure 5.8. The maximum and minimum y-values of the region are indicated by points D and C, where the y-values are given by $y = d$ and $y = c$, respectively. These two points divide the boundary into two curves: one described by $x = x_1(y)$ and the other by $x = x_2(y)$ (see Figure 5.8).

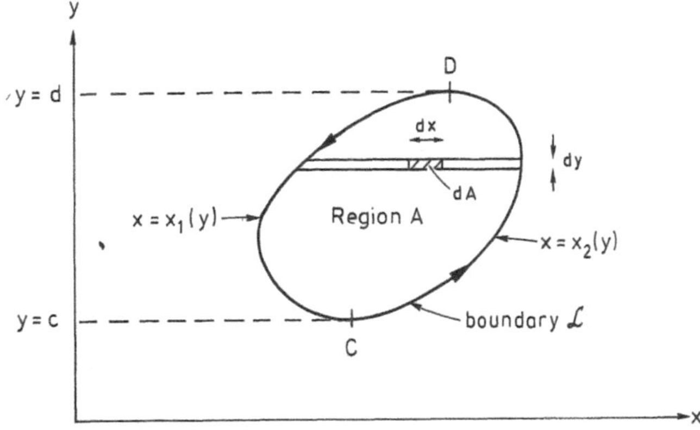

Figure 5.8   Integration over region A

By definition we have

$$\int_A \frac{\partial \phi}{\partial x}\, dA = \iint_A \frac{\partial \phi}{\partial x}\, dx\, dy = \int_c^d \left(\int_{x_1(y)}^{x_2(y)} \frac{\partial \phi}{\partial x}\, dx \right) dy$$

$$= \int_c^d \left[\phi(x, y) \right]_{x=x_1(y)}^{x=x_2(y)}\, dy = \int_c^d \left[\phi(x_2(y), y) - \phi(x_1(y), y) \right] dy$$

$$= \int_c^d \phi(x_2(y), y)\, dy + \int_d^c \phi(x_1(y), y)\, dy \tag{5.13}$$

It appears that the first term on the right-hand side is the integral taken along the boundary $x = x_2(y)$ in the counter-clockwise direction (cf. Figure 5.8), whereas the second term on the right-hand side is the integral taken along the boundary $x = x_1(y)$ in the counter-clockwise direction. By definition, it then follows that (5.13) reduces to

$$\int_A \frac{\partial \phi}{\partial x}\, dA = \oint_{\mathscr{L}} \phi(x, y)\, dy \tag{5.14}$$

where the *boundary integral* is taken in the counter-clockwise direction along the boundary $\mathscr{L}$. In exactly the same manner, it follows that for a function $\psi = \psi(x, y)$ we find

$$\int_A \frac{\partial \psi}{\partial y}\, dA = -\oint_{\mathscr{L}} \psi(x, y)\, dx \tag{5.15}$$

where the boundary integral is again taken in the counter-clockwise direction.

Let us rewrite expressions (5.14) and (5.15) by introducing the *incremental arc length* $d\mathscr{L}$ along the boundary. Recalling that we perform the boundary integration in the counter-clockwise direction, we have the situation sketched in Figure 5.9.

In this figure, **n** denotes the unit vector normal to the boundary and directed

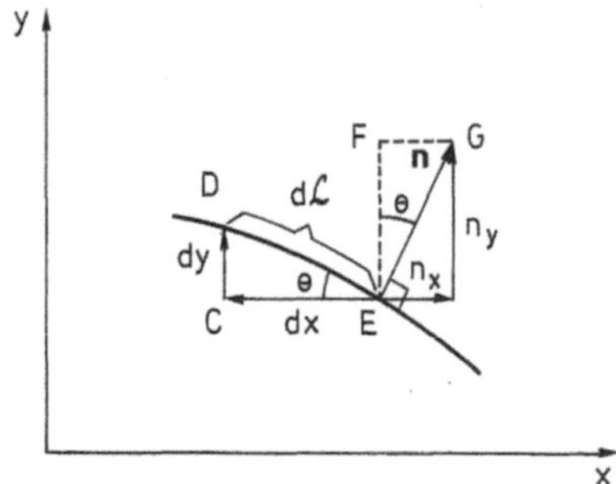

Figure 5.9   Outer unit normal vector **n** and incremental arc length $d\mathscr{L}$

out of the region. This vector is given by

$$\mathbf{n} = \begin{bmatrix} n_x \\ n_y \end{bmatrix}; \quad |\mathbf{n}| = (n_x^2 + n_y^2)^{1/2} = 1 \tag{5.16}$$

where n_x and n_y are the components of the **n**-vector along the x- and y-axes, respectively. By definition, $d\mathscr{L}$ is a *positive quantity* given by

$$\boxed{d\mathscr{L} = (dx^2 + dy^2)^{1/2}} \tag{5.17}$$

where dx and dy indicate the change of position along the boundary, i.e. $d\mathscr{L}$ is the length of the increment along the boundary (cf. Figure 5.9). This expression for $d\mathscr{L}$ may also be written as

$$d\mathscr{L} = |dx|\left[1 + \left(\frac{dy}{dx}\right)^2\right]^{1/2} > 0 \quad \text{or} \quad d\mathscr{L} = |dy|\left[1 + \left(\frac{dx}{dy}\right)^2\right]^{1/2} > 0 \tag{5.18}$$

if $dx \neq 0$ or $dy \neq 0$, respectively.

From Figure 5.9, the angle θ can be identified in the triangles CDE and FGE. That is, for the length of the sides of these triangles we have

$$\cos\theta = \frac{CE}{ED} = \frac{FE}{EG}; \quad \sin\theta = \frac{CD}{ED} = \frac{FG}{EG}$$

From Figure 5.9 these lengths (that are positive quantities) can be identified as

$$-\frac{dx}{d\mathscr{L}} = \frac{n_y}{1}; \quad \frac{dy}{d\mathscr{L}} = \frac{n_x}{1}$$

where $CE = |dx| = -dx$ since dx is negative in the present case. We then conclude that

$$dx = -n_y\,d\mathscr{L}; \quad dy = n_x\,d\mathscr{L} \tag{5.19}$$

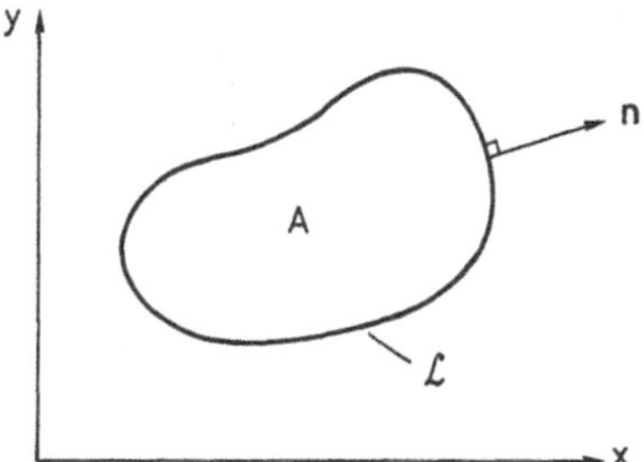

Figure 5.10   Region A with boundary $\mathscr{L}$ and outer unit vector **n**

Using (5.19) in (5.14) and (5.15) we get (see Figure 5.10)

$$\int_A \frac{\partial \phi}{\partial x}\, dA = \oint_{\mathscr{L}} \phi n_x\, d\mathscr{L} \tag{5.20}$$

$$\int_A \frac{\partial \psi}{\partial y}\, dA = \oint_{\mathscr{L}} \psi n_y\, d\mathscr{L} \tag{5.21}$$

In (5.14) and (5.15) the boundary integral should be taken in the counter-clockwise direction. In (5.20) and (5.21), however, as $d\mathscr{L}$, irrespective of the integration direction, is a positive quantity and as the terms ϕn_x and ψn_y are independent of the direction of integration, the direction of the boundary integration is arbitrary.

Now let us define the arbitrary vector **q** by

$$\mathbf{q} = \begin{bmatrix} q_x \\ q_y \end{bmatrix} = \begin{bmatrix} \phi \\ \psi \end{bmatrix} \tag{5.22}$$

where q_x and q_y are the components in the x- and y-directions, respectively. Adding (5.20) and (5.21) and using (5.22), we obtain *Gauss' divergence theorem*

$$\int_A \text{div}\, \mathbf{q}\, dA = \oint_{\mathscr{L}} \mathbf{q}^T \mathbf{n}\, d\mathscr{L} \tag{5.23}$$

where div **q** is the *divergence* of **q** defined by

$$\text{div}\, \mathbf{q} = \frac{\partial q_x}{\partial x} + \frac{\partial q_y}{\partial y} \tag{5.24}$$

The one-dimensional counterpart of (5.23) is given by (4.30).

Assume now that instead of **q** we consider the vector $\phi\mathbf{q}$ where ϕ is a scalar. Using the definition (5.24) we find that

$$\text{div}(\phi\mathbf{q}) = \frac{\partial}{\partial x}(\phi q_x) + \frac{\partial}{\partial y}(\phi q_y) = \phi\frac{\partial q_x}{\partial x} + \phi\frac{\partial q_y}{\partial y} + \frac{\partial \phi}{\partial x}q_x + \frac{\partial \phi}{\partial y}q_y$$

which, using the definition of the gradient $\nabla\phi$ (cf. (5.2)), can be written as

$$\mathrm{div}(\phi\mathbf{q}) = \phi\,\mathrm{div}\,\mathbf{q} + (\nabla\phi)^{\mathrm{T}}\mathbf{q} \tag{5.25}$$

Inserting $\phi\mathbf{q}$ instead of $\mathbf{q}$ in (5.23) and making use of (5.25), the *Green–Gauss theorem* is obtained according to

$$\int_A \phi\,\mathrm{div}\,\mathbf{q}\,\mathrm{d}A = \oint_{\mathscr{L}} \phi\mathbf{q}^{\mathrm{T}}\mathbf{n}\,\mathrm{d}\mathscr{L} - \int_A (\nabla\phi)^{\mathrm{T}}\mathbf{q}\,\mathrm{d}A \tag{5.26}$$

When integrating by parts in the one-dimensional case, the form given by (4.32) is obtained. The Green–Gauss theorem can be seen as the two-dimensional counterpart for integration by parts, and just as for the one-dimensional case, it appears from (5.26) that boundary terms emerge as a result of this integration by parts.

Sometimes, it is important to write the Green–Gauss theorem in a slightly different way. If the arbitrary vector $\mathbf{q}$ is chosen so that $q_x = \psi$ and $q_y = 0$, (5.26) reduces to

$$\int_A \phi\frac{\partial\psi}{\partial x}\,\mathrm{d}A = \oint_{\mathscr{L}} \phi\psi n_x\,\mathrm{d}\mathscr{L} - \int_A \frac{\partial\phi}{\partial x}\psi\,\mathrm{d}A \tag{5.27}$$

Alternatively, if $\mathbf{q}$ is chosen such that $q_x = 0$ and $q_y = \psi$, (5.26) implies that

$$\int_A \phi\frac{\partial\psi}{\partial y}\,\mathrm{d}A = \oint_{\mathscr{L}} \phi\psi n_y\,\mathrm{d}\mathscr{L} - \int_A \frac{\partial\phi}{\partial y}\psi\cdot\mathrm{d}A \tag{5.28}$$

The fundamental results of (5.23) and (5.26) carry over to three dimensions; see for instance Kaplan (1981), Kreyszig (1979) and Sokolnikoff and Redheffer (1958). That is, if the arbitrary vector $\mathbf{q}$ is given by

$$\mathbf{q} = \begin{bmatrix} q_x \\ q_y \\ q_z \end{bmatrix} \tag{5.29}$$

where q_x, q_y and q_z are the components in the x-, y- and z-directions, respectively, then the *divergence* of $\mathbf{q}$ becomes

$$\mathrm{div}\,\mathbf{q} = \frac{\partial q_x}{\partial x} + \frac{\partial q_y}{\partial y} + \frac{\partial q_z}{\partial z} \tag{5.30}$$

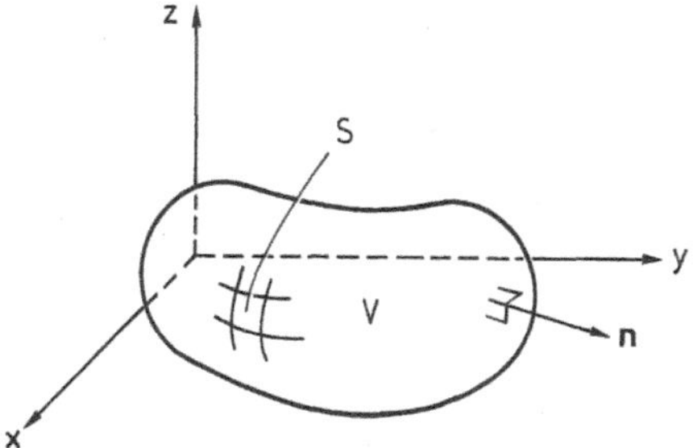

Figure 5.11   Volume V with surface boundary S and outer unit vector **n**

Let the unit vector **n** normal to the boundary and directed outwards be given by

$$\mathbf{n} = \begin{bmatrix} n_x \\ n_y \\ n_z \end{bmatrix}; \quad |\mathbf{n}| = (n_x^2 + n_y^2 + n_z^2)^{1/2} = 1 \tag{5.31}$$

where n_x, n_y, n_z are components in the x-, y-, and z-directions, respectively. *Gauss' divergence theorem* then reads

$$\int_V \operatorname{div} \mathbf{q} \, dV = \int_S \mathbf{q}^T \mathbf{n} \, dS \tag{5.32}$$

where the divergence of **q** is integrated over the volume V and where the quantity $\mathbf{q}^T\mathbf{n}$ is integrated over the surface, i.e. the boundary, of the region V (see Figure 5.11). Finally, the *Green–Gauss theorem* takes the form

$$\int_V \phi \operatorname{div} \mathbf{q} \, dV = \int_S \phi \mathbf{q}^T \mathbf{n} \, dS - \int_V (\nabla \phi)^T \mathbf{q} \, dV \tag{5.33}$$

An alternative form of this theorem is obtained by first choosing the arbitrary vector **q** as $\mathbf{q}^T = [\psi, 0, 0]$, then choosing $\mathbf{q}^T = [0, \psi, 0]$ and finally putting $\mathbf{q}^T = [0, 0, \psi]$. The result is

$$\begin{aligned} \int_V \phi \frac{\partial \psi}{\partial x} \, dV &= \int_S \phi \psi n_x \, dS - \int_V \frac{\partial \phi}{\partial x} \psi \, dV \\[2mm] \int_V \phi \frac{\partial \psi}{\partial y} \, dV &= \int_S \phi \psi n_y \, dS - \int_V \frac{\partial \phi}{\partial y} \psi \, dV \\[2mm] \int_V \phi \frac{\partial \psi}{\partial z} \, dV &= \int_S \phi \psi n_z \, dS - \int_V \frac{\partial \phi}{\partial z} \psi \, dV \end{aligned} \tag{5.34}$$

6

Strong and weak forms – two- and three-dimensional heat flow

In Chapter 4 we derived the one-dimensional heat equation and discussed its strong and weak forms. Moreover, it was demonstrated that various important physical phenomena are controlled by formulations analogous to the heat flow problem. It was emphasized that the FE method is based on the weak formulation. In the present chapter, a similar discussion will be carried out for the heat flow equation in two and three dimensions. It will turn out in this discussion that extensive use is made of Gauss' divergence theorem as well as of the Green–Gauss theorem.

6.1 Heat flux vector–constitutive relation

In one dimension the amount of heat passing through a unit area per unit time was given by the flux, which was related to the derivative of the temperature according to the constitutive relation provided by Fourier's law (cf. (4.3)).

For two- or three-dimensional problems heat may flow in various directions and this flow may conveniently be described by the *heat flux vector* $\mathbf{q}$. This flux vector $\mathbf{q}$ has the direction of the heat flow and its length expresses the heat per unit time which passes through a unit surface area perpendicular to the direction of heat flow. We have

$$\mathbf{q} = \begin{bmatrix} q_x \\ q_y \end{bmatrix}; \quad \mathbf{q} = \begin{bmatrix} q_x \\ q_y \\ q_z \end{bmatrix} \tag{6.1}$$

in two and three dimensions, respectively, where q_x, q_y and q_z are the components of $\mathbf{q}$ in the x-, y- and z-directions, respectively. These components have the dimension $[\text{J}/\text{m}^2\,\text{s}]$.

Referring to Figure 6.1, we now consider the boundary which is characterized

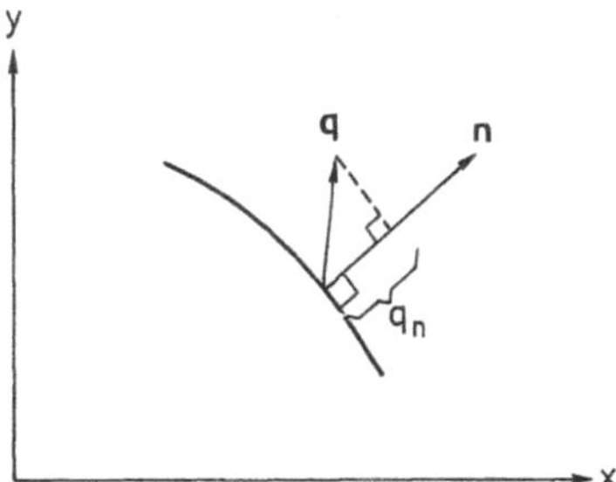

Figure 6.1   Flux q_n at boundary

by the unit vector **n** normal to the boundary and directed outwards, i.e.

$$\mathbf{n} = \begin{bmatrix} n_x \\ n_y \end{bmatrix}; \quad \mathbf{n} = \begin{bmatrix} n_x \\ n_y \\ n_z \end{bmatrix}; \quad |\mathbf{n}| = 1 \tag{6.2}$$

It is of interest to determine the total amount of heat which passes through a unit area of the boundary per unit time. This amount is denoted by q_n and it is called the *flux*. It is essential to distinguish the flux q_n from the flux vector **q**, but clearly the two quantities are related. The flux vector **q**, which indicates the direction of the heat flow, is also illustrated in Figure 6.1. It is obvious that the component of **q** tangential to the boundary does not contribute to the flux q_n which passes through the boundary. Rather, the flux q_n is given by the component of **q** in the direction of the unit vector **n**, i.e. the flux q_n is given by

$$\boxed{q_n = \mathbf{q}^{\mathrm{T}}\mathbf{n}} \tag{6.3}$$

From Figure 6.1 it follows that q_n is positive when heat leaves the boundary at the specific position considered. We are now in a position to evaluate the components q_x, q_y and q_z of **q** (cf. (6.1)). Considering, for instance, the flux q_n in the x-direction, we have $\mathbf{n}^{\mathrm{T}} = (1, 0, 0)$, i.e. (6.3) gives $q_n = q_x$. Similar interpretations apply for the components q_y and q_z.

It is obvious that the heat flux vector **q** is related to the temperature gradient ∇T and this relation is expressed by the *constitutive equation*. For this purpose the generalization of Fourier's law for one-dimensional heat flow, (4.3), to two and three dimensions becomes

$$\boxed{\mathbf{q} = -\mathbf{D}\nabla T} \tag{6.4}$$

The temperature gradient ∇T is given by (5.2) and (5.10), i.e.

$$\nabla T = \begin{bmatrix} \dfrac{\partial T}{\partial x} \\[2mm] \dfrac{\partial T}{\partial y} \end{bmatrix} ; \quad \nabla T = \begin{bmatrix} \dfrac{\partial T}{\partial x} \\[2mm] \dfrac{\partial T}{\partial y} \\[2mm] \dfrac{\partial T}{\partial z} \end{bmatrix} \tag{6.5}$$

in two and three dimensions, respectively. Moreover, in (6.4) $\mathbf{D}$ is the so-called *constitutive matrix* which contains information about the ease with which heat flows in different directions. That is, $\mathbf{D}$ is given by

$$\mathbf{D} = \begin{bmatrix} k_{xx} & k_{xy} \\ k_{yx} & k_{yy} \end{bmatrix} ; \quad \mathbf{D} = \begin{bmatrix} k_{xx} & k_{xy} & k_{xz} \\ k_{yx} & k_{yy} & k_{yz} \\ k_{zx} & k_{zy} & k_{zz} \end{bmatrix} \tag{6.6}$$

in two and three dimensions, respectively. In Fourier's law for one-dimensional heat flow, (4.3), the flux depends linearly on $\partial T / \partial x$. It appears that Fourier's generalized law, (6.4), presents the most general form for which the flux vector $\mathbf{q}$ depends linearly on the derivatives of the temperature T. If the constitutive matrix $\mathbf{D}$ depends on position, i.e. $\mathbf{D} = \mathbf{D}(x, y, z)$ then the material is referred to as *inhomogeneous*; otherwise we speak of a *homogeneous* material.

The implication of the constitutive relation (6.4) is illustrated in Figure 6.2. The temperature gradient ∇T is directed towards hotter regions and physical arguments imply that heat flows from hotter to cooler regions. Consequently, the scalar product of the vectors ∇T and $\mathbf{q}$ is negative, i.e.

$$(\nabla T)^{\mathrm{T}} \mathbf{q} < 0 \tag{6.7}$$

Using relation (6.4) in this expression, it follows that

$$\boxed{(\nabla T)^{\mathrm{T}} \mathbf{D}\, \nabla T > 0} \tag{6.8}$$

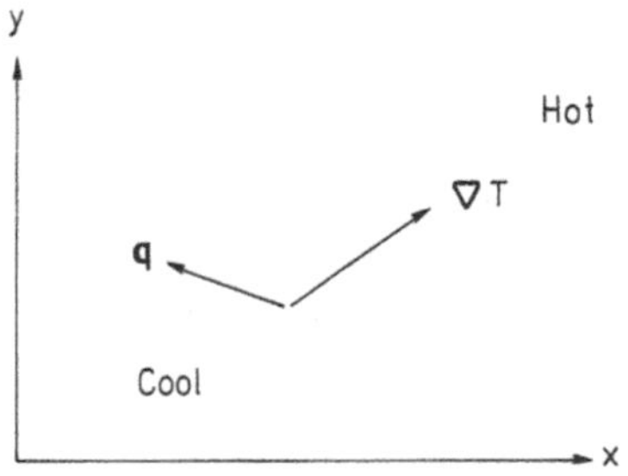

Figure 6.2 Flux vector $\mathbf{q}$ and temperature gradient ∇T

for all $\nabla T \neq \mathbf{0}$, which proves that the constitutive matrix $\mathbf{D}$ is positive definite (cf. (2.66)). From (2.67) we then conclude that $\mathbf{D}$ is non-singular, i.e.

$$\boxed{\det \mathbf{D} \neq 0} \tag{6.9}$$

In the most general form $\mathbf{D}$ may be non-symmetric, but for convenience and because it works well in practice, we assume that $\mathbf{D}$ is symmetric, i.e.

$$\boxed{\mathbf{D} = \mathbf{D}^{\mathrm{T}}} \tag{6.10}$$

Let us investigate the constitutive matrix $\mathbf{D}$ in more detail. Suppose, for instance, that in the two-dimensional case the temperature only changes in the x-direction. The temperature gradient is then given by

$$\nabla T = \begin{bmatrix} \dfrac{\partial T}{\partial x} \\ 0 \end{bmatrix}$$

With (6.4) and (6.6) we then obtain

$$\mathbf{q} = \begin{bmatrix} q_x \\ q_y \end{bmatrix} = \begin{bmatrix} -k_{xx}\dfrac{\partial T}{\partial x} \\ -k_{yx}\dfrac{\partial T}{\partial x} \end{bmatrix}$$

It appears that even though there is no temperature change in the y-direction we still have a heat flow, $q_y \neq 0$, in that direction. In general, this type of coupling effect is caused by the off-diagonal components of the constitutive matrix $\mathbf{D}$. It appears that $\mathbf{D}$ as given by (6.6) applies to very general, so-called *anisotropic* materials.

A somewhat simpler form emerges if $\mathbf{D}$ is taken as

$$\mathbf{D} = \begin{bmatrix} k_{xx} & 0 \\ 0 & k_{yy} \end{bmatrix}; \quad \mathbf{D} = \begin{bmatrix} k_{xx} & 0 & 0 \\ 0 & k_{yy} & 0 \\ 0 & 0 & k_{zz} \end{bmatrix} \tag{6.11}$$

where the thermal conductivities k_{xx}, k_{yy}, k_{zz} in the x-, y- and z-directions may be different. The forms of $\mathbf{D}$ given by (6.11) correspond to an *orthotropic* material and we obtain from (6.4) and (6.11) that

$$\mathbf{q} = \begin{bmatrix} -k_{xx}\dfrac{\partial T}{\partial x} \\ -k_{yy}\dfrac{\partial T}{\partial y} \end{bmatrix}; \quad \mathbf{q} = \begin{bmatrix} -k_{xx}\dfrac{\partial T}{\partial x} \\ -k_{yy}\dfrac{\partial T}{\partial y} \\ -k_{zz}\dfrac{\partial T}{\partial z} \end{bmatrix} \tag{6.12}$$

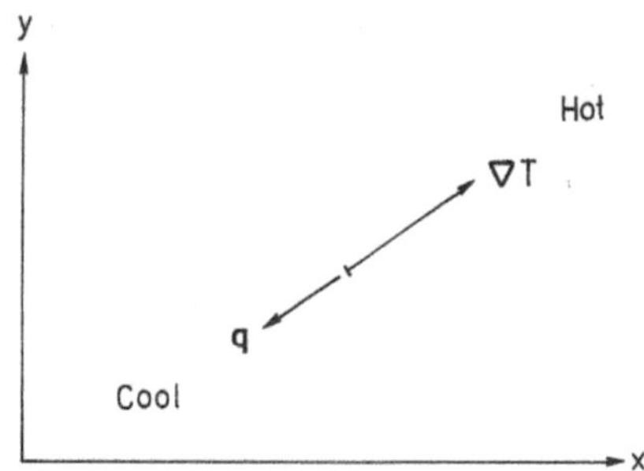

Figure 6.3   Flux vector **q** and temperature gradient ∇T for isotropic materials

For an *isotropic* material, in which the thermal conductivity is the same in all directions, (6.11) reduces to

$$\mathbf{D} = k\begin{bmatrix} 1 & 0 \\ 0 & 1 \end{bmatrix}; \quad \mathbf{D} = k\begin{bmatrix} 1 & 0 & 0 \\ 0 & 1 & 0 \\ 0 & 0 & 1 \end{bmatrix} \tag{6.13}$$

which can be written as

$$\mathbf{D} = k\mathbf{I} \tag{6.14}$$

i.e. we have only one material parameter, the thermal conductivity k. From (6.8) we conclude that k is a positive quantity. In practice, the most common situation is the isotropic one given by (6.14). With (6.14), (6.4) reduces to

$$\mathbf{q} = -k\nabla T \tag{6.15}$$

i.e. the vectors **q** and ∇T have opposite directions as illustrated in Figure 6.3.
With (6.4) we get from (6.3)

$$q_n = -(\nabla T)^{\mathrm{T}}\mathbf{D}\mathbf{n} \tag{6.16}$$

utilizing the fact that **D** is symmetric (cf. (6.10)). If **D** is given by (6.11), we obtain from (6.2) and (6.16)

$$q_n = -k_{xx}\frac{\partial T}{\partial x}n_x - k_{yy}\frac{\partial T}{\partial y}n_y \tag{6.17}$$

for two dimensions and

$$q_n = -k_{xx}\frac{\partial T}{\partial x}n_x - k_{yy}\frac{\partial T}{\partial y}n_y - k_{zz}\frac{\partial T}{\partial z}n_z \tag{6.18}$$

for three dimensions.

6.2 Heat equation for two and three dimensions – strong form

Let us now derive the differential equation which controls heat flow in more dimensions. As in the discussion of the one-dimensional heat equation, we may consider an infinitely small part of the body similar to Figure 4.2. However, in the present case it is much simpler and more elegant to use a derivation analogous to the alternative derivation given in Chapter 4 for one-dimensional heat flow.

The key point here is the fundamental divergence theorem of Gauss. Let us consider a two-dimensional body and let us assume that Q is the amount of heat supplied to the body per unit volume and per unit time; that is, Q is positive when heat is supplied to the body and has the dimension $[J/m^3\,s]$. As we only consider stationary (i.e. time-independent) problems, the balance or conservation principle states that the amount of heat supplied to the body per unit time equals the amount of heat leaving the body per unit time. We then conclude that

$$\int_A Qt\,\mathrm{d}A = \oint_{\mathscr{L}} q_n t\,\mathrm{d}\mathscr{L} \tag{6.19}$$

where $t = t(x, y)$ denotes the thickness in the z-direction of the body located in the xy-plane; see Figure 6.4.

From (6.3) we have that $q_n = \mathbf{q}^{\mathrm{T}}\mathbf{n}$, i.e. Gauss' divergence theorem (5.23) yields

$$\oint_{\mathscr{L}} tq_n\,\mathrm{d}\mathscr{L} = \oint_{\mathscr{L}} t\mathbf{q}^{\mathrm{T}}\mathbf{n}\,\mathrm{d}\mathscr{L} = \oint_{\mathscr{L}} (t\mathbf{q})^{\mathrm{T}}\mathbf{n}\,\mathrm{d}\mathscr{L} = \int_A \mathrm{div}(t\mathbf{q})\,\mathrm{d}A \tag{6.20}$$

With (6.20), (6.19) takes the form

$$\int_A [tQ - \mathrm{div}(t\mathbf{q})]\,\mathrm{d}A = 0$$

In this expression, we did not specify the region A. In fact, this region may be taken as any arbitrary part of the region for the entire body. Therefore, as the region A is arbitrary, we conclude that

$$\boxed{\mathrm{div}(t\mathbf{q}) = tQ} \tag{6.21}$$

which is the balance principle for the two-dimensional body. If the thickness t is

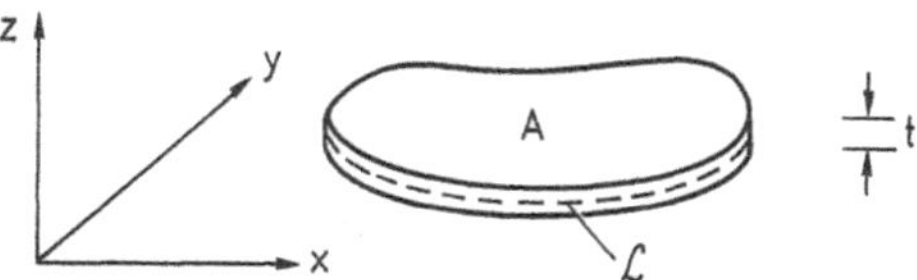

Figure 6.4   Two-dimensional body with thickness t located in the xy-plane

constant, we get

$$\text{div } \mathbf{q} = Q \tag{6.22}$$

where, according to (5.24), the divergence of $\mathbf{q}$ is given by

$$\text{div } \mathbf{q} = \frac{\partial q_x}{\partial x} + \frac{\partial q_y}{\partial y} \tag{6.23}$$

Using the constitutive relation (6.4) in the balance equation (6.21), it follows that

$$\text{div}(t\mathbf{D}\nabla T) + tQ = 0 \quad \text{in region } A \tag{6.24}$$

where region A is the entire region of the body. If the constitutive equation $\mathbf{D}$ is given by (6.11), (6.24) reduces to

$$\frac{\partial}{\partial x}\left(tk_{xx}\frac{\partial T}{\partial x}\right) + \frac{\partial}{\partial y}\left(tk_{yy}\frac{\partial T}{\partial y}\right) + tQ = 0 \tag{6.25}$$

For reasons that will be revealed in a moment, this equation is called the *quasi-harmonic equation*. For isotropic materials, (6.13), we obtain

$$\frac{\partial}{\partial x}\left(tk\frac{\partial T}{\partial x}\right) + \frac{\partial}{\partial y}\left(tk\frac{\partial T}{\partial y}\right) + tQ = 0 \tag{6.26}$$

Furthermore, if the product tk is constant, (6.26) reduces to

$$\frac{\partial^2 T}{\partial x^2} + \frac{\partial^2 T}{\partial y^2} + \frac{Q}{k} = 0 \tag{6.27}$$

This expression is the well-known *Poisson equation* and we shall show later in Chapter 14 that this differential equation also applies for the torsion of non-circular elastic shafts. If the internal heat supply is zero, i.e. $Q = 0$, (6.27) becomes

$$\frac{\partial^2 T}{\partial x^2} + \frac{\partial^2 T}{\partial y^2} = 0 \tag{6.28}$$

which constitutes the well-known *Laplace equation*. Solutions to the Laplace equation are termed *harmonic functions* and this suggests why (6.25) is referred to as the quasi-harmonic equation.

To solve the differential equation (6.24), *boundary conditions* are required. These boundary conditions are typically of the form

$$q_n = \mathbf{q}^{\mathsf{T}}\mathbf{n} = h \quad \text{on } \mathscr{L}_h \tag{6.29}$$

$$T = g \quad \text{on } \mathscr{L}_g \tag{6.30}$$

where h and g are known quantities. $\mathscr{L}_h$ is that part of the boundary $\mathscr{L}$ on which the flux q_n is known, whereas $\mathscr{L}_g$ is that part of the boundary $\mathscr{L}$ on which the temperature T is known. The sum of the boundaries $\mathscr{L}_h$ and $\mathscr{L}_g$ constitutes the entire boundary $\mathscr{L}$ (cf. Figure 6.5).

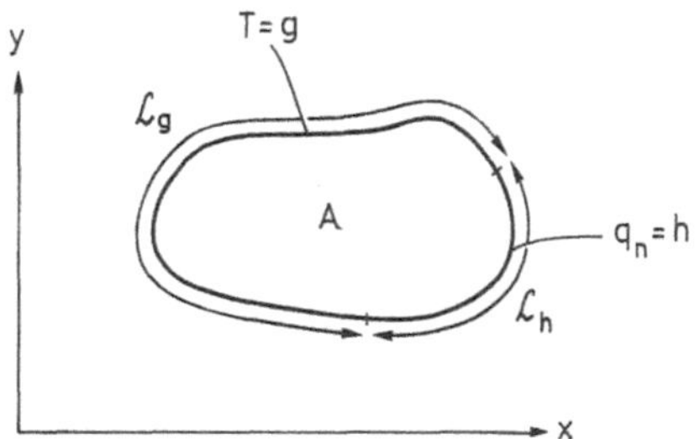

Figure 6.5   Two-dimensional region A with boundary $\mathscr{L} = \mathscr{L}_h + \mathscr{L}_g$

We note that it is not possible to prescribe the flux q_n and the temperature T at the same position.

Whereas the boundary conditions (6.29) and (6.30) are typical for applications, it may also happen that the boundary condition is prescribed in terms of a linear combination of the flux q_n and the temperature T, i.e.

$$q_n + \alpha T = \beta \tag{6.31}$$

where α and β are known quantities. A boundary condition in the form of (6.31) is typical when *convection* occurs along the boundary; see, for instance, Carslaw and Jaeger (1959). Here, we shall mainly be concerned with boundary conditions in the form given by (6.29) and (6.30).

With (6.24), (6.29) and (6.30) we have obtained the strong form of the heat problem:

Strong form of two-dimensional heat flow

$$\operatorname{div}(t\mathbf{D}\,\nabla T) + tQ = 0 \quad \text{in region } A \tag{6.32}$$

$$q_n = \mathbf{q}^{\mathrm{T}}\mathbf{n} = h \qquad\qquad \text{on } \mathscr{L}_h \tag{6.33}$$

$$T = g \qquad\qquad\qquad \text{on } \mathscr{L}_g \tag{6.34}$$

For three-dimensional heat flow and using the conservation principle together with Gauss' divergence theorem for three dimensions (cf. (5.32)) the reader may easily verify that the balance equation (6.22) is again obtained. It then follows directly that

---

Strong form of three-dimensional heat flow

$$\operatorname{div}(\mathbf{D}\,\nabla T) + Q = 0 \quad \text{in region } V \tag{6.35}$$

$$q_n = \mathbf{q}^{\mathrm{T}}\mathbf{n} = h \qquad\qquad \text{on surface } S_h \tag{6.36}$$

$$T = g \qquad\qquad\qquad \text{on surface } S_g \tag{6.37}$$

---

where S_h and S_g are those parts of the boundary S on which the flux q_n and temperature T are known, respectively. The sum of the boundaries S_h and S_g constitutes the entire boundary S. As an example of the differential equation (6.35) assume that $\mathbf{D}$ is given by (6.11). We then obtain the *quasi-harmonic equation* in three dimensions:

$$\frac{\partial}{\partial x}\left(k_{xx}\frac{\partial T}{\partial x}\right) + \frac{\partial}{\partial y}\left(k_{yy}\frac{\partial T}{\partial y}\right) + \frac{\partial}{\partial z}\left(k_{zz}\frac{\partial T}{\partial z}\right) + Q = 0 \tag{6.38}$$

For isotropic materials we conclude that

$$\frac{\partial}{\partial x}\left(k\frac{\partial T}{\partial x}\right) + \frac{\partial}{\partial y}\left(k\frac{\partial T}{\partial y}\right) + \frac{\partial}{\partial z}\left(k\frac{\partial T}{\partial z}\right) + Q = 0 \tag{6.39}$$

Furthermore, if the thermal conductivity k is constant, (6.39) reduces to

$$\frac{\partial^2 T}{\partial x^2} + \frac{\partial^2 T}{\partial y^2} + \frac{\partial^2 T}{\partial z^2} + \frac{Q}{k} = 0 \tag{6.40}$$

i.e. the *Poisson equation* in three dimensions. If the heat input $Q = 0$, (6.40) reduces to

$$\frac{\partial^2 T}{\partial x^2} + \frac{\partial^2 T}{\partial y^2} + \frac{\partial^2 T}{\partial z^2} = 0 \tag{6.41}$$

i.e. the *Laplace equation* in three dimensions.

With these observations we may summarize the equations for three-dimensional heat flow as shown in Figure 6.6.

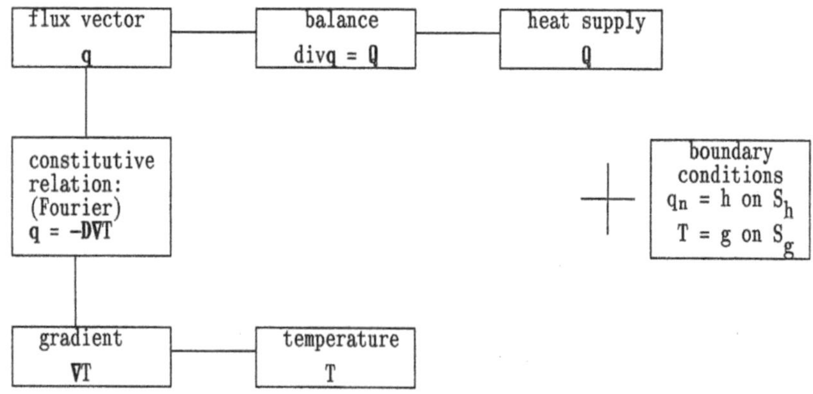

Figure 6.6   Illustration of fundamental equations of three-dimensional heat flow

6.3 Weak form of heat flow in two and three dimensions

We could derive the weak forms directly from the differential equations (6.32) and (6.35). However, it is more instructive first to derive the weak form of the balance equation given by (6.21) and (6.22) for two- and three-dimensional problems, respectively.

Considering two-dimensional problems, we therefore multiply (6.21) by an arbitrary function v – the weight function – and integrate over the region considered, i.e.

$$\int_A v \operatorname{div}(\mathbf{q}t)\, \mathrm{d}A - \int_A vQt\, \mathrm{d}A = 0 \tag{6.42}$$

where $v = v(x, y)$. Integrating by parts the first term using the Green–Gauss theorem of (5.26), we obtain

$$\int_A v \operatorname{div}(\mathbf{q}t)\, \mathrm{d}A = \oint_{\mathscr{L}} vt\mathbf{q}^{\mathrm{T}}\mathbf{n}\, \mathrm{d}\mathscr{L} - \int_A (\nabla v)^{\mathrm{T}}\mathbf{q}t\, \mathrm{d}A \tag{6.43}$$

Expression (6.42) can then be rewritten in the form

$$\int_A (\nabla v)^{\mathrm{T}}\mathbf{q}t\, \mathrm{d}A = \oint_{\mathscr{L}} v\mathbf{q}^{\mathrm{T}}\mathbf{n}t\, \mathrm{d}\mathscr{L} - \int_A vQt\, \mathrm{d}A \tag{6.44}$$

This is the weak form of the balance principle. We may use the boundary condition (6.33) to obtain

$$\int_A (\nabla v)^{\mathrm{T}}\mathbf{q}t\, \mathrm{d}A = \int_{\mathscr{L}_h} vht\, \mathrm{d}\mathscr{L} + \int_{\mathscr{L}_g} vq_n t\, \mathrm{d}\mathscr{L} - \int_A vQt\, \mathrm{d}A \tag{6.45}$$

where h is a known quantity along $\mathscr{L}_h$, whereas the flux q_n is unknown along the boundary $\mathscr{L}_g$. Expression (6.45) is the weak form of the balance equation (6.21).

Insertion of the constitutive relation (6.4) in (6.45) provides the following weak form:

Weak form of two-dimensional heat flow

$$\int_A (\nabla v)^{\mathrm{T}}t\, \mathbf{D}\, \nabla T\, \mathrm{d}A = -\int_{\mathscr{L}_h} vht\, \mathrm{d}\mathscr{L} - \int_{\mathscr{L}_g} vq_n t\, \mathrm{d}\mathscr{L} + \int_A vQt\, \mathrm{d}A \tag{6.46}$$

$$T = g \quad \text{on } \mathscr{L}_g \tag{6.47}$$

where v is any function. This result may also be derived directly from the strong form (6.32)–(6.34), but the advantage of first establishing the weak form of the balance principle (cf. (6.45)) and then introducing the constitutive relation (6.4) is that the assumptions behind the different steps become more apparent. Whereas the balance equation (6.21) or (6.45) is a fundamental physical principle, the constitutive relation (6.4) is an assumption. In principle, constitutive relations other than Fourier's law may be envisioned, but the balance principle always holds. We note that if the thickness t is constant, it may be cancelled in (6.46).

The reader may verify that the weak formulation of the balance principle (6.22) in three dimensions is

$$\int_V (\nabla v)^{\mathrm{T}} \mathbf{q}\, \mathrm{d}V = \int_{S_h} vh\, \mathrm{d}S + \int_{S_g} vq_n\, \mathrm{d}S - \int_V vQ\, \mathrm{d}V \tag{6.48}$$

where the Green–Gauss theorem of (5.33) has been used. In (6.48), the arbitrary weight function v depends on all the coordinates, i.e. $v = v(x, y, z)$. Introducing the constitutive relation (6.4) in (6.48) we obtain the following:

Weak form of three-dimensional heat flow

$$\int_V (\nabla v)^{\mathrm{T}} \mathbf{D}\, \nabla T\, \mathrm{d}V = -\int_{S_h} vh\, \mathrm{d}S - \int_{S_g} vq_n\, \mathrm{d}S + \int_V vQ\, \mathrm{d}V \tag{6.49}$$

$$T = g \quad \text{on } S_g \tag{6.50}$$

where v is any function. Similar to the one-dimensional case, the boundary condition prescribing the derivative of the unknown (i.e. the flux (6.33) and (6.36)) is called a *natural* or *Neumann boundary condition* since it emerges directly from the weak formulation. The boundary conditions prescribing the unknown itself (i.e. (6.34) and (6.37)) is called an *essential* or *Dirichlet boundary condition*. We also note that the convection boundary condition given by (6.31) is occasionally termed a *Newton boundary condition*. Boundary conditions like (6.33) and (6.34) which contain both natural and essential conditions are also called *mixed boundary conditions*.

The weak forms above have been derived from the strong forms. As in the derivation in Chapter 4, it is rather easy to prove that the weak forms also imply the strong forms, i.e. the weak and strong forms are identical. The interested reader may consult Hughes (1987). However, from the discussion in Chapter 4 we recall the advantages of the weak form, namely discontinuities are allowed for and the order of differentiation of the unknown function T has decreased. This decrease has been obtained at the expense of differentiation of the arbitrary weight function v. We also observe that this implies that the order of differentiation for v and T is the same. This indicates a certain symmetry, which is of importance when the FE method is formulated.

6.4 Other analogous physical problems

Just as for one-dimensional heat flow, it follows directly that the formulations for two- and three-dimensional heat flow apply to the diffusion of ions and to electric currents. This is achieved by generalizing Fick's law in one dimension, (4.11), and Ohm's law in one dimension, (4.12), to two and three dimensions, as in the generalized form of Fourier's law (6.4).

We have already mentioned that the Poisson equation in two dimensions, (6.27), also applies to the torsion of non-circular elastic shafts and this will be demonstrated in Chapter 14.

Another important example, which also leads to a formulation similar to that of heat flow, is the flow of fluid through porous media (Bear, 1979). To demonstrate this, consider water flowing in the xy-plane of Figure 6.7, which shows a section of the underground.

A tube is drilled into the ground until point A is reached. At point A the water pressure is p_A and the water will therefore rise inside the tube to a level $p_A/(\rho g)$ above point A. Here ρ is the mass density of the water and g is the acceleration due to gravity. We choose an arbitrary horizontal *datum level* from which the height z is defined, i.e. point A is located z_A above the datum level. The same procedure is adopted for a tube drilled down until point B. An instrument used to measure the pressures p_A and p_B as described above is called a *piezometer*.

At an arbitrary point the so-called *piezometric head* ϕ is defined as

$$\phi = z + \frac{p}{\rho g} \tag{6.51}$$

It appears that $\phi = \phi(x, y)$ has the dimension of length and for the case considered

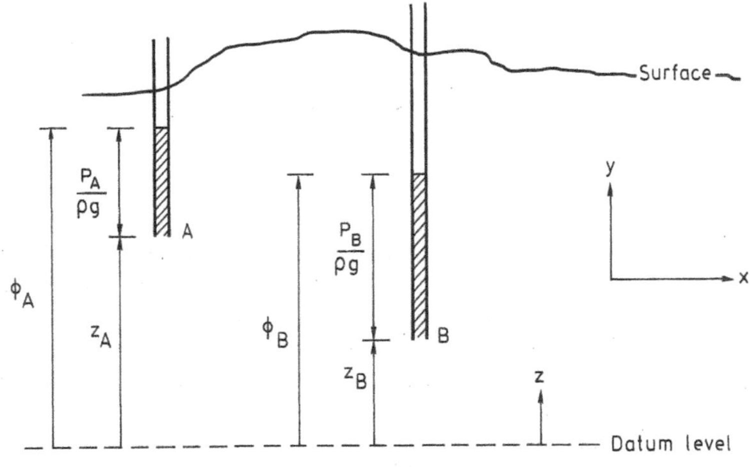

Figure 6.7　Flow of water underground; piezometric head ϕ

Table 6.1 Examples of the heat equation

Differential equation	Physical problem	Quantities	Constitutive law
$\mathrm{div}(\mathbf{D}\nabla T) + Q = 0$	Heat flow	T = temperature $\mathbf{D}$ = constitutive matrix for for thermal conductivities Q = heat supply	Fourier $\mathbf{q} = -\mathbf{D}\nabla T$ $\mathbf{q}$ = heat flux vector
$\mathrm{div}(\mathbf{D}\nabla c) + Q = 0$	Diffusion	c = ion concentration $\mathbf{D}$ = constitutive matrix for diffusion coefficients Q = ion supply	Fick $\mathbf{q} = -\mathbf{D}\nabla c$ $\mathbf{q}$ = ion flux vector
$\mathrm{div}(\mathbf{D}\nabla\phi) + Q = 0$	Fluid flow through porous media	ϕ = piezometric head $\mathbf{D}$ = constitutive matrix for permeability coefficients Q = fluid supply	Darcy $\mathbf{q} = -\mathbf{D}\nabla\phi$ $\mathbf{q}$ = volume flux vector
$\mathrm{div}(\mathbf{D}\nabla V) + Q = 0$	Electric currents in extended bodies	V = voltage $\mathbf{D}$ = constitutive matrix for electric conductivities Q = electric charge supply	Ohm $\mathbf{q} = -\mathbf{D}\nabla V$ $\mathbf{q}$ = electric charge flux vector
$\mathrm{div}(\nabla\phi) + Q = 0$	Irrotational, frictionless and incompressible fluid flow	ϕ = velocity potential Q = fluid supply	$\mathbf{v} = -\nabla\phi$ $\mathbf{v}$ = velocity (not a constitutive law)
$\mathrm{div}\left(\dfrac{1}{G}\nabla\phi\right) + 2\theta = 0$	Saint-Venant torsion of elastic bars	$\phi = \phi(x, y) =$ Prandtl's stress function G = shear modulus θ = rate of twist	Hooke $\sigma_{xz} = G\gamma_{xz}$ $\sigma_{yz} = G\gamma_{yz}$ $\sigma_{xz} = \partial\phi/\partial y;\ \sigma_{yz} = -\partial\phi/\partial x$
$\mathrm{div}(\nabla w) + \dfrac{p}{S} = 0$	Small deflections of membrane	$w = w(x, y)$ = deflection p = lateral pressure S = constant tension force per unit length	

we have $\phi_A > \phi_B$. Occasionally ϕ is called the *hydraulic potential*. The term $p/(\rho g)$ is called the *pressure head*.

Let us imagine that a tube just containing water is connected between points A and B. It is obvious that water will then flow from point A towards point B even though the pressure p_B is greater than the pressure p_A. Therefore, it is not the difference in pressure p that creates the water flow. What really forces the water to flow from point A towards point B is the fact that $\phi_A > \phi_B$. This difference in piezometric head ϕ is the driving force for the water flow even when the imaginary water-filled tube between A and B is removed.

From these arguments we are led to *Darcy's law* from 1856, which for one-dimensional flow is

$$q = -k\frac{d\phi}{dx} \tag{6.52}$$

where q is the flow rate, i.e. the volume of water passing through a unit area per unit time. It appears that the parameter k, the *coefficient of permeability*, has the dimension [m/s]. Moreover, Darcy's law (6.52) is quite analogous with Fourier's law (4.3) for heat flow, Fick's law (4.11) for ion diffusion and Ohm's law (4.12) for electric currents. Just like these constitutive relations (see (6.4)) Darcy's law (6.52) can be generalized to

$$\boxed{\mathbf{q} = -\mathbf{D}\nabla\phi} \tag{6.53}$$

If $\mathbf{D}$, for instance, is given by (6.11), then as in (6.25) the two-dimensional problem of Figure 6.7 leads to the following differential equation:

$$\frac{\partial}{\partial x}\left(tk_{xx}\frac{\partial\phi}{\partial x}\right) + \frac{\partial}{\partial y}\left(tk_{yy}\frac{\partial\phi}{\partial y}\right) + tQ = 0 \tag{6.54}$$

where Q now denotes the water volume supply per unit volume of the body and per unit time and t is the thickness of the section underground considered.

Finally, we mention that the so-called irrotational flow of an incompressible and frictionless fluid is also expressed by the equations given above (see Hughes and Brighton, 1967). In this case the *velocity vector* $\mathbf{v}$ is given as $\mathbf{v} = -\nabla\phi$ where ϕ is the so-called *velocity potential*. It follows that ϕ satisfies the Poisson equation given by (6.27) and (6.40) for two- and three-dimensional flow, respectively.

We have seen that a variety of important physical phenomena leads to the same differential equation and boundary conditions as the heat flow problem, and some examples are collected in Table 6.1. It is with this perspective in mind that we shall evaluate the significance of the FE formulation for the heat flow problem and which will be treated in later chapters.

7
Choice of approximating functions for the FE method – scalar problems

We have emphasized that the FE method is based on the weak formulation of the problem. As a prototype of such a weak formulation we may consider (6.46) for two-dimensional heat flow:

$$\int_A (\nabla v)^{\mathrm{T}} t \mathbf{D} \, \nabla T \, dA = -\int_{\mathscr{L}_h} vht \, d\mathscr{L} - \int_{\mathscr{L}_g} vq_n t \, d\mathscr{L} + \int_A vQt \, dA \qquad (7.1)$$

In this equation, terms involving the unknown temperature T and the arbitrary weight function v appear. Therefore, in order to achieve an FE formulation from (7.1), we need to make certain choices for the approximation of the temperature T as well as for the weight function v.

We can therefore summarize the basic steps in the FE formulation as follows:

Basic steps in the FE formulation

1. Establish the strong formulation of the problem.
2. Obtain the weak form of the problem.
3. Choose approximations for the unknown function – in the present case the unknown temperature T.
4. Choose the weight function v.

Considerable emphasis has already been given to steps 1 and 2. In the present chapter we shall treat the manner in which approximations are made for the unknown function T, i.e. step 3. In the following chapter we shall finally make an evaluation of a suitable choice of the weight function v, i.e. step 4. After these discussions we will eventually be in a position to present the FE method.

In order to establish an approximation for the unknown scalar – the temperature T – we may in principle adopt two basically different routes: (a) the unknown function is approximated directly over the entire region of interest; (b) the region is divided into smaller parts and the approximation is first carried out over these smaller and simpler parts and then, based on these results, it is established for the entire region.

The first route (a) does not present an attractive general approach since the region differs from problem to problem. Therefore, this approach is geared towards the solution of specific problems having in principle the same geometry. What we are looking for is a method applicable to general geometries and this objective is achieved by the second alternative, route (b), described above.

If we split the body, i.e. the region of interest, into smaller parts, i.e. finite elements, for which we can establish an approximation for the unknown temperature T, then we are able to analyze bodies with arbitrary geometries. As already touched upon in Chapter 1, it is a characteristic feature of the FE method that the approximation of the unknown function is carried out 'elementwise'. Actually, the FE method owes its name to this characteristic feature.

In this chapter we shall treat the approximation of a scalar function like the temperature over one-, two- and three-dimensional elements and we shall show how the approximation of the scalar function over the entire region of the body of interest can be established based on this elementwise approximation.

In order to achieve such an approximation we have to make a choice for the type of approximation. For instance, should trigonometric or exponential functions be used in the approximation procedure? It turns out that the use of polynomials is especially advantageous and convenient for establishing an elementwise approximation of the unknown scalar, and in order to understand this important point, it is appropriate to evaluate the criteria for convergence.

7.1 General requirements

7.1.1 *Convergence criteria*

The FE method provides an approximate solution to the problem at hand and it is obvious that the more elements we use (i.e. the more computer power we utilize) the more accurate the approximate solution. In the limit, when the elements are infinitely small, we require that our approximate solution is infinitely close to the exact solution. This is the *convergence* requirement.

Let us consider the inhomogeneous temperature field over the two-dimensional body shown in Figure 7.1. The boundary along AC is insulated and the temperature is prescribed along the boundary ABDC. In Figure 7.1(b), a very coarse element mesh consisting of only two triangular elements is used, whereas many triangular elements are used in the fine element mesh shown in Figure 7.1(c).

Within each element the temperature varies drastically in the coarse mesh and less dramatically in the fine mesh. In the limit, for infinitely small elements, we then have within each element either a constant temperature gradient or a constant temperature. If the approximation of the temperature over each element is to

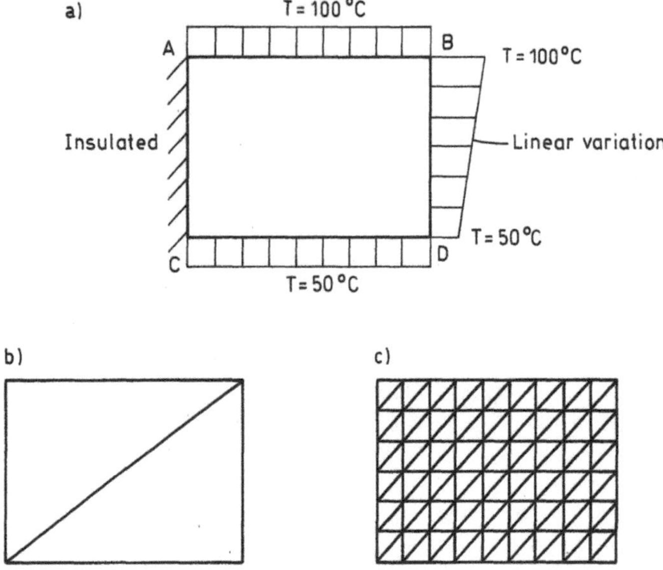

Figure 7.1   (a) Two-dimensional heat problem; (b) coarse element mesh; (c) fine element mesh

approach the exact temperature for decreasing element size, it follows that the approximation must fulfil the so-called *completeness* requirements:

Completeness

- The approximation must be able to represent an arbitrary constant temperature gradient.
- The approximation must be able to represent an arbitrary constant temperature.

The completeness requirements therefore imply that the approximation of the temperature at least must include terms corresponding to an arbitrary linear polynomial. That is, in three dimensions we must have the following approximation:

$$T \simeq \alpha_1 + \alpha_2 x + \alpha_3 y + \alpha_4 z + \text{possibly other terms} \tag{7.2}$$

Consider now two neighbouring elements as shown in Figure 7.2, where triangular elements are used for convenience. The temperature variation along the common element boundary AB is considered. In each element an approximation of the temperature is made and if the temperature variation in element 1 along the dashed line is evaluated, it may differ from the temperature variation in element 2

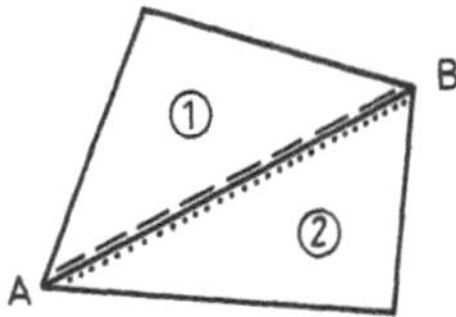

Figure 7.2   Two neighbouring elements

along the dotted line. We are then led to the so-called *compatibility* or *conforming* requirement:

> ## Compatibility or conforming requirement
>
> • The approximation of the temperature over element boundaries must be continuous.

The convergence requirement, stating that for infinitely small elements the approximation approaches the exact temperature field, is clearly fulfilled if both the completeness and compatibility requirements are fulfilled, i.e. we obtain the important statement

> convergence criterion = completeness + compatibility requirements

Here only qualitative arguments have been presented for the fulfilment of convergence, but a more stringent evaluation is presented by Strang and Fix (1973), and in Chapter 9 we shall rederive the compatibility requirement using a different approach. However, it turns out that it *may* even be allowable to relax the compatibility (i.e. the conforming) requirement and still be able to fulfil the convergence criterion. Indeed, even if a discontinuity exists across an element boundary, this discontinuity may approach zero for infinitely small elements. Finite elements fulfilling the compatibility (i.e. the conforming) requirement are called *conforming elements*, otherwise they are termed *non-conforming elements*. Most finite elements are conforming, but even non-conforming elements are used, especially in plate analysis as discussed in Chapter 18.

As indicated above, the fulfilment of both completeness and compatibility is sufficient for convergence. However, whereas the requirement of compatibility *may* be relaxed, the fulfilment of completeness is clearly a necessary condition for convergence that must be satisfied.

As the completeness requirement must be met, one element is always able to reproduce exactly an arbitrary linear temperature variation. Clearly, if an arbitrary assemblage of conforming elements is considered, these elements are also able to reproduce exactly an arbitrary linear temperature variation over the assemblage. This observation leads to the so-called *patch test* for non-conforming elements. To check that non-conforming elements fulfil the convergence requirement they must pass this test. The essence of the patch test is to check that an arbitrary assemblage (i.e. patch) of elements is able to reproduce exactly an arbitrary linear temperature variation. Apart from some plate elements discussed in Chapter 18, in this introductory text we shall only consider conforming elements and we refer to Hughes (1987), Strang and Fix (1973) and Zienkiewicz and Taylor (1989) for a detailed treatment of the patch test and the behaviour of non-conforming elements.

Let us return to the completeness requirements which must be fulfilled in order to achieve convergence. Referring to (7.2) we note that an approximation which at least involves an arbitrary linear polynomial is necessary in order to fulfil the completeness criterion. With this result at hand it will come as no surprise to see later that the elementwise approximations always consist of polynomials possibly containing even higher-order terms than linear ones.

7.1.2 *Rate of convergence – Pascal's triangle*

Having established that the approximation of the unknown function (temperature) is obtained using a polynomial, which at least includes all linear terms, we need information on which higher-order terms it may be advantageous to include. For this purpose we shall discuss the concept of *rate of convergence*, which provides information on how fast the approximate solution converges towards the exact solution when the element size is decreased.

Consider the two-dimensional body shown in Figure 7.3. Assume that the exact temperature distribution T is known and let us make a Taylor expansion of this exact solution about the point (x_0, y_0). It follows that the true solution T can be

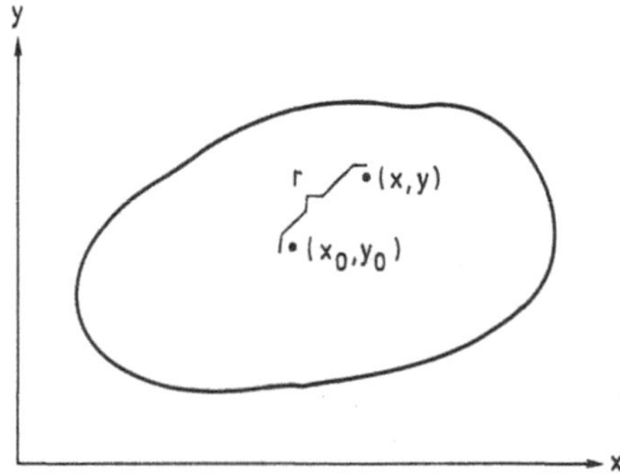

Figure 7.3   Taylor expansion of true solution about point (x_0, y_0)

written as

$$T(x, y) = T_0 + \left(\frac{\partial T}{\partial x}\right)_0 (x - x_0) + \left(\frac{\partial T}{\partial y}\right)_0 (y - y_0)$$

$$+ \frac{1}{2}\left(\frac{\partial^2 T}{\partial x^2}\right)_0 (x - x_0)^2 + \left(\frac{\partial^2 T}{\partial x\,\partial y}\right)_0 (x - x_0)(y - y_0) \tag{7.3}$$

$$+ \frac{1}{2}\left(\frac{\partial^2 T}{\partial y^2}\right)_0 (y - y_0)^2 + \mathrm{O}(r^3)$$

where the term $\mathrm{O}(r^3)$, which is read as 'of order 3', is defined in general as

$$\mathrm{O}(r^n) = Cr^n \quad \text{for } r \to 0 \tag{7.4}$$

In this expression, C is a constant and r is the distance between the current point (x, y) and the fixed point (x_0, y_0) (cf. Figure 7.3). Moreover, in (7.3) T_0 is the exact temperature at point (x_0, y_0) and, likewise, $(\partial T/\partial x)_0$, $(\partial T/\partial y)_0$, $(\partial^2 T/\partial x^2)_0$, $(\partial^2 T/\partial x\,\partial y)_0$ and $(\partial^2 T/\partial y^2)_0$ are the exact derivatives evaluated at the point (x_0, y_0). It appears that (7.3) may be written as

$$T(x, y) = a_1 + a_2 x + a_3 y + a_4 x^2 + a_5 xy + a_6 y^2 + Cr^3 \tag{7.5}$$

where $a_1 \ldots a_6$ are constants.

In a region which contains the point (x_0, y_0) we now assume that the approximation to the true temperature field is taken as

$$T^{\mathrm{app}}(x, y) = a_1 + a_2 x + a_3 y + a_4 x^2 + a_5 xy + a_6 y^2 \tag{7.6}$$

where the notation T^{app} has been used in order to emphasize that (7.6) represents the approximate temperature field. Subtracting (7.6) from (7.5) yields

$$T(x, y) - T^{\mathrm{app}}(x, y) = Cr^3 \tag{7.7}$$

and it appears that $T - T^{\mathrm{app}}$ is the error in the approximation process. Let us now consider a finite element (i.e. a region) which contains the point (x_0, y_0) and let h denote the maximum distance from the point (x_0, y_0) to the element boundaries. Within this finite element the maximum error is written as $|T - T^{\mathrm{app}}|_{\max}$ and from (7.7) we find that

$$|T - T^{\mathrm{app}}|_{\max} = Ch^3$$

We may consider the distance h as representative of the element size and to emphasize this issue, the result above is written as

$$|T - T_h^{\mathrm{app}}|_{\max} = Ch^3 \tag{7.8}$$

where T_h^{app} refers to the approximation for an element of size h.

Let us assume now that instead of an element of size h we use another, geometrically similar element of size αh. In analogy to (7.8) it follows that

$$|T - T_{\alpha h}^{\mathrm{app}}|_{\max} = C\alpha^3 h^3 \tag{7.9}$$

Combining (7.8) and (7.9) results in

$$|T - T^{app}_{\alpha h}|_{max} = \alpha^3 |T - T^{app}_h|_{max} \tag{7.10}$$

If, for example, α is chosen as $\frac{1}{2}$, i.e. the new element size is half its original size, we obtain

$$|T - T^{app}_{h/2}|_{max} = \tfrac{1}{8}|T - T^{app}_h|_{max} \tag{7.11}$$

This implies that with an element size of $h/2$, the maximum error in the approximation is one-eighth of the maximum error using an element size of h. The term α^3 present in (7.10) gives the rate of convergence, which in the present case is of order 3. It follows that the rate of convergence provides information on how fast the approximate solution converges towards the true solution when the element size is decreased. Observe that as the exact solution T is unknown in practice, it is not possible to calculate the value of the error $|T - T^{app}|_{max}$.

In (7.6) we considered all terms up to and including the quadratic terms. Suppose now that only the linear terms are considered in the approximation, i.e.

$$T^{app}(x, y) = a_1 + a_2 x + a_3 y \tag{7.12}$$

As, for instance, the term $(\partial^2 T/\partial x^2)_0 (x - x_0)^2$ is of order $O(r^2)$ we obtain by subtracting (7.12) from (7.3) that

$$T(x, y) - T^{app}(x, y) = C_1 r^2 \tag{7.13}$$

where C_1 is a constant. As in the derivation of (7.10), we obtain from (7.13) that

$$|T - T^{app}_{\alpha h}|_{max} = \alpha^2 |T - T^{app}_h|_{max} \tag{7.14}$$

i.e. the rate of convergence is of order 2 for the approximation given by (7.12). If the element size is halved, i.e. $\alpha = \frac{1}{2}$, (7.14) yields

$$|T - T^{app}_{h/2}|_{max} = \tfrac{1}{4}|T - T^{app}_h|_{max} \tag{7.15}$$

A comparison with (7.11) illustrates that the more higher-order terms we include in the approximation, the faster is the rate of convergence.

However, this conclusion should be used with caution as the following example will show. Consider the approximation given by

$$T^{app}(x, y) = a_1 + a_2 x + a_3 y + a_5 xy \tag{7.16}$$

Even though we include one quadratic term, the true solution (7.3) contains other quadratic terms which are of the order $O(r^2)$. That is, subtracting (7.16) from (7.3) yields

$$T(x, y) - T^{app}(x, y) = C_2 r^2 \tag{7.17}$$

where C_2 is a constant. A comparison of (7.13) and (7.17) shows that the two expressions are similar in structure even though $C_1 \neq C_2$. This means that approximation (7.16) also leads to (7.14) just like approximation (7.12), i.e. we have not improved the rate of convergence. The approximation given by (7.16) is in general more accurate than (7.12), owing to the term $a_5 xy$, and this is reflected by the fact that, in general, $C_2 \leq C_1$ (cf. (7.13) and (7.17)). However, the important point is that the rate of convergence is unchanged. Terms in the approximation which do not improve the rate of convergence are called *parasitic* terms.

Obviously, unless other specific properties of the approximating function are required, one would try to avoid parasitic terms and instead use polynomials which contain terms of lower order before higher-order terms are invoked. A polynomial which contains all terms of a specific order as well as all lower-order terms is called a *complete* polynomial (not to be confused with the completeness requirements, discussed previously). In one dimension, the approximations $a_1 + a_2 x$ and $a_1 + a_2 x + a_3 x^2$ are complete, whereas the approximation $a_1 + a_2 x + a_3 x^3$ is incomplete because it lacks the quadratic term.

In order to keep track of the terms in a polynomial used for two-dimensional approximation it is convenient to make use of *Pascal's triangle*, which lists the terms in a systematic manner:

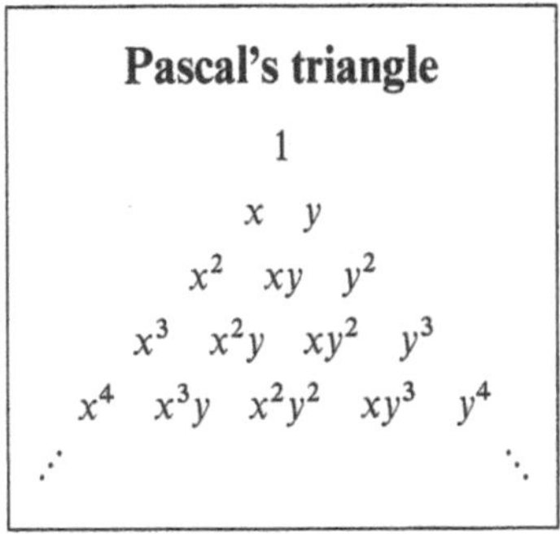

Any polynomial which contains all the terms above a certain horizontal line in Pascal's triangle is a complete polynomial. For one-dimensional problems Pascal's triangle degenerates to $1, x, x^2, x^3, \ldots$ and even for three-dimensional problems it is rather straightforward to generalize the results above.

With these results in mind, we can summarize our discussion for the elementwise approximation procedure:

Elementwise approximation

- A polynomial is used.
- This polynomial must include at least an arbitrary linear polynomial – then the completeness requirements are fulfilled.
- Choose the polynomial so that continuity across element boundaries is achieved – then the compatibility or conforming requirement is fulfilled.
- If possible, avoid parasitic terms.

Having established these general guidelines, we shall now discuss the elementwise approximation of a scalar function like the temperature for one-, two- and three-dimensional problems, and we will show how the approximation of the scalar function over the entire region of the body of interest can be established based on these elementwise approximations. For convenience the scalar function is denoted by T, as it may represent the temperature of a body.

7.2 One-dimensional elements

7.2.1 *Simplest possible one-dimensional element*

Referring to (7.2), it appears that the simplest possible one-dimensional approximation, which fulfils the completeness requirements, is given by

$$T = \alpha_1 + \alpha_2 x \tag{7.18}$$

where α_1 and α_2 are constant parameters for each element. In a moment we will express these parameters in terms of the temperature at the nodal points of the element. As we have two parameters, the element contains two nodal points. Moreover, it is advantageous to locate the nodal points at the ends of the element, i.e. we obtain the element with the length L as shown in Figure 7.4.

Let us rewrite (7.18) in the following matrix form

$$T = \bar{\mathbf{N}}\boldsymbol{\alpha} \tag{7.19}$$

where

$$\bar{\mathbf{N}} = \begin{bmatrix} 1 & x \end{bmatrix}; \quad \boldsymbol{\alpha} = \begin{bmatrix} \alpha_1 \\ \alpha_2 \end{bmatrix} \tag{7.20}$$

It turns out to be advantageous to express the parameters α_1 and α_2 in terms of the

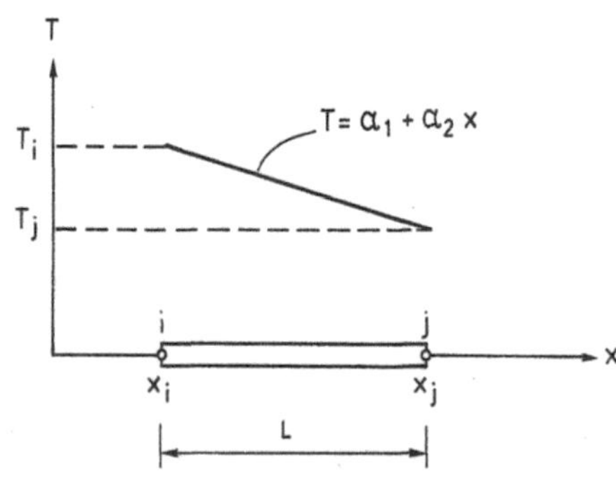

Figure 7.4 Simplest possible one-dimensional element

temperatures T_i and T_j at the nodal points i and j, respectively. From (7.18) and Figure 7.4 we obtain

$$T_i = \alpha_1 + \alpha_2 x_i \quad \text{or} \quad \begin{bmatrix} T_i \\ T_j \end{bmatrix} = \begin{bmatrix} 1 & x_i \\ 1 & x_j \end{bmatrix} \begin{bmatrix} \alpha_1 \\ \alpha_2 \end{bmatrix}$$
$$T_j = \alpha_1 + \alpha_2 x_j$$

which can be written as

$$\mathbf{a}^e = \mathbf{C}\boldsymbol{\alpha} \tag{7.21}$$

where

$$\mathbf{a}^e = \begin{bmatrix} T_i \\ T_j \end{bmatrix}; \quad \mathbf{C} = \begin{bmatrix} 1 & x_i \\ 1 & x_j \end{bmatrix} \tag{7.22}$$

It appears that the column matrix $\mathbf{a}^e$ contains the temperature at the nodal points of the element. From (7.21) it follows that

$$\boldsymbol{\alpha} = \mathbf{C}^{-1}\mathbf{a}^e \tag{7.23}$$

and insertion into (7.19) yields

$$T = \bar{\mathbf{N}}\mathbf{C}^{-1}\mathbf{a}^e \tag{7.24}$$

This expression can be written as

$$\boxed{T = \mathbf{N}^e \mathbf{a}^e} \tag{7.25}$$

where

$$\mathbf{N}^e = [N_i^e \quad N_j^e] = \bar{\mathbf{N}}\mathbf{C}^{-1} \tag{7.26}$$

From (2.46) it is easy to determine the inverse matrix $\mathbf{C}^{-1}$ and we obtain

$$\mathbf{C}^{-1} = \frac{1}{L}\begin{bmatrix} x_j & -x_i \\ -1 & 1 \end{bmatrix} \quad \text{where } L = x_j - x_i \tag{7.27}$$

Using this expression in (7.26) it follows that

$$\boxed{\mathbf{N}^e = [N_i^e \quad N_j^e]} \tag{7.28}$$

where

$$\boxed{\begin{aligned} N_i^e &= -\frac{1}{L}(x - x_j) \\[2mm] N_j^e &= \frac{1}{L}(x - x_i) \end{aligned}} \tag{7.29}$$

The functions N_i^e and N_j^e are called *element shape functions* or *element interpolation*

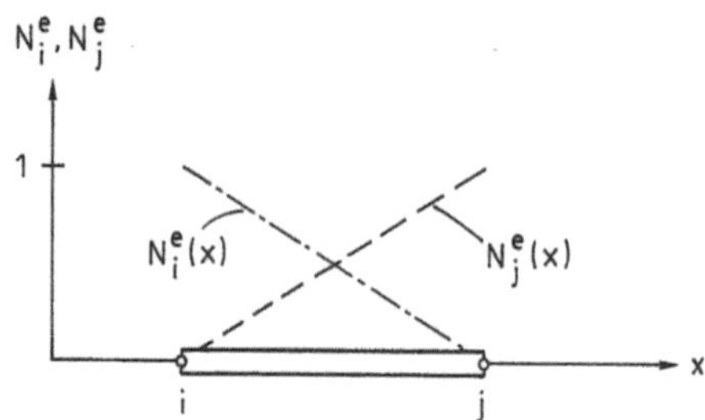

Figure 7.5   Variation of element shape functions with position x

functions and the matrix $\mathbf{N}^e$ is termed the *element shape function matrix*. It will turn out later that different methods exist to establish the element shape functions and the specific method we have adopted here is termed the **C**-*matrix method* (cf. (7.21)).

It is of interest to evaluate how the element shape functions $N_i^e(x)$ and $N_j^e(x)$ vary with position x, and Figure 7.5 shows this dependence. From (7.29) it is obvious that $N_i^e(x)$ and $N_j^e(x)$ vary linearly with position x. Moreover, from (7.29) and Figure 7.5 it follows that the element shape functions possess the following important properties:

$$
\begin{aligned}
N_i^e(x_i) &= 1; & N_i^e(x_j) &= 0 \\
N_j^e(x_i) &= 0; & N_j^e(x_j) &= 1
\end{aligned}
\tag{7.30}
$$

These properties will also turn out to be valid for the more complex elements to be discussed later.

We note that (7.25) can be written as

$$
T(x) = N_i^e(x)\,T_i + N_j^e(x)\,T_j
\tag{7.31}
$$

With expressions (7.25) and (7.31) we have obtained a formulation which separates the influence of geometry via the element shape functions from the influence of physics via the temperatures at the nodal points. This appealing separation is not present in (7.18), where the parameters α_1 and α_2 depend on geometry as well as on nodal point temperatures. It follows that once the geometry of the element is known, the element shape functions can be established directly. This important feature is common for all types of finite elements and it greatly facilitates implementation on a computer.

Expression (7.31) shows clearly that the approximate temperature is expressed as a suitable interpolation between the temperatures at the nodal points. This interpolation is given by the element shape functions, which in the present case are linear functions. Moreover, the approximate temperature T depends linearly on the temperatures T_i and T_j at the nodal points, i.e. the approximation of the temperature can be illustrated as shown in Figure 7.6.

When the fundamental properties (7.30) for the element shape functions were established, use was made of the expressions given by (7.29). However, (7.30) can be

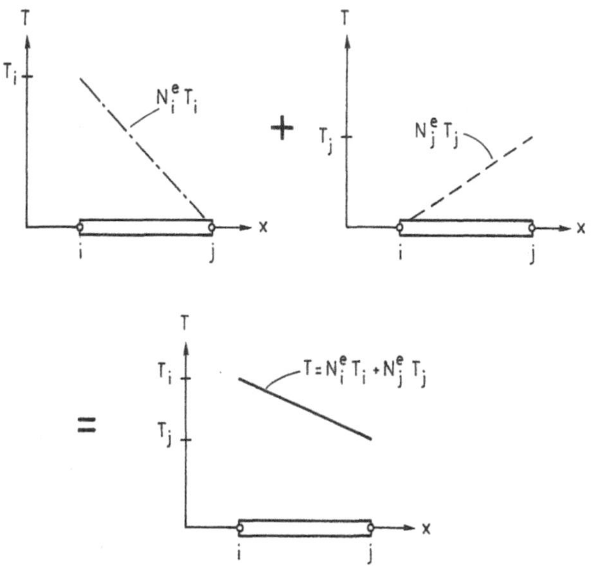

Figure 7.6   Approximation of temperature in linear element

derived directly from (7.31). As an example, for $x = x_i$ we have $T = T_i$, i.e.

$$[N_i^e(x_i) - 1]T_i + N_j^e(x_i)T_j = 0$$

As this expression holds for arbitrary T_i- and T_j-values, we conclude that $N_i^e(x_i) = 1$ and $N_j^e(x_i) = 0$, in accordance with (7.30). Likewise, if $x = x_j$ is used in (7.31) we conclude that $N_i^e(x_j) = 0$ and $N_j^e(x_j) = 1$, also in accordance with (7.30).

Referring to the weak formulation (4.36) of the one-dimensional heat flow problem, it is important to determine the gradient of the temperature T. From (7.25) or (7.31) it follows that

$$\frac{dT}{dx} = \frac{dN^e}{dx}\,a^e = \frac{dN_i^e}{dx}\,T_i + \frac{dN_j^e}{dx}\,T_j \tag{7.32}$$

Let us define the matrix $\mathbf{B}^e$ by

$$\boxed{\mathbf{B}^e = \frac{d\mathbf{N}^e}{dx}} \tag{7.33}$$

Then (7.32) can be written as

$$\boxed{\frac{dT}{dx} = \mathbf{B}^e \mathbf{a}^e} \tag{7.34}$$

Therefore the matrix $\mathbf{B}^e$ relates the temperature gradient to the temperatures at the nodal points of the element. From (7.33) and (7.29) we obtain

$$\mathbf{B}^e = \left[\frac{dN_i^e}{dx} \quad \frac{dN_j^e}{dx} \right] = \left[-\frac{1}{L} \quad \frac{1}{L} \right] \tag{7.35}$$

As $\mathbf{B}^e$ here is a constant matrix, the gradient dT/dx is constant within the element in accordance with the fact that the temperature is assumed to vary linearly within

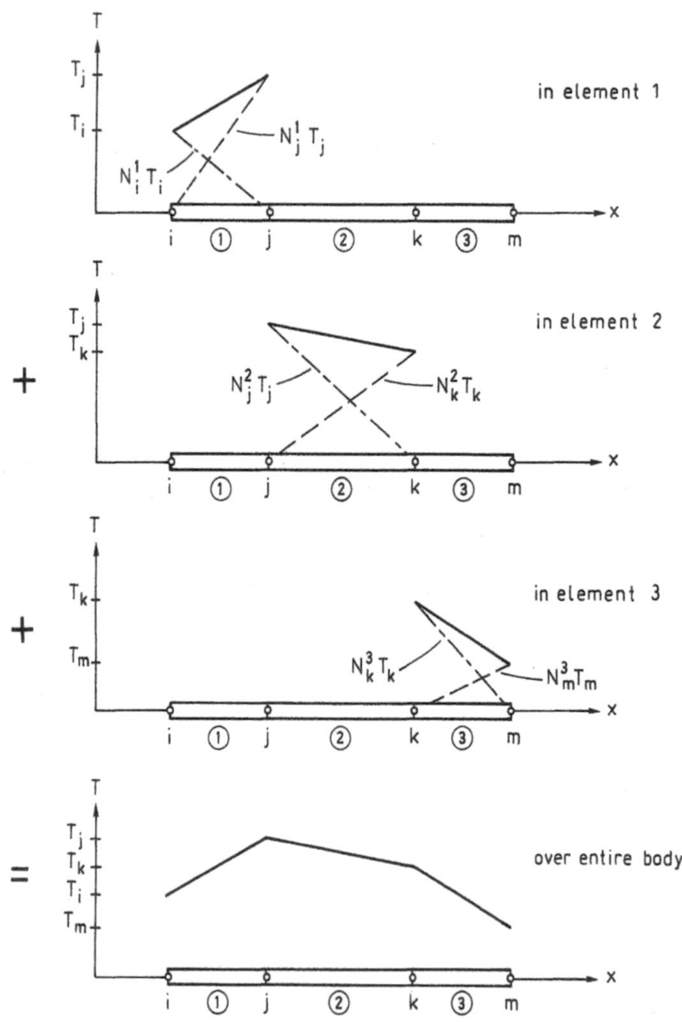

Figure 7.7 Approximation of the temperature over three elements

the element. We note that, due to (7.26) and (7.20), $\mathbf{B}^e$ may be written as

$$\mathbf{B}^e = \frac{d\mathbf{N}^e}{dx} = \frac{d\bar{\mathbf{N}}}{dx}\mathbf{C}^{-1} = [0 \quad 1]\mathbf{C}^{-1} \tag{7.36}$$

where $\mathbf{C}^{-1}$ is given by (7.27).

We have already mentioned that the approximation chosen fulfils the completeness requirements. Let us now check that the element formulation also fulfils the compatibility or conforming requirement stating that the temperature approximation must vary continuously across common element boundaries. For this purpose, consider the body consisting of three elements shown in Figure 7.7, where, for instance, N_j^1 is the element shape function of nodal point j for element 1, whereas N_j^2 is the element shape function of nodal point j for element 2. For each element, we have the approximation shown in Figure 7.6 and as we interpolate between the nodal points, and as, for instance, nodal point j is common for two elements, we clearly have a continuous variation of the temperature over common element boundaries as shown in Figure 7.7. This means that the element is conforming. However, we note that the temperature gradient, in general, varies discontinuously over common element boundaries.

Figure 7.7 shows how the approximate temperature variation over the entire body is established from the approximation over each element. We shall now formulate the approximate temperature variation over the entire body in a slightly different manner. Previously, we determined the element shape functions (cf. (7.29) and Figure 7.7) from which it appears that one element shape function is related to nodal points i and m, whereas two element shape functions are related both to nodal point j and nodal point k. Let us now define the *global shape functions* in such a manner that *one* global shape function is related to each nodal point. The following definitions are adopted:

$$N_i = \begin{cases} N_i^1 & \text{for } x \text{ in element 1} \\ 0 & \text{otherwise} \end{cases}$$

$$N_j = \begin{cases} N_j^1 & \text{for } x \text{ in element 1} \\ N_j^2 & \text{for } x \text{ in element 2} \\ 0 & \text{otherwise} \end{cases}$$

$$N_k = \begin{cases} N_k^2 & \text{for } x \text{ in element 2} \\ N_k^3 & \text{for } x \text{ in element 3} \\ 0 & \text{otherwise} \end{cases} \tag{7.37}$$

$$N_m = \begin{cases} N_m^3 & \text{for } x \text{ in element 3} \\ 0 & \text{otherwise} \end{cases}$$

These global shape functions as well as definitions (7.37) are illustrated in Figure 7.8.

In this figure, it appears that the global shape function for a specific nodal point only differs from zero in those elements which contain this nodal point. This feature

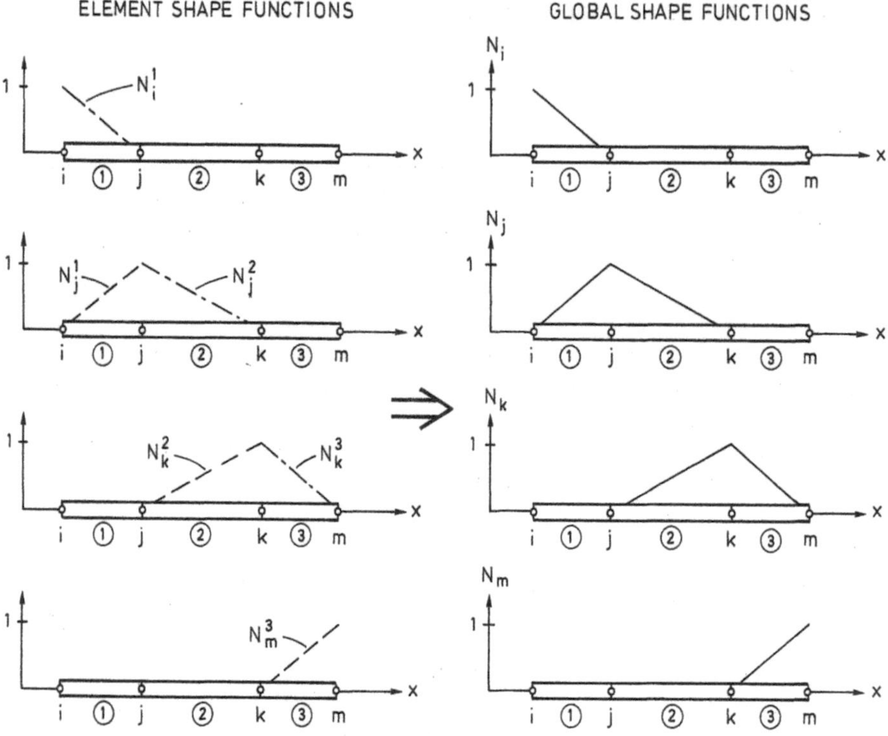

Figure 7.8   Definition of global shape functions from element shape functions

will turn out to be of fundamental importance in the later FE formulation and it is characteristic of all types of finite elements.

With these definitions of the global shape functions, it is obvious that the approximation of the temperature over the entire body can be written as

$$T = N_i T_i + N_j T_j + N_k T_k + N_m T_m \tag{7.38}$$

as illustrated in Figure 7.9. The reader is encouraged to compare Figures 7.7 and 7.9 and check that (7.38) holds.

The temperatures at all nodal points of the body are collected into the column matrix **a** given by

$$\mathbf{a} = \begin{bmatrix} T_i \\ T_j \\ T_k \\ T_m \end{bmatrix} \tag{7.39}$$

Moreover, the *global shape function matrix* **N** is defined by

$$\mathbf{N} = [N_i \ N_j \ N_k \ N_m] \tag{7.40}$$

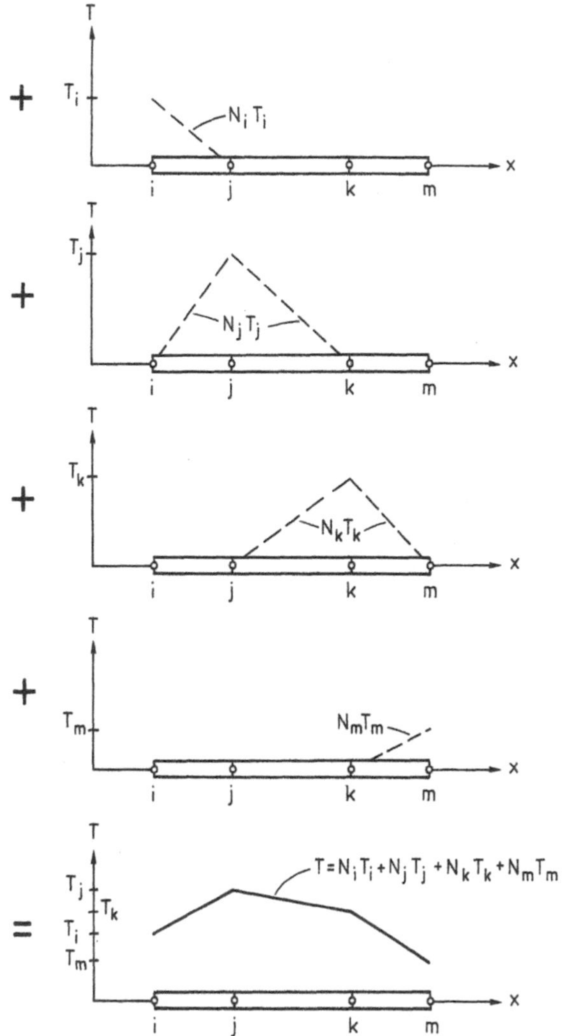

Figure 7.9 Approximation of the temperature over entire body using global shape functions

Then (7.38) can be written as

$$T = \mathbf{Na} \tag{7.41}$$

It follows that the gradient within the body is given by

$$\frac{\mathrm{d}T}{\mathrm{d}x} = \mathbf{Ba} \tag{7.42}$$

where $\mathbf{B}$ is defined as

$$\mathbf{B} = \frac{d\mathbf{N}}{dx} \tag{7.43}$$

i.e.

$$\mathbf{B} = \left[\frac{dN_i}{dx} \quad \frac{dN_j}{dx} \quad \frac{dN_k}{dx} \quad \frac{dN_m}{dx} \right] \tag{7.44}$$

As these results are generalized directly to bodies comprising more elements, we have obtained a complete description of the approximate temperature over an arbitrary one-dimensional body. Note that the elements may be of different lengths.

7.2.2 *Quadratic one-dimensional element*

Instead of the simple linear one-dimensional element presented above we may easily construct higher-order elements where the approximation of the temperature contains terms of higher order than the linear ones. As an example consider the quadratic approximation given by

$$T = \alpha_1 + \alpha_2 x + \alpha_3 x^2 \tag{7.45}$$

Again it is advantageous to express the three parameters α_1, α_2 and α_3 by the temperature at the nodal points. That is, there is room for three nodal points as shown in Figure 7.10. In order to obtain continuity in the temperature across neighbouring elements, nodal points are located at each end of the element and the third nodal point may be located arbitrarily. In practice, however, the third nodal point is located at the middle of the element as shown in Figure 7.10.

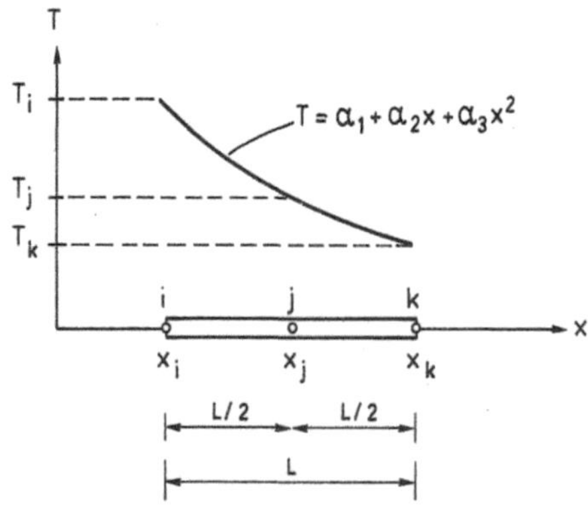

Figure 7.10 Quadratic one-dimensional element

Let us rewrite (7.45) according to

$$T = \bar{\mathbf{N}}\boldsymbol{\alpha} \tag{7.46}$$

where

$$\bar{\mathbf{N}} = \begin{bmatrix} 1 & x & x^2 \end{bmatrix}; \quad \boldsymbol{\alpha} = \begin{bmatrix} \alpha_1 \\ \alpha_2 \\ \alpha_3 \end{bmatrix} \tag{7.47}$$

In order to express the parameters α_1, α_2 and α_3 in terms of the temperatures at the nodal points, we express that (7.45) for $x = x_i$ provides $T = T_i$, etc. This yields

$$\begin{aligned} T_i &= \alpha_1 + \alpha_2 x_i + \alpha_3 x_i^2 \\ T_j &= \alpha_1 + \alpha_2 x_j + \alpha_3 x_j^2 \quad \text{or} \\ T_k &= \alpha_1 + \alpha_2 x_k + \alpha_3 x_k^2 \end{aligned} \qquad \begin{bmatrix} T_i \\ T_j \\ T_k \end{bmatrix} = \begin{bmatrix} 1 & x_i & x_i^2 \\ 1 & x_j & x_j^2 \\ 1 & x_k & x_k^2 \end{bmatrix} \begin{bmatrix} \alpha_1 \\ \alpha_2 \\ \alpha_3 \end{bmatrix}$$

which can be written as

$$\mathbf{a}^e = \mathbf{C}\boldsymbol{\alpha} \tag{7.48}$$

where

$$\mathbf{a}^e = \begin{bmatrix} T_i \\ T_j \\ T_k \end{bmatrix}; \quad \mathbf{C} = \begin{bmatrix} 1 & x_i & x_i^2 \\ 1 & x_j & x_j^2 \\ 1 & x_k & x_k^2 \end{bmatrix} \tag{7.49}$$

As before, the column matrix $\mathbf{a}^e$ contains the temperatures at the nodal points of the element. From (7.48) it follows that

$$\boldsymbol{\alpha} = \mathbf{C}^{-1}\mathbf{a}^e \tag{7.50}$$

and insertion into (7.46) yields

$$T = \bar{\mathbf{N}}\mathbf{C}^{-1}\mathbf{a}^e \tag{7.51}$$

This expression can be written as

$$\boxed{T = \mathbf{N}^e \mathbf{a}^e} \tag{7.52}$$

where

$$\mathbf{N}^e = \begin{bmatrix} N_i^e & N_j^e & N_k^e \end{bmatrix} = \bar{\mathbf{N}}\mathbf{C}^{-1} \tag{7.53}$$

Let us now determine the inverse matrix $\mathbf{C}^{-1}$. First, we determine the determinant of $\mathbf{C}$. From (2.40) and choosing $j = 1$ we obtain

$$\begin{aligned} \det \mathbf{C} &= 1(x_j x_k^2 - x_k x_j^2) - 1(x_i x_k^2 - x_k x_i^2) + 1(x_i x_j^2 - x_j x_i^2) \\ &= x_j x_k(x_k - x_j) - x_i x_k(x_k - x_i) + x_i x_j(x_j - x_i) \end{aligned}$$

Reference to Figure 7.10 gives

$$\det \mathbf{C} = \frac{L}{2}(x_j x_k - 2x_i x_k + x_i x_j) = \frac{L}{2}\left[x_k(x_j - x_i) + x_i(x_j - x_k)\right]$$

$$= \frac{L}{2}\left(x_k\frac{L}{2} - x_i\frac{L}{2}\right) = \frac{L^2}{4}(x_k - x_i) = \frac{L^3}{4}$$

Using (2.46), after a little algebra we then obtain

$$\mathbf{C}^{-1} = \frac{2}{L^2}\begin{bmatrix} x_j x_k & -2x_i x_k & x_i x_j \\ -(x_j + x_k) & 2(x_i + x_k) & -(x_i + x_j) \\ 1 & -2 & 1 \end{bmatrix} \tag{7.54}$$

Inserting (7.47) and (7.54) into (7.53) provides

$$\mathbf{N}^e = \begin{bmatrix} N_i^e & N_j^e & N_k^e \end{bmatrix} \tag{7.55}$$

where

$$N_i^e = \frac{2}{L^2}(x - x_j)(x - x_k)$$

$$N_j^e = -\frac{4}{L^2}(x - x_i)(x - x_k) \tag{7.56}$$

$$N_k^e = \frac{2}{L^2}(x - x_i)(x - x_j)$$

As before N_i^e, N_j^e, N_k^e are called element shape functions, whereas $\mathbf{N}^e$ is termed the element shape function matrix. Just as for the linear element, the specific manner by which we have determined the element shape functions is termed the **C**-matrix method (cf. (7.48)). It appears that the element shape functions are quadratic functions of x, and Figure 7.11 shows their variation with position x.

From (7.56) and Figure 7.11 it follows that the element shape functions possess

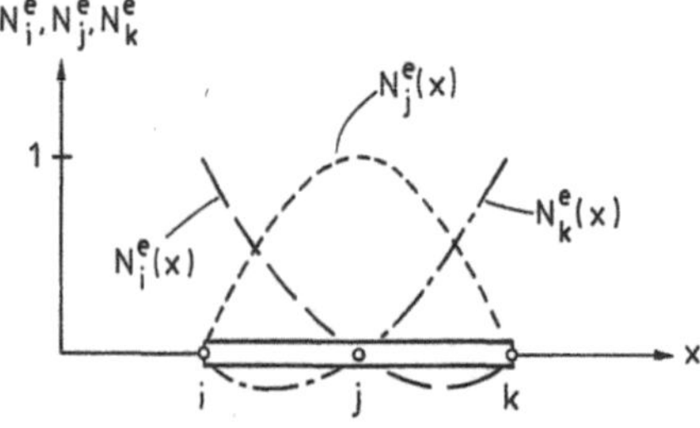

Figure 7.11   Shape functions for quadratic element

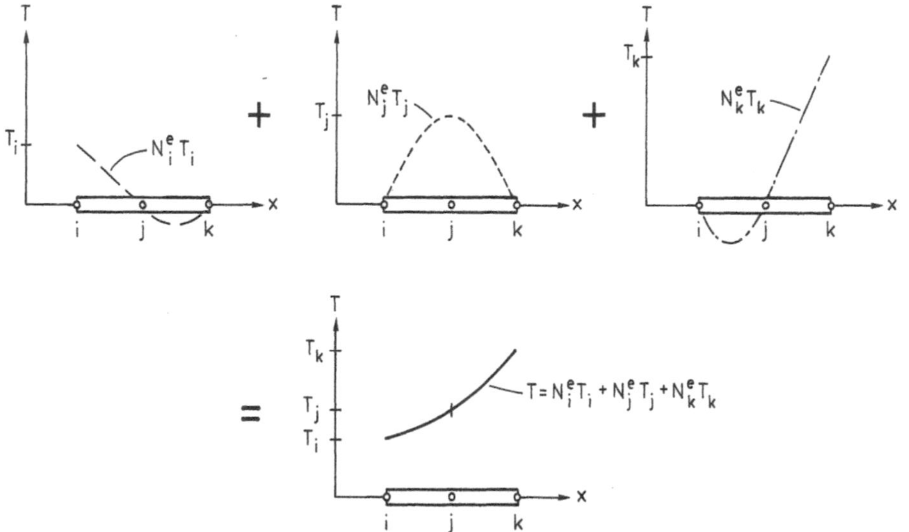

Figure 7.12   Approximation of temperature in quadratic element

the following fundamental property

$$N_i^e = \begin{cases} 1 & \text{at nodal point } i \\ 0 & \text{at all other nodal points} \end{cases} \tag{7.57}$$

in complete analogy to (7.30).

Expression (7.52) may be written as

$$T = N_i^e T_i + N_j^e T_j + N_k^e T_k \tag{7.58}$$

which shows that the temperature T depends linearly on the temperatures T_i, T_j and T_k at the nodal points; that is, the approximation of the temperature can be illustrated as shown in Figure 7.12.

Just as for the linear element, the fundamental property (7.57) for the element shape functions may be derived directly from (7.58). Let us assume that $x = x_i$ for which $T = T_i$, i.e.

$$[N_i^e(x_i) - 1]T_i + N_j^e(x_i)T_j + N_k^e(x_i)T_k = 0$$

As this expression holds for arbitrary T_i-, T_j- and T_k-values, we conclude that $N_i^e(x_i) = 1$, $N_j^e(x_i) = N_k^e(x_i) = 0$ in accordance with (7.57). The remaining properties given by (7.57) are obtained in a similar way by setting $x = x_j$ and $x = x_k$.

In the weak formulation (4.36), the temperature gradient appears. In order to determine this gradient we use (7.52) and (7.58) to obtain

$$\frac{\mathrm{d}T}{\mathrm{d}x} = \frac{\mathrm{d}\mathbf{N}^e}{\mathrm{d}x}\,\mathbf{a}^e = \frac{\mathrm{d}N_i^e}{\mathrm{d}x}\,T_i + \frac{\mathrm{d}N_j^e}{\mathrm{d}x}\,T_j + \frac{\mathrm{d}N_k^e}{\mathrm{d}x}\,T_k \tag{7.59}$$

As with the linear element, the matrix $\mathbf{B}^e$ is defined by

$$\boxed{\mathbf{B}^e = \frac{\mathrm{d}\mathbf{N}^e}{\mathrm{d}x}} \tag{7.60}$$

i.e. we have

$$\boxed{\frac{\mathrm{d}T}{\mathrm{d}x} = \mathbf{B}^e\mathbf{a}^e} \tag{7.61}$$

where the matrix $\mathbf{B}^e$ relates the temperature gradient to the temperatures at the nodal points of the element. With the element shape functions given by (7.56) it follows that

$$\boxed{\mathbf{B}^e = \left[\frac{\mathrm{d}N_i^e}{\mathrm{d}x} \quad \frac{\mathrm{d}N_j^e}{\mathrm{d}x} \quad \frac{\mathrm{d}N_k^e}{\mathrm{d}x}\right] = \frac{2}{L^2}[2x - x_j - x_k \quad -2(2x - x_i - x_k) \quad 2x - x_i - x_j]}$$

$$\tag{7.62}$$

It is no surprise that $\mathbf{B}^e$ and thus the temperature gradient within the element depends linearly on x since the temperature was assumed to vary quadratically with x (cf. (7.45)). We also observe that, due to (7.53) and (7.47), $\mathbf{B}^e$ may be written in the following alternative manner:

$$\mathbf{B}^e = \frac{\mathrm{d}\mathbf{N}^e}{\mathrm{d}x} = \frac{\mathrm{d}\bar{\mathbf{N}}}{\mathrm{d}x}\,\mathbf{C}^{-1} = [0 \quad 1 \quad 2x]\mathbf{C}^{-1} \tag{7.63}$$

where the constant matrix $\mathbf{C}^{-1}$ is given by (7.54).

To illustrate how the temperature variation over the entire body is obtained from the elementwise approximation in a manner analogous to the approach for the linear element, we consider the body shown in Figure 7.13, which is assumed to comprise two elements.

The notation is as before, implying that N_k^1 is the element shape function of nodal point k for element 1, whereas N_k^2 is the element shape function of nodal point k for element 2. Since we interpolate between the temperatures of the nodal points and as, for instance, nodal point k is common to two elements, we clearly have a continuous variation in the temperature over common element boundaries. That is, the element is conforming or compatible. However, the temperature gradient varies, in general, in a discontinuous manner over common element boundaries.

Just as for the linear element and with reference to Figure 7.13, we now define

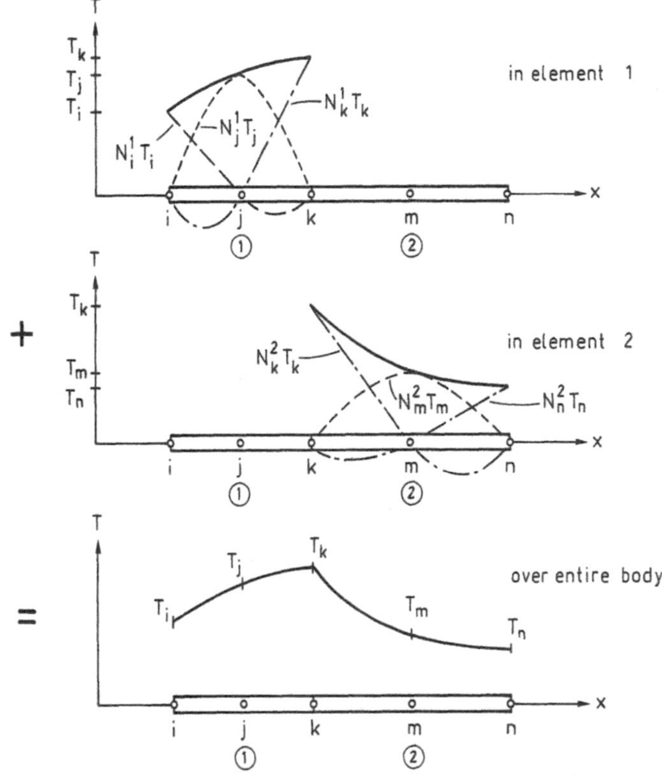

Figure 7.13 Approximation of temperature over two quadratic elements

the global shape functions in such a manner that *one* global shape function is related to each nodal point. The following definitions are made:

$$N_i = \begin{cases} N_i^1 & \text{for } x \text{ in element 1} \\ 0 & \text{otherwise} \end{cases}$$

$$N_j = \begin{cases} N_j^1 & \text{for } x \text{ in element 1} \\ 0 & \text{otherwise} \end{cases}$$

$$N_k = \begin{cases} N_k^1 & \text{for } x \text{ in element 1} \\ N_k^2 & \text{for } x \text{ in element 2} \\ 0 & \text{otherwise} \end{cases} \tag{7.64}$$

$$N_m = \begin{cases} N_m^2 & \text{for } x \text{ in element 2} \\ 0 & \text{otherwise} \end{cases}$$

$$N_n = \begin{cases} N_n^2 & \text{for } x \text{ in element 2} \\ 0 & \text{otherwise} \end{cases}$$

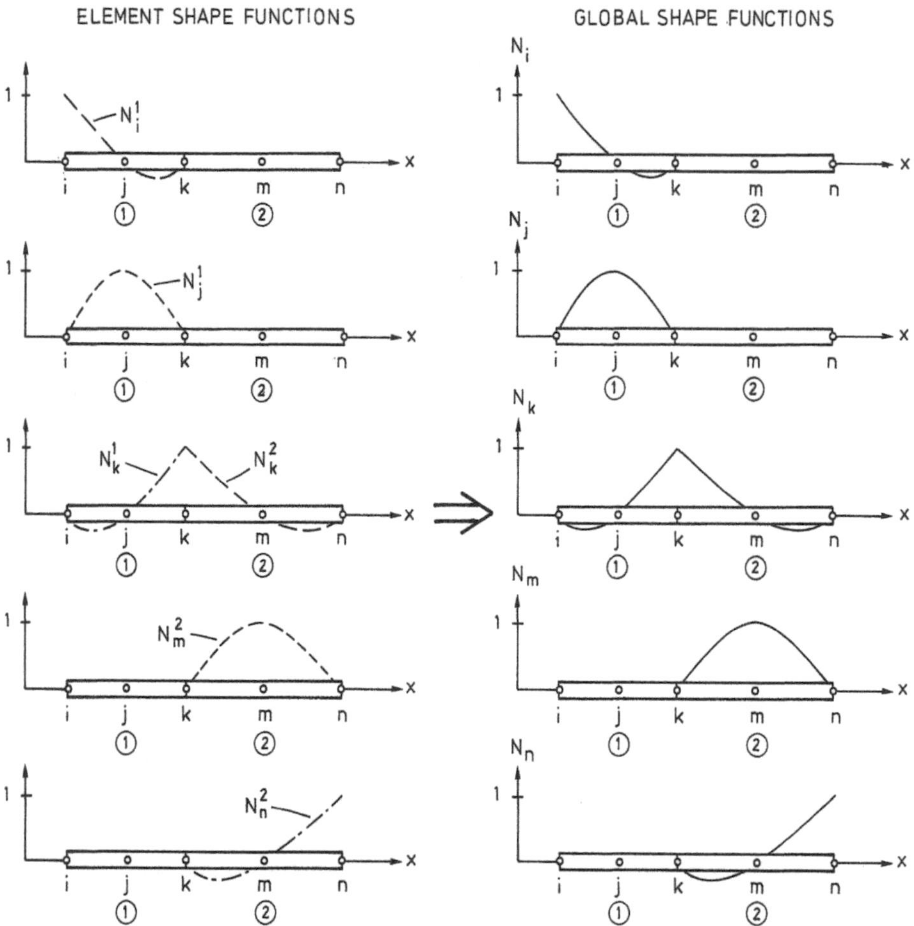

Figure 7.14 Definition of global shape functions from element shape functions

For the body shown in Figure 7.13, these global shape functions take, as well as definitions (7.64), the form shown in Figure 7.14.

In order to write expression (7.64) in a more compact fashion, let us denote a global nodal point by i. For a given position x, we may identify the associated element with number α which contains this position; that is,

$$\boxed{\text{position } x \Rightarrow \text{identification of associated element } \alpha} \qquad (7.65)$$

With this observation, the global shape function $N_i(x)$ can be defined from the element shape function $N_i^\alpha(x)$ according to

$$N_i(x) = \begin{cases} N_i^\alpha(x) & \text{if element } \alpha \text{ contains the global nodal point } i \\ 0 & \text{otherwise} \end{cases} \tag{7.66}$$

The reader is invited to check that (7.64) is contained in this definition and that even (7.37), applicable for the linear element, is contained in the form given by (7.65) and (7.66). In fact, we shall adopt the definitions (7.65) and (7.66) for all types of finite elements. These expressions imply that a global shape function related to a specific nodal point differs from zero only in those elements which contain the nodal point in question.

It appears that the approximation over the entire body given in Figure 7.13 can now be written as

$$T = N_i T_i + N_j T_j + N_k T_k + N_m T_m + N_n T_n \tag{7.67}$$

as illustrated in Figure 7.15. The temperatures at all nodal points of the body are collected into the column matrix **a** given by

$$\mathbf{a} = \begin{bmatrix} T_i \\ T_j \\ T_k \\ T_m \\ T_n \end{bmatrix} \tag{7.68}$$

Moreover, the global shape function matrix **N** is defined by

$$\mathbf{N} = \begin{bmatrix} N_i & N_j & N_k & N_m & N_n \end{bmatrix} \tag{7.69}$$

Then (7.67) can be written as

$$T = \mathbf{Na} \tag{7.70}$$

The gradient within the body becomes

$$\frac{\mathrm{d}T}{\mathrm{d}x} = \mathbf{Ba} \tag{7.71}$$

where **B** is defined as

$$\mathbf{B} = \frac{\mathrm{d}\mathbf{N}}{\mathrm{d}x} \tag{7.72}$$

 Introduction to the finite element method

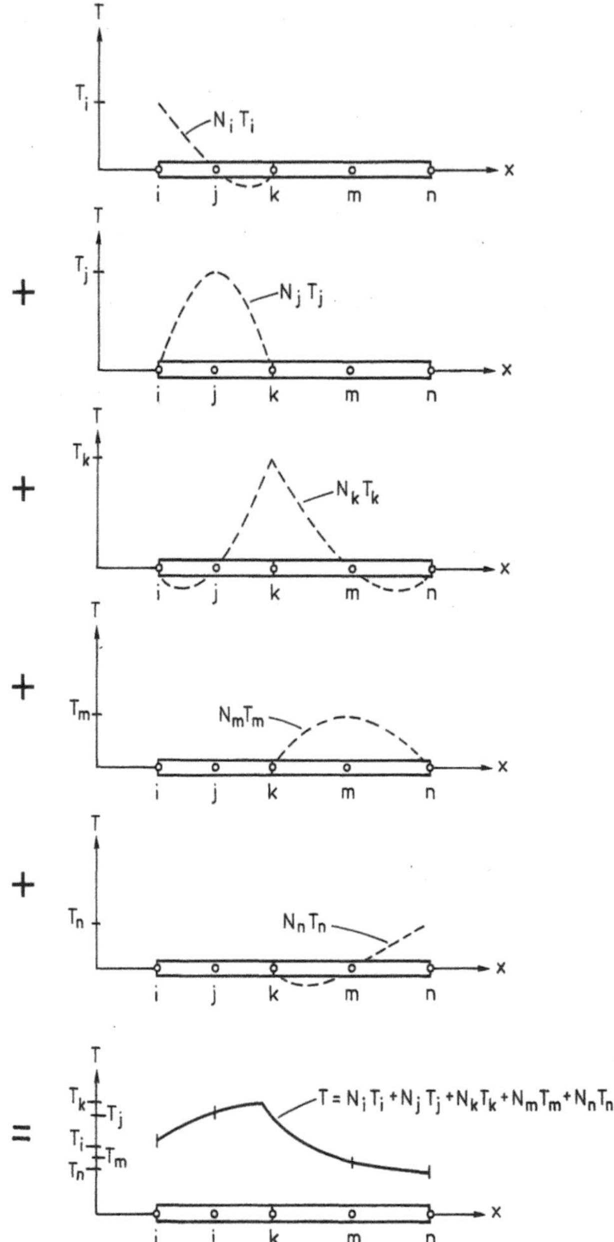

Figure 7.15 Approximation of the temperature over entire body using global shape functions

i.e.

$$\mathbf{B} = \left[\frac{dN_i}{dx} \quad \frac{dN_j}{dx} \quad \frac{dN_k}{dx} \quad \frac{dN_m}{dx} \quad \frac{dN_n}{dx} \right] \tag{7.73}$$

It is obvious that this concept generalizes directly to bodies comprising more elements.

7.2.3 Cubic and quartic one-dimensional elements – Lagrange interpolation

Previously we established the linear and quadratic elements and we may proceed to consider the *cubic* element given by

$$T = \alpha_1 + \alpha_2 x + \alpha_3 x^2 + \alpha_4 x^3 \tag{7.74}$$

In order to express the parameters $\alpha_1 \ldots \alpha_4$ in terms of the temperatures at the nodal point, it is possible again to use the $\mathbf{C}$-matrix concept given by (7.21) and (7.48), which requires that the inverse $\mathbf{C}^{-1}$ is established. For the linear element, this inverse is easily obtained, but for the quadratic element the calculations are rather cumbersome. This obstacle will be much more pronounced when an expression like (7.74) is considered, but it turns out that a general procedure exists by which the element shape functions can be written down directly.

To establish this general procedure, we shall first adopt a suitable numbering of the (local) nodal points belonging to an element. It is emphasized that this local numbering has no effect on the numbers of the global nodal points. With the nodal numbers of the linear element given in Figure 7.16(a), (7.29) provides the element shape functions

$$N_1^e = -\frac{1}{L}(x - x_2); \quad N_2^e = \frac{1}{L}(x - x_1) \tag{7.75}$$

Likewise, with the nodal numbers of the quadratic element given in Figure 7.16(b),

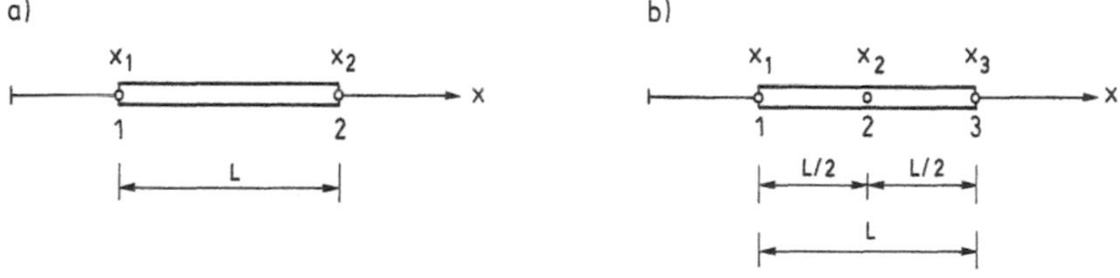

Figure 7.16   (a) Linear element; (b) quadratic element

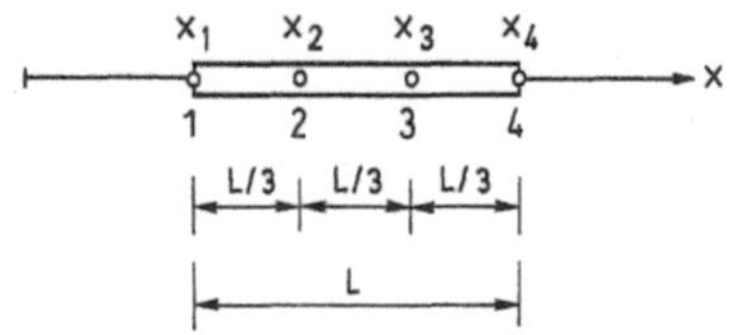

Figure 7.17   Cubic element

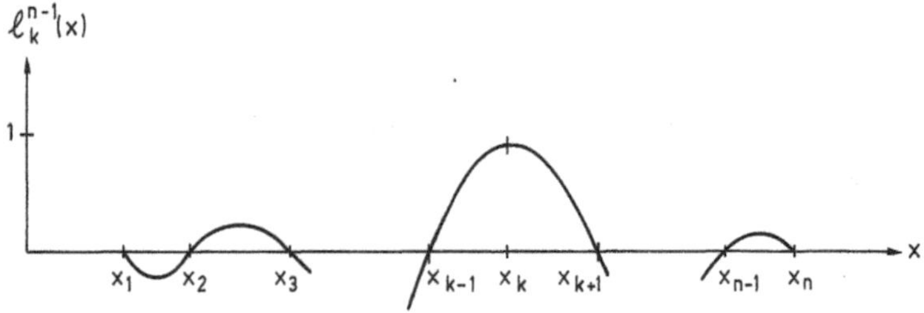

Figure 7.18   Lagrange's interpolation function

(7.56) results in

$$N_1^e = \frac{2}{L^2}(x - x_2)(x - x_3); \quad N_2^e = -\frac{4}{L^2}(x - x_1)(x - x_3);$$

$$N_3^e = \frac{2}{L^2}(x - x_1)(x - x_2) \tag{7.76}$$

Consider now the cubic element in Figure 7.17. It is required that the element shape functions allow us to write

$$T = N_1^e T_1 + N_2^e T_2 + N_3^e T_3 + N_4^e T_4 \tag{7.77}$$

i.e. that the temperature is interpolated between its values at the nodal points, but in order to do so the element shape functions must possess the fundamental property given by (7.57). We also know that the element shape functions must be a polynomial and a polynomial fulfilling (7.57) is given directly by *Lagrange's interpolation formula*. In fact, for n given points this formula provides a polynomial of order $n - 1$ given by

$$l_k^{n-1}(x) = \frac{(x - x_1)(x - x_2) \cdots (x - x_{k-1})(x - x_{k+1}) \cdots (x - x_n)}{(x_k - x_1)(x_k - x_2) \cdots (x_k - x_{k-1})(x_k - x_{k+1}) \cdots (x_k - x_n)};$$

$$k = 1, 2, \ldots, n \tag{7.78}$$

where it should be noted that the term $(x_k - x_k)$ is not present in the denominator and, likewise, the term $(x - x_k)$ is not present in the numerator. It appears that $l_k^{n-1}(x_k) = 1$ and $l_k^{n-1}(x_i) = 0$ for $k \neq i$, as illustrated in Figure 7.18. Moreover, these properties are in accordance with the fundamental property for element shape functions given by (7.57). Therefore, the element shape functions can be constructed

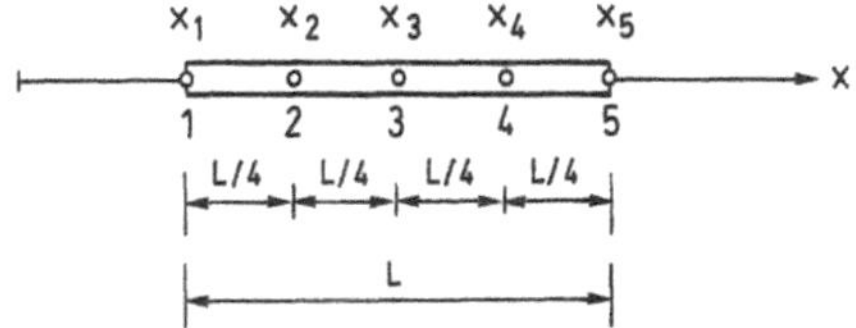

Figure 7.19 Quartic element

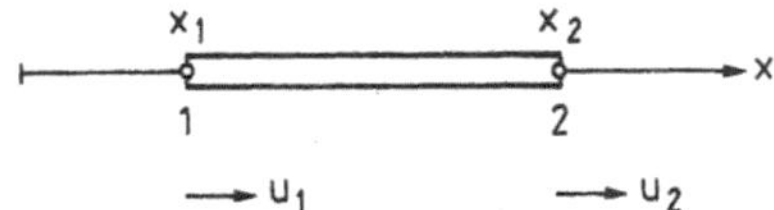

Figure 7.20 Linear element applicable for axially loaded bar

directly by setting

$$\boxed{N_k^e = l_k^{n-1}} \quad k = 1, 2, \ldots, n \qquad (7.79)$$

To illustrate the use of Lagrange's interpolation function, we shall re-establish the element shape functions N_1^e and N_2^e for the linear element shown in Figure 7.16(a). As the element comprises two nodal points, we have $n = 2$ and (7.79) then provides

$$N_k^e = l_k^1; \quad k = 1, 2$$

From (7.78) we get

$$l_1^1 = \frac{x - x_2}{x_1 - x_2}; \quad l_2^1 = \frac{x - x_1}{x_2 - x_1}$$

and as $L = x_2 - x_1$, we have rederived (7.75). In a similar way, the reader may verify that for the quadratic element, i.e. $n = 3$, use of (7.79) yields the results given by (7.76). Therefore, the element shape functions of the cubic element shown in Figure 7.17 are obtained directly from (7.79) by putting $n = 4$, whereas the element shape functions of the *quartic* element given by Figure 7.19 emerge for $n = 5$. This process may be continued for arbitrary n-values. As the element shape functions for all the one-dimensional elements considered may be derived using Lagrange's interpolation formula, these elements are termed *Lagrange elements*.

Having discussed one-dimensional elements for heat flow problems in great detail, we finally mention that these elements obviously apply to axially loaded bars as well. In principle, however, one aspect needs a little attention. Whereas the temperature T is a quantity without any direction, the axial displacement u has direction. When the differential equation (4.18) was derived for an axially loaded elastic bar, the axial displacement u was considered as positive when it was in the direction of the x-axis. Taking account of these remarks, the simple linear element applicable for the axial loading of a bar is illustrated in Figure 7.20, where u_1 and u_2 are the displacements of nodal points 1 and 2, respectively. A similar approach holds for higher-order elements.

7.3 Two-dimensional elements

7.3.1 *Simple triangular element*

After this detailed discussion of one-dimensional elements, it is timely to consider two-dimensional elements. Referring to (7.2), the simplest possible two-dimensional element which fulfils the completeness requirements is given by

$$T = \alpha_1 + \alpha_2 x + \alpha_3 y \tag{7.80}$$

As we have three parameters α_1, α_2 and α_3, which are eventually to be expressed in terms of the temperatures of the nodal points, the element must possess three nodal points. These nodal points are located so that they define the element geometry, i.e. we obtain the triangular element shown in Figure 7.21.

The interpolation given by (7.80) is illustrated in Figure 7.22. As this interpolation is linear in x and y the element is termed the *linear triangular element*. It also follows that contour curves within the element, i.e. curves for which $T = $ constant, are straight lines.

In order to determine the element shape functions corresponding to (7.80) we again use the **C**-matrix method. Therefore, let us write (7.80) as

$$T = \bar{\mathbf{N}}\boldsymbol{\alpha} \tag{7.81}$$

where

$$\bar{\mathbf{N}} = \begin{bmatrix} 1 & x & y \end{bmatrix}; \quad \boldsymbol{\alpha} = \begin{bmatrix} \alpha_1 \\ \alpha_2 \\ \alpha_3 \end{bmatrix} \tag{7.82}$$

Just as for one-dimensional elements, our objective is to reformulate (7.80) so that it clearly appears that the temperature within the element is interpolated between its

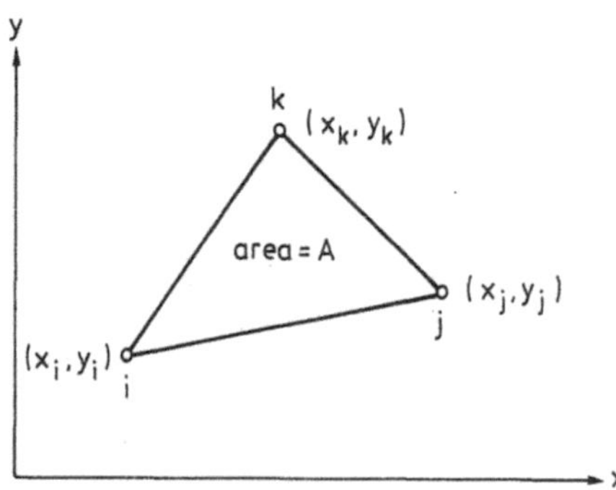

Figure 7.21   Simplest possible two-dimensional element

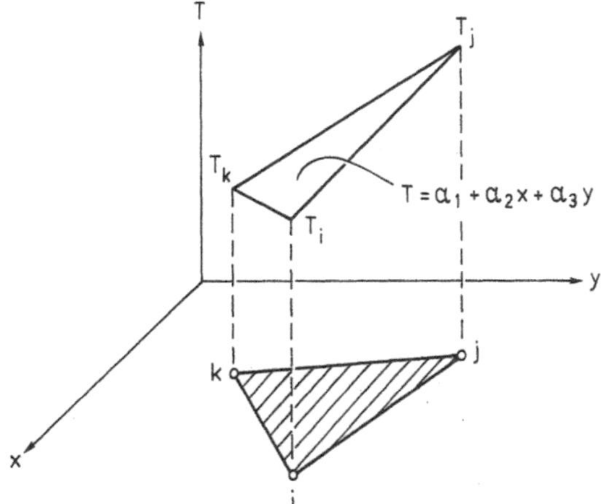

Figure 7.22 Variation of temperature over linear triangular element

values at the nodal points, i.e.

$$T = N_i^e T_i + N_j^e T_j + N_k^e T_k \tag{7.83}$$

where, as before, $N_i^e(x, y)$ is the element shape function at nodal point i and T_i, T_j, T_k are the temperatures at the nodal points. Therefore, inserting the coordinates for the nodal points, we obtain from (7.80) the following system of equations in the same manner as for one-dimensional elements:

$$\mathbf{a}^e = \mathbf{C}\boldsymbol{\alpha} \tag{7.84}$$

where

$$\mathbf{a}^e = \begin{bmatrix} T_i \\ T_j \\ T_k \end{bmatrix}; \quad \mathbf{C} = \begin{bmatrix} 1 & x_i & y_i \\ 1 & x_j & y_j \\ 1 & x_k & y_k \end{bmatrix} \tag{7.85}$$

Moreover, determination of $\boldsymbol{\alpha}$ by means of (7.84) and insertion into (7.81) yields

$$T = \mathbf{N}^e \mathbf{a}^e \tag{7.86}$$

where

$$\mathbf{N}^e = \bar{\mathbf{N}} \mathbf{C}^{-1} \tag{7.87}$$

and

$$\mathbf{N}^e = [N_i^e \quad N_j^e \quad N_k^e] \tag{7.88}$$

It can be observed that (7.86) is in accordance with (7.83).

To determine the inverse matrix $\mathbf{C}^{-1}$, the determinant of $\mathbf{C}$ is calculated first. From (2.40) and choosing $j = 1$ we find

$$\det \mathbf{C} = (x_j y_k - x_k y_j) - (x_i y_k - x_k y_i) + (x_i y_j - x_j y_i) \tag{7.89}$$

Moreover, as the coordinates of the nodal points are known, trivial geometrical arguments show that the area A of the triangle in Figure 7.21 is given by

$$2A = \det \mathbf{C} \tag{7.90}$$

One point needs some attention. In the matrix $\mathbf{a}^e$ given by (7.85) we have listed the nodal points in the sequence i, j and k. Referring to Figure 7.21, it appears that this sequence corresponds to a *counter-clockwise* direction. If a clockwise direction was chosen, as given by the sequence i, k and j, for instance, this implies that the last two rows of $\mathbf{C}$, as given by (7.85), are interchanged. According to property (5) for determinants (see page 18) this would change the sign of the determinant, i.e. $\det \mathbf{C} = -2A$. For convenience, we want (7.90) to apply always, i.e. we adopt the convention that the nodal points should always be taken in the counter-clockwise direction, as expressed through the sequence i, j and k. However, we may equally well take the sequences j, k, i or k, i, j which are obtained from the sequence i, j, k by a so-called *cyclic permutation*. The reason is that each cyclic permutation does not change $\det \mathbf{C}$. To see this, consider the sequence j, k, i. From (7.85) the following is implied: firstly that the first and second rows are interchanged, which changes the sign of $\det \mathbf{C}$; secondly that the second and third rows are interchanged, which again changes the sign of $\det \mathbf{C}$ so that it returns to its original value.

We conclude that for any counter-clockwise direction i, j and k where nodal point i is arbitrary, the relation (7.90) holds.

The inverse matrix $\mathbf{C}^{-1}$ can be established by means of (2.46) and some algebra will show that

$$\mathbf{C}^{-1} = \frac{1}{2A} \begin{bmatrix} x_j y_k - x_k y_j & x_k y_i - x_i y_k & x_i y_j - x_j y_i \\ y_j - y_k & y_k - y_i & y_i - y_j \\ x_k - x_j & x_i - x_k & x_j - x_i \end{bmatrix} \tag{7.91}$$

The element shape functions are obtained from (7.87) and (7.88), which with (7.82) and (7.91) yield

$$\boxed{\begin{aligned} N_i^e &= \frac{1}{2A}[x_j y_k - x_k y_j + (y_j - y_k)x + (x_k - x_j)y] \\[2ex] N_j^e &= \frac{1}{2A}[x_k y_i - x_i y_k + (y_k - y_i)x + (x_i - x_k)y] \\[2ex] N_k^e &= \frac{1}{2A}[x_i y_j - x_j y_i + (y_i - y_j)x + (x_j - x_i)y] \end{aligned}} \tag{7.92}$$

We observe that the element shape functions are linear functions of x and y in accordance with (7.80) and their variation over the element is shown in Figure 7.23.

By inspection it is easily shown that (7.57) again holds, i.e.

$$N_i^e = \begin{cases} 1 & \text{at nodal point } i \\ 0 & \text{at all other nodal points} \end{cases} \tag{7.93}$$

Indeed this is a general property for all element functions and it follows directly from the form (7.83) in a similar manner as in the discussion related to (7.57).

Apart from the fundamental property given by (7.93) we also observe from Figure 7.23 that, for instance, the element shape function $N_k^e(x, y)$ belonging to the nodal point k is zero along the element boundary given by i, j. To prove this formally we recall that $N_k^e(x_i, y_i) = N_k^e(x_j, y_j) = 0$, and as $N_k^e(x, y)$ varies linearly it follows that $N_k^e = 0$ along the element boundary i, j.

Let us return to the interpolation formula (7.83). To appreciate fully how it works, the illustration shown in Figure 7.24 is essential.

In the weak formulation of the two-dimensional heat flow problem (cf. (7.1)), the temperature gradient ∇T enters the expression. Therefore, from (7.86) we derive

$$\nabla T = \begin{bmatrix} \dfrac{\partial T}{\partial x} \\[2ex] \dfrac{\partial T}{\partial y} \end{bmatrix} = \begin{bmatrix} \dfrac{\partial \mathbf{N}^e}{\partial x}\mathbf{a}^e \\[2ex] \dfrac{\partial \mathbf{N}^e}{\partial y}\mathbf{a}^e \end{bmatrix}$$

which can be written as

$$\nabla T = \nabla \mathbf{N}^e \mathbf{a}^e \tag{7.94}$$

where $\nabla \mathbf{N}^e$ is defined according to

$$\nabla \mathbf{N}^e = \begin{bmatrix} \dfrac{\partial \mathbf{N}^e}{\partial x} \\[2ex] \dfrac{\partial \mathbf{N}^e}{\partial y} \end{bmatrix} \tag{7.95}$$

Let us define the matrix $\mathbf{B}^e$ by

$$\mathbf{B}^e = \nabla \mathbf{N}^e = \begin{bmatrix} \dfrac{\partial \mathbf{N}^e}{\partial x} \\[2ex] \dfrac{\partial \mathbf{N}^e}{\partial y} \end{bmatrix} \tag{7.96}$$

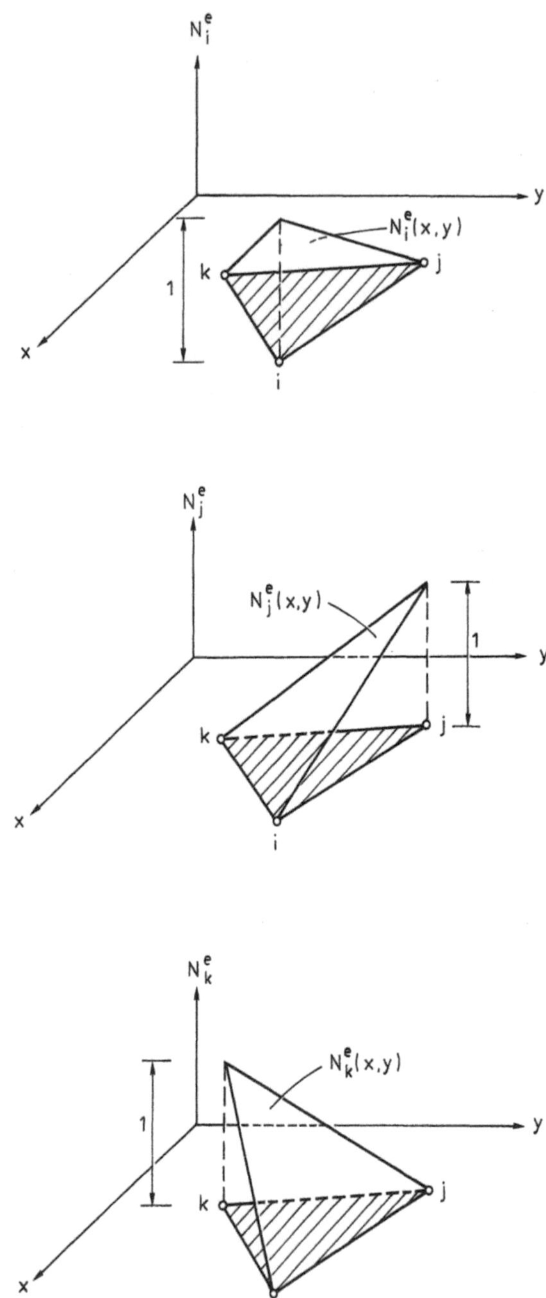

Figure 7.23   Element shape functions for linear triangular element

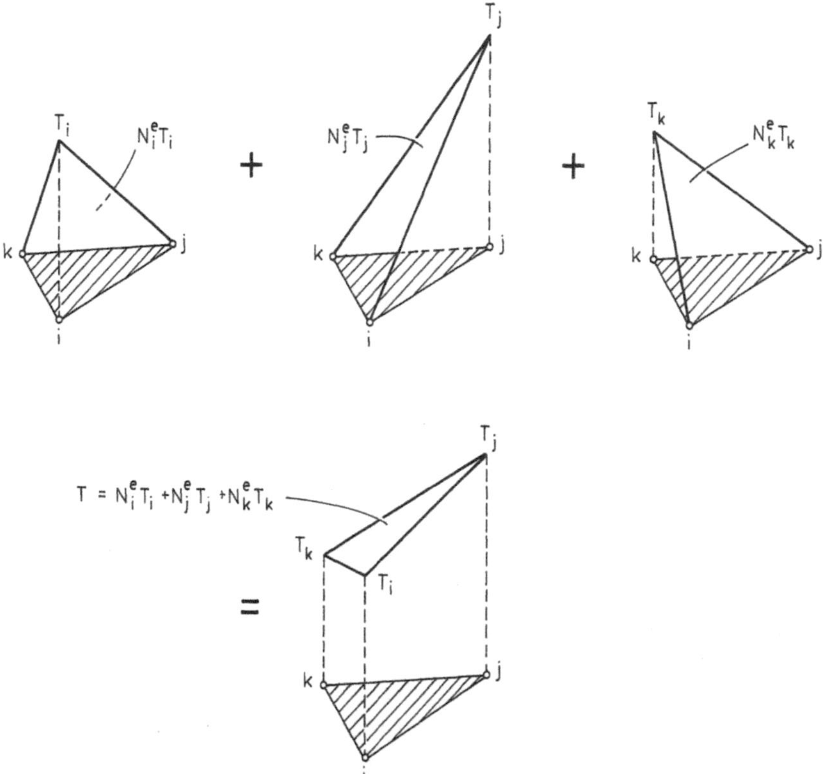

Figure 7.24 Approximation of temperature in triangular element

which implies that (7.94) takes the form

$$\nabla T = \mathbf{B}^e \mathbf{a}^e \qquad (7.97)$$

This expression for the temperature gradient is analogous to the formulations given for one-dimensional elements (cf. (7.34) and (7.61)). The matrix $\mathbf{B}^e$ relates the temperature gradient in the element to the temperatures at the nodal points of the element.

With $\mathbf{N}^e$ given by (7.88) it follows that

$$\mathbf{B}^e = \begin{bmatrix} \dfrac{\partial N_i^e}{\partial x} & \dfrac{\partial N_j^e}{\partial x} & \dfrac{\partial N_k^e}{\partial x} \\[2ex] \dfrac{\partial N_i^e}{\partial y} & \dfrac{\partial N_j^e}{\partial y} & \dfrac{\partial N_k^e}{\partial y} \end{bmatrix} \qquad (7.98)$$

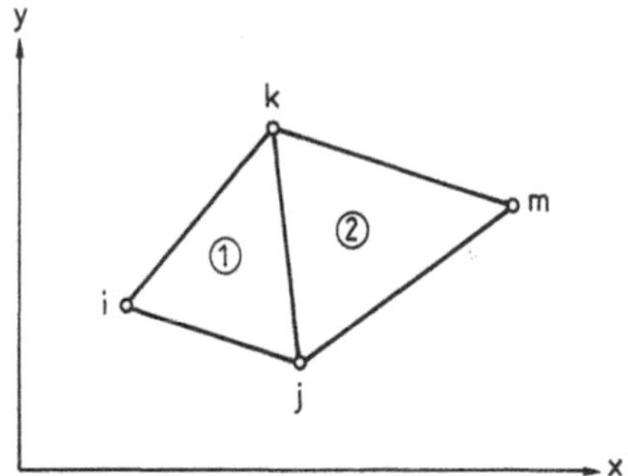

Figure 7.25   Two neighbouring triangular elements

The element shape functions are given by (7.92), i.e.

$$\mathbf{B}^e = \frac{1}{2A}\begin{bmatrix} y_j - y_k & y_k - y_i & y_i - y_j \\ x_k - x_j & x_i - x_k & x_j - x_i \end{bmatrix} \tag{7.99}$$

which shows that $\mathbf{B}^e$ is a constant matrix. Therefore, in accordance with the linear expression (7.80), the temperature gradient ∇T is constant within the element.

Having established the elementwise approximation, our attention is now directed towards the temperature approximation over the entire body. For this purpose we consider two neighbouring triangular elements (Figure 7.25). The element shape functions N_k^1 and N_k^2 both relate to nodal point k, but N_k^1 refers to element 1 and N_k^2 refers to element 2. From (7.93) it follows that $N_k^1(x_k, y_k) = N_k^2(x_k, y_k) = 1$ and $N_k^1(x_j, y_j) = N_k^2(x_j, y_j) = 0$, and as both N_k^1 are N_k^2 are linear functions, we conclude that $N_k^1 = N_k^2$ holds along the common element boundary j, k.

Considering a specific nodal point, let us construct a global shape function related to that point in the same manner as for one-dimensional elements (cf. (7.65) and (7.66)). Therefore, for a given position (x, y), we may identify the associated element α which contains this position, i.e.

$$\text{position } (x, y) \Rightarrow \text{identification of associated element } \alpha \tag{7.100}$$

For a given nodal point i the global shape function N_i is then defined from the element shape functions N_i^α according to

$$N_i(x, y) = \begin{cases} N_i^\alpha(x, y) & \text{if element } \alpha \text{ contains global nodal point } i \\ 0 & \text{otherwise} \end{cases} \tag{7.101}$$

With these definitions the variation of a global shape function is shown in Figure 7.26.

From the global shape functions defined above, the temperature approximation

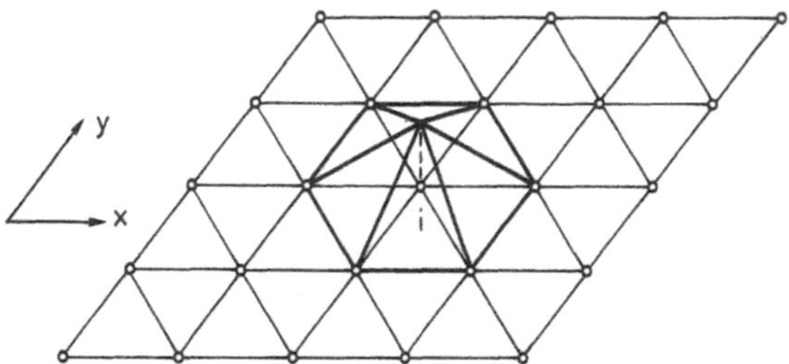

Figure 7.26 Global shape function $N_i(x, y)$ for nodal point i for linear triangular elements

over the entire body can be written as

$$T = \mathbf{Na} \tag{7.102}$$

The column matrix $\mathbf{a}$ contains the temperatures at all nodal points of the body, i.e.

$$\mathbf{a} = \begin{bmatrix} T_1 \\ T_2 \\ \vdots \\ T_n \end{bmatrix} \tag{7.103}$$

where n is the total number of nodal points in the body. The global shape function matrix $\mathbf{N}$ is then given by

$$\mathbf{N} = [N_1 \quad N_2 \quad \cdots \quad N_n] \tag{7.104}$$

The temperature gradient within the body becomes

$$\nabla T = \mathbf{Ba} \tag{7.105}$$

where $\mathbf{B}$ is defined by

$$\mathbf{B} = \nabla \mathbf{N} = \begin{bmatrix} \dfrac{\partial N_1}{\partial x} & \dfrac{\partial N_2}{\partial x} & \cdots & \dfrac{\partial N_n}{\partial x} \\ \dfrac{\partial N_1}{\partial y} & \dfrac{\partial N_2}{\partial y} & \cdots & \dfrac{\partial N_n}{\partial y} \end{bmatrix} \tag{7.106}$$

The approximation of the temperature given by (7.80) clearly fulfils the completeness requirements. Let us now check whether the simple triangular element considered is conforming, i.e. whether the approximate temperature distribution varies continuously over common element boundaries. For this purpose consider the common element boundary given by j, k in Figure 7.25. As the temperatures at nodal points j and k are the same for the two elements and as the temperature varies linearly between nodal points j and k, it follows that it varies in a continuous manner over the common element boundary j, k. This means that the element is conforming.

The linear triangular element dealt with so far is a very simple element. It was suggested by Turner *et al.* (1956), and even though in many respects it marked the beginning of the FE method, it is still widely used. It is essential that the element can take the form of an arbitrary triangle, which implies that the geometry of any irregular two-dimensional body can be approximated as closely as required. This ability to model arbitrary geometries is one of the essential advantages of the FE method.

7.3.2 *Four-node rectangular element – the Lagrange element*

Let us now consider an element which contains four nodal points, i.e. *nodes*. These nodal points are chosen so that they define the geometry of the element and we are then led to the rectangular element shown in Figure 7.27. For reasons that will be revealed later, the sides of the element are parallel to the coordinate axes. As we have four nodal points, (7.2) and Pascal's triangle result in the approximation

$$T = \alpha_1 + \alpha_2 x + \alpha_3 y + \alpha_4 xy \qquad (7.107)$$

The element defined by Figure 7.27 and (7.107) is occasionally referred to as the *Melosh element* (Melosh, 1963), and it involves the parasitic term xy. Instead of this quadratic term xy, one may in principle choose the term x^2 or y^2, but the choice xy is in general preferable because it implies that the dependence on x and y is similar, i.e. the approximation is of the same type in the x- and y-directions. Despite the presence of the quadratic term xy, it is also observed that (7.107) for $y = $ constant implies a linear variation with x and for $x = $ constant a linear variation with y. For this reason, (7.107) is often referred to as a *bilinear* expression.

As before, we want to reformulate (7.107) so that the temperature is interpolated

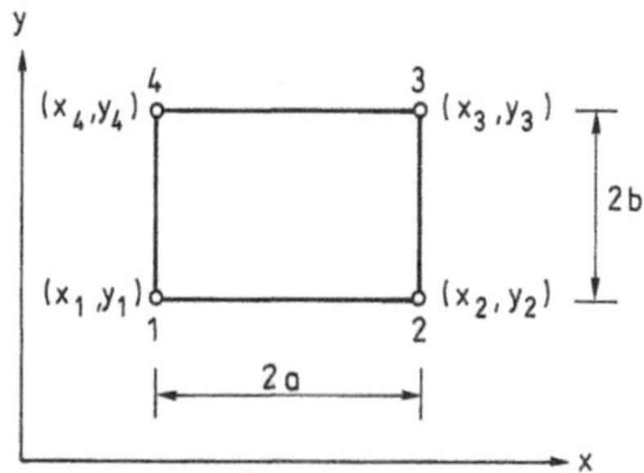

Figure 7.27   Four-node rectangular element

between the temperatures at the nodal points using the element shape functions, i.e.

$$\boxed{T = \mathbf{N}^e \mathbf{a}^e}$$

(7.108)

where

$$\mathbf{N}^e = [N_1^e \quad N_2^e \quad N_3^e \quad N_4^e]; \quad \mathbf{a}^e = \begin{bmatrix} T_1 \\ T_2 \\ T_3 \\ T_4 \end{bmatrix}$$

(7.109)

and where the element shape functions fulfil the usual requirement given by (7.93). This form may be obtained by expressing the parameters $\alpha_1 \ldots \alpha_4$ in terms of the temperatures at the nodal points in the traditional manner using the **C**-matrix method (cf. (7.84)). However, another method is possible using Lagrange's interpolation formula (7.78).

We first observe that in the x-direction, i.e. when $y = $ constant, (7.107) implies that T varies linearly with x. Likewise, in the y-direction, i.e. when $x = $ constant, (7.107) shows that T varies linearly with y. Let us consider next, for instance, nodal point 1 of Figure 7.27. In the x-direction we adopt Lagrange's interpolation formula for $n = 2$ given by

$$l_1^1(x) = \frac{x - x_2}{x_1 - x_2}$$

which fulfils $l_1^1(x_1) = 1$ and $l_1^1(x_2) = 0$. In the y-direction we choose

$$l_1^1(y) = \frac{y - y_4}{y_1 - y_4}$$

i.e. $l_1^1(y_1) = 1$ and $l_1^1(y_4) = 0$. The element shape function N_1^e is then given as

$$N_1^e(x, y) = l_1^1(x) l_1^1(y) = \frac{x - x_2}{x_1 - x_2} \frac{y - y_4}{y_1 - y_4} = \frac{1}{4ab}(x - x_2)(y - y_4)$$

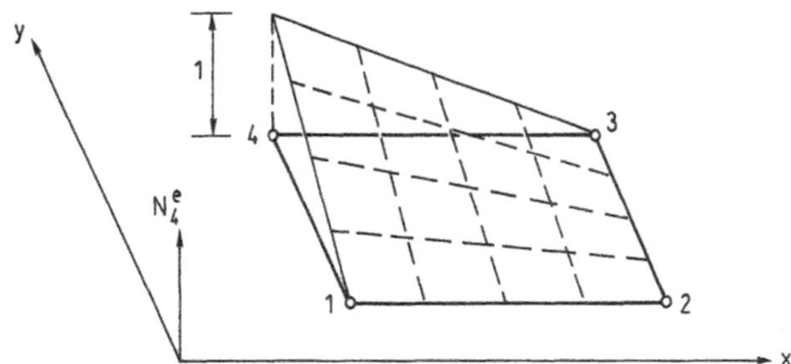

Figure 7.28　Element shape function N_4^e for four-node rectangular element

where $2a = x_2 - x_1 = x_3 - x_4$ and $2b = y_4 - y_1 = y_3 - y_2$ (cf. Figure 7.27). The reader may verify that $N_1^e(x_1, y_1) = 1$ and $N_1^e(x_2, y_2) = N_1^e(x_3, y_3) = N_1^e(x_4, y_4) = 0$ in accordance with the general requirement given by (7.93). Using the same approach for the other nodal points, we find that

$$
\begin{aligned}
N_1^e &= \frac{1}{4ab}(x - x_2)(y - y_4) \\[2mm]
N_2^e &= -\frac{1}{4ab}(x - x_1)(y - y_3) \\[2mm]
N_3^e &= \frac{1}{4ab}(x - x_4)(y - y_2) \\[2mm]
N_4^e &= -\frac{1}{4ab}(x - x_3)(y - y_1)
\end{aligned}
\tag{7.110}
$$

By inserting (7.110) in (7.108) it can easily be checked that the form given by (7.107) is obtained. Since the element shape functions may be derived using Lagrange's interpolation formula, the four-node rectangular element is an example of a *Lagrange element*.

The variation of one element shape function over the element is illustrated in Figure 7.28. The appearance of the other element shape functions is quite similar. Whereas the element shape functions vary linearly for lines parallel to the coordinate axes, the presence of the term xy in (7.107) implies that they vary non-linearly along all other lines. We observe that an element shape function related to a specific nodal point is zero along element boundaries which do not contain the nodal point in question. This property was also found for the triangular linear element and in order to prove this property formally for the four-node element, consider, for instance, the element shape function N_4^e shown in Figure 7.28 along the element boundary 2,3. As $x =$ constant along element boundary 2,3, N_4^e varies linearly with y along this boundary. Moreover, as $N_4^e(x_2, y_2) = N_4^e(x_3, y_3) = 0$, we conclude that $N_4^e = 0$ along element boundary 2,3. The same type of argument implies that if a nodal point is shared by two neighbouring elements, then the two element shape functions belonging to that nodal point are identical along the common element boundary.

As before, the temperature gradient is of interest and similar to (7.94)–(7.98) we obtain from (7.108)

$$
\nabla T = \nabla \mathbf{N}^e \mathbf{a}^e
$$

which can be written as

$$
\nabla T = \mathbf{B}^e \mathbf{a}^e
\tag{7.111}
$$

where

$$\mathbf{B}^e = \nabla \mathbf{N}^e = \begin{bmatrix} \dfrac{\partial \mathbf{N}^e}{\partial x} \\[2ex] \dfrac{\partial \mathbf{N}^e}{\partial y} \end{bmatrix} = \begin{bmatrix} \dfrac{\partial N_1^e}{\partial x} & \dfrac{\partial N_2^e}{\partial x} & \dfrac{\partial N_3^e}{\partial x} & \dfrac{\partial N_4^e}{\partial x} \\[2ex] \dfrac{\partial N_1^e}{\partial y} & \dfrac{\partial N_2^e}{\partial y} & \dfrac{\partial N_3^e}{\partial y} & \dfrac{\partial N_4^e}{\partial y} \end{bmatrix} \qquad (7.112)$$

The element shape functions are given by (7.110), i.e.

$$\mathbf{B}^e = \frac{1}{4ab} \begin{bmatrix} y - y_4 & y_3 - y & y - y_2 & y_1 - y \\ x - x_2 & x_1 - x & x - x_4 & x_3 - x \end{bmatrix} \qquad (7.113)$$

With the global shape functions defined in the usual manner (cf. (7.100) and (7.101)) we obtain similarly to (7.102)–(7.106)

$$T = \mathbf{Na} \qquad (7.114)$$

and

$$\nabla T = \mathbf{Ba} \qquad (7.115)$$

where

$$B = \nabla \mathbf{N} = \begin{bmatrix} \dfrac{\partial N_1}{\partial x} & \dfrac{\partial N_2}{\partial x} & \cdots & \dfrac{\partial N_n}{\partial x} \\[2ex] \dfrac{\partial N_1}{\partial y} & \dfrac{\partial N_2}{\partial y} & \cdots & \dfrac{\partial N_n}{\partial y} \end{bmatrix} \qquad (7.116)$$

The approximation (7.107) fulfils the completeness requirement (cf. (7.2)). To check that the element is conforming, i.e. that the temperature varies continuously across common element boundaries, consider the two elements shown in Figure 7.29.

Along the common boundary 2,5 we have $x = $ constant, i.e. according to (7.107)

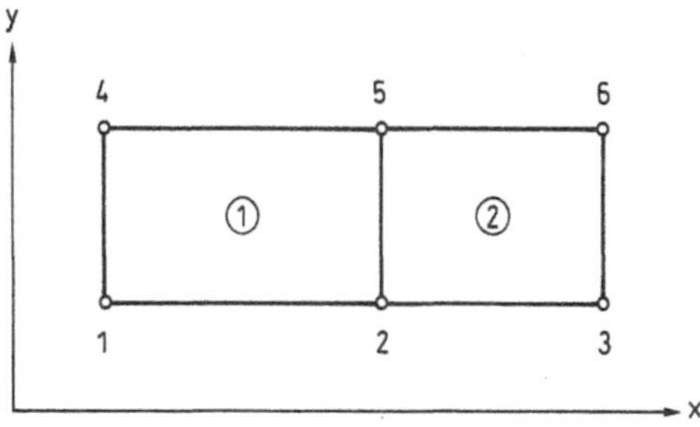

Figure 7.29   Conforming elements

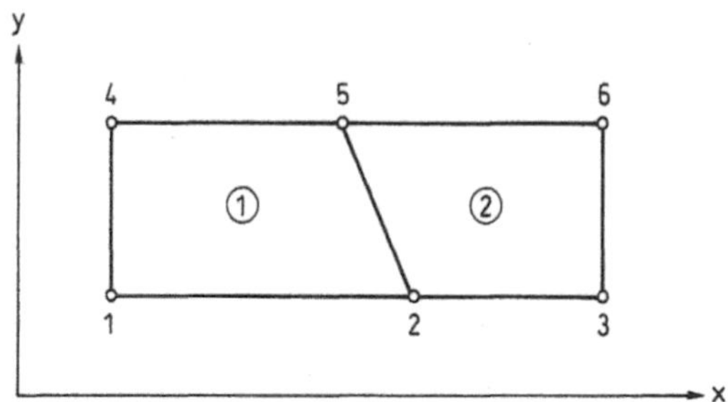

Figure 7.30   Non-conforming elements

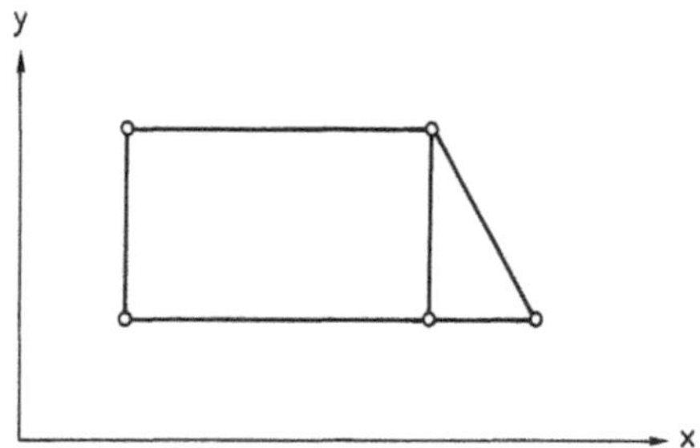

Figure 7.31   Conforming elements

the temperature varies linearly with y for both elements. As the temperatures at nodal points 2 and 5 are identical for the two elements we conclude that continuity exists across the common element boundary; that is, the element is conforming.

However, it is important that this continuity is present only when the element boundaries are parallel to the coordinate axes. To show this consider Figure 7.30.

The common element boundary 2,5 is given by the equation $y = ax + b$. Insertion of this expression into (7.107) implies that

$$T = \alpha_1 + \alpha_3 b + (\alpha_2 + \alpha_3 a + \alpha_4 b)x + \alpha_4 a x^2$$

i.e. T varies quadratically with x along boundary 2,5. This quadratic function is defined uniquely by three temperatures, but as only the two temperatures T_2 and T_5 are known, the quadratic expression above is not uniquely determined. This means that the temperature variation along the common boundary in element 1 differs from that of element 2. Therefore, continuity across common element boundaries is violated implying that the element is non-conforming. In general, we would like to avoid this situation, which means that the four-node element should be used in a form where the element boundaries are parallel to the coordinate axes.

It appears that the four-node rectangular element in itself is unable to model arbitrary geometries of the body. However, if it is combined with the simple triangular element, arbitrary geometries can be modelled, and an example of such a combination is shown in Figure 7.31. Using arguments similar to the above discussion it can easily be shown that the elements in Figure 7.31 are conforming, i.e. continuity across common element boundaries exists.

The restriction that the four-node element must be rectangular with its element boundaries parallel to the coordinate axes has been emphasized. However, we shall see in Chapter 19 that this restriction can be removed by means of the so-called *isoparametric formulation* of finite elements.

7.3.3 *More complicated rectangular and triangular elements – serendipity elements*

It is obvious that the manner in which Lagrange's interpolation formula was used to construct the element shape functions for the four-node rectangular element can be generalized. Without going into detail, consider the rectangular element shown in Figure 7.32. The specific feature of this six-node element is that it contains three nodes in the x-direction and only two nodal points in the y-direction. Using Pascal's triangle the temperature approximation becomes

$$T = \alpha_1 + \alpha_2 x + \alpha_3 y + \alpha_4 x^2 + \alpha_5 xy + \alpha_6 x^2 y \tag{7.117}$$

which implies a quadratic variation in the x-direction and a linear variation in the y-direction, in accordance with Figure 7.32. The element shape functions may be constructed directly in a manner similar to that leading to (7.110), i.e. any element shape function is obtained as a suitable multiplication of a second-order Lagrange polynomial in x and a first-order Lagrange polynomial in y. The element of Figure 7.32 is therefore another example of a Lagrange element and may be of interest for problems where it is known, *a priori*, that the temperature variation is more pronounced in the x-direction than in the y-direction.

Another Lagrange element that is used more often, and which exhibits geometrical isotropy, is the *rectangular nine-node element* shown in Figure 7.33. As we have a quadratic variation in the x- and y-directions and nine nodal points, use of Pascal's triangle yields the following approximation:

$$T = \alpha_1 + \alpha_2 x + \alpha_3 y + \alpha_4 x^2 + \alpha_5 xy + \alpha_6 y^2 + \alpha_7 x^2 y + \alpha_8 xy^2 + \alpha_9 x^2 y^2 \tag{7.118}$$

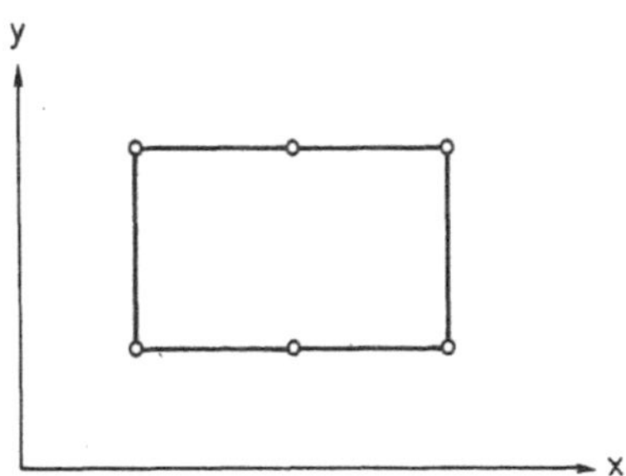

Figure 7.32 Six-node rectangular Lagrange element

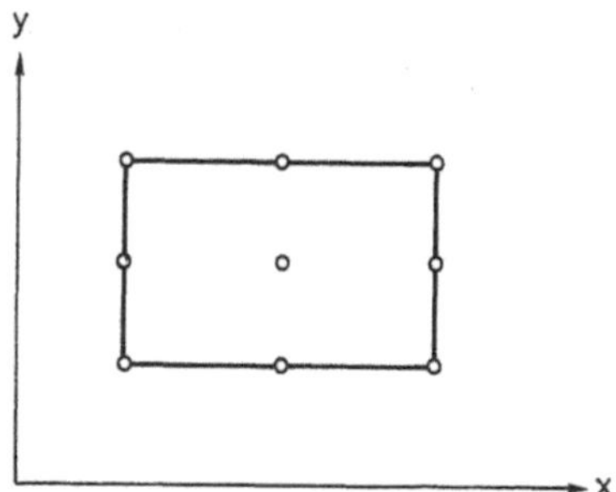

Figure 7.33   Nine-node rectangular Lagrange element

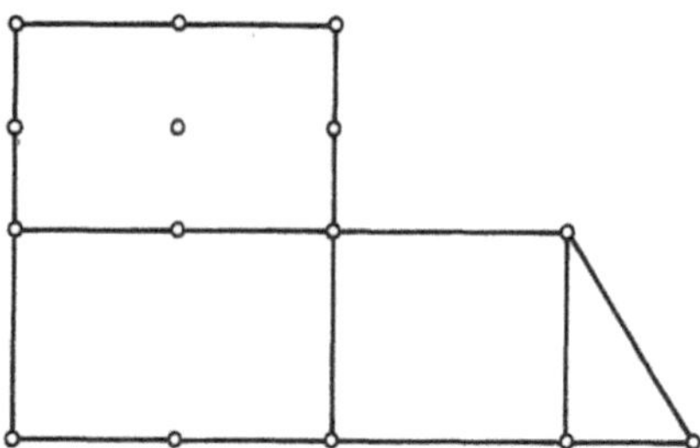

Figure 7.34   Conforming elements

which is symmetric in x and y. It appears that the last three terms are parasitic. Moreover, as all linear and quadratic terms are present, this nine-node element is able to reflect an arbitrary linear variation of the temperature gradient. The element shape functions can be obtained as a suitable multiplication of Lagrange polynomials of second order in the x- and y-direction or using the **C**-matrix method. As the temperature varies parabolically along any element boundary, a combination of rectangular nine-node elements is conforming. However, analogous to the discussion of Figure 7.30, this conforming behaviour requires that the element boundaries are parallel with the coordinate axes. This requirement can be relaxed by use of the isoparametric FE formulation presented in Chapter 19. A conforming combination of the three-node triangle and the four-, six- and nine-node rectangular Lagrange elements is shown in Figure 7.34. Further information on the nine-node Lagrange element may be found, for instance, in Becker *et al.* (1981) and Hughes (1987).

It is also possible to obtain triangular elements of higher order than the simple three-node element treated previously. An example is the six-node triangle shown in Figure 7.35. Referring to Pascal's triangle, we can write the approximation as

$$T = \alpha_1 + \alpha_2 x + \alpha_3 y + \alpha_4 x^2 + \alpha_5 xy + \alpha_6 y^2 \tag{7.119}$$

which forms a complete polynomial suggesting an efficient element. Moreover, any linearly varying temperature gradient can be modelled by (7.119).

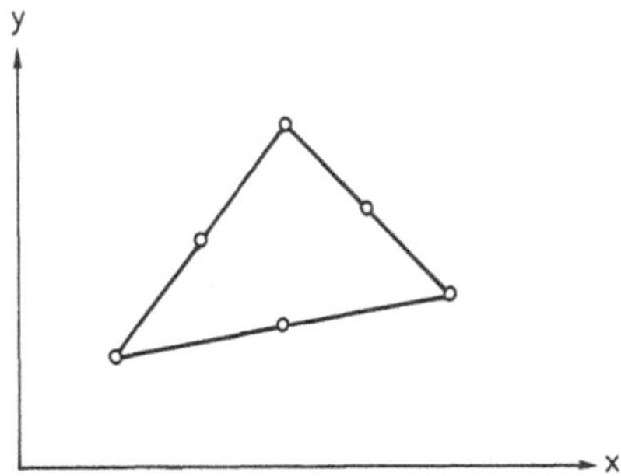

Figure 7.35   Six-node triangular element

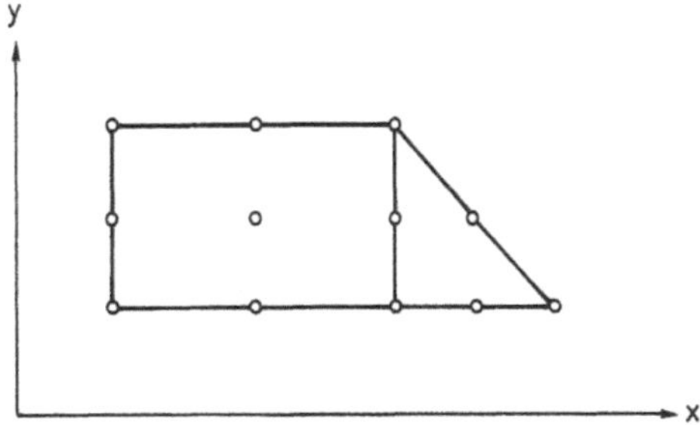

Figure 7.36   Conforming elements

Using arguments similar to those given previously, it can easily be shown that any element mesh consisting of six-node triangular elements is conforming. As the six-node element may form any triangle, we conclude that any two-dimensional geometry can be approximated as closely as requested. As shown in Figure 7.36, the six-node triangle can also be used in combination with the rectangular nine-node element to model bodies with skew boundaries and this combination is easily shown to be conforming. The six-node triangle is treated in detail by, for instance, Zienkiewicz and Taylor (1989) and Gallagher (1975).

At this point, it is no surprise that a variety of elements exists and can be suggested, and we shall conclude the discussion of two-dimensional elements by considering the often used eight-node rectangular element shown in Figure 7.37.

The eight-node element looks a little like that of Figure 7.33, but it is not a Lagrange element because we only have two nodal points along the line given by nodal points 5 and 7. Therefore, Lagrange polynomials cannot be used directly to establish the element shape functions. Nevertheless, it is possible to construct these element shape functions directly. To facilitate this process, we make a coordinate transformation from the xy-system to the $\xi\eta$-system (cf. Figure 7.37). The $\xi\eta$-coordinate system is called a *local coordinate system* and it is devised so that $|\xi| = 1$ and $|\eta| = 1$ hold along the element boundaries. Clearly, it is easier to establish

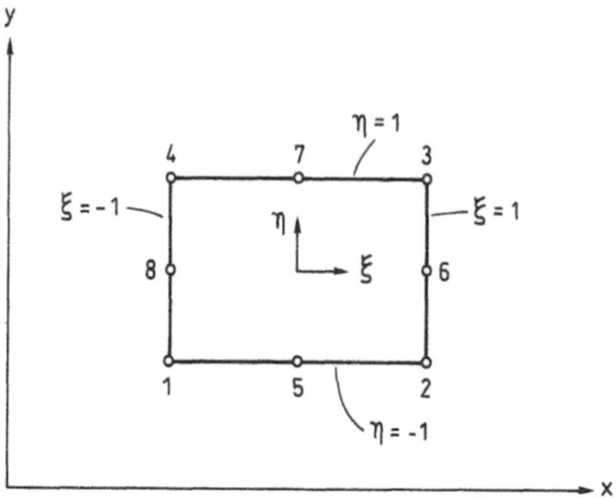

Figure 7.37   Eight-node serendipity element

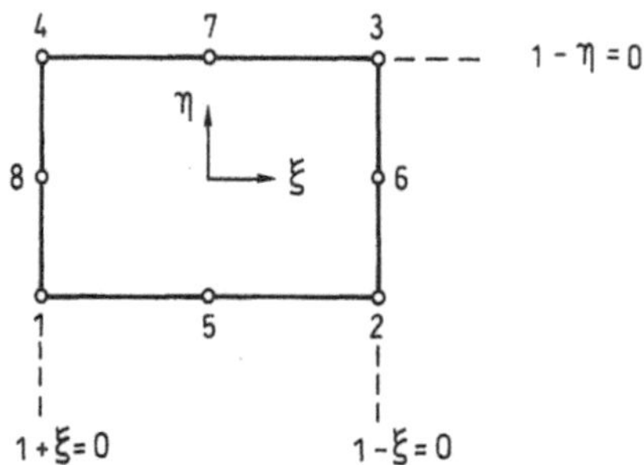

Figure 7.38   Construction of element shape function N_5^e for mid-side node

the element shape functions in the local $\xi\eta$-coordinates, and from these results the element shape functions may be expressed in terms of xy-coordinates, if requested.

Let us first consider a typical mid-size node like nodal point 5 in Figure 7.38. It is required that the element shape function N_5^e is zero along the sides 2,6,3 and 4,7,3 and 1,8,4. From Figure 7.38 we conclude that N_5^e is given by $N_5^e = \frac{1}{2}(1 - \xi)(1 + \xi)(1 - \eta)$, where the factor $\frac{1}{2}$ is determined such that $N_5^e(0, -1) = 1$. By similar arguments for the other mid-side nodes we obtain

$$
\begin{aligned}
N_5^e &= \tfrac{1}{2}(1 - \xi^2)(1 - \eta) \\
N_6^e &= \tfrac{1}{2}(1 + \xi)(1 - \eta^2) \\
N_7^e &= \tfrac{1}{2}(1 - \xi^2)(1 + \eta) \\
N_8^e &= \tfrac{1}{2}(1 - \xi)(1 - \eta^2)
\end{aligned}
\tag{7.120}
$$

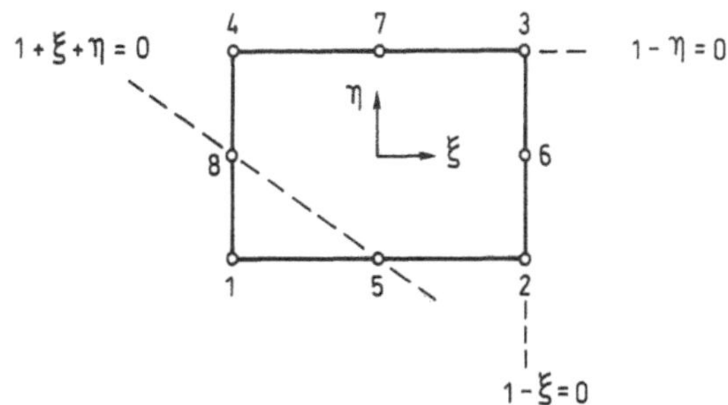

Figure 7.39 Construction of element shape function N_1^e for corner node

Next let us establish the element shape function for a typical corner node like nodal point 1. Referring to Figure 7.39, we require that N_1^e is zero along the sides 2,6,3 and 4,7,3. Moreover, N_1^e is also required to be zero at nodal points 5 and 8. It turns out to be advantageous to assume that N_1^e is zero along the line defined by points 5 and 8. These observations imply that $N_1^e = -\frac{1}{4}(1 - \xi)(1 - \eta)(1 + \xi + \eta)$, where the factor $-\frac{1}{4}$ is determined such that $N_1^e(-1, -1) = 1$. Following the same procedure for the other corner nodes, we obtain

$$
\boxed{
\begin{aligned}
N_1^e &= -\tfrac{1}{4}(1 - \xi)(1 - \eta)(1 + \xi + \eta) \\
N_2^e &= -\tfrac{1}{4}(1 + \xi)(1 - \eta)(1 - \xi + \eta) \\
N_3^e &= -\tfrac{1}{4}(1 + \xi)(1 + \eta)(1 - \xi - \eta) \\
N_4^e &= -\tfrac{1}{4}(1 - \xi)(1 + \eta)(1 + \xi - \eta)
\end{aligned}
}
\tag{7.121}
$$

Even though we have presented a systematic approach for the establishment of the element shape functions, these functions were originally formulated by inspection (Ergatoudis *et al.*, 1968). The eight-node element is therefore called a *serendipity* element after the famous princes of Serendip noted for their chance discoveries (Horace Walpole, 1754). The element shape functions for a typical mid-side and corner node are illustrated in Figure 7.40.

As before, we have that the temperature T is related to the temperature at the nodal points $\mathbf{a}^e$ through $T = \mathbf{N}^e \mathbf{a}^e$, and using (7.120) and (7.121) it follows that

$$
T = \alpha_1 + \alpha_2 x + \alpha_3 y + \alpha_4 x^2 + \alpha_5 xy + \alpha_6 y^2 + \alpha_7 x^2 y + \alpha_8 xy^2
\tag{7.122}
$$

We conclude that the eight-node serendipity element can model any linear variation of the temperature gradient and that (7.122) is symmetric in x and y. The last two terms of (7.122) are parasitic of third order and a comparison with the nine-node Lagrange element (Figure 7.33 and (7.118)), shows that the internal node of the Lagrange element contributes a parasitic term of fourth order. This suggests that for the same number of nodal points (i.e. degrees of freedom) in a body, the eight-node serendipity element is slightly more efficient than the nine-node Lagrange element.

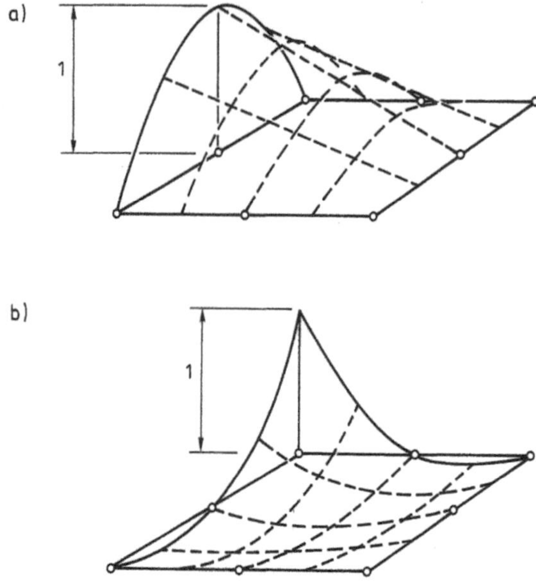

Figure 7.40   Element shape functions for eight-node serendipity element: (a) mid-side node; (b) corner node

As before, the eight-node serendipity element must have element sides parallel to the coordinate axes in order to be conforming. This obstacle can be removed by using the isoparametric formulation presented later.

We emphasize that instead of the derivation prescribed above, it is always possible to establish the element shape functions by means of (7.122) in combination with the C-matrix method.

7.4 Three-dimensional elements

Three-dimensional or *solid elements* can be devised directly from the exposition above. A detailed discussion of three-dimensional elements is given in Hughes (1987), Gallagher (1975) and Zienkiewicz and Taylor (1989), but as no new conceptual matters are involved, we shall confine ourselves here to some rudimentary facts.

The simplest three-dimensional element, the tetrahedral shown in Figure 7.41(a), has four nodal points corresponding to the approximation

$$T = \alpha_1 + \alpha_2 x + \alpha_3 y + \alpha_4 z \tag{7.123}$$

The generalization of the six-node triangle results in the ten-node tetrahedral as

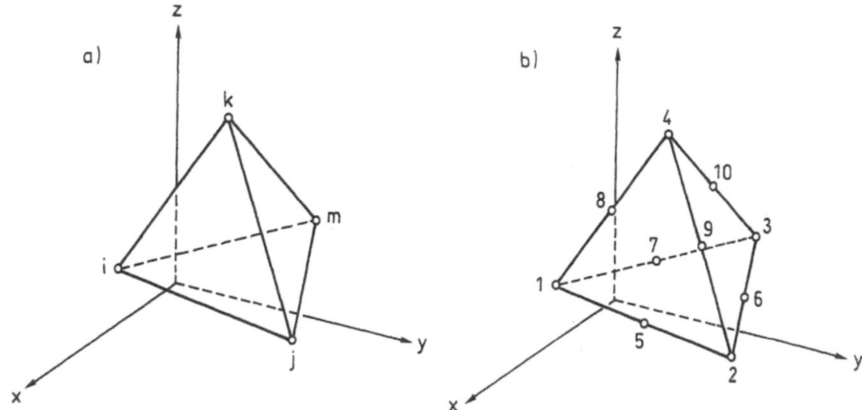

Figure 7.41   (a) Four-node tetrahedral; (b) ten-node tetrahedral

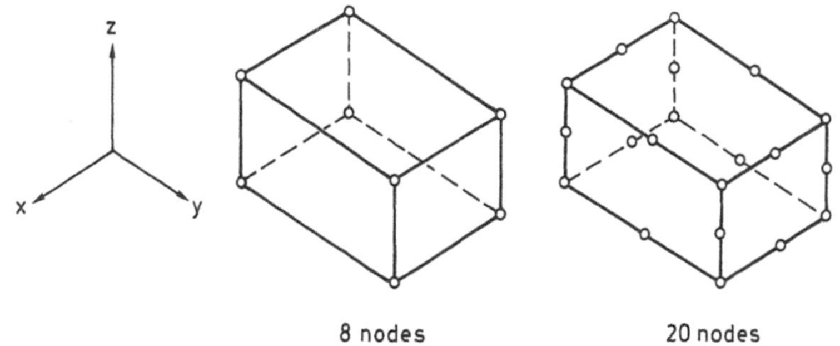

Figure 7.42   Eight- and twenty-node prism element

shown in Figure 7.41(b). The approximation becomes

$$T = \alpha_1 + \alpha_2 x + \alpha_3 y + \alpha_4 z + \alpha_5 x^2 + \alpha_6 y^2 + \alpha_7 z^2 + \alpha_8 xy + \alpha_9 xz + \alpha_{10} yz$$

$$(7.124)$$

which is a complete polynomial in accordance with the approximation for the six-node triangle (cf. (7.119)).

Three-dimensional elements in the form of prisms or *bricks* with element sides parallel to the coordinate axes are easily derived, for instance from the four-node element of Figure 7.27 and the eight-node element of Figure 7.37. These solid elements are illustrated in Figure 7.42 and the temperature approximation for the eight-node prism becomes

$$T = \alpha_1 + \alpha_2 x + \alpha_3 y + \alpha_4 z + \alpha_5 xy + \alpha_6 xz + \alpha_7 yz + \alpha_8 xyz \qquad (7.125)$$

whereas the approximation for the twenty-node isoparametric element is given by

$$T = \alpha_1 + \alpha_2 x + \alpha_3 y + \alpha_4 z + \alpha_5 x^2 + \alpha_6 y^2 + \alpha_7 z^2 + \alpha_8 xy + \alpha_9 xz + \alpha_{10} yz$$
$$+ \alpha_{11} x^2 y + \alpha_{12} xy^2 + \alpha_{13} x^2 z + \alpha_{14} xz^2 + \alpha_{15} y^2 z + \alpha_{16} yz^2 + \alpha_{17} xyz$$
$$+ \alpha_{18} x^2 yz + \alpha_{19} xy^2 z + \alpha_{20} xyz^2 \qquad (7.126)$$

If x, y or z is assumed to be constant, i.e. if one of the sides of the bricks is considered, it follows that (7.125) and (7.126) reduce to (7.107) and (7.122), respectively. Using the isoparametric formulation presented later, these three-dimensional elements may also be used in a conforming manner even when the element sides are not parallel with the coordinate axes.

7.5 Summary

We have seen that the elementwise approximation technique offers a systematic and unified approach of obtaining an approximation for the unknown function – the temperature – over the entire body. The possibility for treating bodies with arbitrary geometries is a hallmark of the FE approximation technique. In addition, the freedom to choose elements with different sizes as well as different types greatly enhances the possibilities offered by this approximation technique.

Irrespective of the particular element chosen, the approximation of the unknown function – the temperature – is carried out as an interpolation between the nodal points of the element. This is expressed through

$$\boxed{T = \mathbf{N}^e \mathbf{a}^e} \qquad (7.127)$$

where $\mathbf{N}^e$ is the element shape function matrix and $\mathbf{a}^e$ contains the temperature at the nodal points. That is,

$$\mathbf{N}^e = [N_1^e \quad N_2^e \quad \cdots \quad N_{n_e}^e]; \quad \mathbf{a}^e = \begin{bmatrix} T_1 \\ T_2 \\ \vdots \\ T_{n_e} \end{bmatrix} \qquad (7.128)$$

where n_e denotes the number of nodal points in the element. The element shape functions, N_i^e, depend on the coordinates. We recall the following fundamental property for the element shape functions:

$$\boxed{N_i^e = \begin{cases} 1 & \text{at nodal point } i \\ 0 & \text{at all other nodal points} \end{cases}} \qquad (7.129)$$

In addition, we found that an element shape function related to a specific nodal point is zero along element boundaries which do not contain the nodal point in question.

Let us now prove another general property for element shape functions. All the elements considered fulfil the completeness requirement, i.e. the approximation is always of the form given by (7.2). By putting all terms except α_1 equal to zero, it appears that all the elements considered are able to describe a constant temperature within the element. If this temperature is called T_c we obtain from (7.127) and (7.128) that

$$T_c = N_1^e T_c + N_2^e T_c + \cdots + N_{n_e}^e T_c$$

i.e.

$$\boxed{\sum_{i=1}^{n_e} N_i^e = 1} \tag{7.130}$$

By inspection we may convince ourselves that this relation holds true for the element shape functions derived previously.

From (7.127) the temperature gradient within the element is given by

$$\boxed{\nabla T = \mathbf{B}^e \mathbf{a}^e} \tag{7.131}$$

For one-dimensional elements we have

$$\nabla T = \frac{\mathrm{d}T}{\mathrm{d}x}; \quad \mathbf{B}^e = \left[\frac{\mathrm{d}N_1^e}{\mathrm{d}x} \quad \frac{\mathrm{d}N_2^e}{\mathrm{d}x} \quad \cdots \quad \frac{\mathrm{d}N_{n_e}^e}{\mathrm{d}x} \right] \tag{7.132}$$

Two-dimensional elements imply that

$$\nabla T = \begin{bmatrix} \dfrac{\partial T}{\partial x} \\[2mm] \dfrac{\partial T}{\partial y} \end{bmatrix}; \quad \mathbf{B}^e = \begin{bmatrix} \dfrac{\partial N_1^e}{\partial x} & \dfrac{\partial N_2^e}{\partial x} & \cdots & \dfrac{\partial N_{n_e}^e}{\partial x} \\[3mm] \dfrac{\partial N_1^e}{\partial y} & \dfrac{\partial N_2^e}{\partial y} & \cdots & \dfrac{\partial N_{n_e}^e}{\partial y} \end{bmatrix} \tag{7.133}$$

whereas three-dimensional elements yield

$$\nabla T = \begin{bmatrix} \dfrac{\partial T}{\partial x} \\[2mm] \dfrac{\partial T}{\partial y} \\[2mm] \dfrac{\partial T}{\partial z} \end{bmatrix}; \quad \mathbf{B}^e = \begin{bmatrix} \dfrac{\partial N_1^e}{\partial x} & \dfrac{\partial N_2^e}{\partial x} & \cdots & \dfrac{\partial N_{n_e}^e}{\partial x} \\[3mm] \dfrac{\partial N_1^e}{\partial y} & \dfrac{\partial N_2^e}{\partial y} & \cdots & \dfrac{\partial N_{n_e}^e}{\partial y} \\[3mm] \dfrac{\partial N_1^e}{\partial z} & \dfrac{\partial N_2^e}{\partial z} & \cdots & \dfrac{\partial N_{n_e}^e}{\partial z} \end{bmatrix} \tag{7.134}$$

To formulate the approximation of the temperature over the entire body we introduce the global shape functions similarly to (7.100) and (7.101). That is, first we observe that any position (x, y, z) identifies the associated element α which contains this position. Therefore

$$\boxed{\text{position } (x, y, z) \Rightarrow \text{identification of associated element } \alpha} \tag{7.135}$$

For a given global nodal point i the global shape function N_i is then defined from the element shape functions N_i^α according to

$$\boxed{N_i(x, y, z) = \begin{cases} N_i^\alpha(x, y, z) & \text{if element } \alpha \text{ contains global nodal point } i \\ 0 & \text{otherwise} \end{cases}}$$

$$\tag{7.136}$$

It is of fundamental importance that these definitions imply that a global shape function related to a specific nodal point differs from zero only in those elements which contain the nodal point in question.

Using these preliminary remarks the approximation of the temperature over the entire body is given by

$$\boxed{T = \mathbf{N}\mathbf{a}} \tag{7.137}$$

where $\mathbf{N}$ is the global shape function matrix and $\mathbf{a}$ contains the temperatures at the nodal points of the entire body. That is,

$$\mathbf{N} = [N_1 \quad N_2 \quad \ldots \quad N_n]; \quad \mathbf{a} = \begin{bmatrix} T_1 \\ T_2 \\ \vdots \\ T_n \end{bmatrix} \tag{7.138}$$

where n denotes the number of nodal points for the entire body. As expression (7.137) is able to reflect a constant temperature over the entire body in an exact manner, we derive in a way similar to (7.130) that

$$\boxed{\sum_{i=1}^{n} N_i = 1} \tag{7.139}$$

From (7.137) the temperature gradient is given by

$$\boxed{\nabla T = \mathbf{B}\mathbf{a}} \tag{7.140}$$

For one-dimensional elements we have

$$\nabla T = \frac{\mathrm{d}T}{\mathrm{d}x}; \quad \mathbf{B} = \begin{bmatrix} \dfrac{\mathrm{d}N_1}{\mathrm{d}x} & \dfrac{\mathrm{d}N_2}{\mathrm{d}x} & \ldots & \dfrac{\mathrm{d}N_n}{\mathrm{d}x} \end{bmatrix} \tag{7.141}$$

Two-dimensional elements imply

$$
\nabla T = \begin{bmatrix} \dfrac{\partial T}{\partial x} \\[2mm] \dfrac{\partial T}{\partial y} \end{bmatrix}; \quad
\mathbf{B} = \begin{bmatrix} \dfrac{\partial N_1}{\partial x} & \dfrac{\partial N_2}{\partial x} & \cdots & \dfrac{\partial N_n}{\partial x} \\[3mm] \dfrac{\partial N_1}{\partial y} & \dfrac{\partial N_2}{\partial y} & \cdots & \dfrac{\partial N_n}{\partial y} \end{bmatrix}
\tag{7.142}
$$

whereas three-dimensional elements results in

$$
\nabla T = \begin{bmatrix} \dfrac{\partial T}{\partial x} \\[2mm] \dfrac{\partial T}{\partial y} \\[2mm] \dfrac{\partial T}{\partial z} \end{bmatrix}; \quad
\mathbf{B} = \begin{bmatrix} \dfrac{\partial N_1}{\partial x} & \dfrac{\partial N_2}{\partial x} & \cdots & \dfrac{\partial N_n}{\partial x} \\[3mm] \dfrac{\partial N_1}{\partial y} & \dfrac{\partial N_2}{\partial y} & \cdots & \dfrac{\partial N_n}{\partial y} \\[3mm] \dfrac{\partial N_1}{\partial z} & \dfrac{\partial N_2}{\partial z} & \cdots & \dfrac{\partial N_n}{\partial z} \end{bmatrix}
\tag{7.143}
$$

A variety of elements has been discussed and it seems natural to ask whether it is possible to nominate the best element. Unfortunately, this is not possible in general because the 'best' element often differs from problem to problem. However, we may state that, as the order of an element increases, the total number of unknowns in a problem can in general be reduced for a given accuracy of the approximation. This is especially true for three-dimensional elements. Use of higher-order elements will also reduce the amount of data preparation for the FE program, but this advantage may be offset by the more involved numerical manipulations. In practice, it is rare to use elements of higher order than those discussed previously.

Apart from these general remarks, we note that in regions where the gradient varies rapidly, more elements are clearly needed in order to capture this variation than in regions where the gradient varies slowly. Finally, we recall that in order to obtain a conforming behaviour of two-dimensional rectangular and three-dimensional brick elements, the element sides must be parallel with the coordinate axes. This essential drawback can be removed by using the isoparametric formulation presented in Chapter 19.

8
Choice of weight function – weighted residual methods

Let us return to our model problem, namely that of heat flow. The strong form of the problem was first established and, based on this result, we derived the corresponding weak formulation. The advantages of the weak form as compared with the strong form have already been discussed and considering, for example, two-dimensional heat flow, the weak form was given by (6.46), which reads

$$\int_A (\nabla v)^{\mathrm{T}} t\, \mathbf{D}\nabla T \, \mathrm{d}A = -\int_{\mathscr{L}_h} vht\, \mathrm{d}\mathscr{L} - \int_{\mathscr{L}_g} vq_n t\, \mathrm{d}\mathscr{L} + \int_A vQt\, \mathrm{d}A \qquad (8.1)$$

The FE method is an approximate method for solving arbitrary differential equations. Therefore, an approximation is required for the unknown function, which in the present case is the temperature T. In the previous chapter, a detailed discussion was devoted to the specific type of approximation for the temperature that is adopted in the FE method. This approximation was found to be characterized as an elementwise approximation over the entire body. Referring to (8.1), it appears that the only topic which remains to be discussed is the choice of the weight function v. We recall that in the weak formulation this weight function is an arbitrary function. At this point we can therefore summarize the basic steps in the FE formulation as follows:

Basic steps in the FE formulation

1. Establish the strong formulation of the problem.
2. Obtain the weak form of the problem.
3. Make an elementwise approximation over the entire body of the unknown function, i.e. the temperature T.
4. Choose the weight function v.

The previous chapters have been devoted to steps 1–3 and in the present chapter we shall establish the choice made for the weight function v, i.e. step 4.

It turns out that a particularly suitable choice of the weight function in the FE method is obtained by the *Galerkin method* (Galerkin, 1915) which is also used in other formulations than the FE approach. More generally, the Galerkin method is an example of a so-called *weighted residual method*, and in order to substantiate our choice of the Galerkin method as an integrated part of the FE formulation, we will discuss different weighted residual methods in this chapter.

General theories for weighted residual methods may be found in Crandall (1956), Finlayson and Scriven (1966), Finlayson (1972) and Zienkiewicz and Taylor (1989). However, our intention is not to provide the reader with comprehensive knowledge of different weighted residual methods, but simply to obtain sufficient motivation for our choice of weight function in terms of the Galerkin method. With this limitation in mind we consider the following one-dimensional differential equation applicable in the range $a \leq x \leq b$:

$$Lu + g = 0; \quad a \leq x \leq b \tag{8.2}$$

where $u(x)$ is the unknown function and $g(x)$ is a known function; L denotes a *differential operator*. When L is chosen, it specifies the actual form of the differential equation given by (8.2). As an example, L may be given by $L = d/dx$, which implies that (8.2) takes the form $du/dx + g = 0$. As another example, for $L = d^2/dx^2 + 1$, (8.2) implies the differential equation $d^2u/dx^2 + u + g = 0$. It appears that by using the format (8.2), it is possible to write quite general one-dimensional differential equations in a compact fashion. In order to simplify the discussion, the boundary conditions for (8.2) are assumed to be given by

$$u(a) = u_a; \quad u(b) = u_b \tag{8.3}$$

where u_a and u_b are known quantities. The specification of two boundary conditions implies that the differential equation (8.2) is of second order.

It is characteristic that the differential equation (8.2) with the boundary conditions (8.3) can only be solved analytically in terms of an exact closed-form solution for certain simple expressions for the differential operator L and the function g. What we are looking for are methods which enable us to solve (8.2) subject to (8.3) for arbitrary expressions for L and g. This gain in generality is paid for in the sense that our solution takes the form of an approximate solution.

For this purpose, we first make a slight reformulation of (8.2). We may multiply (8.2) by an arbitrary function $v(x)$, the weight function, to obtain

$$v(Lu + g) = 0$$

and we may even integrate this expression over the region of interest, i.e.

$$\int_a^b v(Lu + g)\, dx = 0 \tag{8.4}$$

It is obvious that (8.2) implies (8.4) and as v denotes *any* function, it is easily shown that (8.4) implies (8.2), i.e. the two forms are equivalent. To show that (8.4) implies

(8.2), choose the arbitrary weight function v as $v = Lu + g$, i.e. (8.4) becomes

$$\int_a^b (Lu + g)^2 \, dx = 0$$

which can only be true if (8.2) is fulfilled. We have thus proved that (8.2) and (8.4) are equivalent. It is apparent that this discussion has much in common with that of the weak formulation for one-dimensional problems (see Chapter 4). However, it is important to realize that the formulation (8.4) is not a weak formulation. In order to achieve a weak formulation an integration by parts is necessary, whereby the order of differentiation of the unknown function is decreased at the expense of the weight function being differentiated (cf. for instance (4.36)).

With these remarks in mind, we return to the formulation given by (8.4). As we seek a numerical solution, an approximation must be chosen for the unknown function u. Quite generally, we make the following assumption, which is assumed to fulfil the boundary conditions (8.3):

$$u^{\text{app}} = \psi_1 a_1 + \psi_2 a_2 + \cdots + \psi_n a_n \tag{8.5}$$

Here $a_1 \ldots a_n$ are unknown parameters and $\psi_1 \ldots \psi_n$ are functions of x, which we specify in advance. These functions are called *trial* or *basis* functions. Once the parameters $a_1 \ldots a_n$ are known, our approximate solution is given by (8.5), i.e. the task is to determine the parameters $a_1 \ldots a_n$. The trial functions may be chosen arbitrarily, but it is obvious that some knowledge of the physical problem facilitates the choice of suitable trial functions. With reference to Chapter 7, we may even take the approximation (8.5) as an elementwise approximation. In that case $a_1 \ldots a_n$ would be the values of u at the nodal points and the trial functions would become the global shape functions. However, (8.5) may be taken as any approximation and this is the reason why $\psi_1 \ldots \psi_n$ are termed trial functions.

The approximation (8.5) may be written as

$$\boxed{u^{\text{app}} = \boldsymbol{\psi}\mathbf{a}} \tag{8.6}$$

where

$$\boldsymbol{\psi} = [\psi_1 \quad \psi_2 \quad \cdots \quad \psi_n]; \quad \mathbf{a} = \begin{bmatrix} a_1 \\ a_2 \\ \vdots \\ a_n \end{bmatrix} \tag{8.7}$$

We may substitute u in (8.4) by u^{app} and require that

$$\int_a^b v(Lu^{\text{app}} + g) \, dx = 0 \tag{8.8}$$

Likewise in the differential equation (8.2), u may be substituted by u^{app} and it is not surprising that u^{app} will not, in general, satisfy the equation exactly. That is, we obtain

$$Lu^{\text{app}} + g = e \tag{8.9}$$

where $e(x)$ is a measure for the error. This measure is termed the *residual*. With (8.9), (8.8) can be written as

$$\boxed{\int_a^b ve \, \mathrm{d}x = 0} \tag{8.10}$$

Since the residual e depends through (8.9) and (8.6) on the unknown parameters given by **a**, (8.10) is an expression which depends on the parameter $a_1 \ldots a_n$. The objective is to choose the weight function v so that these parameters can be determined, which, in turn, implies that the approximate solution (8.6) is known.

Equation (8.10) may be interpreted so that the residual, $e(x)$, is given a certain weight, $v(x)$, and the integral of this weighted residual, $v(x)e(x)$, over the region of interest is required to be zero. The various weighted residual methods differ in how the weight function v is chosen. Obviously, the specific choice of the weight function influences the values of the parameters $a_1 \ldots a_n$, which are to be determined. Another feature also related to (8.10) is that of *orthogonality*. The orthogonality condition for two column vectors **a** and **b** is expressed through $\mathbf{a}^{\mathrm{T}}\mathbf{b} = 0$. For functions, one can also speak of orthogonality and the function $v(x)$ is said to be orthogonal to the function $e(x)$ in the interval $a \leq x \leq b$ if (8.10) holds.

Equation (8.10) serves as a vehicle to determine the unknowns $a_1 \ldots a_n$, i.e. **a**. To see how this can be accomplished, we write the arbitrary weight function quite generally as

$$v = V_1 c_1 + V_2 c_2 + \cdots + V_n c_n \tag{8.11}$$

where $V_1 \ldots V_n$ are known functions of x, which we specify in advance, and $c_1 \ldots c_n$ are certain parameters. We note that the number of terms in (8.11) is the same as the number of terms in (8.5). With the definitions

$$\mathbf{V} = [V_1 \quad V_2 \quad \ldots \quad V_n]; \quad \mathbf{c} = \begin{bmatrix} c_1 \\ c_2 \\ \vdots \\ c_n \end{bmatrix} \tag{8.12}$$

(8.11) can be written as

$$\boxed{v = \mathbf{Vc}} \tag{8.13}$$

As the weight function is arbitrary and as **V** is known, we conclude that the parameters $c_1 \ldots c_n$, i.e. **c**, are arbitrary. Moreover, as v is one number and as the transpose of a number is equal to the number itself, i.e. $v = v^{\mathrm{T}}$, we may write (8.13) as

$$v = \mathbf{c}^{\mathrm{T}}\mathbf{V}^{\mathrm{T}} \tag{8.14}$$

Inserting (8.14) into (8.10) and observing that $\mathbf{c}^\mathrm{T}$ does not depend on the coordinate x, we obtain

$$\mathbf{c}^\mathrm{T} \int_a^b \mathbf{V}^\mathrm{T} e \, \mathrm{d}x = 0$$

As this expression should hold for arbitrary $\mathbf{c}^\mathrm{T}$-matrices, it is concluded that

$$\boxed{\int_a^b \mathbf{V}^\mathrm{T} e \, \mathrm{d}x = \mathbf{0}} \tag{8.15}$$

It is important to realize that as the column matrix $\mathbf{V}^\mathrm{T}$ has the dimensions $n \times 1$ (cf. (8.12)), expression (8.15) comprises, in fact, n equations. That is, (8.15) may be written as

$$\int_a^b V_1 e \, \mathrm{d}x = 0$$

$$\int_a^b V_2 e \, \mathrm{d}x = 0$$

$$\vdots \tag{8.16}$$

$$\int_a^b V_n e \, \mathrm{d}x = 0$$

The derivation of (8.16) is based on (8.10) and (8.13) and uses the fact that the weight function v is arbitrary, i.e. $\mathbf{c}$ is also arbitrary. Alternatively, in (8.10) one may directly choose n arbitrary weight functions v and first choose $v = V_1$, then $v = V_2$ and so on. Clearly, the result of this approach is again given by (8.16). The approach based on (8.13) is preferred here, as it facilitates a short notation, especially in later FE formulations, but one should be aware of the equivalence of the two approaches both leading to (8.16).

Returning to (8.16), the essential point is that the residual e depends on the n unknowns $a_1 \ldots a_n$ (cf. (8.9) and (8.6)). Therefore (8.16) serves as a system of equations to determine these n unknowns. To see this more clearly, let us insert (8.6) into (8.9) to provide

$$e = L(\boldsymbol{\psi}\mathbf{a}) + g$$

The $\mathbf{a}$-matrix does not depend on the coordinate x, i.e.

$$e = L(\boldsymbol{\psi})\mathbf{a} + g \tag{8.17}$$

Following Chapter 2, differentiation of a matrix means that each component is differentiated, i.e. we obtain from (8.7) that

$$L(\boldsymbol{\psi}) = [L(\psi_1) \quad L(\psi_2) \quad \ldots \quad L(\psi_n)] \tag{8.18}$$

showing that $L(\psi)$ is a matrix with dimension $1 \times n$. Use of (8.17) in (8.15) implies that

$$\left(\int_a^b \mathbf{V}^{\mathrm{T}} L(\psi) \, dx \right) \mathbf{a} = - \int_a^b \mathbf{V}^{\mathrm{T}} g \, dx \tag{8.19}$$

where $\mathbf{a}$ is considered to be independent of x. With the definitions

$$\mathbf{K} = \int_a^b \mathbf{V}^{\mathrm{T}} L(\psi) \, dx; \quad \mathbf{f} = - \int_a^b \mathbf{V}^{\mathrm{T}} g \, dx \tag{8.20}$$

where both $\mathbf{K}$ and $\mathbf{f}$ are independent of $\mathbf{a}$, (8.19) can be written as the system of linear equations

$$\boxed{\mathbf{Ka} = \mathbf{f}} \tag{8.21}$$

As $\mathbf{V}^{\mathrm{T}}$ has the dimension $n \times 1$ and $L(\psi)$ the dimension $1 \times n$ (cf. (8.18)), the expressions for $\mathbf{K}$ and $\mathbf{f}$ may be written as

$$\mathbf{K} = \begin{bmatrix} \int_a^b V_1 L(\psi_1)\,dx & \int_a^b V_1 L(\psi_2)\,dx & \cdots & \int_a^b V_1 L(\psi_n)\,dx \\ \int_a^b V_2 L(\psi_1)\,dx & \int_a^b V_2 L(\psi_2)\,dx & \cdots & \int_a^b V_2 L(\psi_n)\,dx \\ \vdots & \vdots & & \vdots \\ \int_a^b V_n L(\psi_1)\,dx & \int_a^b V_n L(\psi_2)\,dx & \cdots & \int_a^b V_n L(\psi_n)\,dx \end{bmatrix}; \quad \mathbf{f} = - \begin{bmatrix} \int_a^b V_1 g\,dx \\ \int_a^b V_2 g\,dx \\ \vdots \\ \int_a^b V_n g\,dx \end{bmatrix}$$

$$\tag{8.22}$$

showing that $\mathbf{K}$ is a square matrix with dimension $n \times n$. That is, (8.21) consists of n linear equations from which the n unknowns $a_1 \ldots a_n$, i.e. $\mathbf{a}$, can be determined. When $\mathbf{a}$ has been obtained from (8.21), (8.6) provides the required approximate solution.

The procedure described above applies to all weighted residual methods, and a variety of different specific weighted residual methods is obtained depending on our choice of the weight function v, i.e. our choice for $\mathbf{V}$ (cf. (8.13)). Some frequently used weighted residual methods will be described below.

8.1 Point collocation method

In this method, the weight function v is chosen based on *Dirac's delta function* $\delta(x - x_i)$. This function is defined as

$$\delta(x - x_i) = \begin{cases} \infty & \text{if } x = x_i \\ 0 & \text{otherwise} \end{cases} \tag{8.23}$$

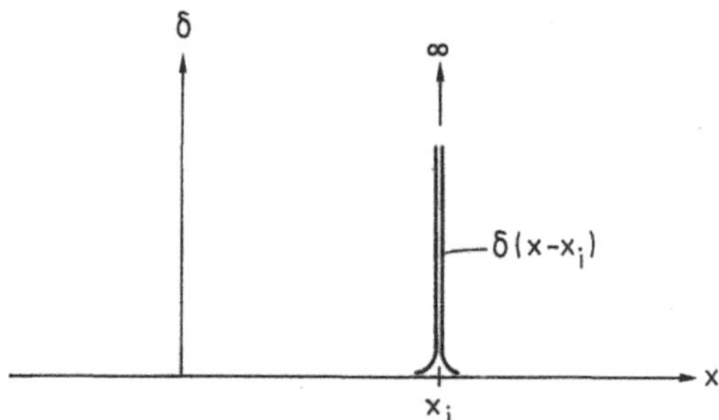

Figure 8.1   Dirac's delta function

and

$$\int_{-\infty}^{\infty} \delta(x - x_i)\,dx = 1 \tag{8.24}$$

where x_i is a given fixed value. Alternatively, due to (8.23), we may write (8.24) as

$$\int_{x_i^-}^{x_i^+} \delta(x - x_i)\,dx = 1 \tag{8.25}$$

where x_i^+ and x_i^- denote x-values slightly larger than and smaller than x_i, respectively. Using these definitions, Dirac's delta function is illustrated in Figure 8.1.

In the *point collocation method*, the weight function v in (8.13) is chosen such that

$$\mathbf{V} = [\delta(x - x_1) \quad \delta(x - x_2) \quad \dots \quad \delta(x - x_n)] \tag{8.26}$$

The fixed points $x_1 \dots x_n$ are chosen arbitrarily within the region $a \leq x \leq b$ and are called *collocation points*. In order to illustrate the consequence of this choice of weight function, we evaluate (8.16), which becomes

$$\int_a^b V_i e(x)\,dx = \int_a^b \delta(x - x_i)e(x)\,dx = 0; \quad i = 1, \dots, n \tag{8.27}$$

As Dirac's delta function is zero unless $x = x_i$ (see (8.23)) we have

$$\int_a^b \delta(x - x_i)e(x)\,dx = \int_{x_i^-}^{x_i^+} \delta(x - x_i)e(x)\,dx = e(x_i)\int_{x_i^-}^{x_i^+} \delta(x - x_i)\,dx = e(x_i) \tag{8.28}$$

where (8.25) has been used. From (8.27) and (8.28) we obtain

$$\int_a^b V_i e(x)\,dx = e(x_i) = 0; \quad i = 1, \dots, n \tag{8.29}$$

That is, the point collocation method forces the residual $e(x)$ to be zero at the collocation points, as illustrated in Figure 8.2.

In order to determine the specific form of the system of equations (8.21), we

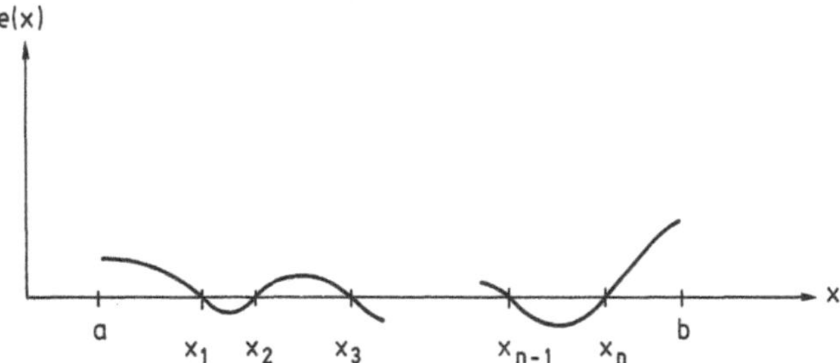

Figure 8.2   Point collocation method

obtain from (8.22) in a manner analogous to (8.28)

$$\mathbf{K} = \begin{bmatrix} L(\psi_1(x_1)) & L(\psi_2(x_1)) & \cdots & L(\psi_n(x_1)) \\ L(\psi_1(x_2)) & L(\psi_2(x_2)) & \cdots & L(\psi_n(x_2)) \\ \vdots & \vdots & & \vdots \\ L(\psi_1(x_n)) & L(\psi_2(x_n)) & \cdots & L(\psi_n(x_n)) \end{bmatrix}; \quad \mathbf{f} = - \begin{bmatrix} g(x_1) \\ g(x_2) \\ \vdots \\ g(x_n) \end{bmatrix} \tag{8.30}$$

The differentiation indicated by the operator L, for instance $L(\psi_1(x_2))$, should be understood as follows: first the differentiation of the function $\psi_1(x)$ is carried out and then the specific x-value, $x = x_2$, is inserted.

8.2 Subdomain collocation method

In the point collocation method n points were chosen. In the *subdomain collocation method*, however, the region of interest is divided into n subregions. Each of these subregions is given by $x_i \leq x \leq x_{i+1}$ where $1 \leq i \leq n$ and both x_i and x_{i+1} are located in the region $a \leq x \leq b$. The matrix $\mathbf{V}$ (cf. (8.12)), which defines the weight function, is chosen such that

$$V_i = \begin{cases} 1 & \text{if } x_i \leq x \leq x_{i+1} \\ 0 & \text{otherwise} \end{cases} \quad i = 1, \dots, n \tag{8.31}$$

as illustrated in Figure 8.3.

This means that (8.16) becomes

$$\int_a^b V_i e(x)\, dx = \int_{x_i}^{x_{i+1}} e(x)\, dx = 0; \quad i = 1, \dots, n \tag{8.32}$$

That is, the subdomain collocation method forces the average of the residual over each subdomain to be equal to zero, as illustrated in Figure 8.4.

With (8.31) and using (8.22), the coefficient matrix $\mathbf{K}$ and the right-hand side $\mathbf{f}$

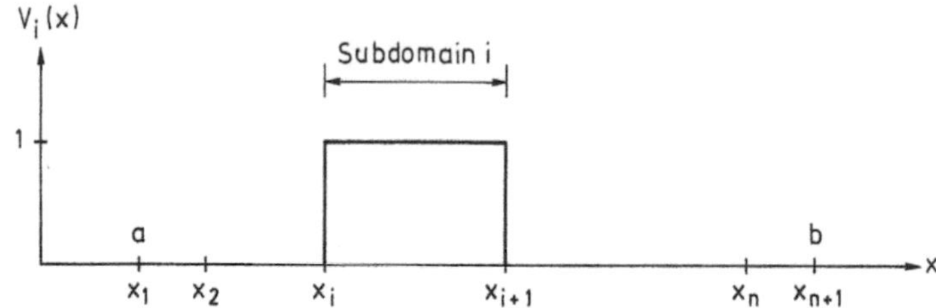

Figure 8.3   Weight function in subdomain collocation method

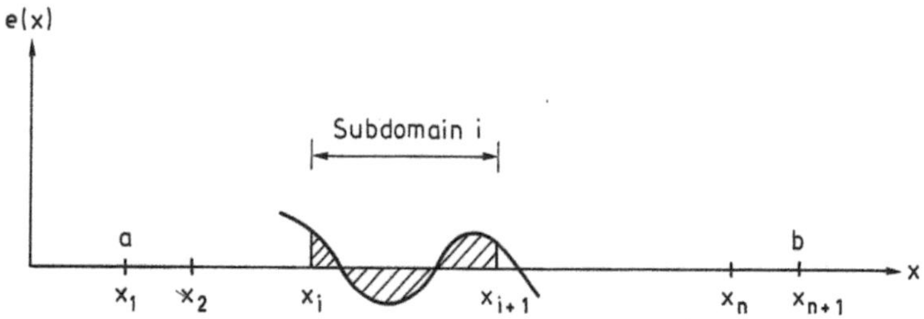

Figure 8.4   Subdomain collocation method

of (8.21) take the forms

$$
\mathbf{K} =
\begin{bmatrix}
\displaystyle\int_{x_1}^{x_2} L(\psi_1)\,dx & \displaystyle\int_{x_1}^{x_2} L(\psi_2)\,dx & \cdots & \displaystyle\int_{x_1}^{x_2} L(\psi_n)\,dx \\[2ex]
\displaystyle\int_{x_2}^{x_3} L(\psi_1)\,dx & \displaystyle\int_{x_2}^{x_3} L(\psi_2)\,dx & \cdots & \displaystyle\int_{x_2}^{x_3} L(\psi_n)\,dx \\[1ex]
\vdots & \vdots & & \vdots \\[1ex]
\displaystyle\int_{x_n}^{x_{n+1}} L(\psi_1)\,dx & \displaystyle\int_{x_n}^{x_{n+1}} L(\psi_2)\,dx & \cdots & \displaystyle\int_{x_n}^{x_{n+1}} L(\psi_n)\,dx
\end{bmatrix};
$$

$$
\mathbf{f} = -
\begin{bmatrix}
\displaystyle\int_{x_1}^{x_2} g\,dx \\[2ex]
\displaystyle\int_{x_2}^{x_3} g\,dx \\[1ex]
\vdots \\[1ex]
\displaystyle\int_{x_n}^{x_{n+1}} g\,dx
\end{bmatrix}
\tag{8.33}
$$

8.3 Least-squares method

Due to (8.17) we observe that the residual e depends on both x and the parameters $a_1 \ldots a_n$, i.e.

$$e = e(x, a_1, \ldots, a_n) \tag{8.34}$$

In the *least-squares method*, a typical component of the matrix $\mathbf{V}$ given by (8.12) is chosen as

$$V_i = \frac{\partial e}{\partial a_i}; \quad i = 1, \ldots, n \tag{8.35}$$

which implies that (8.16) takes the form

$$\int_a^b \frac{\partial e}{\partial a_i} e \, dx = 0; \quad i = 1, \ldots, n \tag{8.36}$$

In order to evaluate this choice, consider the following quantity I:

$$I = \int_a^b e^2(x, a_1, \ldots, a_n) \, dx \tag{8.37}$$

Since we integrate with respect to x, it follows that the quantity I depends on the parameters $a_1, \ldots, a_n$ only, i.e. $I = I(a_1, \ldots, a_n)$. From (8.37) we conclude that

$$\frac{\partial I}{\partial a_i} = 2 \int_a^b e \frac{\partial e}{\partial a_i} \, dx; \quad i = 1, \ldots, n \tag{8.38}$$

An evaluation of (8.37) shows that I is the square of the error, i.e. the residual, integrated over the region of interest. A comparison of (8.36) and (8.38) reveals that

$$\frac{\partial I}{\partial a_i} = 0; \quad i = 1, \ldots, n \tag{8.39}$$

i.e. our choice of weight function given by (8.35) implies that I is stationary. As the residual e, and thus also the quantity I, can be made arbitrarily large, we conclude that the stationarity of I as expressed through (8.39) is a minimum. Consequently, the choice of weight function given by (8.35) implies that the square of the error is a minimum; hence the terminology of the least-squares method.

To determine specifically the coefficient matrix $\mathbf{K}$ and the right-hand side $\mathbf{f}$ of the system of equations (8.21) when the least-squares method is adopted, we first insert (8.17) into (8.35) to obtain

$$V_i = L(\psi_i); \quad i = 1, \ldots, n \tag{8.40}$$

That is, (8.22) yields

$$\mathbf{K} = \begin{bmatrix} \int_a^b L(\psi_1)L(\psi_1)\,dx & \int_a^b L(\psi_1)L(\psi_2)\,dx & \cdots & \int_a^b L(\psi_1)L(\psi_n)\,dx \\ \int_a^b L(\psi_2)L(\psi_1)\,dx & \int_a^b L(\psi_2)L(\psi_2)\,dx & \cdots & \int_a^b L(\psi_2)L(\psi_n)\,dx \\ \vdots & \vdots & & \vdots \\ \int_a^b L(\psi_n)L(\psi_1)\,dx & \int_a^b L(\psi_n)L(\psi_2)\,dx & \cdots & \int_a^b L(\psi_n)L(\psi_n)\,dx \end{bmatrix};$$

$$(8.41)$$

$$\mathbf{f} = -\begin{bmatrix} \int_a^b L(\psi_1)g\,dx \\ \int_a^b L(\psi_2)g\,dx \\ \vdots \\ \int_a^b L(\psi_n)g\,dx \end{bmatrix}$$

which shows that the coefficient matrix $\mathbf{K}$ is symmetric.

8.4 The Galerkin method

In the *Galerkin method* (1915) the components of the matrix $\mathbf{V}$ given by (8.12) are chosen in accordance with

$$\boxed{V_i = \psi_i} \qquad i = 1, \ldots, n \tag{8.42}$$

i.e. the components V_i are equal to the trial functions ψ_i. Loosely speaking we express this by saying that

$$\boxed{\text{weight functions} = \text{trial functions}} \tag{8.43}$$

With (8.42), (8.16) yields

$$\int_a^b \psi_i e\,dx = 0; \quad i = 1, \ldots, n \tag{8.44}$$

Referring to the discussion of (8.10), it follows that in the Galerkin method the trial functions are orthogonal to the residual. To determine explicitly the coefficient matrix $\mathbf{K}$ and the right-hand side $\mathbf{f}$ of the system of equations (8.21), (8.42) is inserted into

(8.22) to provide

$$\mathbf{K} = \begin{bmatrix} \int_a^b \psi_1 L(\psi_1)\,dx & \int_a^b \psi_1 L(\psi_2)\,dx & \cdots & \int_a^b \psi_1 L(\psi_n)\,dx \\ \int_a^b \psi_2 L(\psi_1)\,dx & \int_a^b \psi_2 L(\psi_2)\,dx & \cdots & \int_a^b \psi_2 L(\psi_n)\,dx \\ \vdots & \vdots & & \vdots \\ \int_a^b \psi_n L(\psi_1)\,dx & \int_a^b \psi_n L(\psi_2)\,dx & \cdots & \int_a^b \psi_n L(\psi_n)\,dx \end{bmatrix};$$

$$(8.45)$$

$$\mathbf{f} = -\begin{bmatrix} \int_a^b \psi_1 g\,dx \\ \int_a^b \psi_2 g\,dx \\ \vdots \\ \int_a^b \psi_n g\,dx \end{bmatrix}$$

8.5 Example

To illustrate the weight residual methods described above, we consider the differential equation

$$\frac{d^2 u}{dx^2} + u + x = 0; \quad 0 \le x \le 1 \tag{8.46}$$

In the notation of (8.2) it follows that

$$L = \frac{d^2}{dx^2} + 1; \quad g = x \tag{8.47}$$

The boundary conditions are given as

$$u(0) = u(1) = 0 \tag{8.48}$$

As easily proved by inspection, the exact solution to (8.46) and (8.48) is

$$u = \frac{\sin x}{\sin 1} - x \tag{8.49}$$

In order to obtain an expression for the approximation to u in the form of (8.5), we may explore a number of possibilities. One possibility is given by the trigonometric series

$$u^{\mathrm{app}} = b_0 + (a_1 \sin cx + b_1 \cos cx) + (a_2 \sin 2cx + b_2 \cos 2cx)$$
$$+ \cdots + (a_n \sin ncx + b_n \cos ncx) \tag{8.50}$$

This approximation is required to fulfil the boundary conditions (8.48) and it appears that $u^{\mathrm{app}}(0) = 0$ is fulfilled if

$$b_0 = b_1 = b_2 = \cdots = b_n = 0 \tag{8.51}$$

Moreover, the condition $u^{\mathrm{app}}(1) = 0$ is fulfilled if

$$c = \pi \tag{8.52}$$

With (8.51) and (8.52), (8.50) reduces to

$$u^{\mathrm{app}} = a_1 \sin \pi x + a_2 \sin 2\pi x + \cdots + a_n \sin n\pi x \tag{8.53}$$

As a very simple approximation, we take just one term in this series, i.e.

$$u^{\mathrm{app}} = a_1 \sin \pi x \tag{8.54}$$

and the problem is now to determine the parameter a_1. With reference to (8.6) we have

$$\psi = [\psi_1] = [\sin \pi x] \tag{8.55}$$

Consider first the *point collocation* method. From (8.47) and (8.30) we obtain

$$\mathbf{K} = [L(\psi_1(x_1))] = [-\pi^2 \sin \pi x_1 + \sin \pi x_1]$$
$$\mathbf{f} = -[g(x_1)] = -[x_1] \tag{8.56}$$

Choose the collocation point, for instance, as the midpoint of the interval, i.e. $x_1 = \tfrac{1}{2}$. From (8.21) it then follows that

$$(-\pi^2 + 1)a_1 = -\frac{1}{2} \quad \text{i.e.} \quad a_1 = \frac{1}{2(\pi^2 - 1)} \simeq 0.0564 \tag{8.57}$$

For the *subdomain collocation* method, $\mathbf{K}$ and $\mathbf{f}$ are given by (8.33). In the present example, only one subdomain is involved and this is conveniently chosen as the entire region of interest. This results in

$$\mathbf{K} = \left[\int_0^1 L(\psi_1)\,\mathrm{d}x\right] = \left[\int_0^1 (-\pi^2 \sin \pi x + \sin \pi x)\,\mathrm{d}x\right] = \left[(1 - \pi^2)\frac{2}{\pi}\right]$$
$$\mathbf{f} = -\left[\int_0^1 g\,\mathrm{d}x\right] = -\left[\int_0^1 x\,\mathrm{d}x\right] = -\left[\frac{1}{2}\right] \tag{8.58}$$

Use of these expressions in (8.21) results in

$$(1 - \pi^2)\frac{2}{\pi} a_1 = -\frac{1}{2} \quad \text{i.e.} \quad a_1 = \frac{\pi}{4(\pi^2 - 1)} \simeq 0.0885 \tag{8.59}$$

Next, consider the *least-squares* method, where $\mathbf{K}$ and $\mathbf{f}$ are given by (8.41) which implies that

$$\mathbf{K} = \left[\int_0^1 L(\psi_1)L(\psi_1)\,dx\right] = \left[\int_0^1 (-\pi^2 \sin \pi x + \sin \pi x)^2\,dx\right] = \left[\frac{1}{2}(1-\pi^2)^2\right]$$

$$\mathbf{f} = -\left[\int_0^1 L(\psi_1)g\,dx\right] = -\left[\int_0^1 (-\pi^2 \sin \pi x + \sin \pi x)x\,dx\right] = -\left[\frac{1}{\pi}(1-\pi^2)\right]$$

$$(8.60)$$

and where it was used that

$$\int \sin^2 x\,dx = \frac{x}{2} - \frac{1}{4}\sin 2x; \quad \int x \sin x\,dx = \sin x - x \cos x$$

From (8.21) it follows that

$$\frac{1}{2}(1-\pi^2)^2 a_1 = -\frac{1}{\pi}(1-\pi^2) \quad \text{i.e.} \quad a_1 = \frac{2}{\pi(\pi^2-1)} \simeq 0.0718 \qquad (8.61)$$

Finally, from (8.45) the *Galerkin* method implies that

$$\mathbf{K} = \left[\int_0^1 \psi_1 L(\psi_1)\,dx\right] = \left[\int_0^1 \sin \pi x(-\pi^2 \sin \pi x + \sin \pi x)\,dx\right] = \left[\frac{1}{2}(1-\pi^2)\right]$$

$$\mathbf{f} = -\left[\int_0^1 \psi_1 g\,dx\right] = -\left[\int_0^1 x \sin \pi x\,dx\right] = -\left[\frac{1}{\pi}\right]$$

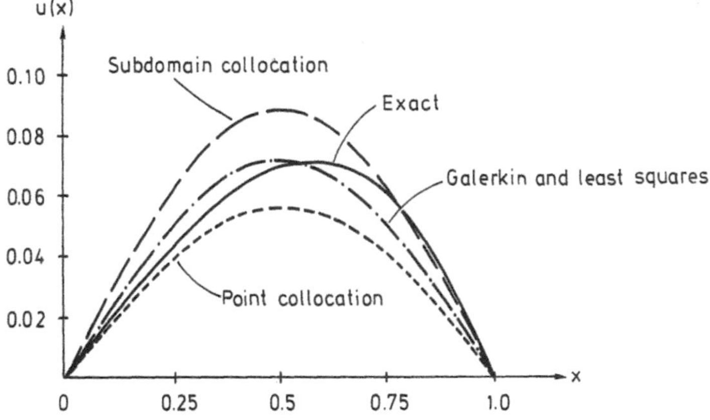

Figure 8.5 Comparison of different weighted residual methods

That is, (8.21) yields

$$\frac{1}{2}(1 - \pi^2)a_1 = -\frac{1}{\pi} \quad \text{i.e.} \quad a_1 = \frac{2}{\pi(\pi^2 - 1)} \simeq 0.0718 \tag{8.62}$$

It appears that the Galerkin and least-squares methods in this example provide the same results. The results of the four weighted residual methods are compared with the exact solution in Figure 8.5 and the superiority of the Galerkin and least-squares methods is obvious for the example considered.

8.6 Concluding remarks

The point collocation method requires that the residual is equal to zero at the collocation points. This condition seems intuitively less exacting than that used in the subdomain method where the average of the residual over each subdomain is required to be zero. Thus one may expect the subdomain method to be more accurate than the point collocation method. The least-squares method and the Galerkin method turn out to be very efficient. The least-squares method always results in a symmetric coefficient matrix (cf. (8.41)), which is clearly an advantage in numerical calculations. In general, the coefficient matrix (8.45) in the Galerkin method is not symmetric, so that the advantage of using this method in the FE formulation may not seem obvious at this point. However, the weighted residual methods discussed in this chapter are based on the integral formulation (8.4) of the differential equation (8.2). We have previously stressed that (8.4) is not a weak formulation. In a weak formulation, an integration by parts is performed whereby the order of differentiation of the unknown function is decreased at the expense of the weight function being differentiated (see for instance (4.36)). As shown in the next chapter, this implies that when the Galerkin method is used in combination with the weak formulation, a symmetric coefficient matrix arises. This aspect as well as the applicability of this method to any differential equation suggest its fundamental importance in FE formulations.

9
FE formulation of one-dimensional heat flow

In the previous chapters we have discussed strong and weak forms of the heat flow problem, the elementwise approximation technique used in the FE formulation for the unknown function and various choices for the weight function. It was emphasized that the FE approach is based on the choice of weight function in accordance with the Galerkin method. The basic steps in the FE method can therefore be summarized as follows:

Basic steps in the FE formulation

1. Establish the strong formulation of the problem.
2. Obtain the weak form of the problem.
3. Make an elementwise approximation over the entire body of the unknown function.
4. Choose the weight function in accordance with the Galerkin method.

In order to illustrate these steps we shall consider the FE formulation of one-dimensional heat flow. Even though this constitutes a simple physical problem, it will turn out that its FE formulation contains all the fundamental features of any FE approach. Therefore, the reader is urged to read the present chapter especially carefully. If the procedures described here are fully comprehended, it will be an easy task to follow the remaining exposition which discusses the FE formulation of more complicated physical problems, but where no essentially new FE concepts will be introduced.

9.1 Global FE formulation

Consider one-dimensional heat flow in the fin shown in Figure 9.1. The strong form of a quite similar problem was given by (4.8)–(4.10) and with the position of the fin

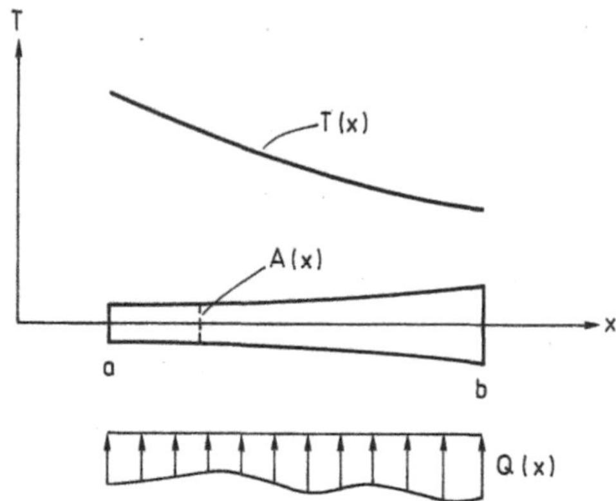

Figure 9.1   Heat conduction in one-dimensional fin

as shown in the figure, the differential equation becomes

$$\frac{d}{dx}\left(Ak\frac{dT}{dx}\right) + Q = 0; \quad a \le x \le b \tag{9.1}$$

where it is recalled that $A(x)$ is the cross-sectional area of the fin, $k(x)$ is the thermal conductivity and $Q(x)$ is the heat supply per unit time and per unit length of the fin. Q may be transferred to the fin across its outer surface or created internally, for instance by electric heating. In order to solve (9.1), boundary conditions must be specified, for instance in the form given by (4.9) and (4.10), but to be able to deal with completely arbitrary boundary conditions, we shall leave the boundary conditions unspecified until later.

To obtain the weak form of (9.1) we multiply by the arbitrary weight function $v(x)$ and integrate over the region of interest, i.e.

$$\int_a^b v\left[\frac{d}{dx}\left(Ak\frac{dT}{dx}\right) + Q\right] dx = 0$$

The first term is integrated by parts in accordance with (4.33) to provide

$$\int_a^b \frac{dv}{dx}Ak\frac{dT}{dx}\,dx = \left[vAk\frac{dT}{dx}\right]_a^b + \int_a^b vQ\,dx \tag{9.2}$$

and using Fourier's law $q = -k\,dT/dx$, we get

$$\int_a^b \frac{dv}{dx}Ak\frac{dT}{dx}\,dx = -[vAq]_a^b + \int_a^b vQ\,dx \tag{9.3}$$

It appears that the integration by parts implies that the order of differentiation for T has decreased at the expense of the weight function v being differentiated.

The approximation for the temperature T is now introduced. Following the summary of Chapter 7, the approximation over the entire region is in general written as

$$\boxed{T = \mathbf{Na}} \tag{9.4}$$

where $\mathbf{N}$ is the global shape function matrix and $\mathbf{a}$ contains the temperatures at the nodal points in the entire body (cf. (7.137) and (7.138)). That is, we have

$$\mathbf{N} = [N_1 \quad N_2 \quad \cdots \quad N_n]; \quad \mathbf{a} = \begin{bmatrix} T_1 \\ T_2 \\ \vdots \\ T_n \end{bmatrix} \tag{9.5}$$

where n denotes the number of nodal points for the entire body and $N_i = N_i(x)$. As $\mathbf{a}$ does not depend on x, (9.4) implies that

$$\boxed{\frac{\mathrm{d}T}{\mathrm{d}x} = \mathbf{Ba}} \quad \text{where} \quad \boxed{\mathbf{B} = \frac{\mathrm{d}\mathbf{N}}{\mathrm{d}x}} \tag{9.6}$$

i.e.

$$\mathbf{B} = \begin{bmatrix} \dfrac{\mathrm{d}N_1}{\mathrm{d}x} & \dfrac{\mathrm{d}N_2}{\mathrm{d}x} & \cdots & \dfrac{\mathrm{d}N_n}{\mathrm{d}x} \end{bmatrix} \tag{9.7}$$

in accordance with (7.140) and (7.141). Inserting (9.6) in (9.3) we obtain

$$\left(\int_a^b \frac{\mathrm{d}v}{\mathrm{d}x} Ak\mathbf{B} \, \mathrm{d}x \right) \mathbf{a} = -[vAq]_a^b + \int_a^b vQ \, \mathrm{d}x \tag{9.8}$$

It is important that the approximation for $\mathrm{d}T/\mathrm{d}x$ as given by (9.6) is not inserted into the boundary terms on the right-hand side of (9.8), which include the flux $q = -k \, \mathrm{d}T/\mathrm{d}x$. The reason is that the boundary conditions either specify the flux or the temperature itself at the boundary and there is no reason to make an approximation for matters that we know beforehand. We shall return to this topic later on.

The last step is the choice of weight function v. In accordance with the Galerkin method, we choose the weight functions to be equal to the trial functions (cf. (8.42) and (8.43)). In the present case the trial functions are the shape functions; that is, (8.13) yields

$$\boxed{v = \mathbf{Nc}} \tag{9.9}$$

Since v is an arbitrary function, the matrix $\mathbf{c}$ is arbitrary. Moreover, as v is one number, we have $v = v^{\mathrm{T}}$, i.e. (9.9) can be written as

$$v = \mathbf{c}^{\mathrm{T}}\mathbf{N}^{\mathrm{T}} \tag{9.10}$$

which implies that

$$\frac{\mathrm{d}v}{\mathrm{d}x} = \mathbf{c}^{\mathrm{T}}\mathbf{B}^{\mathrm{T}} \quad \text{where} \quad \mathbf{B}^{\mathrm{T}} = \frac{\mathrm{d}\mathbf{N}^{\mathrm{T}}}{\mathrm{d}x} \tag{9.11}$$

Inserting (9.10) and (9.11) into (9.8) and using the fact that $\mathbf{c}^{\mathrm{T}}$ is independent of x results in

$$\mathbf{c}^{\mathrm{T}}\left[\left(\int_a^b \mathbf{B}^{\mathrm{T}}Ak\mathbf{B}\,\mathrm{d}x\right)\mathbf{a} + [\mathbf{N}^{\mathrm{T}}Aq]_a^b - \int_a^b \mathbf{N}^{\mathrm{T}}Q\,\mathrm{d}x\right] = 0$$

As this expression should hold for arbitrary $\mathbf{c}^{\mathrm{T}}$-matrices, it is concluded that

$$\left(\int_a^b \mathbf{B}^{\mathrm{T}}Ak\mathbf{B}\,\mathrm{d}x\right)\mathbf{a} = -[\mathbf{N}^{\mathrm{T}}Aq]_a^b + \int_a^b \mathbf{N}^{\mathrm{T}}Q\,\mathrm{d}x \tag{9.12}$$

which is the required FE formulation. Here, we have derived (9.12) by means of (9.8) and (9.9) and using the fact that v is arbitrary, and thus $\mathbf{c}$ is also arbitrary. Alternatively, in (9.8) we may directly choose n arbitrary weight functions v and first choose $v = N_1$, then $v = N_2$ and so on. Clearly, the result of this approach is again given by (9.12). This discussion is similar to that relating to (8.16). Here, we prefer the approach given by (9.9) as it facilitates a clear matrix formulation, especially in the more complicated FE formulations that will be encountered later. However, one should be aware of the equivalence of the two approaches.

In order to write (9.12) in a more compact fashion, we define the following matrices:

$$\boxed{\begin{aligned} \mathbf{K} &= \int_a^b \mathbf{B}^{\mathrm{T}}Ak\mathbf{B}\,\mathrm{d}x \\[2mm] \mathbf{f}_b &= -[\mathbf{N}^{\mathrm{T}}Aq]_a^b \\[2mm] \mathbf{f}_l &= \int_a^b \mathbf{N}^{\mathrm{T}}Q\,\mathrm{d}x \end{aligned}} \tag{9.13}$$

Referring to (9.7) it is obvious that $\mathbf{K}$ is a square matrix with dimension $n \times n$; it is called the *stiffness matrix*. Likewise, both $\mathbf{f}_b$ and $\mathbf{f}_l$ have the dimension $n \times 1$, and they are called the *boundary vector* and *load vector*, respectively, since $\mathbf{f}_b$ refers to conditions at the boundary, whereas $\mathbf{f}_l$ considers the effect of the 'loading' Q. With (9.13), (9.12) takes the form

$$\boxed{\mathbf{Ka} = \mathbf{f}_b + \mathbf{f}_l} \tag{9.14}$$

and defining the *force vector* **f** by

$$\boxed{\mathbf{f} = \mathbf{f}_\mathrm{b} + \mathbf{f}_\mathrm{l}} \tag{9.15}$$

(9.14) can be written as

$$\boxed{\mathbf{K}\mathbf{a} = \mathbf{f}} \tag{9.16}$$

The temperatures at the nodal points given by **a** are obtained by solving this system of linear equations. When **a** is known, the temperature T at an arbitrary point in the fin is given by (9.4) and the temperature gradient at an arbitrary point is given by (9.6). From this temperature gradient, the flux q at any location in the fin is obtained from Fourier's law (4.3). Therefore, when **a** has been determined from (9.16) all quantities of interest can be derived. Let us now evaluate the stiffness matrix **K** and the force vector **f** in more detail.

From (9.13) and (2.30) it follows that

$$\mathbf{K}^\mathrm{T} = \left(\int_a^b \mathbf{B}^\mathrm{T} Ak\mathbf{B} \, \mathrm{d}x \right)^\mathrm{T} = \int_a^b \mathbf{B}^\mathrm{T} Ak\mathbf{B} \, \mathrm{d}x = \mathbf{K} \tag{9.17}$$

i.e. the stiffness matrix **K** is symmetric. It is obvious that this symmetry hinges on the use of the Galerkin method. If instead of this method (9.9) we had chosen $v = \mathbf{V}\mathbf{c}$, where $\mathbf{V} \neq \mathbf{N}$, then the corresponding stiffness matrix would have taken the form

$$\mathbf{K} = \int_a^b \frac{\mathrm{d}\mathbf{V}^\mathrm{T}}{\mathrm{d}x} Ak\mathbf{B} \, \mathrm{d}x \quad \text{where} \quad \frac{\mathrm{d}\mathbf{V}^\mathrm{T}}{\mathrm{d}x} \neq \mathbf{B}^\mathrm{T}$$

which implies that the symmetry of **K** would have been spoiled. This symmetry, the applicability of the Galerkin method to any differential equation and the realistic predictions provided by this method are the primary reasons for its adoption in the FE formulation.

To evaluate the character of the stiffness matrix **K** further we consider its components. From (9.13) and (9.7) it follows that

$$\mathbf{K} = \begin{bmatrix} \int_a^b \dfrac{\mathrm{d}N_1}{\mathrm{d}x} Ak \dfrac{\mathrm{d}N_1}{\mathrm{d}x} \mathrm{d}x & \int_a^b \dfrac{\mathrm{d}N_1}{\mathrm{d}x} Ak \dfrac{\mathrm{d}N_2}{\mathrm{d}x} \mathrm{d}x & \cdots & \int_a^b \dfrac{\mathrm{d}N_1}{\mathrm{d}x} Ak \dfrac{\mathrm{d}N_n}{\mathrm{d}x} \mathrm{d}x \\[2ex] \int_a^b \dfrac{\mathrm{d}N_2}{\mathrm{d}x} Ak \dfrac{\mathrm{d}N_1}{\mathrm{d}x} \mathrm{d}x & \int_a^b \dfrac{\mathrm{d}N_2}{\mathrm{d}x} Ak \dfrac{\mathrm{d}N_2}{\mathrm{d}x} \mathrm{d}x & \cdots & \int_a^b \dfrac{\mathrm{d}N_2}{\mathrm{d}x} Ak \dfrac{\mathrm{d}N_n}{\mathrm{d}x} \mathrm{d}x \\[2ex] \vdots & \vdots & & \vdots \\[2ex] \int_a^b \dfrac{\mathrm{d}N_n}{\mathrm{d}x} Ak \dfrac{\mathrm{d}N_1}{\mathrm{d}x} \mathrm{d}x & \int_a^b \dfrac{\mathrm{d}N_n}{\mathrm{d}x} Ak \dfrac{\mathrm{d}N_2}{\mathrm{d}x} \mathrm{d}x & \cdots & \int_a^b \dfrac{\mathrm{d}N_n}{\mathrm{d}x} Ak \dfrac{\mathrm{d}N_n}{\mathrm{d}x} \mathrm{d}x \end{bmatrix} \tag{9.18}$$

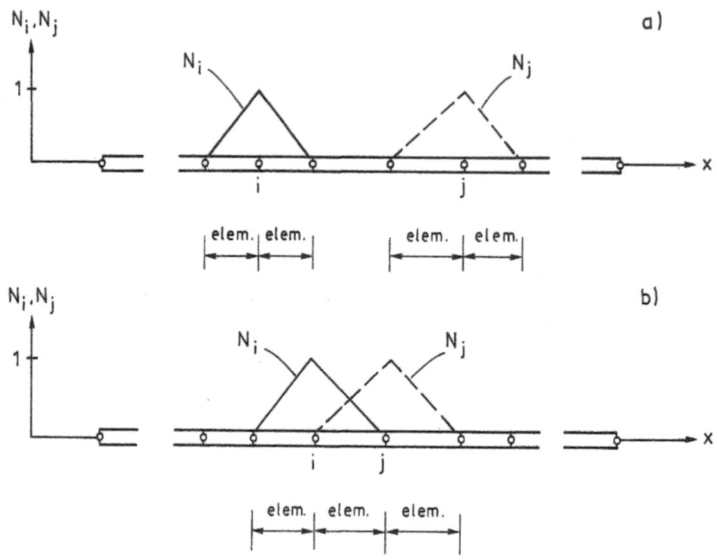

Figure 9.2   Variation of global shape functions N_i and N_j for linear elements

The symmetry of **K** is also obvious from this expression. Moreover, consider a component K_{ij} of **K**, which can be written as

$$K_{ij} = \int_a^b \frac{dN_i}{dx} Ak \frac{dN_j}{dx} dx \tag{9.19}$$

The approximation of the temperature according to (9.4) is an elementwise approximation, and in order to evaluate (9.19) we recall that one global shape function is related to each nodal point. The variation of the global shape functions N_i and N_j is shown in Figure 9.2, where, for simplicity, it has been assumed that the simple linear finite element has been used, i.e. N_i and N_j vary linearly (cf. Figure 7.8). However, the important point, which is common to all types of finite elements, is that a global shape function N_i differs from zero only in those elements which contain the nodal point i. As shown in Figure 9.2, two principally different possibilities exist for the variation of the global shape functions N_i and N_j. In Figure 9.2(a), no elements contain both nodal points i and j, whereas in Figure 9.2(b) one element contains both nodal points i and j. In Figure 9.2(a), N_i is zero when N_j is different from zero and vice versa, i.e. when evaluating (9.19) the component K_{ij} is zero. Only when the two global shape functions N_i and N_j differ from zero at some common regions will K_{ij} be different from zero. Such an example is shown in Figure 9.2(b) where N_i and N_j only differ from zero simultaneously in the element given by nodal points i and j. In conclusion, the component K_{ij} of the stiffness matrix **K** is zero unless both nodal points i and j are present in at least one element. This conclusion is similar to the observation made in Chapter 3 (cf. (3.70) and (3.71)). Moreover, it implies that the

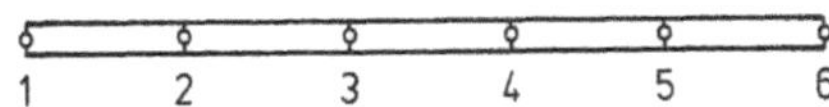

Figure 9.3   Fin modelled by five linear elements

stiffness matrix becomes banded with non-zero components clustered about the diagonal of **K**. This has important numerical advantages as discussed in Chapter 11. We emphasize that the banded structure of the stiffness matrix **K** is a result of the fundamental property that a global shape function differs from zero only in those elements which contain the nodal point in question.

As an illustration of these features, consider the fin shown in Figure 9.3. The fin is modelled by five elements and for convenience we make use of the simple linear element of Chapter 7. With the adopted nodal numbering it follows that the stiffness matrix **K** takes the following form:

$$
\mathbf{K} = \begin{array}{c} \\ 1 \\ 2 \\ 3 \\ 4 \\ 5 \\ 6 \end{array}
\begin{array}{cccccc}
1 & 2 & 3 & 4 & 5 & 6 \\
\end{array}
\begin{bmatrix}
\times & \times & \bigcirc & \bigcirc & \bigcirc & \bigcirc \\
\times & \times & \times & \bigcirc & \bigcirc & \bigcirc \\
\bigcirc & \times & \times & \times & \bigcirc & \bigcirc \\
\bigcirc & \bigcirc & \times & \times & \times & \bigcirc \\
\bigcirc & \bigcirc & \bigcirc & \times & \times & \times \\
\bigcirc & \bigcirc & \bigcirc & \bigcirc & \times & \times \\
\end{bmatrix}
$$

In order to evaluate the force vector **f** given by (9.15), let us first consider the components of the boundary vector $\mathbf{f_b}$ and the load vector $\mathbf{f_l}$. From (9.13) we get

$$
\mathbf{f_b} = - \begin{bmatrix} [N_1 Aq]_a^b \\[6pt] [N_2 Aq]_a^b \\[6pt] \vdots \\[6pt] [N_n Aq]_a^b \end{bmatrix} ; \quad
\mathbf{f_l} = \begin{bmatrix} \int_a^b N_1 Q \, \mathrm{d}x \\[6pt] \int_a^b N_2 Q \, \mathrm{d}x \\[6pt] \vdots \\[6pt] \int_a^b N_n Q \, \mathrm{d}x \end{bmatrix}
\tag{9.20}
$$

A typical term f_{bi} of the boundary vector $\mathbf{f_b}$ is given by

$$
\mathbf{f_{bi}} = -[N_i Aq]_a^b = -(N_i Aq)_{x=b} + (N_i Aq)_{x=a}
\tag{9.21}
$$

The positions $x = a$ and $x = b$, i.e. the ends of the fin, clearly correspond to the position of two nodal points. For the sake of illustration assume that the nodal points 1 and n are located at $x = a$ and $x = b$, respectively. With the properties of a global shape function in mind it follows that

$$N_i(x = a) = \begin{cases} N_1(x = a) = 1 & \text{if } i = 1 \\ 0 & \text{if } i \neq 1 \end{cases} \tag{9.22}$$

and

$$N_i(x = b) = \begin{cases} N_n(x = b) = 1 & \text{if } i = n \\ 0 & \text{if } i \neq n \end{cases} \tag{9.23}$$

With (9.20)–(9.23), the boundary vector becomes

$$\mathbf{f}_b = \begin{bmatrix} (Aq)_{x=a} \\ 0 \\ \vdots \\ 0 \\ -(Aq)_{x=b} \end{bmatrix} \tag{9.24}$$

where q is the flux given by Fourier's law (4.3).

9.1.1 *Example 1*

Having discussed the FE formulation of one-dimensional heat flow in general terms, it is timely to investigate the solution of a specific example. For simplicity, we assume that the fin shown in Figure 9.4 has a constant cross-sectional area A, a constant thermal conductivity k and a constant heat supply Q.

From Figure 9.4 and (9.1) the differential equation is given by

$$\frac{d}{dx}\left(Ak\frac{dT}{dx} \right) + Q = 0; \quad a \leq x \leq b \tag{9.25}$$

where

$$A = 10\,\text{m}^2; \quad k = 5\,\text{J/°C m s}; \quad Q = 100\,\text{J/s m}; \quad a = 2\,\text{m}; \quad b = 8\,\text{m} \tag{9.26}$$

The boundary conditions are assumed to be of the form

$$T(x = a) = g_a; \quad q(x = b) = h_b \tag{9.27}$$

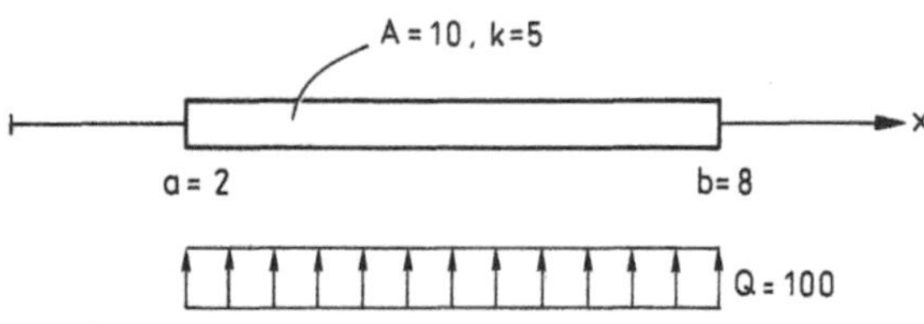

Figure 9.4   Problem definition

Figure 9.5   Three linear elements of equal length

where g_a and h_b are known quantities. These boundary conditions correspond to a prescribed temperature at the left end of the fin and a prescribed flux $q = -k\,\mathrm{d}T/\mathrm{d}x$ at the right end. Moreover, let

$$g_a = 0\,^\circ\mathrm{C}; \quad h_b = 15\,\mathrm{J/m^2\,s} \tag{9.28}$$

Assume that we divide the fin into three elements of equal length and also assume for convenience that we adopt the simple linear element of Chapter 7. The nodal points are indicated in Figure 9.5. The global shape functions together with their derivatives are shown in Figure 9.6.

Let us first determine the stiffness matrix **K**. From (9.19) it follows that

$$K_{11} = \int_a^b \frac{\mathrm{d}N_1}{\mathrm{d}x} Ak \frac{\mathrm{d}N_1}{\mathrm{d}x}\,\mathrm{d}x = \int_2^4 \frac{\mathrm{d}N_1}{\mathrm{d}x} Ak \frac{\mathrm{d}N_1}{\mathrm{d}x}\,\mathrm{d}x$$

$$= \int_2^4 (-0.5)50(-0.5)\,\mathrm{d}x = 25$$

where N_1, i.e. $\mathrm{d}N_1/\mathrm{d}x$, only differs from zero in the interval $2 \le x \le 4$. Likewise

$$K_{12} = \int_a^b \frac{\mathrm{d}N_1}{\mathrm{d}x} Ak \frac{\mathrm{d}N_2}{\mathrm{d}x}\,\mathrm{d}x = \int_2^4 \frac{\mathrm{d}N_1}{\mathrm{d}x} Ak \frac{\mathrm{d}N_2}{\mathrm{d}x}\,\mathrm{d}x$$

$$= \int_2^4 (-0.5)50(0.5)\,\mathrm{d}x = -25$$

Here, we used the fact that N_1 and N_2 only differ from zero simultaneously in the range $2 \le x \le 4$. Since N_1 and N_3 have no common region where they differ from zero, we obtain

$$K_{13} = \int_a^b \frac{\mathrm{d}N_1}{\mathrm{d}x} Ak \frac{\mathrm{d}N_3}{\mathrm{d}x}\,\mathrm{d}x = 0$$

and likewise

$$K_{14} = \int_a^b \frac{\mathrm{d}N_1}{\mathrm{d}x} Ak \frac{\mathrm{d}N_4}{\mathrm{d}x}\,\mathrm{d}x = 0$$

As **K** is symmetric (cf. (9.17)), we next consider component K_{22}, i.e.

$$K_{22} = \int_a^b \frac{\mathrm{d}N_2}{\mathrm{d}x} Ak \frac{\mathrm{d}N_2}{\mathrm{d}x}\,\mathrm{d}x = \int_2^6 \frac{\mathrm{d}N_2}{\mathrm{d}x} Ak \frac{\mathrm{d}N_2}{\mathrm{d}x}\,\mathrm{d}x = \int_2^4 (0.5)50(0.5)\,\mathrm{d}x$$

$$+ \int_4^6 (-0.5)50(-0.5)\,\mathrm{d}x = 50$$

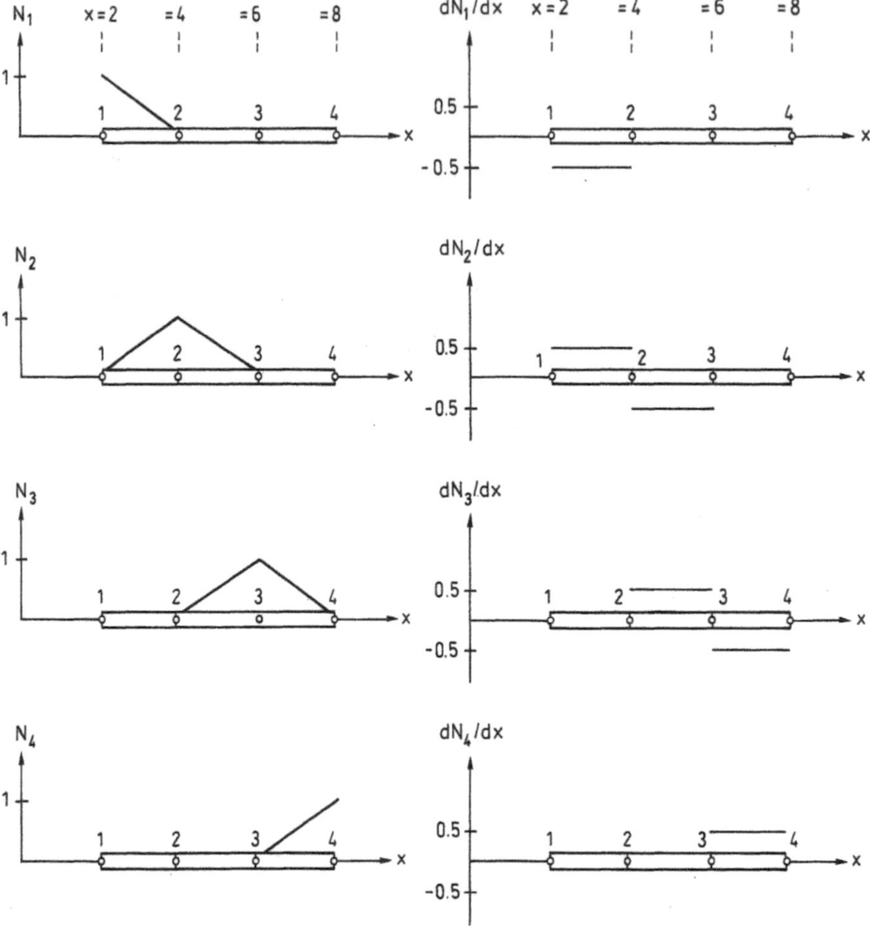

Figure 9.6 Global shape functions and their derivatives

Moreover, proceeding in the same manner we get

$$K_{23} = \int_a^b \frac{dN_2}{dx} Ak \frac{dN_3}{dx}\, dx = \int_4^6 \frac{dN_2}{dx} Ak \frac{dN_3}{dx}\, dx$$

$$= \int_4^6 (-0.5)50(0.5)\, dx = -25$$

and

$$K_{24} = \int_a^b \frac{dN_2}{dx} Ak \frac{dN_4}{dx}\, dx = 0$$

$$K_{33} = \int_a^b \frac{dN_3}{dx} Ak \frac{dN_3}{dx} dx = \int_4^6 (0.5)10 \times 5(0.5) dx$$

$$+ \int_6^8 (-0.5)10 \times 5(-0.5) dx = 50$$

$$K_{34} = \int_a^b \frac{dN_3}{dx} Ak \frac{dN_4}{dx} dx = \int_6^8 (-0.5)10 \times 5(0.5) dx = -25$$

$$K_{44} = \int_a^b \frac{dN_4}{dx} Ak \frac{dN_4}{dx} dx = \int_6^8 (0.5)10 \times 5(0.5) dx = 25$$

Collecting these results together, we obtain the stiffness matrix $\mathbf{K}$:

$$\mathbf{K} = \begin{bmatrix} 25 & -25 & 0 & 0 \\ -25 & 50 & -25 & 0 \\ 0 & -25 & 50 & -25 \\ 0 & 0 & -25 & 25 \end{bmatrix} \tag{9.29}$$

The banded structure of the stiffness matrix is obvious. Next, consider the boundary vector $\mathbf{f}_b$. From (9.24) and (9.26)–(9.28) we obtain

$$\mathbf{f}_b = \begin{bmatrix} (Aq)_{x=a} \\ 0 \\ 0 \\ -(Aq)_{x=b} \end{bmatrix} = \begin{bmatrix} 10q(x=2) \\ 0 \\ 0 \\ -150 \end{bmatrix} \tag{9.30}$$

The load vector $\mathbf{f}_l$ is given by (9.20), and from (9.26) and Figure 9.6 we find

$$\mathbf{f}_l = \begin{bmatrix} \int_a^b N_1 Q \, dx \\ \int_a^b N_2 Q \, dx \\ \int_a^b N_3 Q \, dx \\ \int_a^b N_4 Q \, dx \end{bmatrix} = Q \begin{bmatrix} \int_2^4 N_1 \, dx \\ \int_2^4 N_2 \, dx + \int_4^6 N_2 \, dx \\ \int_4^6 N_3 \, dx + \int_6^8 N_3 \, dx \\ \int_6^8 N_4 \, dx \end{bmatrix} = \begin{bmatrix} 100 \\ 200 \\ 200 \\ 100 \end{bmatrix} \tag{9.31}$$

where again we utilized the fact that a global shape function N_i differs from zero only in those elements which contain the nodal point i.

From (9.29)–(9.31), the final equation system (9.16) can be established and since the prescribed temperature T_1 is $T_1 = T(x = a) = g_a = 0$ (cf. (9.27) and (9.28)), this implies that

$$\begin{bmatrix} 25 & -25 & 0 & 0 \\ -25 & 50 & -25 & 0 \\ 0 & -25 & 50 & -25 \\ 0 & 0 & -25 & 25 \end{bmatrix}\begin{bmatrix} 0 \\ T_2 \\ T_3 \\ T_4 \end{bmatrix} = \begin{bmatrix} 10q_{x=2} \\ 0 \\ 0 \\ -150 \end{bmatrix} + \begin{bmatrix} 100 \\ 200 \\ 200 \\ 100 \end{bmatrix} \tag{9.32}$$

It can be noted that the largest computational effort is involved in determining the stiffness matrix $\mathbf{K}$ and that other flux boundary conditions and loads Q only influence the right-hand side of (9.32). Considering the first row and the last three rows of (9.32) separately, we obtain

$$-25T_2 = 10q(x = 2) + 100 \tag{9.33}$$

and

$$\begin{bmatrix} 50 & -25 & 0 \\ -25 & 50 & -25 \\ 0 & -25 & 25 \end{bmatrix}\begin{bmatrix} T_2 \\ T_3 \\ T_4 \end{bmatrix} = \begin{bmatrix} 200 \\ 200 \\ -50 \end{bmatrix} \tag{9.34}$$

By using the Gauss elimination technique presented in Chapter 2, it can easily be shown that the solution of (9.34) becomes

$$\begin{bmatrix} T_2 \\ T_3 \\ T_4 \end{bmatrix} = \begin{bmatrix} 14 \\ 20 \\ 18 \end{bmatrix} \,^\circ\mathrm{C} \tag{9.35}$$

With the known T_2-value, the flux q at $x = a = 2$ is determined from (9.33) to provide

$$q_{x=2} = -\left(k\frac{\mathrm{d}T}{\mathrm{d}x}\right)_{x=2} = -45 \text{ J/m}^2\text{ s} \tag{9.36}$$

i.e. heat must be extracted at the left end of the fin to keep the temperature at the required value.

To evaluate this solution, we make comparisons with the exact solution. It can easily be checked that

$$T = -x^2 + 13x - 22 \tag{9.37}$$

is the exact solution which satisfies the differential equation (9.25) and the boundary conditions (9.27) with the specific data given by (9.26) and (9.28).

The FE solution and the exact solution are compared in Figure 9.7, where due to the simple element adopted, the temperature determined by the FE method varies linearly between its values at the nodal points. It is of interest that the FE solution is exact at the nodal points.

While this is certainly not a general conclusion, we note that this fortunate situation occurs for some one-dimensional problems (a detailed discussion is presented

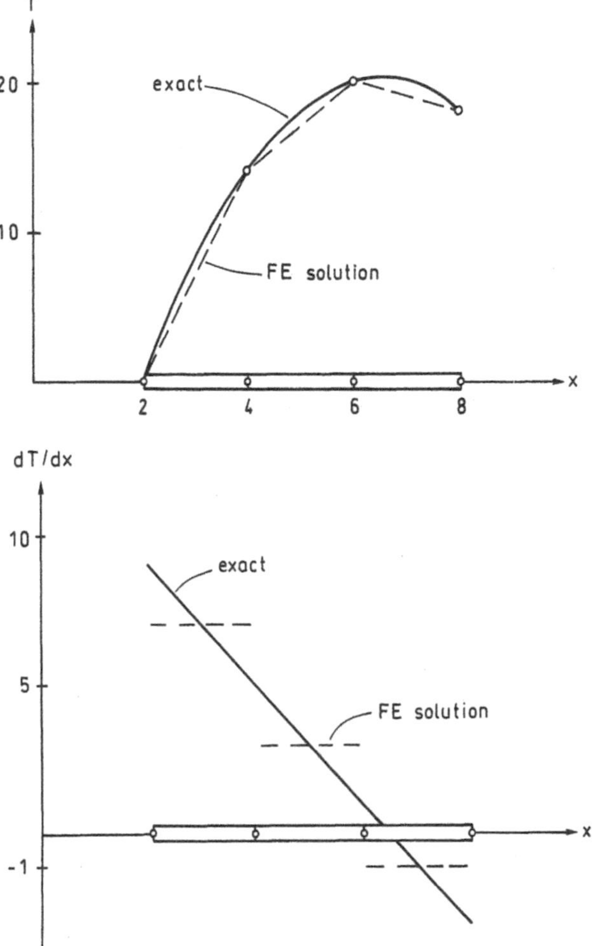

Figure 9.7   Comparison of exact and FE solution

by Hughes (1987, p. 27)). Besides, a fair agreement exists between the exact and the FE solution and we may draw the general conclusion that, even though the temperatures of the FE solution are close to the exact values, the temperature gradient of the FE solution is in much less agreement with the exact temperature gradient. In recognition of this, it may be of some surprise that the flux q at the left end of the fin as determined by (9.36) is in perfect agreement with the exact solution, as is easily checked from (9.37). The reason is because the temperature gradient dT/dx present in (9.36) is not determined from $dT/dx = \mathbf{Ba}$, which would lead to an inferior answer, but rather from the boundary vector $\mathbf{f}_b$ where the exact q-value enters. In a later section we shall prove that the balance principle for the entire body is fulfilled exactly by the FE solution, and as the loading due to Q is given, as is also the flux at the right end of the fin, the FE solution is forced to provide the exact flux at the left end of the fin.

9.1.2 *Further properties of the stiffness matrix*

Apart from the problem discussed in Example 1, the boundary conditions relating to the differential equation (9.1) have not been discussed, i.e. the system of equations (9.16) holds for arbitrary boundary conditions. Clearly, the boundary conditions must be considered in order to solve a specific physical problem and in order to do so in a systematic manner we first need to establish some important properties of the stiffness matrix $\mathbf{K}$.

We shall first show that

$$\boxed{\det \mathbf{K} = 0} \tag{9.38}$$

To prove this, consider the following homogeneous equation system:

$$\mathbf{Ka} = \mathbf{0} \tag{9.39}$$

Referring to (2.55), we recall that a non-trivial $\mathbf{a}$-solution exists if and only if (9.38) is true. As $\mathbf{a}$ is independent of x, use of (9.13) in (9.39) implies that

$$\int_a^b \mathbf{B}^T A k \mathbf{Ba} \, \mathrm{d}x = \mathbf{0} \tag{9.40}$$

All the finite elements considered are able to reflect a zero temperature gradient exactly (cf. (7.2)). Therefore, if all the temperatures at the nodal points are equal then $\mathrm{d}T/\mathrm{d}x = \mathbf{Ba} = \mathbf{0}$ even though $\mathbf{a} \neq \mathbf{0}$, i.e. (9.40) and thus also (9.39) possess a non-trivial $\mathbf{a}$-solution. In turn, this proves that (9.38) is correct.

Next, consider the quantity I defined by the following quadratic form:

$$\boxed{I = \mathbf{a}^T \mathbf{Ka}} \tag{9.41}$$

With (9.13) and (9.6) we get

$$I = \int_a^b \mathbf{a}^T \mathbf{B}^T A k \mathbf{Ba} \, \mathrm{d}x = \int_a^b \frac{\mathrm{d}T}{\mathrm{d}x} A k \frac{\mathrm{d}T}{\mathrm{d}x} \, \mathrm{d}x \geq 0 \tag{9.42}$$

where $Ak > 0$ and $(\mathrm{d}T/\mathrm{d}x)^2 \geq 0$. As $I \geq 0$ for arbitrary $\mathbf{a}$-matrices different from zero, we conclude from (2.68) that the stiffness matrix $\mathbf{K}$ is a positive semi-definite matrix. It also follows from (9.42) that

$$\boxed{\begin{aligned} I &> 0 \quad \text{if } \mathrm{d}T/\mathrm{d}x \neq 0 \\ I &= 0 \quad \text{if } \mathrm{d}T/\mathrm{d}x = 0 \end{aligned}} \tag{9.43}$$

To evaluate the properties of the stiffness matrix $\mathbf{K}$ further, we make a partitioning of $\mathbf{K}$ and $\mathbf{a}$ (cf. Chapter 2). As an example we split $\mathbf{K}$ and $\mathbf{a}$ into the following submatrices:

$$\mathbf{K} = \begin{bmatrix} \mathbf{A}_1 & \mathbf{A}_2 \\ \mathbf{A}_2^T & \underset{\sim}{\mathbf{K}} \end{bmatrix}; \quad \mathbf{a} = \begin{bmatrix} \mathbf{g} \\ \underset{\sim}{\mathbf{a}} \end{bmatrix} \tag{9.44}$$

In this partitioning of $\mathbf{K}$, use is made of the fact that $\mathbf{K}$ is symmetric and it also follows that the submatrix $\underset{\sim}{\mathbf{K}}$ is symmetric, i.e.

$$\boxed{\underset{\sim}{\mathbf{K}} = \underset{\sim}{\mathbf{K}}^T} \tag{9.45}$$

Using (9.44) the quadratic form (9.41) can be written as

$$I = \begin{bmatrix} \mathbf{g}^T & \underset{\sim}{\mathbf{a}}^T \end{bmatrix} \begin{bmatrix} \mathbf{A}_1 & \mathbf{A}_2 \\ \mathbf{A}_2^T & \underset{\sim}{\mathbf{K}} \end{bmatrix} \begin{bmatrix} \mathbf{g} \\ \underset{\sim}{\mathbf{a}} \end{bmatrix} \geq 0 \tag{9.46}$$

where (9.43) has been used. We emphasize that the submatrices given by (9.44) must have proper dimensions so that the matrix multiplications indicated in (9.46) are permissible. Evaluating (9.46) we obtain

$$I = \mathbf{g}^T \mathbf{A}_1 \mathbf{g} + \mathbf{g}^T \mathbf{A}_2 \underset{\sim}{\mathbf{a}} + \underset{\sim}{\mathbf{a}}^T \mathbf{A}_2^T \mathbf{g} + \underset{\sim}{\mathbf{a}}^T \underset{\sim}{\mathbf{K}} \underset{\sim}{\mathbf{a}} \geq 0 \tag{9.47}$$

Taking account of (9.43), we observe that $I = 0$ if and only if all components of the a-matrix are identical; otherwise we have $I > 0$. Therefore, choosing $\mathbf{a}^T = \begin{bmatrix} \mathbf{g}^T & \underset{\sim}{\mathbf{a}}^T \end{bmatrix} = \begin{bmatrix} \mathbf{0}^T & \underset{\sim}{\mathbf{a}}^T \end{bmatrix}$ it follows from (9.47) that

$$\boxed{\underset{\sim}{\mathbf{a}}^T \underset{\sim}{\mathbf{K}} \underset{\sim}{\mathbf{a}} > 0} \tag{9.48}$$

for $\underset{\sim}{\mathbf{a}} \neq 0$. This proves that $\underset{\sim}{\mathbf{K}}$ is positive definite and from (2.67) it then follows that

$$\boxed{\det \underset{\sim}{\mathbf{K}} \neq 0} \tag{9.49}$$

The submatrix $\underset{\sim}{\mathbf{K}}$ considered above can be viewed as obtained from the stiffness matrix $\mathbf{K}$ by deleting some rows and corresponding columns in the manner indicated by the partitioning shown in (9.44). It then turns out that $\underset{\sim}{\mathbf{K}}$ possesses the three fundamental properties given by (9.45), (9.48) and (9.49). However, by using exactly the same type of arguments as presented above, it follows that *any* submatrix $\underset{\sim}{\mathbf{K}}$ obtained from the stiffness matrix $\mathbf{K}$ by deleting one or more rows and the corresponding columns possesses these fundamental properties. This conclusion has important consequences when considering the boundary conditions, as shown next.

9.1.3 *Systematic consideration of boundary conditions*

When the system of equations (9.16) was established, no use was made of the boundary conditions. The differential equation (9.1) contains only the temperature gradient

$\mathrm{d}T/\mathrm{d}x$. Let us assume that the boundary conditions are prescribed in terms of given fluxes at both ends of the fin, i.e.

$$q(x = a) = h_a; \quad q(x = b) = h_b \tag{9.50}$$

where h_a and h_b are known quantities and where it is recalled that $q = -k\,\mathrm{d}T/\mathrm{d}x$. In this case, both the differential equation and the boundary conditions involve only the temperature gradient $\mathrm{d}T/\mathrm{d}x$. This means that if $T(x)$ is a solution to the problem, then $T(x) + C$, where C is an arbitrary constant, is also a solution to the problem. In this situation, we would expect that a unique solution to (9.16) does not exist. Referring to (2.56), we would therefore expect the determinant of the stiffness matrix **K** to be zero and this is precisely the situation indicated by (9.38).

Consequently, in order to obtain a unique solution from (9.16) we must specify at least one nodal temperature, and valid boundary conditions could therefore be of the form

$$q(x = a) = h_a; \quad T(x = b) = g_b \tag{9.51}$$

$$T(x = a) = g_a; \quad q(x = b) = h_b \tag{9.52}$$

$$T(x = a) = g_a; \quad T(x = b) = g_b \tag{9.53}$$

where h_a, h_b and g_a, g_b are given values of the fluxes and temperatures, respectively. We shall now prove that these boundary conditions lead to a modification of the system of equations (9.16), which ensures a unique solution. The proof follows the same lines irrespective of whether (9.51), (9.52) or (9.53) is given and, as an example, we assume boundary conditions in the form given by (9.52). For convenience, we also assume that nodal points 1 and n are located at $x = a$ and $x = b$, respectively; that is, at $x = a$ we have $T_1 = g_a$. With (9.15) and (9.24), the system of equations (9.16) can be written as

$$
\begin{bmatrix}
K_{11} & K_{12} & K_{13} & \cdots & K_{1n} \\
K_{21} & K_{22} & K_{23} & \cdots & K_{2n} \\
K_{31} & K_{32} & K_{33} & \cdots & K_{3n} \\
\vdots & \vdots & \vdots & & \vdots \\
K_{n1} & K_{n2} & K_{n3} & \cdots & K_{nn}
\end{bmatrix}
\begin{bmatrix}
g_a \\
T_2 \\
T_3 \\
\vdots \\
T_n
\end{bmatrix}
=
\begin{bmatrix}
(Aq)_{x=a} \\
0 \\
0 \\
\vdots \\
-A(x=b)h_b
\end{bmatrix}
+
\begin{bmatrix}
f_{l1} \\
f_{l2} \\
f_{l3} \\
\vdots \\
f_{ln}
\end{bmatrix}
=
\begin{bmatrix}
f_1 \\
f_2 \\
f_3 \\
\vdots \\
f_n
\end{bmatrix}
$$

$$\tag{9.54}$$

All the components of the stiffness matrix are known and the same holds for the components of the load vector $\mathbf{f}_l$. However, the total amount of heat flow at $x = a$, i.e. $(Aq)_{x=a}$, is unknown, which means that the force component f_1 is unknown, whereas the components $f_2 \ldots f_n$ are known. Moreover, component g_a of **a** is known, whereas the components $T_2 \ldots T_n$ are unknown. A partitioning of the matrices is indicated in (9.54). Let us define the following submatrices of the stiffness matrix $\mathbf{K}$:

$$\mathbf{A}_1 = [K_{11}]; \quad \mathbf{A}_2 = [K_{12} \quad K_{13} \quad \cdots \quad K_{1n}]$$

$$\mathbf{A}_2^\mathsf{T} = \begin{bmatrix} K_{21} \\ K_{31} \\ \vdots \\ K_{n1} \end{bmatrix}; \quad \underset{\sim}{\mathbf{K}} = \begin{bmatrix} K_{22} & K_{23} & \cdots & K_{2n} \\ K_{32} & K_{33} & \cdots & K_{3n} \\ \vdots & \vdots & & \vdots \\ K_{n2} & K_{n3} & \cdots & K_{nn} \end{bmatrix} \tag{9.55}$$

where the symmetry of $\mathbf{K}$ was used in the definitions of $\mathbf{A}_2$ and $\mathbf{A}_2^\mathsf{T}$. Likewise, we define the following submatrices of $\mathbf{a}$ and $\mathbf{f}$:

$$\mathbf{g} = [g_a]; \quad \underset{\sim}{\mathbf{a}} = \begin{bmatrix} T_2 \\ T_3 \\ \vdots \\ T_n \end{bmatrix}; \quad \mathbf{r} = [f_1]; \quad \underset{\sim}{\mathbf{f}} = \begin{bmatrix} f_2 \\ f_3 \\ \vdots \\ f_n \end{bmatrix} \tag{9.56}$$

This means that the system of equations (9.54) can be written as

$$\begin{bmatrix} \mathbf{A}_1 & \mathbf{A}_2 \\ \mathbf{A}_2^\mathsf{T} & \underset{\sim}{\mathbf{K}} \end{bmatrix} \begin{bmatrix} \mathbf{g} \\ \underset{\sim}{\mathbf{a}} \end{bmatrix} = \begin{bmatrix} \mathbf{r} \\ \underset{\sim}{\mathbf{f}} \end{bmatrix} \tag{9.57}$$

It appears that the only unknowns in this equation system are $\underset{\sim}{\mathbf{a}}$ and $\mathbf{r}$. Performing the matrix multiplications indicated in (9.57) we obtain

$$\mathbf{A}_1 \mathbf{g} + \mathbf{A}_2 \underset{\sim}{\mathbf{a}} = \mathbf{r}$$

$$\mathbf{A}_2^\mathsf{T} \mathbf{g} + \underset{\sim}{\mathbf{K}}\,\underset{\sim}{\mathbf{a}} = \underset{\sim}{\mathbf{f}}$$

which can be restated as

$$\boxed{\underset{\sim}{\mathbf{K}}\,\underset{\sim}{\mathbf{a}} = \underset{\sim}{\mathbf{f}} - \mathbf{A}_2^\mathsf{T} \mathbf{g}} \tag{9.58}$$

$$\boxed{\mathbf{r} = \mathbf{A}_1 \mathbf{g} + \mathbf{A}_2 \underset{\sim}{\mathbf{a}}} \tag{9.59}$$

The right-hand side of (9.58) is known and as $\underset{\sim}{\mathbf{K}}$ is obtained from the stiffness matrix $\mathbf{K}$ by deleting a row and the corresponding column, $\underset{\sim}{\mathbf{K}}$ possesses the properties given by (9.45), (9.48) and (9.49). In particular, $\det \underset{\sim}{\mathbf{K}} \neq 0$, i.e. a unique solution $\underset{\sim}{\mathbf{a}}$ is obtained from (9.58) and this solution does not depend on the unknown $\mathbf{r}$-matrix. By inserting the $\underset{\sim}{\mathbf{a}}$-solution into (9.59), we can determine the unknown $\mathbf{r}$-matrix. Therefore, all quantities of interest have been calculated in a unique manner. We also take the opportunity to draw attention to the similar observations made in Chapter 3 regarding the properties of $\mathbf{K}$ and $\underset{\sim}{\mathbf{K}}$ as well as the introduction of boundary conditions (cf. the discussion related to (3.25)–(3.32)).

The systematic consideration of the boundary conditions illustrated above proves that, when at least one nodal temperature is prescribed, a unique solution of the system of equations (9.16) follows. Moreover, this systematic approach lends itself directly to a computational implementation of boundary conditions in an FE program.

With the discussion above we are even able to find all solutions to (9.16) when the boundary conditions are given in terms of prescribed fluxes alone (cf. (9.50)). For this purpose we prescribe an arbitrary temperature at one end of the fin, for instance at $x = a$; that is, we set $T(x = a) = g_a$, where g_a is arbitrary. Assuming that nodal point 1 corresponds to $x = a$, we have $T(x = a) = g_a = T_1$, i.e. we again obtain the system of equations (9.57). Now, the only unknown matrix is $\mathbf{a}$, but this matrix can be found by solving (9.58), i.e. all nodal temperatures are now available. If we denote this solution by T, it is obvious that $T + C$ is also a solution, where C denotes an arbitrary constant.

9.1.4 *Example 2*

In order to illustrate the systematic consideration of boundary conditions as described above, we again consider the problem of Example 1. According to (9.14) we have seen in general that

$$\mathbf{Ka} = \mathbf{f}_b + \mathbf{f}_l$$

For the problem of Example 1 this system of equations takes the form (9.32), i.e.

$$\begin{bmatrix} 25 & -25 & 0 & 0 \\ -25 & 50 & -25 & 0 \\ 0 & -25 & 50 & -25 \\ 0 & 0 & -25 & 25 \end{bmatrix} \begin{bmatrix} 0 \\ T_2 \\ T_3 \\ T_4 \end{bmatrix} = \begin{bmatrix} 10q_{x=2} \\ 0 \\ 0 \\ -150 \end{bmatrix} + \begin{bmatrix} 100 \\ 200 \\ 200 \\ 100 \end{bmatrix} \tag{9.60}$$

where the specific boundary conditions were introduced. The unknowns of this equation are T_2, T_3, T_4 and $10q(x = 2)$. To solve it in a systematic manner, we make the partitioning as shown above, which is analogous to that indicated by (9.54)–(9.57). That is,

$$\mathbf{A}_1 = [25]; \quad \mathbf{A}_2 = [-25 \quad 0 \quad 0] \tag{9.61}$$

$$\mathbf{A}_2^{\mathrm{T}} = \begin{bmatrix} -25 \\ 0 \\ 0 \end{bmatrix}; \quad \underset{\sim}{\mathbf{K}} = \begin{bmatrix} 50 & -25 & 0 \\ -25 & 50 & -25 \\ 0 & -25 & 25 \end{bmatrix}$$

and

$$\mathbf{g} = [0]; \quad \mathbf{a} = \begin{bmatrix} T_2 \\ T_3 \\ T_4 \end{bmatrix} \tag{9.62}$$

$$\mathbf{r} = [10q(x=2) + 100]; \quad \mathbf{f} = \begin{bmatrix} 200 \\ 200 \\ -50 \end{bmatrix} \tag{9.63}$$

From (9.58) we then obtain

$$\begin{bmatrix} 50 & -25 & 0 \\ -25 & 50 & -25 \\ 0 & -25 & 25 \end{bmatrix} \begin{bmatrix} T_2 \\ T_3 \\ T_4 \end{bmatrix} = \begin{bmatrix} 200 \\ 200 \\ -50 \end{bmatrix} - \begin{bmatrix} -25 \\ 0 \\ 0 \end{bmatrix} [0] = \begin{bmatrix} 200 \\ 200 \\ -50 \end{bmatrix}$$

This system of equations is identical to (9.34), i.e. the solution becomes

$$\mathbf{a} = \begin{bmatrix} T_2 \\ T_3 \\ T_4 \end{bmatrix} = \begin{bmatrix} 14 \\ 20 \\ 18 \end{bmatrix} \tag{9.64}$$

From (9.59) and (9.61)–(9.64) we get

$$\mathbf{r} = [10q(x=2) + 100] = [25][0] + [-25 \quad 0 \quad 0] \begin{bmatrix} 14 \\ 20 \\ 18 \end{bmatrix} = -350 \tag{9.65}$$

i.e.

$$q(x=2) = -45 \tag{9.66}$$

The solutions (9.64) and (9.66) are evidently in accordance with (9.35) and (9.36) of Example 1. Even though the analysis above may seem a little involved, the important point is that we have illustrated how a systematic consideration of the boundary conditions can be carried out. Moreover, this systematic approach applies to general situations, i.e. it lends itself directly to an efficient computer implementation of the FE method.

9.1.5 *Evaluation of the force vector – fulfilment of the global balance principle*

It is an important characteristic of the FE method that even though it is an approximate method, the balance principle for the entire body is fulfilled exactly. In Figure 9.8 all heat quantities which enter and leave the one-dimensional fin are shown. As the flux q given by Fourier's law (4.3) is positive in the x-direction, $(Aq)_{x=a}$ is the heat per unit time which enters the fin at end point $x=a$ and $(Aq)_{x=b}$ is the heat per unit time which leaves the fin at end point $x=b$. We recall that Q is the heat input per unit time and per unit length of the fin and is measured as positive if heat is supplied to the fin.

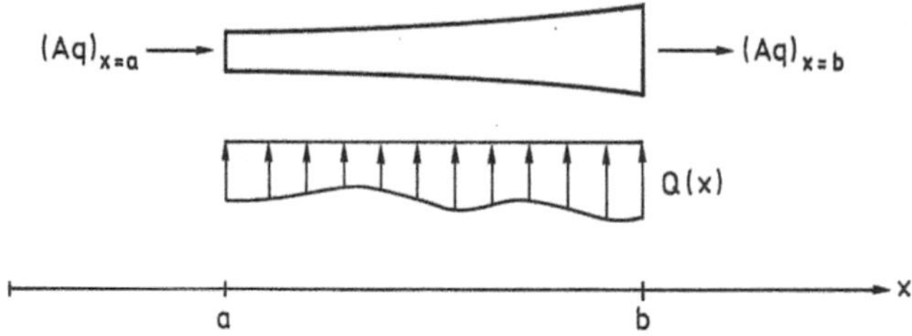

Figure 9.8   Heat supply and removal over the entire body

As we are considering time-independent problems, the balance principle for the entire body states that the heat entering the body equals the heat leaving the body, i.e.

$$(Aq)_{x=a} + \int_a^b Q \, dx = (Aq)_{x=b}$$

which can be written as

$$-[Aq]_a^b + \int_a^b Q \, dx = 0 \tag{9.67}$$

This equation is the balance principle for the entire body. Let us consider now the force vector **f** which, according to (9.15), is given by

$$f_i = f_{bi} + f_{li}; \quad i = 1, \ldots, n$$

This implies that

$$\sum_{i=1}^n f_i = \sum_{i=1}^n f_{bi} + \sum_{i=1}^n f_{li}$$

From (9.20) we get

$$\sum_{i=1}^n f_i = - \sum_{i=1}^n [N_i Aq]_a^b + \sum_{i=1}^n \int_a^b N_i Q \, dx$$

$$= - \left[\left(\sum_{i=1}^n N_i \right) Aq \right]_a^b + \int_a^b \left(\sum_{i=1}^n N_i \right) Q \, dx$$

Use of (7.139) yields

$$\sum_{i=1}^n f_i = -[Aq]_a^b + \int_a^b Q \, dx$$

and from (9.67) we conclude that

$$\boxed{\sum_{i=1}^n f_i = 0} \tag{9.68}$$

This means that the exact fulfilment of the balance principle for the entire body is expressed by the fact that, in accordance with (9.68), the sum of all the components of the force vector **f** is equal to zero. As an illustration consider the solution obtained in Examples 1 and 2. From (9.60) and (9.66) it follows directly that the force vector fulfils the balance principle of (9.68). Recalling that our FE equations were derived from the balance principle itself, the property (9.68) is indeed not surprising.

9.1.6 *Evaluation of the load vector – point sources*

Let us continue the evaluation of the force vector **f**. For this purpose we will provide a physical interpretation of the boundary vector $\mathbf{f}_b$ and the load vector $\mathbf{f}_l$. We observe that they have the same dimension, and to establish this dimension it is convenient to consider $\mathbf{f}_l$ as given by (9.13). The global shape function matrix **N** is dimensionless, and as Q is the heat supply per unit time and per unit length of the fin, then $\mathbf{f}_l$ and thus also $\mathbf{f}_b$ have the dimension $[\text{J/s}]$, i.e. heat per unit time.

In order to obtain a physical interpretation of the load vector $\mathbf{f}_l$ we evaluate a component f_{li} of $\mathbf{f}_l$. From (9.20) such a component is given by

$$f_{li} = \int_a^b N_i Q \, dx \tag{9.69}$$

It follows that if the load $Q(x)$ is prescribed over a region where N_i is different from zero then $f_{li} \neq 0$, otherwise $f_{li} = 0$.

Next let us consider the influence of a load Q in terms of a so-called *point source*. This means that the heat supply is concentrated to a point and at this point a heat supply Q_s per unit time is prescribed. Q_s is called the *strength* of the point source and it has the dimension $[\text{J/s}]$. Using Dirac's delta function (cf. (8.23)–(8.25)), we may express the load Q as

$$Q = Q_s \delta(x - c) = \begin{cases} \infty & \text{if } x = c \\ 0 & \text{otherwise} \end{cases} \tag{9.70}$$

where c is the position of the point source. By definition, we have that

$$\int_{-\infty}^{\infty} Q \, dx = \int_{-\infty}^{\infty} Q_s \delta(x - c) \, dx = Q_s \tag{9.71}$$

Due to (9.70), (9.71) may also be written as

$$\int_{c^-}^{c^+} Q_s \delta(x - c) \, dx = Q_s \tag{9.72}$$

where c^+ and c^- denote x-values slightly larger than and smaller than c, respectively. For such a point source, the component f_{li} given by (9.69) becomes

$$f_{li} = \int_a^b N_i Q_s \delta(x - c) \, dx = \int_{c^-}^{c^+} N_i Q_s \delta(x - c) \, dx = N_i(c) \int_{c^-}^{c^+} Q_s \delta(x - c) \, dx$$

i.e.

$$\boxed{f_{li} = N_i(c)Q_s}$$

(9.73)

If the position c of the point source coincides with the position of nodal point i, then $N_i(c) = N_i(x_i) = 1$, i.e. we have

$$\boxed{f_{li} = Q_s \quad \text{if the point source is located at nodal point } i}$$

(9.74)

Referring to (9.69) for an arbitrary load Q, it follows that it is always possible to devise point sources located at the nodal points and with such strengths that f_{li} obtained from (9.74) equals f_{li} obtained from (9.69), i.e.

$$f_{li} = \int_a^b N_i Q \, dx = Q_{si}$$

(9.75)

where Q_{si} is the strength of the nodal point source located at nodal point i. We may therefore envision the components of the load vector $\mathbf{f}_l$ as given by the strengths of point sources located at the nodal points. This equivalence is shown schematically in Figure 9.9. As the load Q is positive when heat is supplied to the fin, the same holds for the strengths Q_{si} (cf. Figure 9.9). Occasionally, the components of the load vector $\mathbf{f}_l$ from a distributed load Q are obtained by intuitive physical considerations. However, when determining $\mathbf{f}_l$ using (9.69) one speaks of a *consistent load vector*, since this determination is consistent with the Galerkin method.

Let us now return to the boundary vector $\mathbf{f}_b$ given by (9.24). In accordance with Fourier's law (4.3), the flux q is measured as positive in the direction of the x-axis. That is, the components of $\mathbf{f}_b$ given by (9.24) can be illustrated as in Figure 9.10.

With Figures 9.9 and 9.10 we have now illustrated all the heat supplied to the

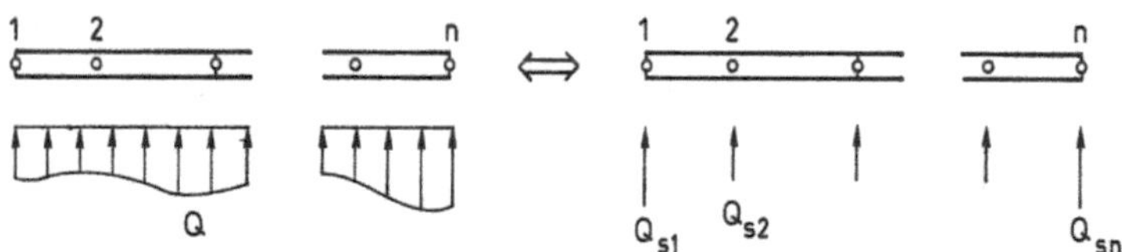

Figure 9.9   Equivalence between distributed load Q and point sources Q_{si}

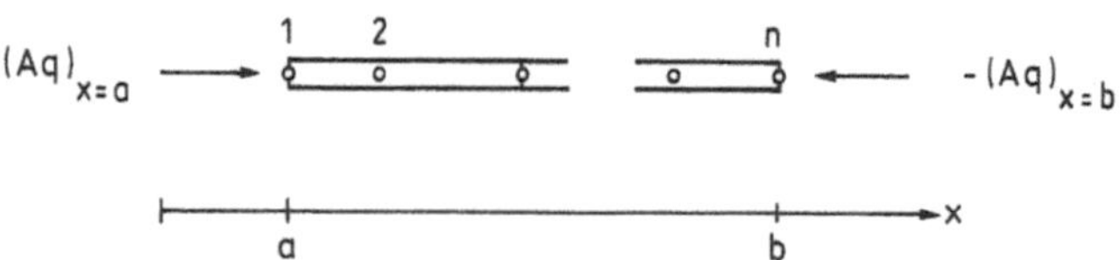

Figure 9.10   Illustration of boundary vector $\mathbf{f}_b$

fin. From (9.15) we have

$$\sum_{i=1}^{n} f_i = \sum_{i=1}^{n} f_{bi} + \sum_{i=1}^{n} f_{li} \tag{9.76}$$

Using (9.75) and Figures 9.9 and 9.10, it is obvious that the balance principle states that

$$\sum_{i=1}^{n} f_i = 0$$

in accordance with (9.68).

9.2 Expanded FE formulation of one element – assembling process

Previously, we established the FE formulation for the entire body, but for computer implementations it will turn out to be important also to know the FE formulation of one element. The number of such an element is denoted by α and the ends of the element are located at $x = x_i$ and $x = x_j$, as shown in Figure 9.11. For the linear element the only nodal points are those at $x = x_i$ and $x = x_j$, but for higher-order elements we also have interior nodal points. The space which element α occupies is denoted by L_α.

The approximation of the temperature given by (9.4) holds for the entire body, and therefore it also holds in the region L_α. By treating the region L_α as a separate body, we can make an FE formulation for this region and will clearly end up with a formulation similar to (9.12). Noting that the region L_α of element α is $x_i \leq x \leq x_j$ (cf. Figure 9.11), it follows that

$$\left(\int_{x_i}^{x_j} \mathbf{B}^T A k \mathbf{B} \, dx \right) \mathbf{a} = - [\mathbf{N}^T A q]_{x_i}^{x_j} + \int_{x_i}^{x_j} \mathbf{N}^T Q \, dx \tag{9.77}$$

When performing integrations over a region L_α which extends from x_i to x_j, it is

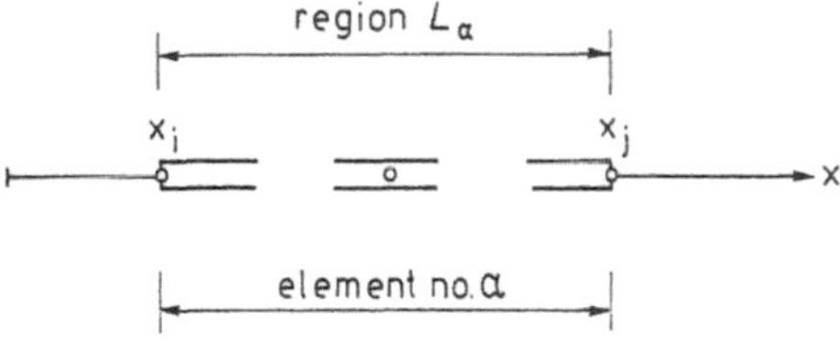

Figure 9.11   One element with ends at x_i and x_j and with possible interior nodal points

convenient to write this as

$$\int_{x_i}^{x_j} H(x)\,\mathrm{d}x = \int_{L_\alpha} H(x)\,\mathrm{d}x \tag{9.78}$$

where $H(x)$ denotes any function of x. Likewise, we adopt the notation

$$[\mathbf{N}^{\mathrm{T}} A q]_{x_i}^{x_j} = [\mathbf{N}^{\mathrm{T}} A q]_{L_\alpha} \tag{9.79}$$

If we define the following matrices:

$$\boxed{\begin{aligned}
\mathbf{K}^{ee} &= \int_{L_\alpha} \mathbf{B}^{\mathrm{T}} A k \mathbf{B}\,\mathrm{d}x \\[2mm]
\mathbf{f}_{\mathrm{b}}^{ee} &= -[\mathbf{N}^{\mathrm{T}} A q]_{L_\alpha} \\[2mm]
\mathbf{f}_{\mathrm{l}}^{ee} &= \int_{L_\alpha} \mathbf{N}^{\mathrm{T}} Q\,\mathrm{d}x
\end{aligned}} \tag{9.80}$$

then (9.77) can be written as

$$\boxed{\mathbf{K}^{ee}\mathbf{a} = \mathbf{f}_{\mathrm{b}}^{ee} + \mathbf{f}_{\mathrm{l}}^{ee}} \tag{9.81}$$

For reasons that will be explained later, superscript ee denotes 'expanded element'. That is, $\mathbf{K}^{ee}$ is the *expanded element stiffness matrix* for element α, $\mathbf{f}_{\mathrm{b}}^{ee}$ is the *expanded element boundary vector* for element α and $\mathbf{f}_{\mathrm{l}}^{ee}$ is the *expanded element load vector* for element α. Moreover, if we define the *expanded element force vector* $\mathbf{f}^{ee}$ for element α according to

$$\boxed{\mathbf{f}^{ee} = \mathbf{f}_{\mathrm{b}}^{ee} + \mathbf{f}_{\mathrm{l}}^{ee}} \tag{9.82}$$

(9.81) may be written as

$$\boxed{\mathbf{K}^{ee}\mathbf{a} = \mathbf{f}^{ee}} \tag{9.83}$$

This system of equations constitutes the FE formulation of one element. We remark that as $\mathbf{a}$ in (9.83) is the same as that present in the system of equations (9.16), which applies to the entire body, the dimension of $\mathbf{a}$ is $n \times 1$, where n is the total number of nodal points in the body. This means that the matrices $\mathbf{K}^{ee}$ and $\mathbf{f}^{ee}$ have the same dimensions as those of $\mathbf{K}$ and $\mathbf{f}$ in (9.16), respectively. As (9.83) has been formulated in terms of all the degrees of freedom for the entire body and not only those related to the specific element, (9.83) is referred to as the *expanded FE formulation of one element*. Examples of such a formulation are given by (3.15) and (3.17).

Now let n_{el} denote the total number of elements for the entire body which occupies the region $a \leqslant x \leqslant b$; cf. Figure 9.1. An integration over this region can always be expressed as a sum of integrations over each element. With element α occupying the region L_α, we therefore have

$$\int_a^b H(x)\,\mathrm{d}x = \sum_{\alpha=1}^{n_{el}} \left(\int_{L_\alpha} H(x)\,\mathrm{d}x \right) \tag{9.84}$$

where, again, $H(x)$ is an arbitrary function. With (9.13), (9.80) and (9.84), we obtain

$$\boxed{\begin{aligned} \mathbf{K} &= \sum_{\alpha=1}^{n_{el}} \mathbf{K}_\alpha^{ee} \\[2mm] \mathbf{f}_l &= \sum_{\alpha=1}^{n_{el}} \mathbf{f}_{l_\alpha}^{ee} \end{aligned}} \tag{9.85}$$

where $\mathbf{K}_\alpha^{ee}$ denotes the expanded element stiffness matrix for element α and $\mathbf{f}_{l_\alpha}^{ee}$ is the expanded load vector for element α.

Therefore, the (global) stiffness matrix $\mathbf{K}$ is the sum of all the expanded element stiffnesses $\mathbf{K}_\alpha^{ee}$. Likewise, the (global) load vector $\mathbf{f}_l$ is the sum of all the expanded element load vectors $\mathbf{f}_{l_\alpha}^{ee}$. We emphasize that $\mathbf{K}$ and $\mathbf{K}_\alpha^{ee}$ have the same dimension $n \times n$, where n is the total number of nodal points in the body. Likewise $\mathbf{f}_l$ and $\mathbf{f}_{l_\alpha}^{ee}$ have the same dimension $n \times 1$.

With reference to Figure 9.12, let us consider two neighbouring elements with numbers α and β. It appears that

$$-[\mathbf{N}^{\mathrm{T}} A q]_{x_i}^{x_j} - [\mathbf{N}^{\mathrm{T}} A q]_{x_j}^{x_k} = -[\mathbf{N}^{\mathrm{T}} A q]_{x_i}^{x_k} \tag{9.86}$$

where $q = -k\,\mathrm{d}T/\mathrm{d}x$. We have seen that, in general, the temperature gradient $\mathrm{d}T/\mathrm{d}x$ varies in a discontinuous manner over element boundaries; see for instance Figure 9.7. Considering point x_j of Figure 9.12, it may therefore be somewhat surprising that the boundary term from region L_α cancels the corresponding boundary term from region L_β. However, the important fact is that the boundary fluxes q entering the boundary terms in (9.86) are the *exact* ones, since no approximation of the temperature gradient $\mathrm{d}T/\mathrm{d}x$ was used for these boundary fluxes (cf. the discussion of (9.8)). Let us now determine the sum of the boundary terms $-[\mathbf{N}^{\mathrm{T}} A q]_{x_i}^{x_j}$ for all the elements of the body.

From (9.86) and with the notation of (9.79), we obtain

$$-[\mathbf{N}^{\mathrm{T}} A q]_a^b = -\sum_{\alpha=1}^{n_{el}} [\mathbf{N}^{\mathrm{T}} A q]_{L_\alpha} \tag{9.87}$$

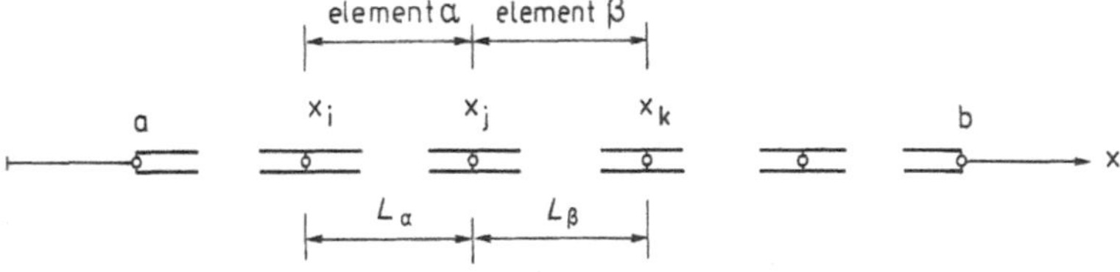

Figure 9.12 Two neighbouring elements with regions L_α and L_β

Using (9.80) and (9.13), (9.87) can be written as

$$\boxed{\mathbf{f}_b = \sum_{\alpha=1}^{n_{el}} \mathbf{f}_{b_\alpha}^{ee}} \tag{9.88}$$

where $\mathbf{f}_{b_\alpha}^{ee}$ denotes the element boundary vector for element α.

From (9.82) it follows that

$$\sum_{\alpha=1}^{n_{el}} \mathbf{f}_\alpha^{ee} = \sum_{\alpha=1}^{n_{el}} \mathbf{f}_{b_\alpha}^{ee} + \sum_{\alpha=1}^{n_{el}} \mathbf{f}_{l_\alpha}^{ee}$$

i.e. (9.85), (9.88) and (9.15) imply that

$$\boxed{\mathbf{f} = \sum_{\alpha=1}^{n_{el}} \mathbf{f}_\alpha^{ee}} \tag{9.89}$$

We are now in a position to establish a fundamental property for the FE formulation. Based on the expanded FE formulation of one element, expression (9.83), we sum over all elements of the body to obtain

$$\sum_{\alpha=1}^{n_{el}} \mathbf{K}_\alpha^{ee}\mathbf{a} = \sum_{\alpha=1}^{n_{el}} \mathbf{f}_\alpha^{ee}$$

and with (9.85) and (9.89) this expression becomes

$$\mathbf{Ka} = \mathbf{f} \tag{9.90}$$

which is precisely the FE formulation of the entire body (cf. (9.16)). This shows that the FE equations for the entire body can be established as the sum of the contributions of all elements and the process of making this summation is called *assembling*. This essential result is analogous to the assembling procedure discussed in Chapter 3 (see for instance the discussion of (3.23) and (3.26)).

We emphasize that we *need not* establish the FE equations for the entire body as a sum of the contributions of all elements. We can just as well use the formulation (9.16) directly, but it will turn out later that a slight reformulation of the process given by (9.85) and (9.89) provides essential advantages when constructing a computer program for FE analysis. We also note that obtaining the boundary vector $\mathbf{f}_b$ as the sum of the expanded element boundary vectors $\mathbf{f}_{b_\alpha}^{ee}$ in accordance with (9.88) does not provide any advantages. In fact, from (9.24) it follows that the boundary vector $\mathbf{f}_b$ only possesses two non-zero components and it is therefore much more advantageous to establish the boundary vector $\mathbf{f}_b$ directly from (9.24). Moreover, we emphasize that the boundary conditions of an element are in fact unknown, except when the element boundary coincides with the boundary of the body.

9.2.1 *Example 3*

In order to illustrate how the FE formulation of the entire body can be established as a sum of the contributions of all elements in accordance with (9.85), we consider again the problem described in Example 1.

With this procedure we have to identify the number of the elements; thus Figure 9.5 is redrawn in Figure 9.13. As we have three elements, i.e. $n_{el} = 3$, the element numbers become 1, 2 and 3, but apart from that the specific numbering of the elements if of course immaterial.

We recall that the boundary vector $\mathbf{f}_b$ is established in the same manner as before; that is, from (9.30) we get

$$\mathbf{f}_b = \begin{bmatrix} 10q(x = 2) \\ 0 \\ 0 \\ -150 \end{bmatrix} \tag{9.91}$$

Next let us establish the expanded element stiffness matrix and expanded element load vector for each element. From (9.80), we have

$$\mathbf{K}^{ee} = \int_{L_\alpha} \mathbf{B}^{\mathrm{T}} Ak\mathbf{B} \, \mathrm{d}x = \int_{L_\alpha} \frac{\mathrm{d}\mathbf{N}^{\mathrm{T}}}{\mathrm{d}x} Ak \frac{\mathrm{d}\mathbf{N}}{\mathrm{d}x} \, \mathrm{d}x$$

i.e. the component $(\mathbf{K}^{ee})_{ij}$ of $\mathbf{K}^{ee}$ is

$$(\mathbf{K}^{ee})_{ij} = \int_{L_\alpha} \frac{\mathrm{d}N_i}{\mathrm{d}x} Ak \frac{\mathrm{d}N_j}{\mathrm{d}x} \, \mathrm{d}x \tag{9.92}$$

Likewise, we have from (9.80) that a component $(\mathbf{f}_1^{ee})_i$ of $\mathbf{f}_1^{ee}$ is

$$(\mathbf{f}_1^{ee})_i = \int_{L_\alpha} N_i Q \, \mathrm{d}x \tag{9.93}$$

Consider first element 1, where $2 \leq x \leq 4$ corresponds to the region L_1, i.e.

$$(\mathbf{K}_1^{ee})_{ij} = \int_2^4 \frac{\mathrm{d}N_i}{\mathrm{d}x} Ak \frac{\mathrm{d}N_j}{\mathrm{d}x} \, \mathrm{d}x$$

$$(\mathbf{f}_{11}^{ee})_i = \int_2^4 N_i Q \, \mathrm{d}x$$

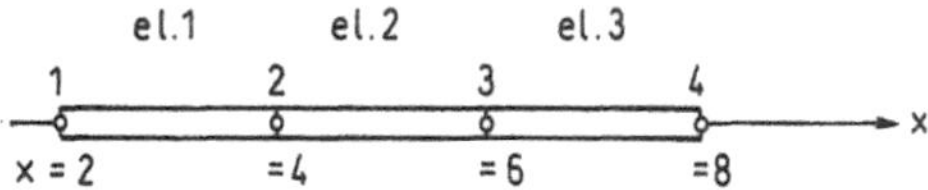

Figure 9.13 Numbering of elements

According to (9.26) we have $Ak = 50$ and $Q = 100$. Moreover, referring to Figure 9.6, the only shape functions which differ from zero in the region $2 \le x \le 4$ are N_1 and N_2. It follows trivially that

$$\mathbf{K}_1^{ee} = \begin{bmatrix} 25 & -25 & 0 & 0 \\ -25 & 25 & 0 & 0 \\ 0 & 0 & 0 & 0 \\ 0 & 0 & 0 & 0 \end{bmatrix}; \quad \mathbf{f}_{l1}^{ee} = \begin{bmatrix} 100 \\ 100 \\ 0 \\ 0 \end{bmatrix} \tag{9.94}$$

In the same manner for element 2, where $4 \le x \le 6$ corresponds to the region L_2, we obtain

$$\mathbf{K}_2^{ee} = \begin{bmatrix} 0 & 0 & 0 & 0 \\ 0 & 25 & -25 & 0 \\ 0 & -25 & 25 & 0 \\ 0 & 0 & 0 & 0 \end{bmatrix}; \quad \mathbf{f}_{l2}^{ee} = \begin{bmatrix} 0 \\ 100 \\ 100 \\ 0 \end{bmatrix} \tag{9.95}$$

For element 3 with the region L_3 corresponding to $6 \le x \le 8$, we find that

$$\mathbf{K}_3^{ee} = \begin{bmatrix} 0 & 0 & 0 & 0 \\ 0 & 0 & 0 & 0 \\ 0 & 0 & 25 & -25 \\ 0 & 0 & -25 & 25 \end{bmatrix}; \quad \mathbf{f}_{l3}^{ee} = \begin{bmatrix} 0 \\ 0 \\ 100 \\ 100 \end{bmatrix} \tag{9.96}$$

From (9.85), we obtain with (9.94)–(9.96)

$$\mathbf{K} = \begin{bmatrix} 25 & -25 & 0 & 0 \\ -25 & 50 & -25 & 0 \\ 0 & -25 & 50 & -25 \\ 0 & 0 & -25 & 25 \end{bmatrix}; \quad \mathbf{f}_l = \begin{bmatrix} 100 \\ 200 \\ 200 \\ 100 \end{bmatrix} \tag{9.97}$$

As expected, $\mathbf{K}$ and $\mathbf{f}_l$ established in this manner correspond exactly to the expressions given in Example 1 (cf. (9.29) and (9.31)). Again we note the analogy of the assembling process used above to that discussed in Chapter 3 (see for instance (3.23) and (3.26)).

9.3 FE formulation of one element

We have seen that the FE equation for the entire body can be established as the sum of the FE equations for each element. We also noted that this summation was made only for the stiffness matrix $\mathbf{K}$ and the load vector $\mathbf{f}_l$, since it is more convenient to establish the boundary vector $\mathbf{f}_b$ directly from (9.24).

The expanded element stiffness matrix $\mathbf{K}^{ee}$ and the expanded element load vector $\mathbf{f}_1^{ee}$ are of dimensions $n \times n$ and $n \times 1$, respectively, where n is the total number of nodal points. The nodal temperatures, which influence the temperature approximation within an element, are those related to the nodal points belonging to the element. Therefore, if nodal point j does not belong to the element, the nodal temperature T_j has no influence on the results and, consequently, the component $(\mathbf{K}^{ee})_{ij}$ of the expanded element stiffness matrix is zero. As $\mathbf{K}^{ee}$ is symmetric, we conclude that a component $(\mathbf{K}^{ee})_{ij}$ is zero unless both nodal point i and nodal point j belong to the element α considered. Therefore, if the element contains n_e nodal points then only $n_e \times n_e$ components of the expanded element stiffness matrix $\mathbf{K}^{ee}$ will be non-zero. As $\mathbf{K}^{ee}$ has the dimension $n \times n$, this implies that many components of $\mathbf{K}^{ee}$ are zero (see for instance (9.94)–(9.96)). This will be especially pronounced if the body is divided into many elements.

As an example, assume that simple linear elements are used, i.e. $n_e = 2$; then the number of non-zero components in each expanded element stiffness matrix is $2 \times 2 = 4$. Assume that the total number of nodal points in the body is $n = 30$. As the dimension of $\mathbf{K}^{ee}$ is $n \times n = 30 \times 30 = 900$, the number of zero components in $\mathbf{K}^{ee}$ is $900 - 4 = 896$. This implies that a computer program based on the use of expanded element stiffness matrices would be highly inefficient, as it would perform a sequence of operations with numbers equal to zero. The same arguments hold for the expanded element load vector.

It appears that it would be much more efficient to establish the stiffness matrix and the load vector for an element using only those degrees of freedom that are related to the specific element in question.

When (9.83) was established for element α, use was made of the approximation for the temperature T given by (9.4) and valid for the entire body. Obviously, as we only consider one element we may equally well consider the approximation for T applicable for this specific element only. This approximation is given by (7.127), i.e.

$$\boxed{T = \mathbf{N}^e \mathbf{a}^e} \tag{9.98}$$

where $\mathbf{N}^e$ is the element shape function matrix and $\mathbf{a}^e$ contains the temperatures at the nodal points of the element. Thus

$$\mathbf{N}^e = [N_1^e \quad N_2^e \quad \dots \quad N_{n_e}^e]; \quad \mathbf{a}^e = \begin{bmatrix} T_1 \\ T_2 \\ \vdots \\ T_{n_e} \end{bmatrix} \tag{9.99}$$

where n_e denotes the number of nodal points for the element. From (9.98) we obtain

$$\boxed{\frac{\mathrm{d}T}{\mathrm{d}x} = \mathbf{B}^e \mathbf{a}^e} \quad \text{where} \quad \boxed{\mathbf{B}^e = \frac{\mathrm{d}\mathbf{N}^e}{\mathrm{d}x}} \tag{9.100}$$

i.e.

$$\mathbf{B}^e = \left[\frac{dN_1^e}{dx} \quad \frac{dN_2^e}{dx} \quad \cdots \quad \frac{dN_{n_e}^e}{dx} \right] \tag{9.101}$$

Use of (9.100) in the weak form (9.3) and adoption of the Galerkin method evidently result in

$$\left(\int_{L_a} \mathbf{B}^{eT} Ak\mathbf{B}^e \, dx \right) \mathbf{a}^e = -[\mathbf{N}^{eT} Aq]_{L_a} + \int_{L_a} \mathbf{N}^{eT} Q \, dx \tag{9.102}$$

where, as before, L_α is the region of the element α considered and where the notations (9.78) and (9.79) have been utilized. If we define the following matrices:

$$\begin{aligned}
\mathbf{K}^e &= \int_{L_a} \mathbf{B}^{eT} Ak\mathbf{B}^e \, dx \\[2mm]
\mathbf{f}_b^e &= -[\mathbf{N}^{eT} Aq]_{L_a} \\[2mm]
\mathbf{f}_l^e &= \int_{L_a} \mathbf{N}^{eT} Q \, dx
\end{aligned} \tag{9.103}$$

then (9.102) can be written as

$$\mathbf{K}^e \mathbf{a}^e = \mathbf{f}_b^e + \mathbf{f}_l^e \tag{9.104}$$

$\mathbf{K}^e$ is the *element stiffness matrix* for element α, $\mathbf{f}_b^e$ is the *element boundary vector* for element α and $\mathbf{f}_l^e$ is the *element load vector* for element α. Moreover, if we define the *element force vector* $\mathbf{f}^e$ for element α as

$$\mathbf{f}^e = \mathbf{f}_b^e + \mathbf{f}_l^e \tag{9.105}$$

then (9.104) may be written as

$$\mathbf{K}^e \mathbf{a}^e = \mathbf{f}^e \tag{9.106}$$

This system of equations constitutes the FE formulation of one element. Contrary to its expanded form, (9.106) contains only those degrees of freedom which belong to the element considered and it follows that the non-zero components of $\mathbf{K}^{ee}$ and $\mathbf{f}_l^{ee}$ are those given by $\mathbf{K}^e$ and $\mathbf{f}_l^e$, respectively. We recall that the dimensions of $\mathbf{K}^e$ and $\mathbf{f}_l^e$ are $n_e \times n_e$ and $n_e \times 1$, respectively.

The three different FE formulations – global formulation, expanded element formulation and element formulation – are illustrated in Figure 9.14. We recall that, in the global formulation, the entire body is considered and the approximation $T = \mathbf{N}\mathbf{a}$, applicable to the entire body, is adopted. In the expanded element

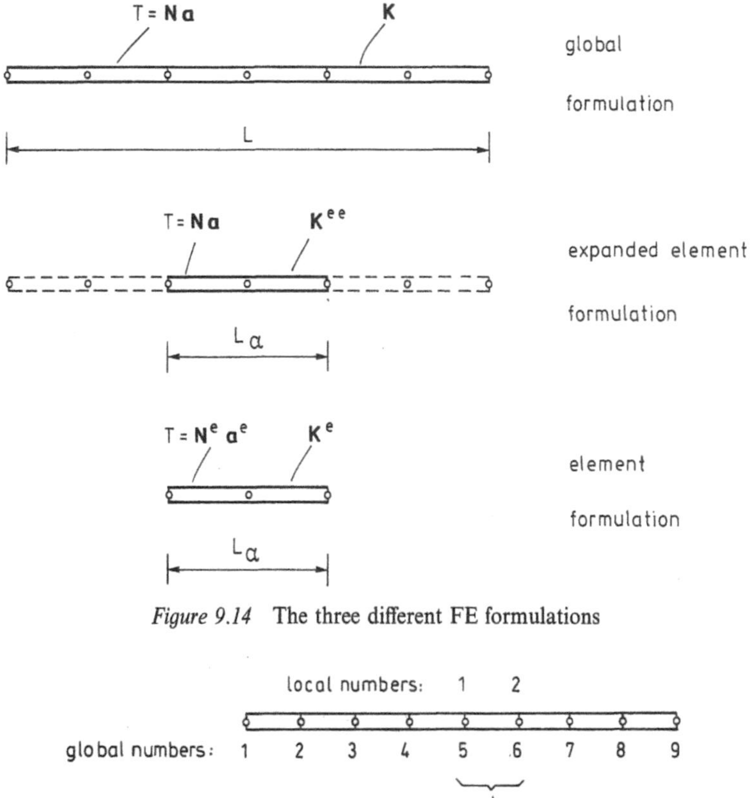

Figure 9.14   The three different FE formulations

Figure 9.15   Local and global nodal point numbering

formulation, one element is considered, but the global approximation $T = \mathbf{Na}$ is still used. Finally, in the element formulation, one element is considered and the approximation $T = \mathbf{N}^e\mathbf{a}^e$, applicable to this particular element, is used.

Therefore, to obtain an efficient FE computer program the element stiffness matrix $\mathbf{K}^e$ is derived for each element. Then, instead of using the inefficient formulation (9.85), each term of $\mathbf{K}^e$ is added directly to the proper component of the (global) stiffness matrix $\mathbf{K}$. In a similar manner, each term of the element load vector $\mathbf{f}_i^e$ is added directly to the proper component of the (global) load vector $\mathbf{f}_i$. To describe this procedure, we first note that in order to obtain a standardized determination of the FE equations (9.104) for an element, it is advantageous to use a *local nodal point numbering* which should be distinguished from the *global nodal point numbering*.

To firm up these ideas, consider the body shown in Figure 9.15, where, for convenience, simple linear elements are used. The element given by the global numbers 5 and 6 is considered. Using the local numbering, the FE equations (9.104) take the form

$$\begin{bmatrix} K^e_{11} & K^e_{12} \\ K^e_{21} & K^e_{22} \end{bmatrix} \begin{bmatrix} T_1 \\ T_2 \end{bmatrix} = \begin{bmatrix} f^e_{b1} \\ f^e_{b2} \end{bmatrix} + \begin{bmatrix} f^e_{l1} \\ f^e_{l2} \end{bmatrix} \tag{9.107}$$

As local point number 1 corresponds to global point number 5 and local point number 2 corresponds to global point number 6, we obtain the following procedure:

$$
\begin{matrix}
& 5 & 6 & & K^e_{11} \text{ is added to } K_{55} \\
5 & \begin{bmatrix} K^e_{11} & K^e_{12} \\ K^e_{21} & K^e_{22} \end{bmatrix} & \Rightarrow & & K^e_{12} \text{ is added to } K_{56} \\
6 & & & & K^e_{21} \text{ is added to } K_{65} \\
& & & & K^e_{22} \text{ is added to } K_{66}
\end{matrix}
\tag{9.108}
$$

$$
\begin{matrix}
5 & \begin{bmatrix} f^e_{l1} \\ f^e_{l2} \end{bmatrix} \Rightarrow & & f^e_{l1} \text{ is added to } f_{l5} \\
6 & & & f^e_{l2} \text{ is added to } f_{l6}
\end{matrix}
$$

It appears that the identification of proper components of $\mathbf{K}$ and $\mathbf{f}_l$ as given by (9.108) is purely a geometrical or topological scheme. In a general situation, the information on how local numbers communicate with global numbers is given by the so-called *topology data*.

Let us summarize the discussion above. The FE equations can always be established directly from (9.16), (9.15) and (9.13). The stiffness matrix $\mathbf{K}$ and the load vector $\mathbf{f}_l$ may also be derived according to (9.85) using the expanded form of the FE equations for the individual elements. However, in order to construct an efficient FE computer program, the stiffness matrix $\mathbf{K}$ and the load vector $\mathbf{f}_l$ are established using the element stiffness $\mathbf{K}^e$ and element load vector $\mathbf{f}^e_l$ given by (9.103), where the components of $\mathbf{K}^e$ and $\mathbf{f}^e_l$ are added to $\mathbf{K}$ and $\mathbf{f}_l$ using the topology data. In any case, the boundary vector $\mathbf{f}_b$ is most conveniently obtained by using (9.13).

9.3.1 *Example 4*

We shall now show the manner in which the FE equations for the entire body can be established using the element stiffness $\mathbf{K}^e$ and element load vector $\mathbf{f}^e_l$ given by (9.103). Again the problem discussed in Examples 1, 2 and 3 is considered. For convenience, the element numbering and global nodal point numbering are shown again in Figure 9.16.

We consider *element 1* first with the local nodal points as shown in Figure 9.17.

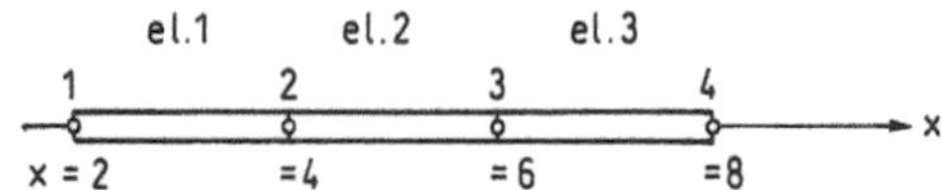

Figure 9.16   Global nodal points and element numbering

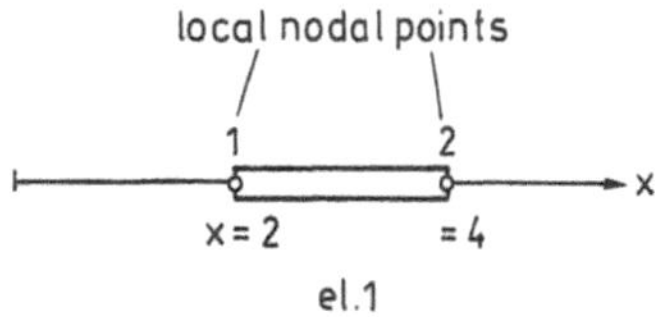

Figure 9.17   Local nodal point numbering of element 1

From (9.103) the element stiffness matrix $\mathbf{K}^e$ of element 1 becomes

$$\mathbf{K}^e = \int_2^4 \mathbf{B}^{eT} Ak\mathbf{B}^e \, dx \tag{9.109}$$

According to (9.26) we have $Ak = 50$ and using the $\mathbf{B}^e$-matrix as given by (7.35), we obtain

$$\mathbf{K}^e = \int_2^4 \begin{bmatrix} -1/L \\ 1/L \end{bmatrix} Ak \begin{bmatrix} -\dfrac{1}{L} & \dfrac{1}{L} \end{bmatrix} dx = \frac{Ak}{L^2} \int_2^4 \begin{bmatrix} 1 & -1 \\ -1 & 1 \end{bmatrix} dx$$

$$= \frac{50}{2^2} \begin{bmatrix} 2 & -2 \\ -2 & 2 \end{bmatrix} \tag{9.110}$$

i.e.

$$\mathbf{K}^e = \begin{bmatrix} 25 & -25 \\ -25 & 25 \end{bmatrix} \tag{9.111}$$

The element load vector $\mathbf{f}_1^e$ is given by (9.103) and as the load $Q = 100$ (cf. (9.26)) we obtain for *element* 1

$$\mathbf{f}_1^e = \int_2^4 \mathbf{N}^{eT} Q \, dx = \frac{Q}{L} \int_2^4 \begin{bmatrix} -x + 4 \\ x - 2 \end{bmatrix} dx = \frac{100}{2} \int_2^4 \begin{bmatrix} -x + 4 \\ x - 2 \end{bmatrix} dx \tag{9.112}$$

where (7.29) has been used. The result is that

$$\mathbf{f}_1^e = \begin{bmatrix} 100 \\ 100 \end{bmatrix} \tag{9.113}$$

According to Figures 9.16 and 9.17 the local nodal point 1 corresponds to the global nodal point 1 and the local nodal point 2 corresponds to the global nodal point 2. The contributions of (9.111) and (9.113) are added to the (global) stiffness matrix $\mathbf{K}$ and the (global) load vector $\mathbf{f}_1$ and as these are originally zero and of dimension 4×4 and 4×1, respectively, we obtain the intermediate result

$$\mathbf{K} = \begin{bmatrix} 25 & -25 & 0 & 0 \\ -25 & 25 & 0 & 0 \\ 0 & 0 & 0 & 0 \\ 0 & 0 & 0 & 0 \end{bmatrix} ; \quad \mathbf{f}_1 = \begin{bmatrix} 100 \\ 100 \\ 0 \\ 0 \end{bmatrix} \tag{9.114}$$

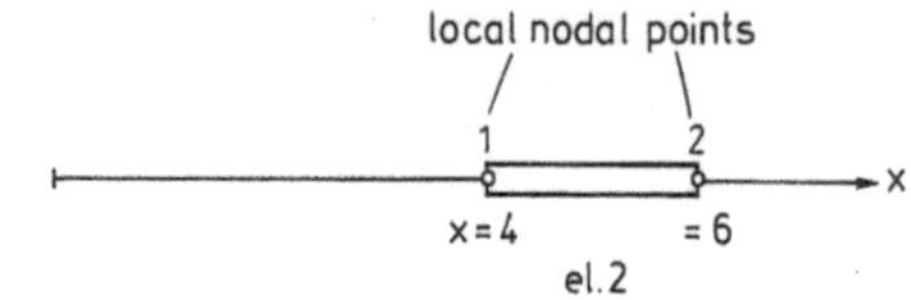

Figure 9.18   Local nodal point numbering of element 2

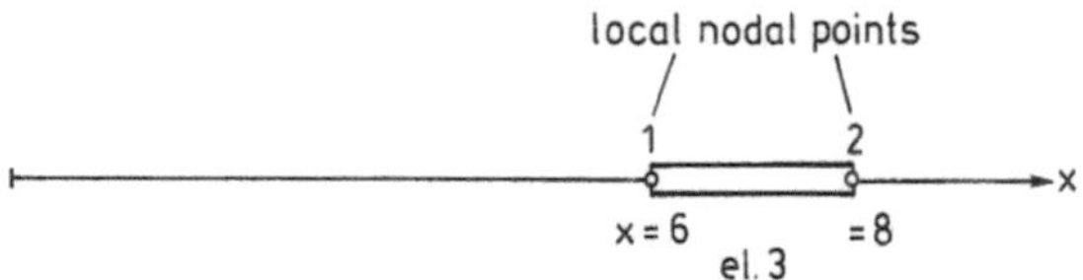

Figure 9.19   Local nodal point numbering of element 3

We next consider *element 2* with the local nodal points as shown in Figure 9.18. From (9.103) the element stiffness matrix $\mathbf{K}^e$ of element 2 is

$$\mathbf{K}^e = \int_4^6 \mathbf{B}^{eT} A k \mathbf{B}^e \, dx$$

With (7.35) and as $Ak = 50$ we obtain

$$\mathbf{K}^e = \int_4^6 \begin{bmatrix} -1/L \\ 1/L \end{bmatrix} Ak \begin{bmatrix} -\dfrac{1}{L} & \dfrac{1}{L} \end{bmatrix} dx$$

and a comparison with (9.110) shows that the element stiffnesses $\mathbf{K}^e$ for elements 1 and 2 are identical. This is certainly not surprising, since Ak and L are identical for the two elements. Moreover, it can easily be shown that the element load vector $\mathbf{f}_1^e$ for element 2 is equal to that of element 1, i.e. it is given by (9.113). From Figures 9.16 and 9.18 the topology data of element 2 is that local nodal point 1 corresponds to global nodal point 2 and local nodal point 2 corresponds to global nodal point 3. Therefore, adding (9.111) and (9.113) to the proper components of (9.114) results in

$$\mathbf{K} = \begin{bmatrix} 25 & -25 & 0 & 0 \\ -25 & 50 & -25 & 0 \\ 0 & -25 & 25 & 0 \\ 0 & 0 & 0 & 0 \end{bmatrix}; \quad \mathbf{f}_1 = \begin{bmatrix} 100 \\ 200 \\ 100 \\ 0 \end{bmatrix} \tag{9.115}$$

We then consider *element 3* with the local nodal points as shown in Figure 9.19.

It can easily be shown that the element stiffness matrix $\mathbf{K}^e$ and the element load vector $\mathbf{f}_1^e$ are again given by (9.111) and (9.113), respectively. According to Figures 9.16 and 9.19 the topology data for element 3 is that local point 1 corresponds to global nodal point 3 and local point 2 corresponds to global nodal point 4. Therefore, adding (9.111) and (9.113) to the proper components of (9.115) gives the following final result for the global stiffness matrix **K** and the global load vector f_l:

$$\mathbf{K} = \begin{bmatrix} 25 & -25 & 0 & 0 \\ -25 & 50 & -25 & 0 \\ 0 & -25 & 50 & -25 \\ 0 & 0 & -25 & 25 \end{bmatrix}; \quad \mathbf{f}_l = \begin{bmatrix} 100 \\ 200 \\ 200 \\ 100 \end{bmatrix} \tag{9.116}$$

As expected, these **K**- and f_l-matrices correspond exactly to those obtained in Examples 1 and 2 (cf. (9.29), (9.31) and (9.97)), respectively. Finally, we recall that the boundary vector $\mathbf{f}_b$ is obtained as in Example 1, i.e. it is given by (9.30).

9.3.2 *Use of local coordinate systems*

In order to standardize further the determination of the element stiffness matrix $\mathbf{K}^e$ and the element load vector $\mathbf{f}_l^e$, we may use a convenient *local coordinate system*. Since $\mathbf{K}^e$ and $\mathbf{f}_l^e$ are obtained by integration (cf. (9.103)), it is obvious that we may choose any coordinate transformation when these integrations are carried out.

To show this, we adopt the linear transformation

$$x = c\xi + d \tag{9.117}$$

where c and d are parameters and ξ is the new variable. As shown in Figure 9.20, let the coordinates of the left and right ends of the element be given by x_i and x_j, respectively, i.e. the length L of the element becomes $L = x_j - x_i$.

Now we choose the coordinate transformation (9.117), for instance, such that $\xi = -1$ and $\xi = 1$ at the left end and right end of the element, respectively. That is, in the local ξ-coordinate system, the element of Figure 9.20 takes the form shown in Figure 9.21.

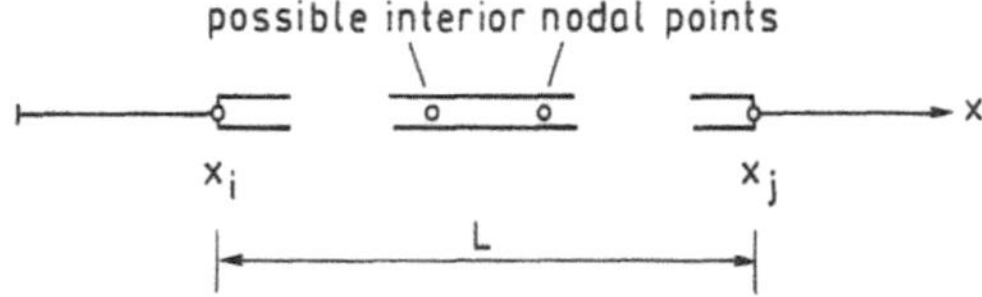

Figure 9.20   Geometry of one element in global coordinate system

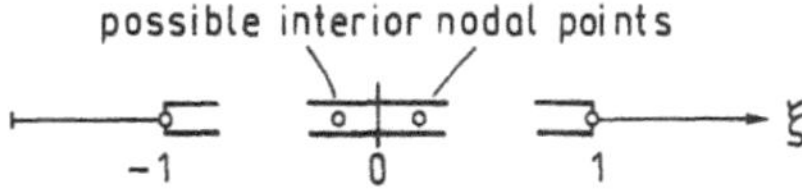

Figure 9.21   Geometry of one element in local ξ-coordinate system

This transformation is easily shown to be given by

$$x = \frac{L}{2}\xi + \frac{1}{2}(x_i + x_j) \tag{9.118}$$

which gives

$$dx = \frac{L}{2}d\xi \tag{9.119}$$

As an example, consider the simple linear element of Chapter 7. From (7.28) and (7.29) we have

$$\mathbf{N}^e = [N_i^e \quad N_j^e] \tag{9.120}$$

where

$$N_i^e(x) = -\frac{1}{L}(x - x_j) = -\frac{1}{L}\left[\frac{L}{2}\xi + \frac{1}{2}(x_i + x_j) - x_j\right] = \frac{1}{2}(1 - \xi)$$

$$N_j^e(x) = \frac{1}{L}(x - x_i) = \frac{1}{L}\left[\frac{L}{2}\xi + \frac{1}{2}(x_i + x_j) - x_i\right] = \frac{1}{2}(1 + \xi) \tag{9.121}$$

Moreover, from (7.33) we have that

$$\mathbf{B}^e = \frac{d\mathbf{N}^e}{dx} = \frac{d\mathbf{N}^e}{d\xi}\frac{d\xi}{dx} = \left[-\frac{1}{2} \quad \frac{1}{2}\right]\frac{2}{L} = \left[-\frac{1}{L} \quad \frac{1}{L}\right] \tag{9.122}$$

in accordance with (7.35). From (9.119) and (9.122) it follows that the element stiffness matrix $\mathbf{K}^e$ given by (9.103) is

$$\mathbf{K}^e = \int_{x_i}^{x_j} \mathbf{B}^{eT} Ak\mathbf{B}^e \, dx = \int_{-1}^{1}\begin{bmatrix} -1/L \\ 1/L \end{bmatrix} Ak \begin{bmatrix} -\frac{1}{L} & \frac{1}{L} \end{bmatrix}\frac{L}{2}d\xi$$

$$= \frac{1}{2L}\int_{-1}^{1} Ak\begin{bmatrix} 1 & -1 \\ -1 & 1 \end{bmatrix}d\xi$$

i.e.

$$\mathbf{K}^e = \frac{1}{2L}\begin{bmatrix} 1 & -1 \\ -1 & 1 \end{bmatrix}\int_{-1}^{1} Ak \, d\xi \tag{9.123}$$

If $Ak = $ constant we get

$$\boxed{\mathbf{K}^e = \frac{Ak}{L}\begin{bmatrix} 1 & -1 \\ -1 & 1 \end{bmatrix}} \tag{9.124}$$

From (9.119) and (9.121) it follows that the element load vector $\mathbf{f}_i^e$ given by (9.103) can be written as

$$\mathbf{f}_l^e = \int_{x_i}^{x_j} \mathbf{N}^{eT} Q \, dx = \frac{1}{2} \int_{-1}^{1} \begin{bmatrix} 1 - \xi \\ 1 + \xi \end{bmatrix} Q \frac{L}{2} \, d\xi \tag{9.125}$$

and if $Q = $ constant we get

$$\mathbf{f}_l^e = \frac{QL}{2} \begin{bmatrix} 1 \\ 1 \end{bmatrix} \tag{9.126}$$

It appears from (9.124) and (9.126) that for constant Ak- and Q-values the element stiffness matrix $\mathbf{K}^e$ and the element load vector $\mathbf{f}_l^e$ can be established directly using closed-form expressions. In an FE program such expressions clearly facilitate computational efficiency and similar closed-form solutions may be established for a sequence of typical forms of Ak and Q. Here we have only considered the simple linear element, but similar closed-form expressions may be derived for higher-order elements.

9.4 Axially loaded elastic bar

Consider the axially loaded elastic bar shown in Figure 9.22, where $A(x)$ is the cross-sectional area, $E(x)$ is Young's modulus and $b(x)$ is the body force per unit axial length.

With u being the displacement measured positive in the x-direction we have

$$u = \mathbf{Na} \tag{9.127}$$

where $\mathbf{a}$ now contains the displacements at the nodal points. The strain ε is given by

$$\varepsilon = \frac{du}{dx} = \mathbf{Ba} \tag{9.128}$$

In Chapter 4 it was shown that the equations for such a problem are similar to those of heat flow; cf. (4.8)–(4.10) with (4.18) and (4.20), (4.21). Referring to (9.12) and

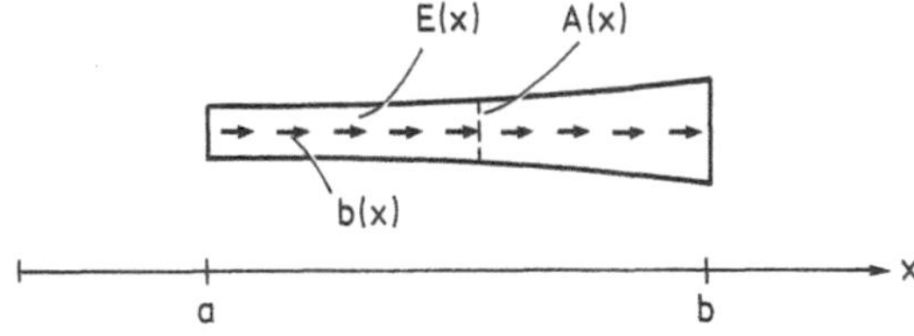

Figure 9.22   Axially loaded elastic bar

Table 4.1 we can therefore write the FE formulation directly as

$$\left(\int_a^b \mathbf{B}^T AE\mathbf{B} \, dx \right) \mathbf{a} = [\mathbf{N}^T A\sigma]_a^b + \int_a^b \mathbf{N}^T b \, dx \tag{9.129}$$

where σ is the normal stress. Care should be taken not to confuse the axial load b with the position $x = b$ at the right end of the bar. The force N in the bar is given by

$$N = A\sigma; \quad \sigma = E\varepsilon = E\frac{du}{dx} \tag{9.130}$$

N is seen to be positive if a tensile stress exists in the bar. Let us define the following matrices:

$$\boxed{\begin{aligned} \mathbf{K} &= \int_a^b \mathbf{B}^T AE\mathbf{B} \, dx \\ \mathbf{f}_b &= [\mathbf{N}^T A\sigma]_a^b \\ \mathbf{f}_l &= \int_a^b \mathbf{N}^T b \, dx \end{aligned}} \tag{9.131}$$

as well as

$$\boxed{\mathbf{f} = \mathbf{f}_b + \mathbf{f}_l} \tag{9.132}$$

Then (9.129) can be written as

$$\boxed{\mathbf{Ka} = \mathbf{f}} \tag{9.133}$$

where it is easily seen that the force vector $\mathbf{f}$ has the dimension of force, i.e. [N].

Next let us consider the FE formulation of one element. The element stiffness matrix $\mathbf{K}^e$ and the element load vector $\mathbf{f}_l^e$ now take the form

$$\boxed{\begin{aligned} \mathbf{K}^e &= \int_{L_e} \mathbf{B}^{eT} AE\mathbf{B}^e \, dx \\ \mathbf{f}_l^e &= \int_{L_e} \mathbf{N}^{eT} b \, dx \end{aligned}} \tag{9.134}$$

For constant AE- and b-values, these expression can, analogous to (9.124) and (9.126), be written as

$$\boxed{\mathbf{K}^e = k\begin{bmatrix} 1 & -1 \\ -1 & 1 \end{bmatrix}; \quad k = \frac{AE}{L}} \tag{9.135}$$

$$\boxed{\mathbf{f}_1^e = \frac{bL}{2}\begin{bmatrix} 1 \\ 1 \end{bmatrix}} \tag{9.136}$$

where k is the stiffness of the bar (cf. (3.38)).

For the heat flow problem, the global stiffness matrix $\mathbf{K}$ is singular (cf. (9.38)), whereas any submatrix $\underset{\sim}{\mathbf{K}}$ of $\mathbf{K}$ is non-singular (cf. (9.49)). The introduction of the boundary conditions by which at least one nodal temperature is specified implies that $\underset{\sim}{\mathbf{K}}$ could be established from $\mathbf{K}$. The present case of an axially loaded bar is quite similar and the stiffness matrix $\mathbf{K}$ of (9.131) is also singular. The introduction of boundary conditions in terms of the specification of at least one nodal displacement is now the tool which enables us to obtain a non-singular submatrix $\underset{\sim}{\mathbf{K}}$ from $\mathbf{K}$. This can be proved formally just as for heat flow, but the need to specify at least one nodal displacement is physically obvious since it removes possible rigid-body motions. We recall that a rigid-body motion does not create any strains.

9.4.1 *Example 5*

With the discussion of the element stiffness matrix $\mathbf{K}^e$ and the element load vector $\mathbf{f}_i^e$ and the introduction of a convenient local coordinate system for the establishment of these matrices, we are in a position to illustrate the details of the efficient implementation of the FE equations into a computer program. For this purpose consider the axially loaded elastic bar shown in Figure 9.23. The bar consists of two parts, 1 and 2, each having a constant Young's modulus E, constant cross-sectional area A and constant external load b.

This problem is analyzed using the two simple linear elements shown in Figure 9.24.

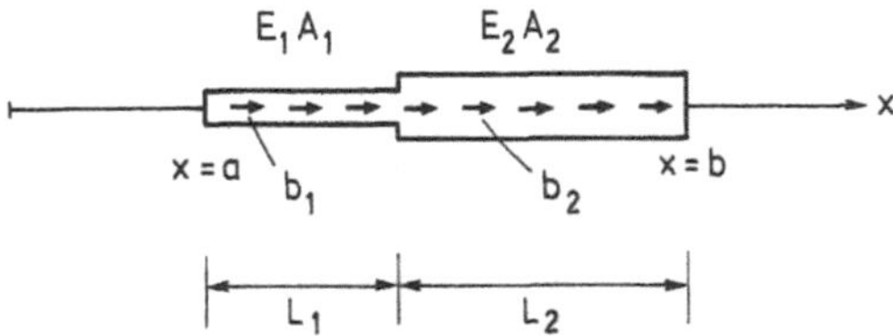

Figure 9.23 Configuration of axially loaded elastic bar

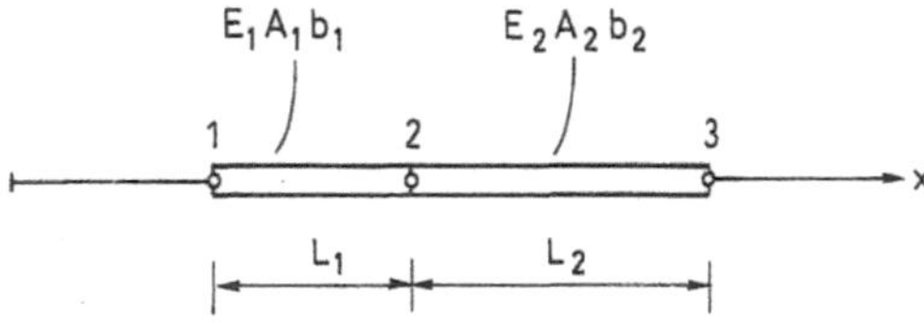

Figure 9.24 Two simple linear elements used to model the problem of Figure 9.23

Let us consider first the boundary vector $\mathbf{f}_b$ given by (9.131). As in (9.24) we obtain

$$\mathbf{f}_b = \begin{bmatrix} -(A\sigma)_{x=a} \\ 0 \\ (A\sigma)_{x=b} \end{bmatrix} \tag{9.137}$$

Next we evaluate the element with length L_1. From (9.135) and (9.136) we have

$$\mathbf{K}^e = k_1 \begin{bmatrix} 1 & -1 \\ -1 & 1 \end{bmatrix}; \quad \mathbf{f}_l^e = \frac{b_1 L_1}{2} \begin{bmatrix} 1 \\ 1 \end{bmatrix}$$

where $k_1 = A_1 E_1 / L_1$. The topology data follows from Figure 9.24; that is, after the contributions for the element with length L_1, $\mathbf{K}$ and $\mathbf{f}_l$ are given by

$$\mathbf{K} = \begin{bmatrix} k_1 & -k_1 & 0 \\ -k_1 & k_1 & 0 \\ 0 & 0 & 0 \end{bmatrix}; \quad \mathbf{f}_l = \frac{1}{2} \begin{bmatrix} b_1 L_1 \\ b_1 L_1 \\ 0 \end{bmatrix}$$

For the element with length L_2 we obtain

$$\mathbf{K}^e = k_2 \begin{bmatrix} 1 & -1 \\ -1 & 1 \end{bmatrix}; \quad \mathbf{f}_l^e = \frac{b_2 L_2}{2} \begin{bmatrix} 1 \\ 1 \end{bmatrix}$$

where $k_2 = A_2 E_2 / L_2$. With Figure 9.24, after the contributions of this element we obtain

$$\mathbf{K} = \begin{bmatrix} k_1 & -k_1 & 0 \\ -k_1 & k_1 + k_2 & -k_2 \\ 0 & -k_2 & k_2 \end{bmatrix}; \quad \mathbf{f}_l = \frac{1}{2} \begin{bmatrix} b_1 L_1 \\ b_1 L_1 + b_2 L_2 \\ b_2 L_2 \end{bmatrix} \tag{9.138}$$

From (9.137) and (9.138) the final FE equations are

$$\begin{bmatrix} k_1 & -k_1 & 0 \\ -k_1 & k_1 + k_2 & -k_2 \\ 0 & -k_2 & k_2 \end{bmatrix} \begin{bmatrix} u_1 \\ u_2 \\ u_3 \end{bmatrix} = \begin{bmatrix} -(A\sigma)_{x=a} \\ 0 \\ (A\sigma)_{x=b} \end{bmatrix} + \frac{1}{2} \begin{bmatrix} b_1 L_1 \\ b_1 L_1 + b_2 L_2 \\ b_2 L_2 \end{bmatrix} \tag{9.139}$$

The boundary conditions are now introduced. We assume that the bar is fixed at $x = a$ and that the right end of the bar is free, i.e.

$$u_1 = u_{x=a} = 0; \quad \sigma_{x=b} = 0 \tag{9.140}$$

The solution becomes

$$u_2 = \frac{1}{2k_1}(b_1 L_1 + 2b_2 L_2); \quad u_3 = \frac{1}{2k_1 k_2}[k_2 b_1 L_1 + (k_1 + 2k_2)b_2 L_2] \tag{9.141}$$

$$N_{x=a} = (A\sigma)_{x=a} = b_1 L_1 + b_2 L_2 \tag{9.142}$$

It appears that $N_{x=a}$ is that tensile force – namely, the reaction – which must be applied at $x = a$ in order to maintain equilibrium of the bar.

9.5 Basic features of an FE computer program

We are now in a position to sketch the *computer flow diagram* for an FE program as follows:

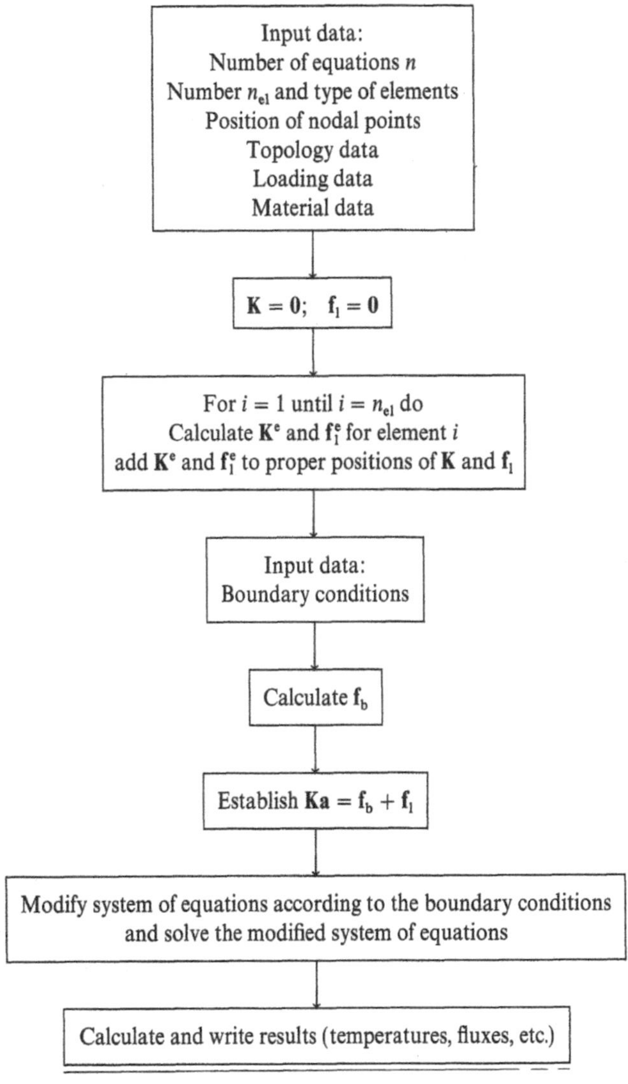

First a sequence of input data is read. Then the total stiffness matrix $\mathbf{K}$ and the total load vector $\mathbf{f}_l$ are *initialized*, i.e. the matrix dimensions are identified and all their components are set to zero. For each element, the element stiffness matrix $\mathbf{K}^e$ and the element load vector $\mathbf{f}_l^e$ are calculated and added to the proper components of $\mathbf{K}$ and $\mathbf{f}_l$ using the topology data. This is the assembling process. The boundary conditions are read and the boundary vector $\mathbf{f}_b$ is calculated. Those components of $\mathbf{f}_b$ that are unknown (i.e. the 'reactions') are included in the matrix $\mathbf{r}$ of (9.57). The entire system of equations is established, and after the introduction of the boundary conditions, the modified system of equations is solved according to (9.58). The component $\mathbf{r}$ of $\mathbf{f}$ which was originally unknown is then calculated using (9.59), i.e. the 'reactions' can now also be calculated. Finally, the results in terms of temperature, fluxes and 'reactions' are calculated and written out.

In practice, a so-called *pre-processor* is often used in connection with the input to the program. This pre-processor may perform an automatic division into elements following some rules specified by the user, and it may also provide graphical information on the element mesh, nodal points, etc. This facilitates the possibilities for ensuring that correct input data is used. Likewise, a so-called *post-processor* is often used to provide graphical information on the results, for instance in terms of contour curves for the temperature etc.

9.6 Heat flow with convection

Generally, fluid motion around a body is a very effective means of transporting heat between the body and the fluid. This type of heat transport is called *convection*. Assume that the fluid far away from the body has the uniform temperature T_∞. The temperature at the surface of the body is given by T and it is natural to expect that the temperature difference $T - T_\infty$ controls the amount of heat flow between the body and the fluid. This leads to *Newton's convection boundary condition* expressed through

$$\boxed{q_n = \alpha(T - T_\infty)} \tag{9.143}$$

where q_n is the flux through the boundary with the outer unit normal vector $\mathbf{n}$ (cf. Figure 6.1 and (6.3)) and α is a parameter, the so-called *convection coefficient* with the dimension $[\mathrm{J/m^2\,s\,°C}]$. We note the similarity of (9.143) and (6.31). The convection coefficient α depends on a variety of circumstances such as fluid velocity, type of fluid, surface conditions, etc. but for our present purpose we assume that the convection coefficient α is known. Moreover, α is a positive quantity, implying that when the temperature difference $T - T_\infty$ is positive, the flux q_n as given by (9.143) becomes positive. This is in accordance with the definition of the flux q_n (cf. Figure 6.1), since a positive q_n means that heat leaves the body.

As an example of (9.143) assume that convection occurs at the right end of the

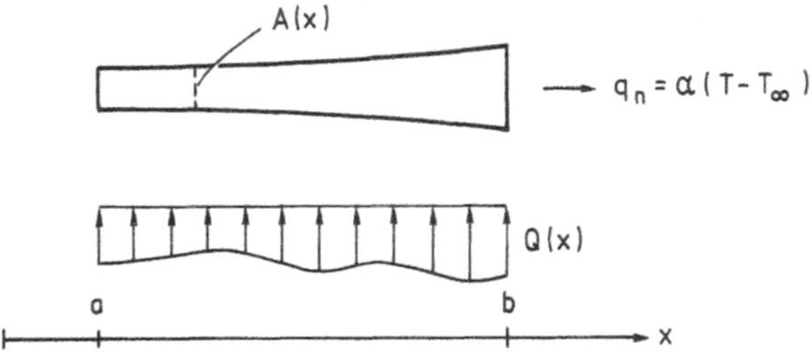

Figure 9.25   Convection at the right end of the fin

fin as shown in Figure 9.25. According to (9.14) the FE formulation is

$$\mathbf{Ka} = \mathbf{f}_b + \mathbf{f}_l \tag{9.144}$$

The boundary vector $\mathbf{f}_b$ given by (9.13) can be written as

$$\mathbf{f}_b = -[\mathbf{N}^T Aq]_a^b = -(\mathbf{N}^T Aq)_{x=b} + (\mathbf{N}^T Aq)_{x=a} \tag{9.145}$$

As we have one-dimensional heat flow, the flux q_n of Figure 9.25 is $q_n = q(x = b)$, i.e. (9.143) takes the following form of a boundary condition:

$$(q)_{x=b} = \alpha(T_{x=b} - T_\infty) \tag{9.146}$$

We note that if convection occurs at the left end of the fin, then $q_n = -q(x = a)$. In general, we have the approximation $T = \mathbf{Na}$, i.e. we may write the unknown temperature $T_{x=b}$ as

$$T_{x=b} = \mathbf{N}_{x=b}\mathbf{a}$$

This implies that (9.146) becomes

$$(q)_{x=b} = \alpha\mathbf{N}_{x=b}\mathbf{a} - \alpha T_\infty$$

The boundary vector $\mathbf{f}_b$ given by (9.145) can then be written as

$$\mathbf{f}_b = -\alpha(A\mathbf{N}^T\mathbf{N})_{x=b}\mathbf{a} + \alpha T_\infty(A\mathbf{N}^T)_{x=b} + (\mathbf{N}^T Aq)_{x=a}$$

Use of this expression in (9.144) gives

$$\boxed{\begin{aligned} (\mathbf{K} + \mathbf{K}_c)\mathbf{a} &= \alpha T_\infty(A\mathbf{N}^T)_{x=b} + (\mathbf{N}^T Aq)_{x=a} + \mathbf{f}_l \\ \text{where} \quad \mathbf{K}_c &= \alpha(A\mathbf{N}^T\mathbf{N})_{x=b} \end{aligned}} \tag{9.147}$$

It may be of interest to evaluate the expression for $\mathbf{K}_c$. The number of nodal points for the entire body is given by n. Assuming that nodal point n is located at $x = b$ we obtain

$$\mathbf{N}_{x=b} = [0 \quad 0 \quad \dots \quad 1]$$

i.e.

$$\mathbf{K}_c = \alpha(A\mathbf{N}^\mathrm{T}\mathbf{N})_{x=b} = \alpha A_b \begin{bmatrix} 0 & 0 & \cdots & 0 \\ 0 & 0 & \cdots & 0 \\ \vdots & \vdots & & \vdots \\ 0 & 0 & \cdots & 1 \end{bmatrix} \tag{9.148}$$

Expression (9.147) shows that the convection condition implies a modification of the stiffness matrix. The new stiffness matrix is symmetric just like **K**, but it is important to observe that in contrast to **K** the modified stiffness matrix is non-singular, i.e.

$$\boxed{\det(\mathbf{K} + \mathbf{K}_c) \neq 0} \tag{9.149}$$

To show this, consider the quadratic form I given by

$$I = \mathbf{a}^\mathrm{T}(\mathbf{K} + \mathbf{K}_c)\mathbf{a} = \mathbf{a}^\mathrm{T}\mathbf{k}\mathbf{a} + \mathbf{a}^\mathrm{T}\mathbf{K}_c\mathbf{a} \tag{9.150}$$

Referring to (9.41) and (9.43) we have $\mathbf{a}^\mathrm{T}\mathbf{Ka} = 0$ if and only if $dT/dx = 0$, i.e. if the components of **a** are equal; otherwise $\mathbf{a}^\mathrm{T}\mathbf{Ka} > 0$. From (9.150) and (9.148) it therefore follows that

$$I > 0 \quad \text{for all } \mathbf{a} \neq 0 \tag{9.151}$$

That is, the modified stiffness matrix is positive definite and, according to (2.67), it is then non-singular. This proves that (9.149) is correct and implies that the system of equations (9.147) can be solved directly irrespective of the boundary condition at the left end of the fin.

As another important engineering problem of convection, assume that lateral convection occurs along the outer surface of the fin as shown in Figure 9.26. This type of convection may cause conceptual difficulties because it can only occur if heat flows in directions perpendicular to the x-axis and this means that the heat flow is

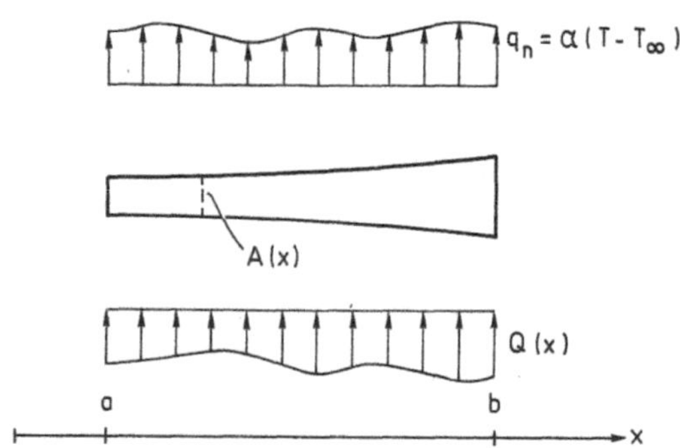

Figure 9.26   Lateral convection along the outer surface of the fin

not truly one-dimensional. Actually, this situation is similar to the discussion related to the heat supply Q, and referring to the discussion on page 48 we accept the engineering approximation of a one-dimensional heat flow in the x-direction provided that the temperature variations along the fin are much larger than temperature variations normal to the fin. This assumption is usually realistic if the fin is thin.

With this discussion it is obvious that we may consider the heat flux q_n due to convection as a modification of the heat supply given by Q. Now Q is the heat supply per unit time and per unit length of the fin. Moreover, Q is positive when heat is supplied to the fin. In contrast, q_n is the heat per unit time and per unit area and q_n is positive when heat leaves the fin. Let P be the perimeter of the fin; that is, $q_n P$ is the heat per unit time and per unit length of the fin. It appears that we may treat the convection as a modification of Q according to the scheme

$$Q \circlearrowright Q - q_n P \tag{9.152}$$

which means that Q should be replaced by the term $Q - q_n P$. The FE formulation is again given by (9.14), i.e.

$$\mathbf{Ka} = \mathbf{f}_b + \mathbf{f}_l \tag{9.153}$$

where the load vector $\mathbf{f}_l$ is given by (9.13), i.e.

$$\mathbf{f}_l = \int_a^b \mathbf{N}^T Q \, dx \tag{9.154}$$

Combining (9.152) and (9.143) gives

$$Q \circlearrowright Q - \alpha P T + \alpha P T_\infty$$

The unknown temperature T can be replaced by $T = \mathbf{Na}$ to obtain

$$Q \circlearrowright Q - \alpha P \mathbf{Na} + \alpha P T_\infty \tag{9.155}$$

With (9.153)–(9.155) we obtain

$$\boxed{\begin{aligned} (\mathbf{K} + \mathbf{K}_c)\mathbf{a} &= \mathbf{f}_b + \mathbf{f}_l + \int_a^b \alpha P T_\infty \mathbf{N}^T \, dx \\[2mm] \text{where} \quad \mathbf{K}_c &= \int_a^b \alpha P \mathbf{N}^T \mathbf{N} \, dx \end{aligned}} \tag{9.156}$$

and $\mathbf{f}_b$ and $\mathbf{f}_l$ are given in the usual manner by (9.13). Again it appears that the convection results in a modification of the stiffness matrix. The modified stiffness matrix is symmetric, and in a manner similar to (9.150) and (9.151) it can easily be proved that this matrix is non-singular, i.e.

$$\boxed{\det(\mathbf{K} + \mathbf{K}_c) \neq 0} \tag{9.157}$$

Hence the system of equations (9.156) can be solved directly irrespective of the boundary conditions imposed on the ends of the fin.

We finally observe that with the results above it is straightforward to combine the convection cases of Figures 9.25 and 9.26.

9.7 C^0-continuity

We have already established criteria in Chapter 7 which ensure that the FE formulation converges towards the exact solution for decreasing element size. According to page 93, these convergence criteria consist of the fulfilment of completeness and compatibility. The compatibility or conforming requirement stated that the approximation of the temperature also varies in a continuous manner over element boundaries.

Let us now derive this continuity requirement in a different manner. It is not surprising that the fulfilment of the convergence criteria hinges on the expressions chosen for the approximation, i.e. on the global shape function matrix $\mathbf{N}$. Referring to the global stiffness matrix $\mathbf{K}$, the boundary vector $\mathbf{f_b}$ and the load vector $\mathbf{f_l}$ given by (9.13), it appears that $\mathbf{K}$ places the most severe conditions on $\mathbf{N}$ since the derivative $\mathbf{B} = d\mathbf{N}/dx$ enters the expression for $\mathbf{K}$. A component K_{ij} of $\mathbf{K}$ is given by (9.19), i.e.

$$K_{ij} = \int_a^b \frac{dN_i}{dx} Ak \frac{dN_j}{dx} dx \tag{9.158}$$

If the terms Ak and dN_i/dx are continuous functions, the integration given by (9.158) can be carried out directly. However, if discontinuities occur we may run into trouble. In order to evaluate this situation, assume that the global shape function N_i varies in the manner shown in Figure 9.27. It appears that N_i is continuous whereas dN_i/dx is discontinuous over the element boundary. Moreover, even though dN_i/dx varies discontinuously, an integration which involves dN_i/dx is still well defined.

To investigate whether it is also possible to make a well-defined integration of $d^2 N_i/dx^2$, we replace N_i by the smooth dashed curve shown in Figure 9.27(a). The corresponding variations of dN_i/dx and $d^2 N_i/dx^2$ are shown in Figure 9.27(b) and (c), respectively, and it follows that an integration involving $d^2 N_i/dx^2$ is not, in general, well defined. We may draw the conclusion that in order for the stiffness matrix to be established in a well-defined manner, we are led to the compatibility requirement stating that the approximation of the temperature must vary in a continuous manner over the element boundaries. A function which is continuous but which has a discontinuous first derivative is termed a C^0-*continuous function*. In general, we have the following statement:

C^n-continuity means that the n-derivative is continuous.

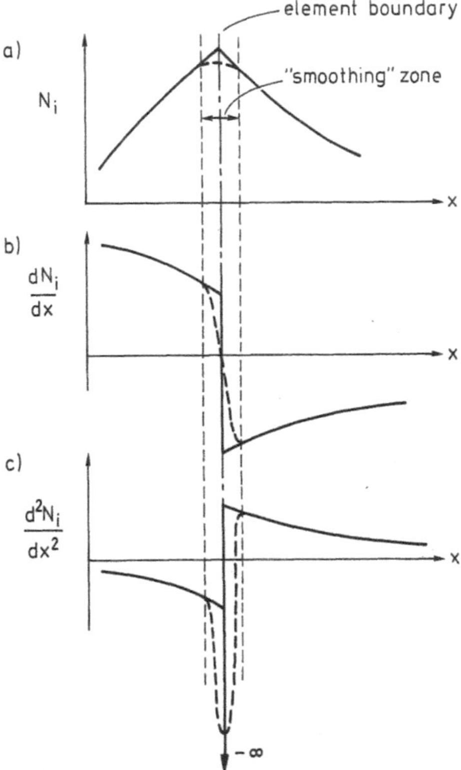

Figure 9.27   Differentiation of shape function with C^0-continuity

Here we conclude that C^0-continuity is sufficient for the global shape functions of heat flow problems, but it is obvious that the continuity required depends on the differential equations involved. We shall return to this topic in later chapters.

9.8 Concluding remarks

In principle, even though this chapter has been devoted to simple physical problems, it contains all the ingredients of an FE formulation. In later chapters more complicated physical problems will be treated, but the fundamental FE results are analogous to those discussed here. It is for this reason that we have presented a rather detailed discussion of various typical issues related to the FE formulation. As these matters are easily generalized, we will adopt in later chapters a much more compact method of presentation for establishing the different FE formulations.

Let us therefore review some of the fundamental facts. The FE formulation can always be given in the form of (9.16), (9.15) and (9.13) and this is typically the format

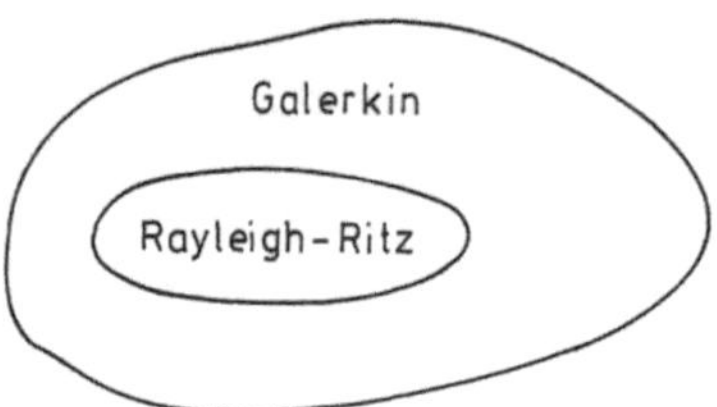

Figure 9.28   Relation between the Rayleigh–Ritz method and the Galerkin method

chosen when discussing theoretical aspects. Alternatively, the global stiffness matrix $\mathbf{K}$ and the global load vector $\mathbf{f}_l$ may be derived as the sum of the expanded element stiffness matrix $\mathbf{K}^{ee}$ and the expanded element load vector $\mathbf{f}_l^{ee}$, respectively, for all the individual elements (cf. (9.85)). Conceptually (9.85) is of fundamental importance, but when establishing a computer program for the FE method, these expanded forms are extremely inconvenient to work with, so, instead, the element stiffness matrix $\mathbf{K}^e$ and the element load vector $\mathbf{f}_l^e$ are derived using only those degrees of freedom related to each individual element (cf. (9.103)). The contributions of $\mathbf{K}^e$ and $\mathbf{f}_l^e$ are then added to the global stiffness matrix $\mathbf{K}$ and the global load vector $\mathbf{f}_l$, respectively, using the topology data. We also note that the boundary vector $\mathbf{f}_b$ is always derived using the form given by (9.13).

It appears that the FE method is extremely well suited for efficient implementation on a computer, which enables us to deal with a variety of problems.

As a global shape function N_i only differs from zero in those elements which contain the nodal point i, the global stiffness matrix $\mathbf{K}$ was found to be banded. Moreover, the Galerkin method implies that $\mathbf{K}$ is symmetric. We found that $\mathbf{K}$ is singular and positive semi-definite; see (9.38), (9.41) and (9.43). We proved that any submatrix $\underset{\sim}{\mathbf{K}}$ obtained from $\mathbf{K}$ by deleting one or more rows and the corresponding columns is symmetric, non-singular and positive definite; see (9.45), (9.48) and (9.49). These aspects have important consequences when the boundary conditions are prescribed. We also showed that, apart from problems involving convection, the specification of at least one nodal temperature is a necessity in order to obtain a unique solution of the FE equations. Moreover, the balance equation for the entire body is fulfilled through relation (9.68).

In this textbook, the Galerkin method is used universally, but we remark that, as an alternative, one may also adopt the so-called *Rayleigh–Ritz method*. This method relies on the construction of certain functionals for which a minimum is sought using the *calculus of variations*. If such a functional can be established, it turns out that the Rayleigh–Ritz and Galerkin methods result in identical FE formulations. For a discussion of these topics the reader is referred to Hughes (1987), Strang and Fix (1973) and Zienkiewicz and Taylor (1989). However, the important point is that whereas the Rayleigh–Ritz method can be used only for certain types of differential equations, the Galerkin method is applicable to arbitrary differential equations. This means that the Rayleigh–Ritz method can be viewed as a subset of the Galerkin method and this fact is illustrated in Figure 9.28. We also note that, in addition to being more general, the Galerkin method often provides a more straightforward derivation of the FE equations than the Rayleigh–Ritz method.

We finally draw the reader's attention to the comprehensive textbooks by Becker *et al.* (1981), Hughes (1987) and Stasa (1985) that also treat one-dimensional heat flow.

10

FE formulation of two- and three-dimensional heat flow

Following the detailed discussion of the FE formulation for one-dimensional heat flow, it is an easy task to treat two- and three-dimensional heat flow. We emphasize that even though we focus on heat flow here, a sequence of other important physical problems is governed by similar equations (see Table 6.1). Therefore, the FE formulation presented in this chapter has significance in a variety of other fields.

10.1 Two-dimensional heat flow

Considering two-dimensional heat flow first, the strong form was given by (6.32)–(6.34) and the corresponding weak form by

$$\int_A (\nabla v)^{\mathrm{T}} t\, \mathbf{D} \nabla T\, \mathrm{d}A = -\int_{\mathscr{L}_h} vht\, \mathrm{d}\mathscr{L} - \int_{\mathscr{L}_g} vq_n t\, \mathrm{d}\mathscr{L} + \int_A vQt\, \mathrm{d}A \tag{10.1}$$

$$T = g \quad \text{on } \mathscr{L}_g \tag{10.2}$$

(cf. (6.46) and (6.47)). In (10.1) and (10.2) we recall that v is the arbitrary weight function, $\mathbf{D}$ the constitutive matrix, T the temperature, t the thickness and Q the heat supply per unit time and per unit volume of the body. The flux vector $\mathbf{q}$ is determined from Fourier's law (6.4), i.e.

$$\mathbf{q} = -\mathbf{D}\nabla T \tag{10.3}$$

where the temperature gradient ∇T is

$$\nabla T = \begin{bmatrix} \dfrac{\partial T}{\partial x} \\[2ex] \dfrac{\partial T}{\partial y} \end{bmatrix} \tag{10.4}$$

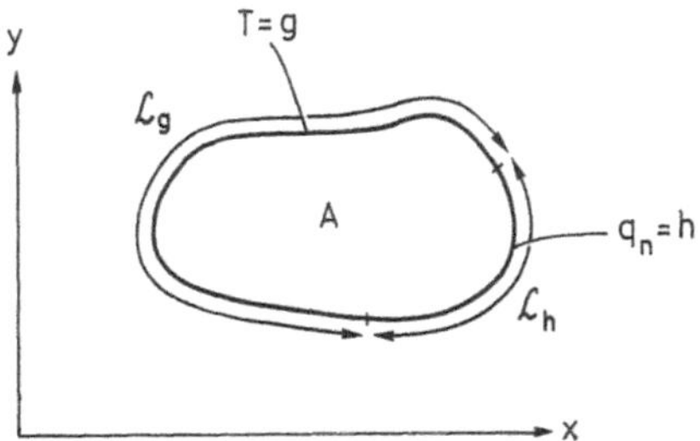

Figure 10.1　Two-dimensional region A with boundary $\mathcal{L} = \mathcal{L}_h + \mathcal{L}_g$

Moreover, the flux q_n at the boundary of the region is given by Figure 6.1 and (6.3), i.e.

$$q_n = \mathbf{q}^\mathrm{T}\mathbf{n} \tag{10.5}$$

and q_n is positive when heat leaves the body. As shown in Figure 10.1, the boundary $\mathcal{L}$ of the region A consists of $\mathcal{L}_h$ and $\mathcal{L}_g$. On $\mathcal{L}_h$ we have $q_n = h$ where h is a known quantity and on $\mathcal{L}_g$ we have $T = g$, where g is a known quantity. With these preliminary remarks we are in a position to obtain the FE formulation.

The temperature T is approximated by means of (7.137), i.e.

$$T = \mathbf{Na} \tag{10.6}$$

where $\mathbf{N}$ is the global shape function matrix and $\mathbf{a}$ contains the temperatures at the nodal points in the entire body. This means that

$$\mathbf{N} = [N_1 \quad N_2 \quad \ldots \quad N_n]; \quad \mathbf{a} = \begin{bmatrix} T_1 \\ T_2 \\ \vdots \\ T_n \end{bmatrix} \tag{10.7}$$

where n is the number of nodal points for the entire body and a component N_i depends on x and y, i.e. $N_i = N_i(x, y)$. From (10.6) we obtain

$$\nabla T = \mathbf{Ba} \quad \text{where} \quad \mathbf{B} = \nabla\mathbf{N} \tag{10.8}$$

which implies that

$$\mathbf{B} = \begin{bmatrix} \dfrac{\partial N_1}{\partial x} & \dfrac{\partial N_2}{\partial x} & \cdots & \dfrac{\partial N_n}{\partial x} \\[2mm] \dfrac{\partial N_1}{\partial y} & \dfrac{\partial N_2}{\partial y} & \cdots & \dfrac{\partial N_n}{\partial y} \end{bmatrix} \tag{10.9}$$

in accordance with (7.140) and (7.142). Inserting (10.8) into (10.1) gives

$$\left(\int_A (\nabla v)^\mathrm{T}\mathbf{DB}t \, \mathrm{d}A \right)\mathbf{a} = -\int_{\mathcal{L}_h} vht \, \mathrm{d}\mathcal{L} - \int_{\mathcal{L}_g} vq_n t \, \mathrm{d}\mathcal{L} + \int_A vQt \, \mathrm{d}A \tag{10.10}$$

The final step is to choose the arbitrary weight function v. In accordance with the Galerkin method we set

$$v = \mathbf{Nc} \tag{10.11}$$

Since v is arbitrary, the matrix $\mathbf{c}$ is arbitrary. From (10.11) we obtain

$$\nabla v = \mathbf{Bc} \tag{10.12}$$

As $v = v^{\mathrm{T}}$, (10.11) can also be written as

$$v = \mathbf{c}^{\mathrm{T}}\mathbf{N}^{\mathrm{T}} \tag{10.13}$$

Inserting (10.12) and (10.13) into (10.10), and noting that $\mathbf{c}$ is independent of position, gives

$$\mathbf{c}^{\mathrm{T}}\left[\left(\int_A \mathbf{B}^{\mathrm{T}}\mathbf{DB}t\,\mathrm{d}A\right)\mathbf{a} + \int_{\mathscr{L}_h} \mathbf{N}^{\mathrm{T}}ht\,\mathrm{d}\mathscr{L} + \int_{\mathscr{L}_g} \mathbf{N}^{\mathrm{T}}q_n t\,\mathrm{d}\mathscr{L} - \int_A \mathbf{N}^{\mathrm{T}}Qt\,\mathrm{d}A\right] = 0$$

As this expression should hold for arbitrary $\mathbf{c}^{\mathrm{T}}$-matrices, we conclude that

$$\left(\int_A \mathbf{B}^{\mathrm{T}}\mathbf{DB}t\,\mathrm{d}A\right)\mathbf{a} = -\int_{\mathscr{L}_h} \mathbf{N}^{\mathrm{T}}ht\,\mathrm{d}\mathscr{L} - \int_{\mathscr{L}_g} \mathbf{N}^{\mathrm{T}}q_n t\,\mathrm{d}\mathscr{L} + \int_A \mathbf{N}^{\mathrm{T}}Qt\,\mathrm{d}A \tag{10.14}$$

which is the FE formulation sought.

To write (10.14) in a more compact fashion, we define the following matrices:

$$\boxed{\begin{aligned}
\mathbf{K} &= \int_A \mathbf{B}^{\mathrm{T}}\mathbf{DB}t\,\mathrm{d}A \\[2mm]
\mathbf{f}_b &= -\int_{\mathscr{L}_h} \mathbf{N}^{\mathrm{T}}ht\,\mathrm{d}\mathscr{L} - \int_{\mathscr{L}_g} \mathbf{N}^{\mathrm{T}}q_n t\,\mathrm{d}\mathscr{L} \\[2mm]
\mathbf{f}_l &= \int_A \mathbf{N}^{\mathrm{T}}Qt\,\mathrm{d}A
\end{aligned}} \tag{10.15}$$

As $\mathbf{D}$ has the dimension 2×2 and $\mathbf{B}$ the dimension $2 \times n$, it follows that $\mathbf{K}$ is a square matrix with dimension $n \times n$ and it is the *stiffness matrix*. Likewise, both $\mathbf{f}_b$ and $\mathbf{f}_l$ have the dimension $n \times 1$ and they are termed the *boundary vector* and *load vector*, respectively. With (10.15), (10.14) can be written as

$$\boxed{\mathbf{Ka} = \mathbf{f}_b + \mathbf{f}_l} \tag{10.16}$$

We define the *force vector* $\mathbf{f}$ by

$$\boxed{\mathbf{f} = \mathbf{f}_b + \mathbf{f}_l} \tag{10.17}$$

i.e. (10.16) becomes

$$\boxed{\mathbf{Ka} = \mathbf{f}} \tag{10.18}$$

Just like one-dimensional heat flow, the force vector $\mathbf{f}$ has the dimension of [J/s] and a comparison with the corresponding formulation for one-dimensional heat flow (cf. (9.13)–(9.16)) shows that a complete similarity exists.

Owing to the symmetry of the constitutive matrix $\mathbf{D}$ (cf. (6.10)) it follows from (10.15) that $\mathbf{K}$ is symmetric

$$\boxed{\mathbf{K} = \mathbf{K}^\mathrm{T}} \tag{10.19}$$

Similar to the evaluation of (9.38)–(9.40) it follows that $\mathbf{K}$ is singular, i.e.

$$\boxed{\det \mathbf{K} = 0} \tag{10.20}$$

Moreover, since $\mathbf{D}$ is positive definite (cf. (6.8)), we derive in a similar manner to (9.41)–(9.43) that $\mathbf{K}$ is positive semi-definite, i.e.

$$\boxed{\mathbf{a}^\mathrm{T}\mathbf{Ka} \geq 0} \tag{10.21}$$

for all $\mathbf{a} \neq 0$ and that only if the temperature gradient is zero, i.e. $\nabla T = \mathbf{Ba} = \mathbf{0}$, is the quadratic form above equal to zero. Finally, using the approach discussed in relation to (9.44)–(9.49) it can easily be shown that any submatrix $\underset{\sim}{\mathbf{K}}$ obtained from $\mathbf{K}$ by deleting one or more rows and the corresponding columns is symmetric, non-singular and positive definite. With reference to Chapter 9, we conclude that at least one nodal temperature has to be prescribed in order to obtain a unique solution of the FE equations. A systematic consideration of the boundary conditions is again given by the approach outlined in (9.57)–(9.59).

Let us now prove that the components of the force vector again fulfil relation (9.68). As we consider a region A with thickness t, the balance principle states that

$$\int_A Qt \, \mathrm{d}A = \oint_{\mathscr{L}} q_n t \, \mathrm{d}\mathscr{L} \tag{10.22}$$

(cf. (6.19)). Let us write (10.17) as

$$f_i = f_{\mathrm{b}i} + f_{\mathrm{l}i}; \quad i = 1, \ldots, n$$

which leads to

$$\sum_{i=1}^{n} f_i = \sum_{i=1}^{n} f_{\mathrm{b}i} + \sum_{i=1}^{n} f_{\mathrm{l}i} \tag{10.23}$$

According to (10.15) we have that

$$f_{\mathrm{b}i} = -\int_{\mathscr{L}_h} N_i ht \, \mathrm{d}\mathscr{L} - \int_{\mathscr{L}_g} N_i q_n t \, \mathrm{d}\mathscr{L} \tag{10.24}$$

We recall that the boundary conditions specify the flux $q_n = h$ along $\mathscr{L}_h$, whereas the flux q_n along $\mathscr{L}_g$ is unspecified beforehand. Therefore, (10.24) may be rewritten as

$$f_{\mathrm{b}i} = -\oint_{\mathscr{L}} N_i q_n t \, \mathrm{d}\mathscr{L} \tag{10.25}$$

From (10.15) a component of the load vector is given by

$$f_{\mathrm{l}i} = \int_A N_i Q t \, \mathrm{d}A \tag{10.26}$$

Using (10.25) and (10.26) in (10.23) leads to

$$\sum_{i=1}^{n} f_i = -\oint_{\mathscr{L}} \left(\sum_{i=1}^{n} N_i \right) q_n t \, \mathrm{d}\mathscr{L} + \int_A \left(\sum_{i=1}^{n} N_i \right) Q t \, \mathrm{d}A$$

which with (7.139) results in

$$\sum_{i=1}^{n} f_i = -\oint_{\mathscr{L}} q_n t \, \mathrm{d}\mathscr{L} + \int_A Q t \, \mathrm{d}A \tag{10.27}$$

A comparison with (10.22) shows that

$$\boxed{\sum_{i=1}^{n} f_i = 0} \tag{10.28}$$

in accordance with (9.68). This means that the balance principle for the body is expressed by the fact that the sum of the components of the force vector $\mathbf{f}$ is equal to zero. We emphasize that (10.28) holds *exactly* even though the FE method is an approximate approach.

The components of the load vector $\mathbf{f}_{\mathrm{l}}$ are given by (10.26). Let us assume now that the load Q is given in terms of a *line source*, where the heat supply is concentrated to a point in the xy-plane and at this point a heat supply Q_{s} per unit time and per unit thickness of the region is prescribed. Q_{s} is the strength of the line source and has the dimension $[\mathrm{J/s\,m}]$. In such a situation, we may express Q using Dirac's delta function in a similar way to (9.70)–(9.72), i.e.

$$Q = Q_{\mathrm{s}}\delta(x - a)\delta(y - b) = \begin{cases} \infty & \text{if } (x, y) = (a, b) \\ 0 & \text{otherwise} \end{cases} \tag{10.29}$$

where (a, b) is the position of the line source. By definition, we have that

$$\int_{-\infty}^{\infty} \int_{-\infty}^{\infty} Q \, \mathrm{d}x \, \mathrm{d}y = \int_{-\infty}^{\infty} \int_{-\infty}^{\infty} Q_{\mathrm{s}}\delta(x - a)\delta(y - b) \, \mathrm{d}x \, \mathrm{d}y = Q_{\mathrm{s}} \tag{10.30}$$

As a result of (10.29), (10.30) may also be written as

$$\int_{b-}^{b+} \int_{a-}^{a+} Q_{\mathrm{s}}\delta(x - a)\delta(y - b) \, \mathrm{d}x \, \mathrm{d}y = Q_{\mathrm{s}} \tag{10.31}$$

For such a line source, the component f_{1i} of the load vector given by (10.26) becomes

$$f_{1i} = \int_A N_i(x, y)Qt \, \mathrm{d}A = \int_{b^-}^{b^+} \int_{a^-}^{a^+} N_i(x, y)Q_s\delta(x - a)\delta(y - b)t \, \mathrm{d}x \, \mathrm{d}y$$

$$= N_i(a, b)t(a, b)Q_s \tag{10.32}$$

If the position (a, b) of the line source coincides with the position of nodal point i, then $N_i(a, b) = N_i(x_i, y_i) = 1$, i.e. we obtain

$$\boxed{f_{1i} = Q_s t(x_i, y_i) \quad \text{if the line source is located at nodal point } i} \tag{10.33}$$

As in the one-dimensional case of Chapter 9, it follows that any distributed load Q may be replaced by line sources located at the nodal points.

The global stiffness matrix $\mathbf{K}$ and the global load vector $\mathbf{f}_1$ given by (10.15) are obtained by integration over the entire region A. These integrations may be obtained as a summation of integrations over each element. In this way we are led to the *expanded element stiffness matrix* $\mathbf{K}^{ee}$ for element α and the *expanded element load vector* $\mathbf{f}_1^{ee}$ for element α given by

$$\boxed{\begin{aligned} \mathbf{K}^{ee} &= \int_{A_\alpha} \mathbf{B}^\mathrm{T}\mathbf{D}\mathbf{B}t \, \mathrm{d}A \\[2mm] \mathbf{f}_1^{ee} &= \int_{A_\alpha} \mathbf{N}^\mathrm{T}Qt \, \mathrm{d}A \end{aligned}} \tag{10.34}$$

where A_α is the region of element α. Moreover, it follows directly that

$$\boxed{\begin{aligned} \mathbf{K} &= \sum_{\alpha=1}^{n_{el}} \mathbf{K}_\alpha^{ee} \\[2mm] \mathbf{f}_1 &= \sum_{\alpha=1}^{n_{el}} \mathbf{f}_{1\alpha}^{ee} \end{aligned}} \tag{10.35}$$

where n_{el} denotes the total number of elements and where $\mathbf{K}_\alpha^{ee}$ and $\mathbf{f}_{1\alpha}^{ee}$ refer to the pertinent quantities for element α. Conceptually, relations (10.34) and (10.35) are of fundamental importance, but just as for the one-dimensional FE formulation these relations lead to a very inefficient computer implementation. In order to identify the non-zero components of $\mathbf{K}^{ee}$ and $\mathbf{f}_1^{ee}$ directly, we consider the FE formulation for an element using only those degrees of freedom which belong to that specific element.

The approximation of the temperature over each element is given by

$$T = \mathbf{N}^e\mathbf{a}^e \tag{10.36}$$

where $\mathbf{N}^e$ is the element shape function matrix and $\mathbf{a}^e$ contains the temperatures at the nodal points of the element, i.e.

$$\mathbf{N}^e = [N_1^e \quad N_2^e \quad \cdots \quad N_{n_e}^e]; \quad \mathbf{a}^e = \begin{bmatrix} T_1 \\ T_2 \\ \vdots \\ T_{n_e} \end{bmatrix} \tag{10.37}$$

where n_e denotes the number of nodal points for the element. From (10.36) we obtain

$$\nabla T = \mathbf{B}^e \mathbf{a}^e \quad \text{where} \quad \mathbf{B}^e = \nabla \mathbf{N}^e \tag{10.38}$$

i.e.

$$\mathbf{B}^e = \begin{bmatrix} \dfrac{\partial N_1^e}{\partial x} & \dfrac{\partial N_2^e}{\partial x} & \cdots & \dfrac{\partial N_{n_e}^e}{\partial x} \\[2ex] \dfrac{\partial N_1^e}{\partial y} & \dfrac{\partial N_2^e}{\partial y} & \cdots & \dfrac{\partial N_{n_e}^e}{\partial y} \end{bmatrix} \tag{10.39}$$

A comparison with (10.14)–(10.18) shows directly that the FE formulation for one element is given by

$$\boxed{\mathbf{K}^e \mathbf{a}^e = \mathbf{f}^e} \tag{10.40}$$

where the *element force vector* $\mathbf{f}^e$ is given by

$$\boxed{\mathbf{f}^e = \mathbf{f}_b^e + \mathbf{f}_l^e} \tag{10.41}$$

Moreover, the *element stiffness matrix* $\mathbf{K}^e$, the *element boundary vector* $\mathbf{f}_b^e$ and the *element load vector* $\mathbf{f}_l^e$ for element α are given by

$$\boxed{\begin{aligned} \mathbf{K}^e &= \int_{A_\alpha} \mathbf{B}^{e\mathrm{T}} \mathbf{D} \mathbf{B}^e t \, \mathrm{d}A \\[2ex] \mathbf{f}_b^e &= -\int_{\mathscr{L}_{h\alpha}} \mathbf{N}^{e\mathrm{T}} h t \, \mathrm{d}\mathscr{L} - \int_{\mathscr{L}_{g\alpha}} \mathbf{N}^{e\mathrm{T}} q_n t \, \mathrm{d}\mathscr{L} \\[2ex] \mathbf{f}_l^e &= \int_{A_\alpha} \mathbf{N}^{e\mathrm{T}} Q t \, \mathrm{d}A \end{aligned}} \tag{10.42}$$

where A_α is the region of element α, $\mathscr{L}_{h\alpha}$ is that part of the element boundary on which the flux $q_n = h$ is known, whereas $\mathscr{L}_{g\alpha}$ is the remaining part of the element boundary. We recall that the boundary conditions along an element boundary are unknown, except when an element boundary coincides with the boundary of the body.

Similar to Figure 9.14, the three different FE formulations – global formulation, expanded element formulation and element formulation – are illustrated in Figure 10.2. We recall that in the global formulation the entire body is considered and the

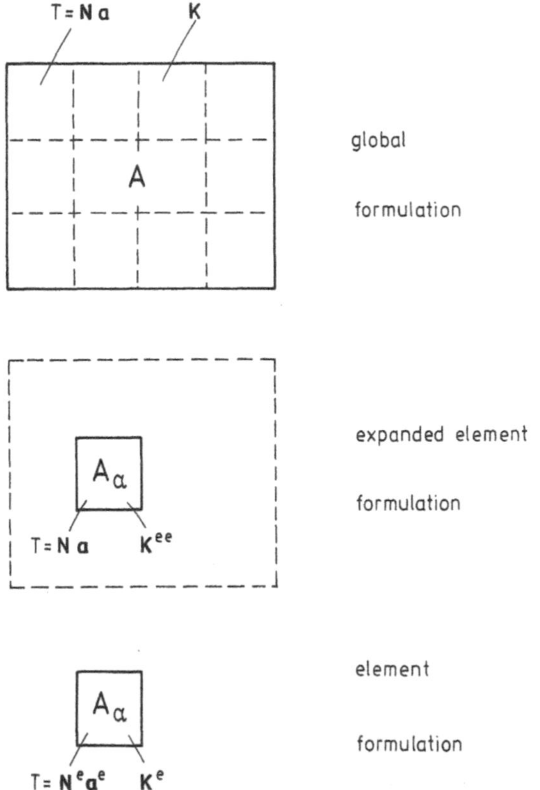

Figure 10.2 The three different FE formulations

approximation $T = \mathbf{N}\mathbf{a}$, applicable to the entire body, is adopted. In the expanded element formulation, one element is considered, but the global approximation $T = \mathbf{N}\mathbf{a}$ is used. Finally, in the element formulation, one element is considered and the approximation $T = \mathbf{N}^e\mathbf{a}^e$, applicable to this particular element, is used.

In a computer program, the element stiffness $\mathbf{K}^e$ and the element load vector $\mathbf{f}_l^e$ are determined by (10.42) and the contributions to the global stiffness matrix $\mathbf{K}$ and the global load vector $\mathbf{f}_l$ are obtained using the topology data. The boundary vector $\mathbf{f}_b$ is always derived using the form given by (10.15). It appears that the approach described is completely similar to that discussed for the FE formulation of one-dimensional heat flow. We also observe that just as for one-dimensional heat flow, the stiffness matrix $\mathbf{K}$ given by (10.15) contains first derivatives of the global shape functions, i.e. we again require that these shape functions fulfil C^0-continuity.

When the nodal temperature $\mathbf{a}$ has been determined by (10.18), the temperature T at an arbitrary point in an element is given by (10.36) and the temperature gradient ∇T is given by (10.38). From the temperature gradient, the flux vector $\mathbf{q}$ at any place in the body is obtained from Fourier's law (10.3). Therefore, when **a** has been determined from (10.18) all quantities of interest can be derived.

10.1.1 *Example*

Consider the square panel shown in Figure 10.3, where the boundary along the y-axis is insulated (i.e. $q_n = h = 0$) and a constant flux $q_n = h = 30$ J/m² s is prescribed along the boundaries $y = 1$ m and $y = -1$ m. A constant temperature $T = 10\,°C$ is prescribed along $x = 2$ m and a constant heat supply $Q = 45$ J/m² s is applied all over the panel. The thickness t is constant and is equal to 1 m. The material is assumed to be isotropic and homogeneous, i.e. according to (6.13) and (6.14) the constitutive matrix **D** can be written as

$$\mathbf{D} = k \begin{bmatrix} 1 & 0 \\ 0 & 1 \end{bmatrix} = k\mathbf{I} \tag{10.43}$$

where the thermal conductivity is given by $k = 4$ J/°C m s.

It is obvious that the heat flow in the panel is symmetric about the x-axis, i.e. no heat flows across the x-axis. We may therefore reformulate the problem as shown in Figure 10.4(a).

In order to solve this problem by the FE method using hand calculations only, we adopt the extremely simple FE model consisting of two three-node triangles as shown in Figure 10.4(b). The element numbers and the global nodal numbers are also indicated on this figure.

The element stiffness matrix $\mathbf{K}^e$ and the element load vector $\mathbf{f}_i^e$ are given by (10.42). With (10.43) and noting that $\mathbf{B}^e$ is a constant matrix for the three-node

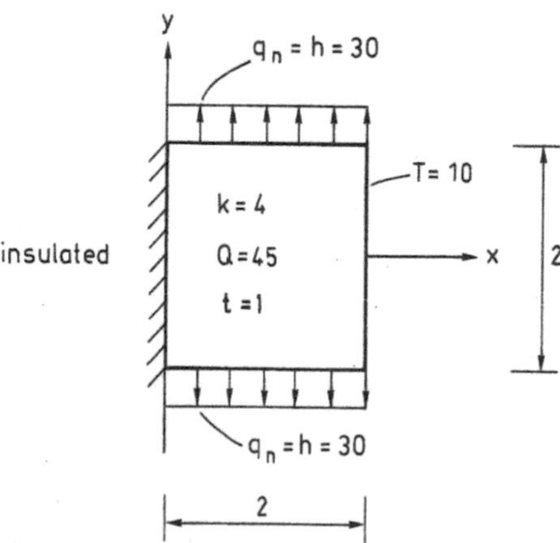

Figure 10.3   Square panel with heat flow

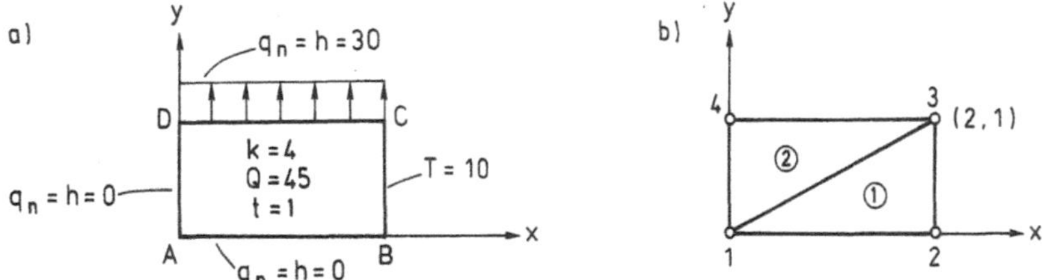

Figure 10.4   (a) Problem reformulation; (b) finite element mesh

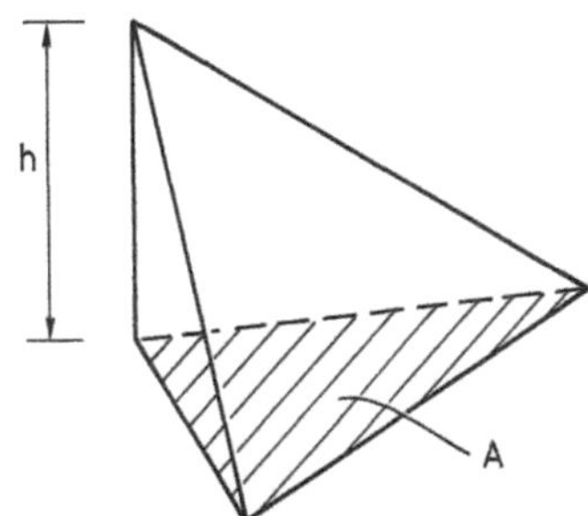

Figure 10.5   Body with triangular base and height h

triangle (cf. (7.99)), we obtain

$$\mathbf{K}^e = k\mathbf{B}^{eT}\mathbf{B}^e t A_\alpha \tag{10.44}$$

where

$$\mathbf{B}^e = \frac{1}{2A_\alpha}\begin{bmatrix} y_j - y_k & y_k - y_i & y_i - y_j \\ x_k - x_j & x_i - x_k & x_j - x_i \end{bmatrix} \tag{10.45}$$

As both the load Q and the thickness t are constants, (10.42) yields

$$\mathbf{f}_1^e = Qt \int_{A_\alpha} \mathbf{N}^{eT}\, dA \tag{10.46}$$

The element shape function matrix $\mathbf{N}^e$ is given by (7.92) and (10.46) can be derived using standard rules for area integration. In the present case, however, the shape functions vary linearly, which means that (10.46) may be evaluated in a very simple manner. For this purpose consider the body shown in Figure 10.5. The volume V of this body is known to be given by

$$V = \tfrac{1}{3}hA \tag{10.47}$$

where h is the height and A the area of the triangle which forms the base of the body.

Reference to Figure 7.23 shows that the element shape functions vary in the

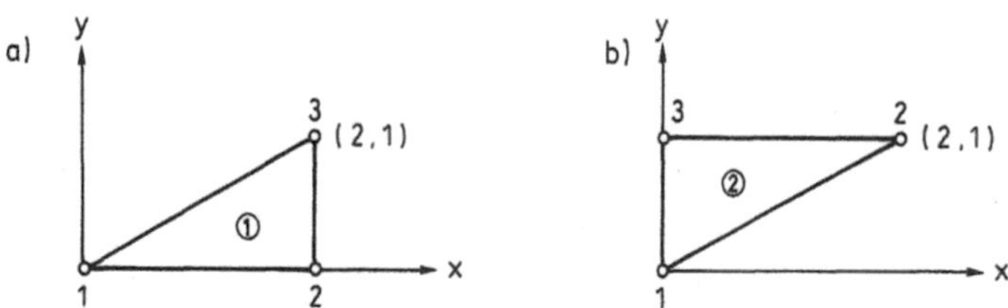

Figure 10.6   Local nodal points for element 1 and 2

manner given by Figure 10.5, i.e. from (10.46) we obtain with (10.47)

$$\mathbf{f}_i^e = \frac{Q}{3} t A_\alpha \begin{bmatrix} 1 \\ 1 \\ 1 \end{bmatrix} \tag{10.48}$$

This result is certainly not surprising as it shows that a constant load Q contributes equal amounts to each nodal point.

The local nodal points of the two elements are shown in Figure 10.6 and in accordance with Chapter 7 these nodal points are listed in the counter-clockwise direction 1, 2, 3. We first consider *element 1* where $i = 1$, $j = 2$ and $k = 3$, i.e.

$$x_i = y_i = 0; \quad x_j = 2,\ y_j = 0; \quad x_k = 2,\ y_k = 1$$

With these values and as the area $A_1 = 1$, $\mathbf{B}^e$ as given by (10.45) becomes

$$\mathbf{B}^e = \frac{1}{2} \begin{bmatrix} -1 & 1 & 0 \\ 0 & -2 & 2 \end{bmatrix} \tag{10.49}$$

From (10.44) and (10.48) and as the thickness $t = 1$, the results for the first element are given by

$$\mathbf{K}^e = \begin{bmatrix} 1 & -1 & 0 \\ -1 & 5 & -4 \\ 0 & -4 & 4 \end{bmatrix}; \quad \mathbf{f}_i^e = 15 \begin{bmatrix} 1 \\ 1 \\ 1 \end{bmatrix}$$

Comparing Figures 10.4(b) and 10.6(a), local nodal points 1, 2, 3 correspond to the global nodal points 1, 2, 3 respectively. That is, after the contributions from the first element, the global stiffness matrix $\mathbf{K}$ and the global load vector $\mathbf{f}_i$ are given by

$$\mathbf{K} = \mathbf{K}_i^{ee} = \begin{bmatrix} 1 & -1 & 0 & 0 \\ -1 & 5 & -4 & 0 \\ 0 & -4 & 4 & 0 \\ 0 & 0 & 0 & 0 \end{bmatrix}; \quad \mathbf{f}_i = \mathbf{f}_{ii}^{ee} = 15 \begin{bmatrix} 1 \\ 1 \\ 1 \\ 0 \end{bmatrix} \tag{10.50}$$

Next we consider *element 2* of Figure 10.6(b) where $i = 1, j = 2$ and $k = 3$, i.e.

$$x_i = y_i = 0; \quad x_j = 2, \ y_j = 1; \quad x_k = 0, \ y_k = 1$$

The area A_2 is $A_2 = 1$ and $\mathbf{B}^e$ as given by (10.45) becomes

$$\mathbf{B}^e = \frac{1}{2}\begin{bmatrix} 0 & 1 & -1 \\ -2 & 0 & 2 \end{bmatrix} \tag{10.51}$$

From (10.44) and (10.48) the results for the second element are then given by

$$\mathbf{K}^e = \begin{bmatrix} 4 & 0 & -4 \\ 0 & 1 & -1 \\ -4 & -1 & 5 \end{bmatrix}; \quad \mathbf{f}_1^e = 15\begin{bmatrix} 1 \\ 1 \\ 1 \end{bmatrix}$$

Comparing Figures 10.4(b) and 10.6(b), the topology data for this element is that the local nodal points 1, 2, 3 correspond to the global nodal points 1, 3, 4 respectively. For clarity, and using this topology data, the expanded element stiffness matrix $\mathbf{K}^{ee}$ and the expanded element load vector $\mathbf{f}_1^{ee}$ are

$$\mathbf{K}_2^{ee} = \begin{bmatrix} 4 & 0 & 0 & -4 \\ 0 & 0 & 0 & 0 \\ 0 & 0 & 1 & -1 \\ -4 & 0 & -1 & 5 \end{bmatrix}; \quad \mathbf{f}_{12}^{ee} = 15\begin{bmatrix} 1 \\ 0 \\ 1 \\ 1 \end{bmatrix}$$

Adding these results to (10.50) we then obtain

$$\mathbf{K} = \begin{bmatrix} 5 & -1 & 0 & -4 \\ -1 & 5 & -4 & 0 \\ 0 & -4 & 5 & -1 \\ -4 & 0 & -1 & 5 \end{bmatrix}; \quad \mathbf{f}_1 = 15\begin{bmatrix} 2 \\ 1 \\ 2 \\ 1 \end{bmatrix} \tag{10.52}$$

Let us now evaluate the boundary vector $\mathbf{f}_b$ given by (10.15). From Figure 10.4 we conclude that

$$\mathbf{f}_b = -\int_{\mathscr{L}_{AB}} \mathbf{N}^T ht \, d\mathscr{L} - \int_{\mathscr{L}_{DC}} \mathbf{N}^T ht \, d\mathscr{L} - \int_{\mathscr{L}_{AD}} \mathbf{N}^T ht \, d\mathscr{L} - \int_{\mathscr{L}_{BC}} \mathbf{N}^T q_n t \, d\mathscr{L}$$

i.e.

$$\mathbf{f}_b = -t\int_{\mathscr{L}_{DC}} \mathbf{N}^T h \, d\mathscr{L} - t\int_{\mathscr{L}_{BC}} \mathbf{N}^T q_n d\mathscr{L} \tag{10.53}$$

The only shape functions which differ from zero along the boundary DC are N_3 and

N_4. Moreover, as they vary linearly and as $h = 30$ along DC we have

$$\int_{\mathscr{L}_{DC}} \mathbf{N}^T h \, d\mathscr{L} = 30 \int_{\mathscr{L}_{DC}} \begin{bmatrix} 0 \\ 0 \\ N_3 \\ N_4 \end{bmatrix} d\mathscr{L} = 30 \begin{bmatrix} 0 \\ 0 \\ 1 \\ 1 \end{bmatrix} \tag{10.54}$$

Likewise, it follows that

$$\int_{\mathscr{L}_{BC}} \mathbf{N}^T q_n \, d\mathscr{L} = \begin{bmatrix} 0 \\ \displaystyle\int_{\mathscr{L}_{BC}} N_2 q_n \, d\mathscr{L} \\ \displaystyle\int_{\mathscr{L}_{BC}} N_3 q_n \, d\mathscr{L} \\ 0 \end{bmatrix} \tag{10.55}$$

Using (10.53)–(10.55) and as the thickness $t = 1$, the boundary vector $\mathbf{f}_b$ is given by

$$\mathbf{f}_b = - \begin{bmatrix} 0 \\ \displaystyle\int_{\mathscr{L}_{BC}} N_2 q_n \, d\mathscr{L} \\ 30 + \displaystyle\int_{\mathscr{L}_{BC}} N_3 q_n \, d\mathscr{L} \\ 30 \end{bmatrix} \tag{10.56}$$

The FE equations for the body are obtained from (10.52) and (10.56), i.e.

$$\begin{bmatrix} 5 & -1 & 0 & -4 \\ -1 & 5 & -4 & 0 \\ 0 & -4 & 5 & -1 \\ -4 & 0 & -1 & 5 \end{bmatrix} \begin{bmatrix} T_1 \\ 10 \\ 10 \\ T_4 \end{bmatrix} = - \begin{bmatrix} 0 \\ \displaystyle\int_{\mathscr{L}_{BC}} N_2 q_n \, d\mathscr{L} \\ 30 + \displaystyle\int_{\mathscr{L}_{BC}} N_3 q_n \, d\mathscr{L} \\ 30 \end{bmatrix} + 15 \begin{bmatrix} 2 \\ 1 \\ 2 \\ 1 \end{bmatrix} \tag{10.57}$$

where $T_2 = T_3 = 10$. From the first and fourth row we obtain

$$\begin{bmatrix} 5 & -4 \\ -4 & 5 \end{bmatrix} \begin{bmatrix} T_1 \\ T_4 \end{bmatrix} = \begin{bmatrix} 10 - 0 + 30 \\ 10 - 30 + 15 \end{bmatrix} = \begin{bmatrix} 40 \\ -5 \end{bmatrix}$$

with the solution

$$\begin{bmatrix} T_1 \\ T_4 \end{bmatrix} = \begin{bmatrix} 20 \\ 15 \end{bmatrix} {}^\circ\text{C} \tag{10.58}$$

Use of this solution in the second and third row of (10.57) implies that

$$\int_{\mathscr{L}_{\mathrm{BC}}} N_2 q_n \, \mathrm{d}\mathscr{L} = 25; \qquad \int_{\mathscr{L}_{\mathrm{BC}}} N_3 q_n \, \mathrm{d}\mathscr{L} = 5 \tag{10.59}$$

As N_2 and N_3 are positive functions, (10.59) implies that q_n is positive along the boundary BC, i.e. we conclude that in order to maintain the temperature at $T = 10$ along the boundary BC, heat must be extracted from the body along this boundary. With the result of (10.59) it follows from (10.57) that the balance principle for the body is satisfied in accordance with (10.28).

Let us finally determine the flux vector $\mathbf{q}$ in each element. From (10.3) and (10.38) we get

$$\mathbf{q} = -\mathbf{D}\nabla T = -k\nabla T = -k\mathbf{B}^{\mathrm{e}}\mathbf{a}^{\mathrm{e}}$$

where $k = 4$. For element 1 we obtain with (10.49) that

$$\mathbf{q} = -4 \times \tfrac{1}{2} \begin{bmatrix} -1 & 1 & 0 \\ 0 & -2 & 2 \end{bmatrix} \begin{bmatrix} 20 \\ 10 \\ 10 \end{bmatrix} = \begin{bmatrix} 20 \\ 0 \end{bmatrix} \mathrm{J/m^2\,s}$$

whereas for element 2, where (10.51) holds, we get

$$\mathbf{q} = -4 \times \tfrac{1}{2} \begin{bmatrix} 0 & 1 & -1 \\ -2 & 0 & 2 \end{bmatrix} \begin{bmatrix} 20 \\ 10 \\ 15 \end{bmatrix} = \begin{bmatrix} 10 \\ 20 \end{bmatrix} \mathrm{J/m^2\,s}$$

10.1.2 *Two-dimensional heat flow with convection*

Previously, we assumed that the boundary $\mathscr{L}$ consists of the two parts, $\mathscr{L}_h$ and $\mathscr{L}_g$, where the flux q_n and the temperature T are prescribed, respectively. Let us now also assume that convection occurs on part of the boundary and let $\mathscr{L}_c$ denote that part. Referring to (10.15) this affects only the boundary vector $\mathbf{f}_b$, which now becomes

$$\mathbf{f}_{\mathrm{b}} = -\int_{\mathscr{L}_h} \mathbf{N}^{\mathrm{T}} h t \, \mathrm{d}\mathscr{L} - \int_{\mathscr{L}_g} \mathbf{N}^{\mathrm{T}} q_n t \, \mathrm{d}\mathscr{L} - \int_{\mathscr{L}_c} \mathbf{N}^{\mathrm{T}} q_n t \, \mathrm{d}\mathscr{L} \tag{10.60}$$

where $\mathscr{L}_h$, $\mathscr{L}_g$ and $\mathscr{L}_c$ now comprise the entire boundary, as shown in Figure 10.7. The convection along $\mathscr{L}_c$ is again given by Newton's convection boundary condition (9.143), i.e.

$$q_n = \alpha(T - T_\infty) \tag{10.61}$$

To consider the influence of convection, the procedure discussed in Chapter 9 is followed, i.e. the unknown temperature T in (10.61) is replaced by $T = \mathbf{N}\mathbf{a}$. This gives

$$q_n = \alpha \mathbf{N}\mathbf{a} - \alpha T_\infty$$

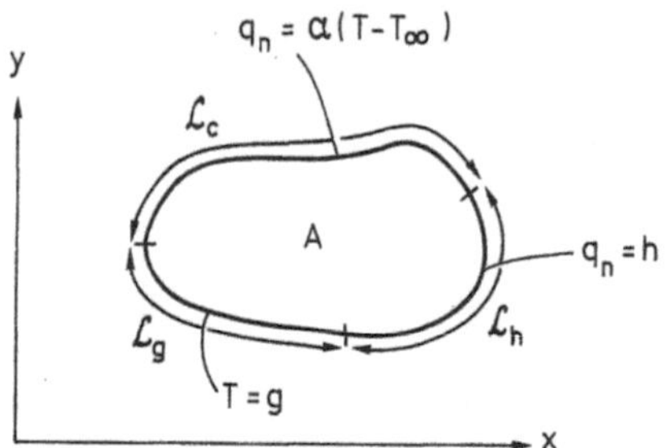

Figure 10.7   Two-dimensional region with boundary $\mathscr{L} = \mathscr{L}_h + \mathscr{L}_g + \mathscr{L}_c$

and use of this expression in (10.60) results in

$$\mathbf{f}_b = -\int_{\mathscr{L}_h} \mathbf{N}^T h t \, \mathrm{d}\mathscr{L} - \int_{\mathscr{L}_g} \mathbf{N}^T q_n t \, \mathrm{d}\mathscr{L} - \left(\int_{\mathscr{L}_c} \alpha \mathbf{N}^T \mathbf{N} t \, \mathrm{d}\mathscr{L} \right) \mathbf{a} + T_\infty \int_{\mathscr{L}_c} \mathbf{N}^T \alpha t \, \mathrm{d}\mathscr{L}$$

With this expression, the FE formulation of (10.16) takes the form

$$\boxed{\begin{aligned}(\mathbf{K} + \mathbf{K}_c)\mathbf{a} &= -\int_{\mathscr{L}_h} \mathbf{N}^T h t \, \mathrm{d}\mathscr{L} - \int_{\mathscr{L}_g} \mathbf{N}^T q_n t \, \mathrm{d}\mathscr{L} + T_\infty \int_{\mathscr{L}_c} \mathbf{N}^T \alpha t \, \mathrm{d}\mathscr{L} + \mathbf{f}_l \\ \text{where} \quad \mathbf{K}_c &= \int_{\mathscr{L}_c} \alpha \mathbf{N}^T \mathbf{N} t \, \mathrm{d}\mathscr{L}\end{aligned}}$$

(10.62)

It appears that a modification of the stiffness matrix occurs. This new stiffness matrix is symmetric, just like $\mathbf{K}$, and similar to the discussion of (9.149) it follows that the modified stiffness matrix is non-singular, i.e.

$$\boxed{\det(\mathbf{K} + \mathbf{K}_c) \neq 0}$$

(10.63)

We finally mention that we are able to consider convection occurring along the lateral surfaces of the region A, i.e. convection normal to the xy-plane. This situation can be dealt with by a modification of the load Q which leads to a modification of the load vector $\mathbf{f}_l$ as in the procedure for one-dimensional heat flow (cf. (9.152), (9.154) and (9.155)). This modification is straightforward and is left as an exercise for the interested reader.

10.2 Three-dimensional heat flow

The weak form of three-dimensional heat flow is given by (6.49) and (6.50), i.e.

$$\int_V (\nabla v)^T \mathbf{D} \nabla T \, \mathrm{d}V = -\int_{S_h} v h \, \mathrm{d}S - \int_{S_g} v q_n \, \mathrm{d}S + \int_V v Q \, \mathrm{d}V$$
$$T = g \quad \text{on } S_g$$

(10.64)

where V is the volume of the body and S is the boundary surface which consists of S_h and S_g, where the flux $q_n = h$ and the temperature $T = g$ are prescribed, respectively.

The temperature is again approximated by means of

$$T = \mathbf{Na} \tag{10.65}$$

where

$$\mathbf{N} = [N_1 \quad N_2 \quad \dots \quad N_n]; \quad \mathbf{a} = \begin{bmatrix} T_1 \\ T_2 \\ \vdots \\ T_n \end{bmatrix} \tag{10.66}$$

and n is the number of nodal points for the entire body and a component N_i now depends on x, y and z, i.e. $N_i = N_i(x, y, z)$. From (10.65) we obtain

$$\nabla T = \mathbf{Ba} \quad \text{where} \quad \mathbf{B} = \nabla \mathbf{N} \tag{10.67}$$

i.e.

$$\mathbf{B} = \begin{bmatrix} \dfrac{\partial N_1}{\partial x} & \dfrac{\partial N_2}{\partial x} & \dots & \dfrac{\partial N_n}{\partial x} \\[2mm] \dfrac{\partial N_1}{\partial y} & \dfrac{\partial N_2}{\partial y} & \dots & \dfrac{\partial N_n}{\partial y} \\[2mm] \dfrac{\partial N_1}{\partial z} & \dfrac{\partial N_2}{\partial z} & \dots & \dfrac{\partial N_n}{\partial z} \end{bmatrix} \tag{10.68}$$

The derivation of the FE formulation from (10.64) using the approximation above in combination with the Galerkin method follows the same procedure as for two-dimensional heat flow. We may therefore write the result directly as

$$\boxed{\mathbf{Ka} = \mathbf{f}} \tag{10.69}$$

where

$$\boxed{\mathbf{f} = \mathbf{f}_b + \mathbf{f}_l} \tag{10.70}$$

and

$$\boxed{\begin{aligned} \mathbf{K} &= \int_V \mathbf{B}^\mathrm{T} \mathbf{D} \mathbf{B} \, \mathrm{d}V \\[2mm] \mathbf{f}_b &= -\int_{S_h} \mathbf{N}^\mathrm{T} h \, \mathrm{d}S - \int_{S_g} \mathbf{N}^\mathrm{T} q_n \, \mathrm{d}S \\[2mm] \mathbf{f}_l &= \int_V \mathbf{N}^\mathrm{T} Q \, \mathrm{d}V \end{aligned}} \tag{10.71}$$

Just like one- and two-dimensional heat flow, the force vector **f** has the dimension [J/s]. Moreover, it is evident that the discussion for two-dimensional heat flow also carries over directly to that of three-dimensional heat flow, and we simply mention that convection now can occur only through a part S_c of the boundary surface. The corresponding modification of the boundary vector $\mathbf{f}_b$ is similar to that performed for two-dimensional heat flow.

We have seen that once the FE formulation for one-dimensional heat flow is comprehended, it is a straightforward task to generalize these results to two- and three-dimensional heat flow. It follows that we are now in a position to treat heat flow for arbitrary geometries and with arbitrary material properties. By now it should also be obvious why the FE method provides a unique tool for engineers and physicists. We mention finally that two- and three-dimensional heat flow is also treated in the textbooks by Bathe (1982), Becker *et al.* (1981), Hughes (1987), Johnson (1987), Stasa (1985) and Zienkiewicz and Taylor (1989).

11

Guidelines for element meshes and global nodal numbering

We shall now discuss some general guidelines for the establishment of FE meshes and for choosing the numbering of global nodal points. Also, the equation solution process will be investigated in more detail.

11.1 FE mesh

The first step in an FE analysis is to select the type of elements and the corresponding FE mesh. There are no fixed rules on how to make these decisions. Clearly, for a given type of element the accuracy increases with decreasing element size and, in general, one will use small elements in regions where the unknown function – the temperature, say – varies rapidly. This means that a sound physical understanding of the problem considered is of fundamental importance for a realistic analysis. However, the decision on element types and size is more delicate than that. Every analysis involves the use of resources, whether they are measured in terms of money or manpower, and even though we aim at an accurate analysis, we do not want it to be more accurate than required. For some problems, we are interested in detailed information on the behaviour even in local regions, while for others we only want to obtain a rather general and crude indication of the overall response. Moreover, for some problems, simplifications may have already been introduced when the problem was defined. For instance, we may have only vague ideas about the loading, the boundary conditions and the material data. As engineers we must therefore use our judgement in order to obtain that optimum choice for element type and element mesh which balances the requirement of reliable results with that of cost effectiveness. It is not surprising, therefore, that such a choice presents a delicate problem, and we may refer to Cook *et al.* (1989) and Zienkiewicz and Taylor (1989) for further information. One often starts with a simple FE model, and these results may form the basis for the establishment of a more refined model.

All finite elements are based on rather simple polynomial interpolations of the unknown function within the element. For a given type of element this means that

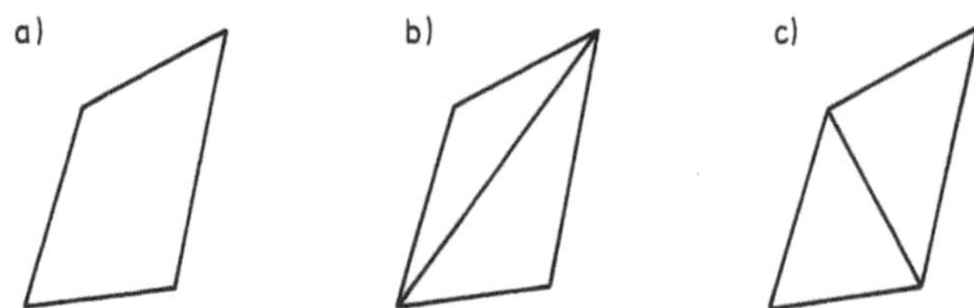

Figure 11.1   (a) Quadrilateral; (b) inferior division; (c) desirable division

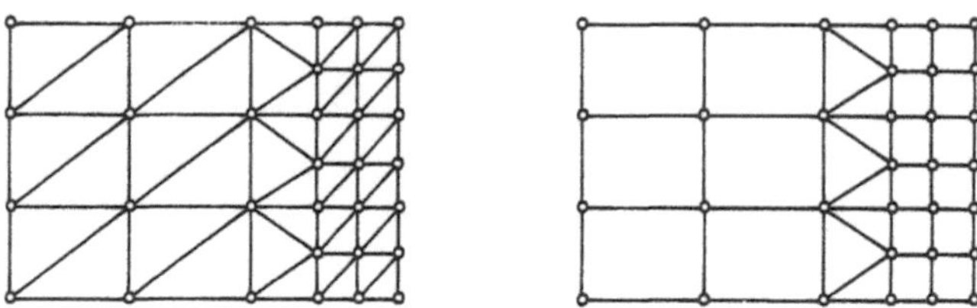

Figure 11.2   Examples of mesh refinement

the smaller the elements, the greater the accuracy. However, it also implies that *any* dimension of an element should be kept as small as possible, i.e. it is not only the size, but also the form of the element which is of importance. As an example, assume that the quadrilateral shown in Figure 11.1(a) is to be divided into two triangular elements.

It is obvious that the division in Figure 11.1(c) is better than that of 11.1(b), since the largest dimension of the elements in 11.1(c) is smaller than that given by 11.1(b). The ratio between the largest and smallest dimension of an element is called the *aspect ratio* and in a good FE mesh, the aspect ratio is as close as possible to unity.

In order to obtain an efficient solution scheme, we want to use few elements in regions where the unknown function varies slowly, but many elements in regions where it varies rapidly. Two possibilities, which allow for such a mesh refinement and which fulfil the continuity requirement, are illustrated in Figure 11.2 for the three-node triangular element and four-node rectangle.

To increase computational efficiency, symmetry properties should be used whenever possible. An example was given in Chapter 10 on page 214. We recall that symmetry involves not only geometry, but also loading and material data. As another example of symmetry, consider a heating cable placed in the isotropic body shown in Figure 11.3(a). It is clearly advantageous to utilize the symmetry about the centre line (cf. Figure 11.3(b)) and thereby reduce the number of elements. The boundary condition along the symmetry line is that the flux $q_n = 0$.

It is also often possible to make use of symmetry properties for more complicated situations. In Figure 11.4(a) we assume that heating cables are located at constant distances and that the isotropic body extends infinitely far away in the horizontal direction. Due to symmetry, the flux q_n is zero along the planes AB and CD, which

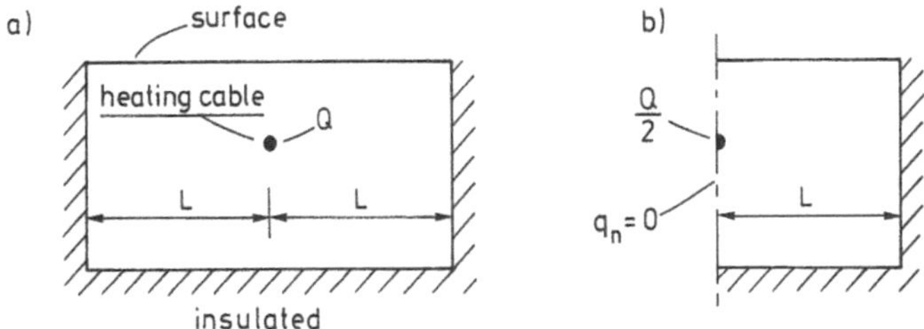

Figure 11.3   (a) Heat flow problem; (b) use of symmetry

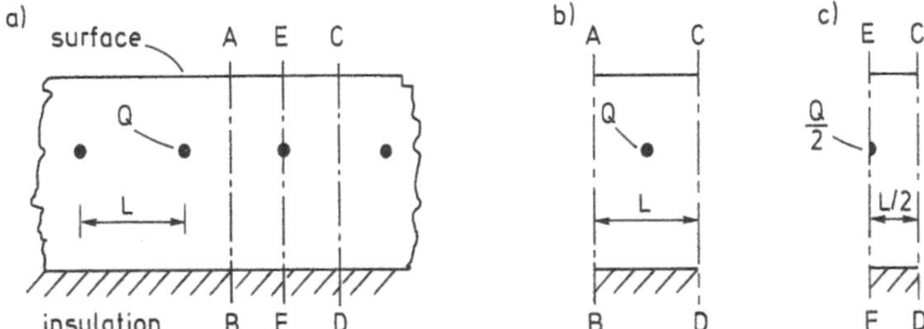

Figure 11.4   (a) Heat flow problem; (b) use of symmetry; (c) use of symmetry

provides the model of Figure 11.4(b). We also note that the flux q_n is zero along the plane EF, which leads to the even simpler model of Figure 11.4(c).

11.2 Methods of solution

11.2.1 *Bandwidth*

We have seen that the FE method results in a symmetric global stiffness matrix **K** which is banded, i.e. with non-zero components clustered about the diagonal of **K** and zero components far away from the diagonal. With a cross indicating non-zero components, this situation is illustrated in Figure 11.5 where all non-zero components fall between two lines parallel to the diagonal.

Including the diagonal, the distance from the diagonal to one of these lines is termed the *bandwidth* (see Figure 11.5). In an efficient FE program, advantage is taken not only of the symmetry, but also of the banded structure of the **K**-matrix so that only the components of **K** within the bandwidth are considered when solving the system of equations **Ka** = **f**. Therefore, the bandwidth plays a significant role in the computational cost of solving this system of equations and we shall now investigate this aspect when the Gauss elimination technique is used.

bandwidth (=5)

$$\mathbf{K} = \begin{bmatrix} \times & \times & \times & 0 & \times & 0 & 0 & 0 & 0 \\ \times & \times & \times & \times & \times & \times & 0 & 0 & 0 \\ \times & \times & \times & \times & 0 & \times & \times & 0 & 0 \\ 0 & \times & \times & \times & \times & \times & \times & \times & 0 \\ \times & \times & 0 & \times & \times & \times & \times & 0 & \times \\ 0 & \times & \times & \times & \times & \times & \times & \times & \times \\ 0 & 0 & \times & \times & \times & \times & \times & \times & 0 \\ 0 & 0 & 0 & \times & 0 & \times & \times & \times & \times \\ 0 & 0 & 0 & 0 & \times & \times & 0 & \times & \times \end{bmatrix}$$

Figure 11.5   Illustration of bandwidth for symmetric matrix $\mathbf{K}$

11.2.2 *Systematic Gauss elimination*

The principal ingredients of the Gauss elimination technique were described in Chapter 2. However, we shall now present a formulation of this technique, which facilitates its implementation in a computer program. We recall that Gauss elimination comprises a systematic elimination of unknowns so that a triangularized form is achieved (cf. (2.56)–(2.58)).

Consider the equation system $\mathbf{Ka} = \mathbf{f}$ and perform the partitioning shown below

$$\begin{bmatrix} K_{11} & K_{12} & K_{13} & \cdots & K_{1n} \\ K_{21} & K_{22} & K_{23} & \cdots & K_{2n} \\ K_{31} & K_{32} & K_{33} & \cdots & K_{3n} \\ \vdots & \vdots & \vdots & & \vdots \\ K_{n1} & K_{n2} & K_{n3} & \cdots & K_{nn} \end{bmatrix} \begin{bmatrix} a_1 \\ a_2 \\ a_3 \\ \vdots \\ a_n \end{bmatrix} = \begin{bmatrix} f_1 \\ f_2 \\ f_3 \\ \vdots \\ f_n \end{bmatrix} \tag{11.1}$$

With the following definitions

$$\mathbf{A}^{\mathrm{T}} = \begin{bmatrix} K_{21} \\ K_{31} \\ \vdots \\ K_{n1} \end{bmatrix}; \quad \underset{\sim}{\mathbf{K}} = \begin{bmatrix} K_{22} & K_{23} & \cdots & K_{2n} \\ K_{32} & K_{33} & \cdots & K_{3n} \\ \vdots & \vdots & & \vdots \\ K_{n2} & K_{n3} & \cdots & K_{nn} \end{bmatrix}; \quad {}_1\mathbf{a} = \begin{bmatrix} a_2 \\ a_3 \\ \vdots \\ a_n \end{bmatrix}; \quad \underset{\sim}{\mathbf{f}} = \begin{bmatrix} f_2 \\ f_3 \\ \vdots \\ f_n \end{bmatrix}$$

the system of equations (11.1) can be written as

$$\begin{bmatrix} K_{11} & \mathbf{A} \\ \mathbf{A}^{\mathrm{T}} & \underset{\sim}{\mathbf{K}} \end{bmatrix} \begin{bmatrix} a_1 \\ {}_1\mathbf{a} \end{bmatrix} = \begin{bmatrix} f_1 \\ \underset{\sim}{\mathbf{f}} \end{bmatrix}$$

Carrying out the matrix multiplications we obtain

$$K_{11}a_1 + \mathbf{A}_1\mathbf{a} = f_1; \quad \mathbf{A}^{\mathrm{T}}a_1 + \underset{\sim}{\mathbf{K}}_1\mathbf{a} = \underset{\sim}{\mathbf{f}}$$

Elimination of a_1 in the second system of equations yields

$$K_{11}a_1 + \mathbf{A}_1\mathbf{a} = f_1; \quad {}_1\mathbf{K}_1\mathbf{a} = {}_1\mathbf{f} \tag{11.2}$$

where

$$ {}_1\mathbf{K} = \mathbf{\underset{\sim}{K}} - K_{11}^{-1}\mathbf{A}^{\mathrm{T}}\mathbf{A} \quad \text{and} \quad {}_1\mathbf{f} = \mathbf{\underset{\sim}{f}} - K_{11}^{-1}f_1\mathbf{A}^{\mathrm{T}} \tag{11.3}$$

This elimination process is called a *static condensation*. The derivations above require that $K_{11} \neq 0$. However, the system of equations $\mathbf{Ka} = \mathbf{f}$ refers here to that system which is obtained after proper boundary conditions have been invoked. This implies that $\mathbf{K}$ is positive definite and, consequently, all diagonal terms are positive (cf. page 24).

It appears that the two systems given by (11.2) may be expressed as

$$\begin{bmatrix} K_{11} & \mathbf{A} \\ \mathbf{0} & {}_1\mathbf{K} \end{bmatrix}\begin{bmatrix} a_1 \\ {}_1\mathbf{a} \end{bmatrix} = \begin{bmatrix} f_1 \\ {}_1\mathbf{f} \end{bmatrix} \tag{11.4}$$

where $\mathbf{0}^{\mathrm{T}} = [0 \quad 0 \quad \dots \quad 0]$. It is evident that (11.4) consists of the first step in the Gauss elimination process in which the unknown a_1 is eliminated from all the equations except the first.

Considering now the system of equations ${}_1\mathbf{K}_1\mathbf{a} = {}_1\mathbf{f}$ we perform the same kind of elimination process as indicated by (11.1)–(11.4), i.e. the unknown a_2 is eliminated from all the equations of ${}_1\mathbf{K}_1\mathbf{a} = {}_1\mathbf{f}$ except the first. This procedure is continued until the triangularized form of the system of equations is obtained. It appears that this approach to Gauss elimination in reality consists of a sequence of static condensations.

To illustrate this approach, we consider the following system of equations:

$$\begin{bmatrix} 2 & -2 & -2 & 0 \\ -2 & 4 & -2 & -2 \\ -2 & -2 & 12 & -2 \\ 0 & -2 & -2 & 22 \end{bmatrix}\begin{bmatrix} a_1 \\ a_2 \\ a_3 \\ a_4 \end{bmatrix} = \begin{bmatrix} 1 \\ 0 \\ -5 \\ 7 \end{bmatrix} \tag{11.5}$$

With (11.3) we obtain

$$ {}_1\mathbf{K} = \begin{bmatrix} 4 & -2 & -2 \\ -2 & 12 & -2 \\ -2 & -2 & 22 \end{bmatrix} - \frac{1}{2}\begin{bmatrix} -2 \\ -2 \\ 0 \end{bmatrix}[-2 \quad -2 \quad 0] = \begin{bmatrix} 2 & -4 & -2 \\ -4 & 10 & -2 \\ -2 & -2 & 22 \end{bmatrix}$$

$$ {}_1\mathbf{f} = \begin{bmatrix} 0 \\ -5 \\ 7 \end{bmatrix} - \frac{1}{2} \times 1\begin{bmatrix} -2 \\ -2 \\ 0 \end{bmatrix} = \begin{bmatrix} 1 \\ -4 \\ 7 \end{bmatrix}$$

The analogue to (11.4) then reads

$$\begin{bmatrix} 2 & -2 & -2 & 0 \\ 0 & 2 & -4 & -2 \\ 0 & -4 & 10 & -2 \\ 0 & -2 & -2 & 22 \end{bmatrix} \begin{bmatrix} a_1 \\ a_2 \\ a_3 \\ a_4 \end{bmatrix} = \begin{bmatrix} 1 \\ 1 \\ -4 \\ 7 \end{bmatrix} \tag{11.6}$$

Using (11.3) gives

$$_2\mathbf{K} = \begin{bmatrix} 10 & -2 \\ -2 & 22 \end{bmatrix} - \frac{1}{2}\begin{bmatrix} -4 \\ -2 \end{bmatrix}[-4 \quad -2] = \begin{bmatrix} 2 & -6 \\ -6 & 20 \end{bmatrix}$$

$$_2\mathbf{f} = \begin{bmatrix} -4 \\ 7 \end{bmatrix} - \frac{1}{2} \times 1 \begin{bmatrix} -4 \\ -2 \end{bmatrix} = \begin{bmatrix} -2 \\ 8 \end{bmatrix}$$

i.e. the following system of equations can now be established:

$$\begin{bmatrix} 2 & -2 & -2 & 0 \\ 0 & 2 & -4 & -2 \\ 0 & 0 & 2 & -6 \\ 0 & 0 & -6 & 20 \end{bmatrix} \begin{bmatrix} a_1 \\ a_2 \\ a_3 \\ a_4 \end{bmatrix} = \begin{bmatrix} 1 \\ 1 \\ -2 \\ 8 \end{bmatrix} \tag{11.7}$$

From (11.3) it follows that

$$_3\mathbf{K} = [20] - \tfrac{1}{2}[-6][-6] = 2; \quad _3\mathbf{f} = [8] - \tfrac{1}{2}(-2)[-6] = 2$$

i.e. the final triangularized form becomes

$$\begin{bmatrix} 2 & -2 & -2 & 0 \\ 0 & 2 & -4 & -2 \\ 0 & 0 & 2 & -6 \\ 0 & 0 & 0 & 2 \end{bmatrix} \begin{bmatrix} a_1 \\ a_2 \\ a_3 \\ a_4 \end{bmatrix} = \begin{bmatrix} 1 \\ 1 \\ -2 \\ 2 \end{bmatrix} \tag{11.8}$$

in accordance with (2.57) and (2.58). Indeed, the reader is encouraged to compare the steps indicated by (11.5)–(11.8) with those arising from the elimination process described in Chapter 2. It appears that a complete similarity exists, but the advantage of the method indicated here by (11.5)–(11.8) is its general and systematic formulation.

We mentioned previously that the size of the bandwidth of the system of equations plays a significant role in the computational costs of solving the system $\mathbf{Ka} = \mathbf{f}$. Let us now investigate this topic. As the triangularization of $\mathbf{K}$ is much more demanding than the operations on the right-hand side $\mathbf{f}$, we shall for convenience only consider the triangularization process of $\mathbf{K}$.

Let the number of unknowns be denoted by n and the bandwidth by B. Then, referring to Figure 11.5 and (11.4), the dimension of $\mathbf{A}$ is $1 \times (B - 1)$. We then have that the establishment of

$$K_{11}^{-1}\mathbf{A}^{\mathrm{T}} \quad \text{requires } B - 1 \text{ divisions} \tag{11.9}$$

Referring to (11.3) we need to establish the matrix $K_{11}^{-1}\mathbf{A}^{\mathrm{T}}\mathbf{A}$. The matrix $\mathbf{A}^{\mathrm{T}}\mathbf{A}$ consists of $(B-1)^2$ components, but as $\mathbf{A}^{\mathrm{T}}\mathbf{A}$ is symmetric, we only need to compute $\frac{1}{2}B(B-1)$ components. That is, knowing the quantity $K_{11}^{-1}\mathbf{A}^{\mathrm{T}}$, the determination of

$$K_{11}^{-1}\mathbf{A}^{\mathrm{T}}\mathbf{A} \quad \text{requires } \tfrac{1}{2}B(B-1) \text{ multiplications} \tag{11.10}$$

Knowing this quantity, the determination of

$$\underset{\sim}{\mathbf{K}} - K_{11}^{-1}\mathbf{A}^{\mathrm{T}}\mathbf{A} \quad \text{requires } \tfrac{1}{2}B(B-1) \text{ subtractions} \tag{11.11}$$

Let us now define an operation as

$$\text{one operation} = \text{one multiplication} + \text{one subtraction} \tag{11.12}$$

Moreover, let us disregard the computational effort related to the determination of $\mathbf{K}_{11}^{-1}\mathbf{A}^{\mathrm{T}}$ and given by (11.9). It follows that the total computational effort for the determination of

$$\underset{\sim}{\mathbf{K}} - K_{11}^{-1}\mathbf{A}\mathbf{A}^{\mathrm{T}} \quad \text{requires } \tfrac{1}{2}B(B-1) \text{ operations.} \tag{11.13}$$

This effort relates to the elimination of one known. As we have n unknowns, we need to make $n - 1$ eliminations, i.e. the total effort in the triangularization of **K** becomes

$$\text{total effort} = (n-1)\tfrac{1}{2}B(B-1) \text{ operations} \tag{11.14}$$

This expression is not completely true. We have assumed that the dimension of **A** is $1 \times (B - 1)$, but in the later stages of the elimination process the dimension of **A** decreases, and at the last stage its dimension is only 1×1 (cf. (11.7)). However, when $n \gg B$ we can ignore this aspect and accept the estimate given by (11.14). Moreover, if $B \gg 1$, (11.14) can approximately be written as

$$\boxed{\text{computational cost} = \alpha n B^2} \tag{11.15}$$

where α is some parameter that depends on the computer. It is obvious that a small bandwidth B enhances the computational efficiency. We emphasize that (11.15) is valid for the Gauss elimination technique and that the computational cost for other solution procedures cannot be estimated using (11.15).

In addition, we mention that whereas the solution procedure presented above is based on a constant value of the bandwidth B – where B in reality is the maximum bandwidth – other Gauss elimination techniques exist by which the variation of the bandwidth is considered. This variation forms a 'skyline' as indicated in Figure 11.6 and solution methods that take advantage of this aspect are called *skyline* or *profile solvers*. Such methods of solution as well as other approaches to the solution of systems of equations are treated in depth by Bathe (1982).

$$B = 5$$

$$
\mathbf{K} = \begin{bmatrix}
\times & \times & 0 & 0 & 0 & 0 & 0 & 0 & 0 \\
 & \times & 0 & \times & 0 & \times & 0 & 0 & 0 \\
 & & \times & \times & 0 & 0 & 0 & 0 & 0 \\
 & & & \times & 0 & \times & 0 & 0 & 0 \\
 & & & & \times & \times & 0 & \times & 0 \\
 & & & & & \times & \times & 0 & 0 \\
 & & & & & & \times & \times & 0 \\
 & & & & & & & \times & \times \\
 & & & & & & & & \times
\end{bmatrix}
$$

skyline

Figure 11.6   Bandwidth $B = 5$ (in reality the maximum bandwidth). The variation of bandwidth is indicated by the 'skyline'

11.2.3 *Global nodal point numbering*

The global stiffness matrix $\mathbf{K}$ and the global load vector $\mathbf{f}_i$ are obtained as a summation of all element stiffness matrices $\mathbf{K}^e$ and all element load vectors $\mathbf{f}_i^e$, respectively, using the topology data. This implies that the element numbers have no influence whatsoever on the result or on computational efficiency. Likewise, as the topology data uniquely relates the local nodal numbers to the global nodal numbers, the local nodal numbers that we choose affect neither the result nor the computational efficiency. When we obtained the FE equations, we did not prescribe the global nodal numbers in any way. Therefore, the choice of global nodal numbers has no effect on the FE results as such, but it does play a major role in computational efficiency because it influences the bandwidth B.

Let us illustrate this important fact by the axially loaded bar shown in Figure 11.7, where simple linear elements are used. Referring to (9.135), the element stiffness matrix $\mathbf{K}^e$ can be written as

$$
\mathbf{K}^e = k \begin{bmatrix} 1 & -1 \\ -1 & 1 \end{bmatrix}
\tag{11.16}
$$

where the stiffness k is given by $k = AE/L$ and where it was assumed that AE is constant along the element. The stiffnesses of the elements are also indicated in Figure 11.7. Following the procedures for establishing the global stiffness matrix $\mathbf{K}$ from the element stiffness matrices, and using the global nodal numbering of Figure 11.7, we

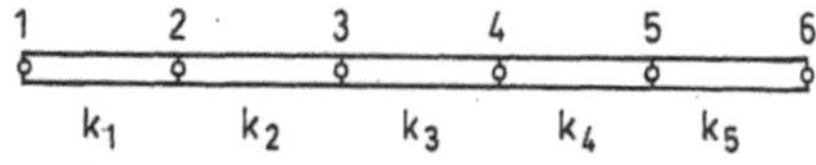

Figure 11.7   Axially loaded bar modelled with simple linear elements

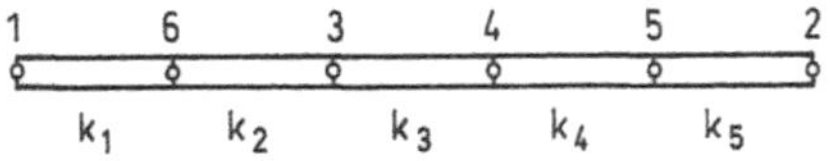

1   6   3   4   5   2

k_1 k_2 k_3 k_4 k_5

Figure 11.8   Axially loaded bar modelled with simple linear elements

obtain

$$
\mathbf{K} = \begin{array}{c} \\ 1 \\ 2 \\ 3 \\ 4 \\ 5 \\ 6 \end{array}
\begin{array}{cccccc}
1 & 2 & 3 & 4 & 5 & 6 \\
\left[\begin{array}{cccccc}
k_1 & -k_1 & 0 & 0 & 0 & 0 \\
-k_1 & k_1 + k_2 & -k_2 & 0 & 0 & 0 \\
0 & -k_2 & k_2 + k_3 & -k_3 & 0 & 0 \\
0 & 0 & -k_3 & k_3 + k_4 & -k_4 & 0 \\
0 & 0 & 0 & -k_4 & k_4 + k_5 & -k_5 \\
0 & 0 & 0 & 0 & -k_5 & k_5
\end{array}\right]
\end{array}
\tag{11.17}
$$

Instead of the numbering of the global nodal points as shown in Figure 11.7, we now choose the numbering shown in Figure 11.8. Using (11.16) again, we now obtain

$$
\mathbf{K} = \begin{array}{c} \\ 1 \\ 2 \\ 3 \\ 4 \\ 5 \\ 6 \end{array}
\begin{array}{cccccc}
1 & 2 & 3 & 4 & 5 & 6 \\
\left[\begin{array}{cccccc}
k_1 & 0 & 0 & 0 & 0 & -k_1 \\
0 & k_5 & 0 & 0 & -k_5 & 0 \\
0 & 0 & k_2 + k_3 & -k_3 & 0 & -k_2 \\
0 & 0 & -k_3 & k_3 + k_4 & -k_4 & 0 \\
0 & -k_5 & 0 & -k_4 & k_4 + k_5 & 0 \\
-k_1 & 0 & -k_2 & 0 & 0 & k_1 + k_2
\end{array}\right]
\end{array}
\tag{11.18}
$$

It appears that whereas the bandwidth of (11.17) is $B = 2$, the bandwidth of (11.18) is $B = 6$. In the present case, where n is a small number, we cannot use (11.15) directly, but it is obvious that the computational efficiency of the global nodal numbering of Figure 11.7 is much more efficient than that of Figure 11.8. We emphasize that, corresponding to Figures 11.7 and 11.8, not only the global stiffness matrices, but also the force vectors are changed in such a way that the two sets of FE results become identical.

11.3 Estimation of bandwidth

With this background, let us now identify a method of determining the bandwidth B for an FE model. For one-dimensional heat flow, a component K_{ij} of the global stiffness matrix $\mathbf{K}$ is given by (9.19), i.e.

$$
K_{ij} = \int_a^b \frac{dN_i}{dx} Ak \frac{dN_j}{dx} dx
\tag{11.19}
$$

With reference to the discussion of this expression, we recall that K_{ij} is zero unless both nodal points i and j are present in at least one element. Considering all elements in the body, let R denote the largest difference between the global nodal points i and j found in an element; that is,

$$\boxed{R = |i - j|_{\max} \quad \text{found in an element}} \tag{11.20}$$

where $|i - j|$ is the numerical value of the difference $i - j$ and where R refers to the largest difference found in all elements of the body. Considering the component K_{ij}, it follows that if $|i - j| > R$ then the global nodal points i and j are never present in the same element. Thus

$$K_{ij} = 0 \quad \text{if } |i - j| > R \tag{11.21}$$

Referring to Figure 11.5, we conclude that the bandwidth B is given by

$$B = R + 1 \tag{11.22}$$

This result was obtained for one-dimensional heat flow and it is also of interest to study two-dimensional heat flow. For this purpose consider a component K_{ij} of $\mathbf{K}$ as given by (10.15). With (10.9) and (6.6) it is easily shown that

$$K_{ij} = \int_A \left[\frac{\partial N_i}{\partial x} \left(k_{xx} \frac{\partial N_j}{\partial x} + k_{xy} \frac{\partial N_j}{\partial y} \right) + \frac{\partial N_i}{\partial y} \left(k_{yx} \frac{\partial N_j}{\partial x} + k_{yy} \frac{\partial N_j}{\partial y} \right) \right] t \, dA \tag{11.23}$$

Just as for (11.19), we draw the conclusion that K_{ij} is zero unless both nodal points i and j are present in at least one element; that is, we are again led to (11.21) and thus to (11.22).

Indeed, this result is not surprising. We recall that a global shape function N_i differs from zero only in those elements which contain nodal point i. Therefore, if we increase the temperature in nodal point i and leave all other nodal temperatures unchanged, this increase only affects those elements that contain nodal point i. Therefore, the temperatures in nodal points i and j are uncoupled unless i and j belong to the same element. Consequently, $K_{ij} = 0$ unless i and j belong to the same element. We conclude that (11.21) and thus (11.22) apply not only to one- and two-dimensional heat flow, but also to three-dimensional heat flow.

Up until now we have studied heat problems where one degree of freedom, i.e. one unknown, is coupled to each nodal point. We shall later encounter problems where more unknowns are related to each nodal point and we immediately generalize (11.22) to

$$\boxed{B = (R + 1)N_{\mathrm{DOF}}} \tag{11.24}$$

where N_{DOF} is the number of unknowns (degrees of freedom) for each node.

In order to illustrate the applicability of (11.24), consider the two different nodal point numberings shown in Figure 11.9. The largest difference R between two nodal numbers in an element is found to be $R = 9$ in Figure 11.9(a) and $R = 35$ in 11.9(b).

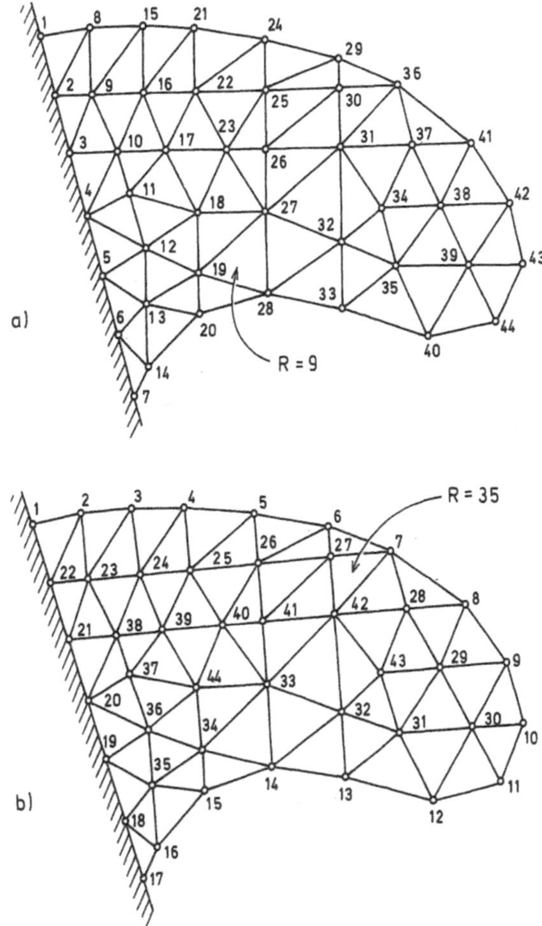

Figure 11.9 Different global nodal point numbering and effect on R

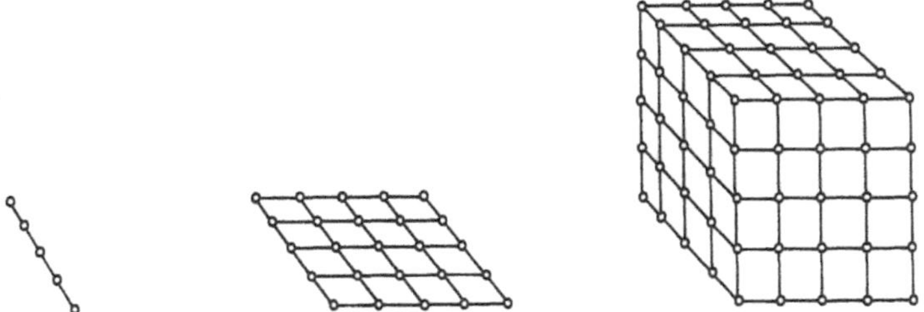

Figure 11.10 One-, two- and three-dimensional models

Table 11.1

	No. of equations, n	Bandwidth, B	Computational cost, αnB^2
1-D	5	2	$\alpha 20$
2-D	25	~ 7	$\alpha 1225$
3-D	125	~ 32	$\alpha 128\,000$

Using (11.24) the respective bandwidths become 10 and 36, if there is one unknown (N_{DOF}) at each node, and 20 and 72 if there are two unknowns at each node.

Finally, let us illustrate that an increase of dimension for a problem increases the computational costs dramatically. Considering heat flow, Figure 11.10 shows one-, two- and three-dimensional models with $n = 5$, $n = 5 \times 5$ and $n = 5 \times 5 \times 5$ nodal points, respectively. The bandwidth B for these problems is given approximately by Table 11.1. This table clearly shows that apparently simple three-dimensional problems present a challenge even for large computers.

12

Stresses and strains

In the previous chapters, the FE method was formulated for problems having one unknown function – the temperature. Apart from their technical importance, such problems present the most straightforward method of introducing the concepts of FE analysis. We shall now shift our focus of interest towards solid mechanics problems where, in general, we have more unknown functions, namely the displacements in the x-, y- and z-directions. To obtain a firm grasp of the differential equations that we will encounter, we will start with some introductory remarks on stresses and strains. For a more comprehensive treatment of stresses and strains, we may refer to Fung (1965), Malvern (1969), Spencer (1980) and Timoshenko and Goodier (1970).

12.1 Stresses

The body is assumed to be continuous and two kinds of forces are assumed: *body forces* (i.e. force per unit volume) and *surface forces* (i.e. force per unit area).

Consider a surface of the body as shown in Figure 12.1. This surface can be an external surface or an internal surface obtained from a section of the body. The vector **n** is a unit vector normal to the surface and directed out of the body. The incremental force vector d**P** acts on the infinitesimal surface area dA. When dA approaches zero, it is assumed that the ratio $d\mathbf{P}/dA$ approaches a value given by

$$\boxed{\mathbf{t} = \frac{d\mathbf{P}}{dA}} \qquad dA \to 0; \quad \mathbf{t} = \begin{bmatrix} t_x \\ t_y \\ t_z \end{bmatrix} \tag{12.1}$$

The vector **t** with components t_x, t_y and t_z in the x-, y- and z-directions, respectively, is termed the *traction vector* and has the dimension $[\text{N/m}^2]$. The traction vector **t** defined above is related to a surface with the outer unit normal vector **n**. It is obvious that the traction vector will, in general, be different when other sections through the same point are considered. What we look for is a quantity – the *stress tensor* – which

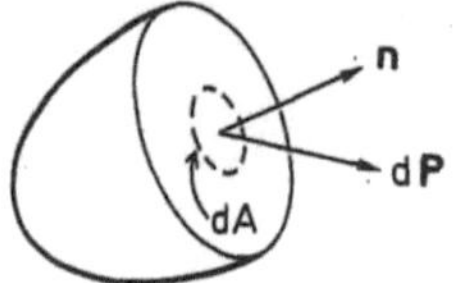

Figure 12.1   Force d**P** on area dA with outer unit normal vector **n**

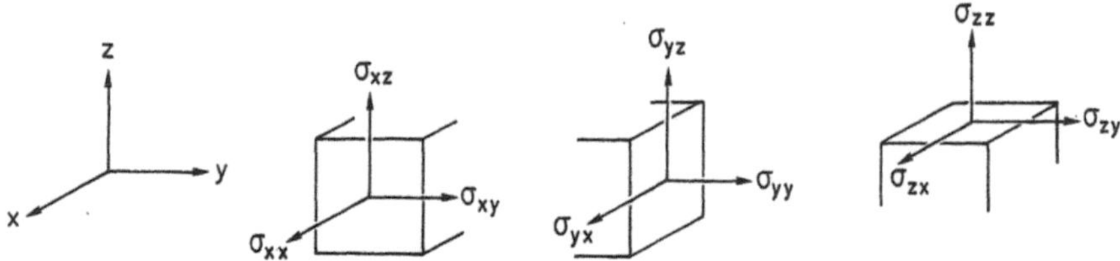

Figure 12.2   Illustration of stress components

for a particular point contains all the information necessary to determine the traction vector for arbitrary sections through the point.

Let us now consider some special traction vectors, namely those obtained when sections perpendicular to the coordinate axes are considered. Assume that the outer normal vector **n** (see Figure 12.1) is in the direction of the x-axis. The corresponding traction vector is denoted by $\mathbf{s}_x$ and we can resolve this vector into its components along the coordinate axes, i.e.

$$\mathbf{s}_x = \begin{bmatrix} \sigma_{xx} \\ \sigma_{xy} \\ \sigma_{xz} \end{bmatrix} \tag{12.2}$$

where σ_{xx}, σ_{xy} and σ_{xz} denote the components of $\mathbf{s}_x$ in the x-, y- and z-directions, respectively. These components are illustrated in Figure 12.2.

Likewise, if the outer normal unit vector **n** is taken in the direction of the y-axis, we denote the corresponding traction vector by $\mathbf{s}_y$, i.e.

$$\mathbf{s}_y = \begin{bmatrix} \sigma_{yx} \\ \sigma_{yy} \\ \sigma_{yz} \end{bmatrix} \tag{12.3}$$

where σ_{yx}, σ_{yy} and σ_{yz} denote the components of $\mathbf{s}_y$ in the x-, y- and z-directions, respectively (cf. Figure 12.2). Finally, if the outer normal unit vector **n** is taken in the direction of the z-axis, we denote the corresponding traction vector by $\mathbf{s}_z$, i.e.

$$\mathbf{s}_z = \begin{bmatrix} \sigma_{zx} \\ \sigma_{zy} \\ \sigma_{zz} \end{bmatrix} \tag{12.4}$$

where σ_{zx}, σ_{zy} and σ_{zz} denote the components of $\mathbf{s}_y$ in the x-, y- and z-directions, respectively (cf. Figure 12.2).

The components given by (12.2)–(12.4) are termed the *stress components* and σ_{xx}, σ_{yy}, σ_{zz} are called *normal stresses*, whereas σ_{xy}, σ_{xz}, σ_{yx}, σ_{yz}, σ_{zx}, σ_{zy} are referred to as *shear stresses*. We observe the consistent notation of the stress components where, for instance, σ_{yz} is the z-component of the traction vector for a surface with the outer unit vector in the y-direction, whereas σ_{xx} is the x-component of the traction vector for a surface with the outer unit vector in the x-direction.

Using the special traction vectors considered above, we define the matrix $\mathbf{S}$ as

$$\mathbf{S} = \begin{bmatrix} \mathbf{s}_x^T \\ \mathbf{s}_y^T \\ \mathbf{s}_z^T \end{bmatrix} = \begin{bmatrix} \sigma_{xx} & \sigma_{xy} & \sigma_{xz} \\ \sigma_{yx} & \sigma_{yy} & \sigma_{yz} \\ \sigma_{zx} & \sigma_{zy} & \sigma_{zz} \end{bmatrix} \tag{12.5}$$

This matrix is called the *stress tensor*. Here we need not probe into the concepts of tensors, but simply mention that tensors are quantities that change in a particular manner when the coordinate system is changed. The only topic of interest here is that $\mathbf{S}$ contains all the stress components.

Let us now prove that $\mathbf{S}$ is symmetric. From the body we cut a small parallelepiped with planes parallel to the coordinate planes. We then consider the moment equilibrium about an axis through the centre E of this parallelepiped and parallel to the z-axis (cf. Figure 12.3). It appears that body forces do not provide a moment about this axis. It is also obvious that only forces acting on planes parallel to the moment axis can contribute to the moment equilibrium. On these planes, only shear stresses normal to the moment axis can give rise to moments; see Figure 12.3.

Referring to this figure, the positive direction of the shear stresses along BC and DC is in accordance with the previous interpretation of the stress components (cf. Figure 12.2). The positive direction of the shear stresses along AB and AD follows from the law of action and reaction. Taking moments as positive in the

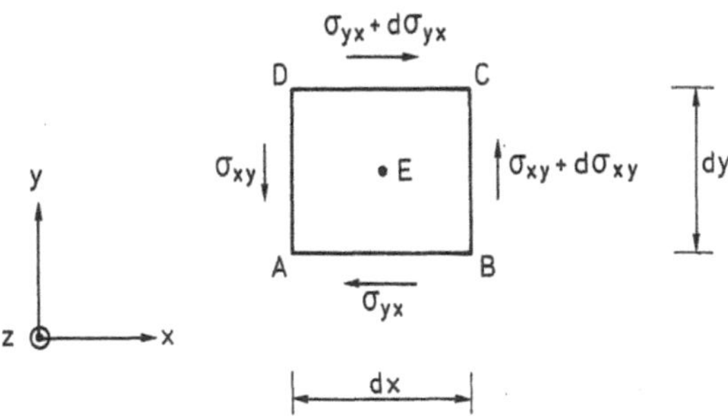

Figure 12.3 Moment about an axis through the centre E and parallel to the z-axis

counter-clockwise direction, moment equilibrium about point E yields

$$(\sigma_{xy} + d\sigma_{xy})\, dy\, dz\tfrac{1}{2}dx - (\sigma_{yx} + d\sigma_{yx})\, dx\, dz\tfrac{1}{2}dy$$
$$+\, \sigma_{xy}\, dy\, dz\tfrac{1}{2}dx - \sigma_{yx}\, dx\, dz\tfrac{1}{2}dy = 0$$

i.e.

$$2\sigma_{xy} - 2\sigma_{yx} + d\sigma_{xy} - d\sigma_{yx} = 0$$

Letting dx, dy and dz approach zero, both $d\sigma_{xy}$ and $d\sigma_{yx}$ also approach zero; that is, moment equilibrium requires that $\sigma_{xy} = \sigma_{yx}$. Likewise, considering moment equilibrium about axes parallel to the x- and y-axes implies that $\sigma_{yz} = \sigma_{zy}$ and $\sigma_{xz} = \sigma_{zx}$, respectively. In conclusion, moment equilibrium requires that

$$\sigma_{xy} = \sigma_{yx}; \quad \sigma_{xz} = \sigma_{zx}; \quad \sigma_{yz} = \sigma_{zy} \tag{12.6}$$

Referring to (12.5) we conclude that **S** is symmetric, i.e.

$$\boxed{\mathbf{S} = \mathbf{S}^{\mathrm{T}}} \tag{12.7}$$

Our aim was to establish a quantity – the stress tensor – which contains all the information necessary to determine the traction vector **t** for arbitrary sections through the point in question. We shall now prove that **S** contains this information.

Consider the infinitesimal tetrahedron shown in Figure 12.4(a). At the surface ABC with the outer unit normal vector **n**, we have the traction vector **t**. On the planes parallel to the coordinate planes the traction vectors are $-\mathbf{s}_x$, $-\mathbf{s}_y$ and $-\mathbf{s}_z$ (cf. (12.2)–(12.4); minus signs appear because of the law of action and reaction and because the outer normal vectors are in the negative direction of the coordinate axes).

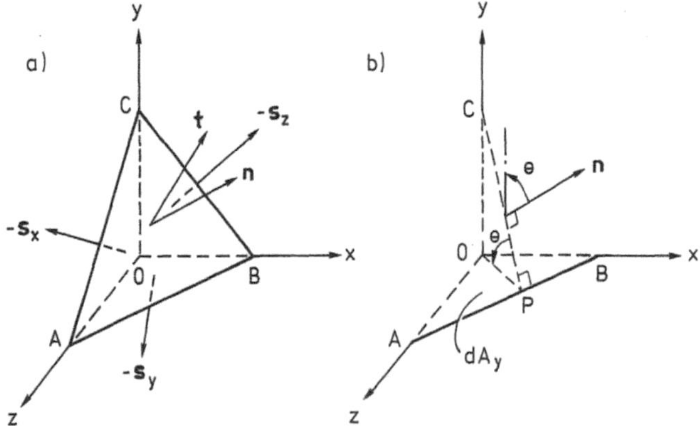

Figure 12.4   (a) Traction vectors on tetrahedron: t acts on ABC, $-\mathbf{s}_x$ on AOC, $-\mathbf{s}_y$ on AOB and $-\mathbf{s}_z$ on BOC; (b) determination of dA_y by geometrical arguments. Vector n is located in the plane OCP

The components of the vector **n** are given by

$$\mathbf{n} = \begin{bmatrix} n_x \\ n_y \\ n_z \end{bmatrix} \tag{12.8}$$

Denoting the angle between **n** and the x-axis by $(\mathbf{n}, x)$ we find, for instance, that $\cos(\mathbf{n}, x) = n_x$. The area ABC is denoted by $\mathrm{d}A$, the area AOC by $\mathrm{d}A_x$, the area AOB by $\mathrm{d}A_y$ and the area BOC by $\mathrm{d}A_z$. In Figure 12.4(b), the line CP is orthogonal to the line AB. However, as **n** is perpendicular to the surface $\mathrm{d}A$, it follows that **n** is located in the plane defined by OCP. Consequently

$$\mathrm{d}A_y = \tfrac{1}{2}\mathrm{OP.AB}; \quad \mathrm{OP} = \mathrm{CP}\cos\theta; \quad \mathrm{d}A = \tfrac{1}{2}\mathrm{CP.AB}$$

whereby

$$\mathrm{d}A_y = \mathrm{d}A\cos\theta; \quad \cos\theta = n_y; \quad \text{i.e. } \mathrm{d}A_y = n_y\,\mathrm{d}A$$

With this and analogous arguments we find that

$$\mathrm{d}A_x = n_x\,\mathrm{d}A; \quad \mathrm{d}A_y = n_y\,\mathrm{d}A; \quad \mathrm{d}A_z = n_z\,\mathrm{d}A \tag{12.9}$$

The condition of force equilibrium of the tetrahedron of Figure 12.4(a) requires that

$$\mathbf{t}\,\mathrm{d}A - \mathbf{s}_x\,\mathrm{d}A_x - \mathbf{s}_y\,\mathrm{d}A_y - \mathbf{s}_z\,\mathrm{d}A_z + \mathbf{b}\,\mathrm{d}V = \mathbf{0} \tag{12.10}$$

where the body force **b** is the force per unit volume and $\mathrm{d}V$ is the volume of the infinitesimal tetrahedron. The body force **b** has the components

$$\mathbf{b} = \begin{bmatrix} b_x \\ b_y \\ b_z \end{bmatrix} \tag{12.11}$$

Using (12.9) in (12.10) gives

$$\mathbf{t} - \mathbf{s}_x n_x - \mathbf{s}_y n_y - \mathbf{s}_z n_z + \mathbf{b}\,\frac{\mathrm{d}V}{\mathrm{d}A} = \mathbf{0}$$

Letting the size of the tetrahedron shrink towards zero, we have $\mathrm{d}V/\mathrm{d}A \to 0$, i.e. we obtain

$$\mathbf{t} = \mathbf{s}_x n_x + \mathbf{s}_y n_y + \mathbf{s}_z n_z \tag{12.12}$$

This may be written as

$$\mathbf{t} = \begin{bmatrix} \mathbf{s}_x & \mathbf{s}_y & \mathbf{s}_z \end{bmatrix} \begin{bmatrix} n_x \\ n_y \\ n_z \end{bmatrix} = \mathbf{S}^\mathrm{T}\mathbf{n}$$

where (12.5) and (12.8) were used. As **S** is symmetric (cf. (12.7)) we conclude that

$$\boxed{\mathbf{t} = \mathbf{S}\mathbf{n}}$$

(12.13)

This expression proves that knowledge of the stress tensor **S** provides sufficient information for the traction vector **t** to be derived for any direction **n**. Equation (12.13) is occasionally referred to as *Cauchy's formula* from 1822. It should be observed that on the exterior surface of the body, (12.13) represents a *boundary condition* expressing a relation between the forces on the external surface and the stress tensor. With (12.1), (12.5) and (12.8), (12.13) can be written as

$$\boxed{\begin{aligned}
t_x &= \sigma_{xx}n_x + \sigma_{xy}n_y + \sigma_{xz}n_z \\
t_y &= \sigma_{yx}n_x + \sigma_{yy}n_y + \sigma_{yz}n_z \\
t_z &= \sigma_{zx}n_x + \sigma_{zy}n_y + \sigma_{zz}n_z
\end{aligned}}$$

(12.14)

Moreover, use of (12.2)–(12.4) yields the alternative formulation

$$t_x = \mathbf{s}_x^{\mathrm{T}}\mathbf{n}; \quad t_y = \mathbf{s}_y^{\mathrm{T}}\mathbf{n}; \quad t_z = \mathbf{s}_z^{\mathrm{T}}\mathbf{n}$$

(12.15)

Instead of the infinitesimal tetrahedron of Figure 12.4, we now consider equilibrium for an arbitrary part of the body. The forces acting on this arbitrary body are given by the traction vector **t** along the boundary surface S and the body force **b** in the region V. Equilibrium requires that

$$\int_S \mathbf{t}\, dS + \int_V \mathbf{b}\, dV = \mathbf{0}$$

(12.16)

This expression comprises three equations which become

$$\int_S t_x\, dS + \int_V b_x\, dV = 0$$
$$\int_S t_y\, dS + \int_V b_y\, dV = 0$$
$$\int_S t_z\, dS + \int_V b_z\, dV = 0$$

(12.17)

With t_x given by (12.15), the first equation of (12.17) may be written as

$$\int_S \mathbf{s}_x^{\mathrm{T}}\mathbf{n}\, dS + \int_V b_x\, dV = 0$$

The first term can be reformulated using Gauss' divergence theorem (cf. (5.32)) to obtain

$$\int_V (\operatorname{div} \mathbf{s}_x + b_x)\, dV = 0$$

As this expression holds for arbitrary regions V, it is concluded that

$$\text{div } \mathbf{s}_x + b_x = 0 \tag{12.18}$$

Inserting the expression for $\mathbf{s}_x$ as given by (12.2) and using the definition of the divergence of a vector (cf. (5.30)) give

$$\frac{\partial \sigma_{xx}}{\partial x} + \frac{\partial \sigma_{xy}}{\partial y} + \frac{\partial \sigma_{xz}}{\partial z} + b_x = 0$$

Treating the two last equations of (12.17) in the same manner, we obtain the following differential equations:

$$\boxed{\begin{aligned}
\frac{\partial \sigma_{xx}}{\partial x} + \frac{\partial \sigma_{xy}}{\partial y} + \frac{\partial \sigma_{xz}}{\partial z} + b_x &= 0 \\[2mm]
\frac{\partial \sigma_{yx}}{\partial x} + \frac{\partial \sigma_{yy}}{\partial y} + \frac{\partial \sigma_{yz}}{\partial z} + b_y &= 0 \\[2mm]
\frac{\partial \sigma_{zx}}{\partial x} + \frac{\partial \sigma_{zy}}{\partial y} + \frac{\partial \sigma_{zz}}{\partial z} + b_z &= 0
\end{aligned}} \tag{12.19}$$

These differential equations express the *equilibrium condition* (i.e. the balance principle) for the body.

Let us define the following matrices:

$$\tilde{\nabla}^{\mathrm{T}} = \begin{bmatrix}
\dfrac{\partial}{\partial x} & 0 & 0 & \dfrac{\partial}{\partial y} & \dfrac{\partial}{\partial z} & 0 \\[3mm]
0 & \dfrac{\partial}{\partial y} & 0 & \dfrac{\partial}{\partial x} & 0 & \dfrac{\partial}{\partial z} \\[3mm]
0 & 0 & \dfrac{\partial}{\partial z} & 0 & \dfrac{\partial}{\partial x} & \dfrac{\partial}{\partial y}
\end{bmatrix} ; \quad \boldsymbol{\sigma} = \begin{bmatrix}
\sigma_{xx} \\ \sigma_{yy} \\ \sigma_{zz} \\ \sigma_{xy} \\ \sigma_{xz} \\ \sigma_{yz}
\end{bmatrix} \tag{12.20}$$

where $\boldsymbol{\sigma}$ contains all the stress components and $\tilde{\nabla}$ is a *matrix differential operator*. With these definitions (12.19) can be written in the compact matrix form

$$\boxed{\tilde{\nabla}^{\mathrm{T}}\boldsymbol{\sigma} + \mathbf{b} = \mathbf{0}} \tag{12.21}$$

12.1.1 *Plane stress*

Plane stress is defined as a stress state where the only non-zero stresses are σ_{xx}, σ_{yy} and σ_{xy}, i.e. the stress tensor $\mathbf{S}$ becomes

$$\mathbf{S} = \begin{bmatrix}
\sigma_{xx} & \sigma_{xy} & 0 \\
\sigma_{yx} & \sigma_{yy} & 0 \\
0 & 0 & 0
\end{bmatrix} \tag{12.22}$$

The boundary conditions (12.14) then take the form

$$\boxed{\begin{aligned} t_x &= \sigma_{xx}n_x + \sigma_{xy}n_y \\ t_y &= \sigma_{yx}n_x + \sigma_{yy}n_y \end{aligned}}$$

(12.23)

and

$$t_z = 0$$

(12.24)

It is important that plane stress conditions require that the component $t_z = 0$. Moreover, we may write

$$\mathbf{t} = \begin{bmatrix} t_x \\ t_y \end{bmatrix}; \quad \mathbf{n} = \begin{bmatrix} n_x \\ n_y \end{bmatrix}$$

(12.25)

Finally, the equilibrium conditions (12.19) become

$$\boxed{\begin{aligned} \frac{\partial \sigma_{xx}}{\partial x} + \frac{\partial \sigma_{xy}}{\partial y} + b_x &= 0 \\ \frac{\partial \sigma_{yx}}{\partial x} + \frac{\partial \sigma_{yy}}{\partial y} + b_y &= 0 \end{aligned}}$$

(12.26)

and it is required that $b_z = 0$. Defining the matrices

$$\tilde{\nabla}^{\mathrm{T}} = \begin{bmatrix} \dfrac{\partial}{\partial x} & 0 & \dfrac{\partial}{\partial y} \\ 0 & \dfrac{\partial}{\partial y} & \dfrac{\partial}{\partial x} \end{bmatrix}; \quad \sigma = \begin{bmatrix} \sigma_{xx} \\ \sigma_{yy} \\ \sigma_{xy} \end{bmatrix}; \quad \mathbf{b} = \begin{bmatrix} b_x \\ b_y \end{bmatrix}$$

(12.27)

(12.26) can be written as

$$\boxed{\tilde{\nabla}^{\mathrm{T}}\sigma + \mathbf{b} = \mathbf{0}}$$

(12.28)

It was observed that plane stress requires that $t_z = 0$ and $b_z = 0$, i.e. both the traction vector $\mathbf{t}$ along the boundary of the body and the body force $\mathbf{b}$ must be forces

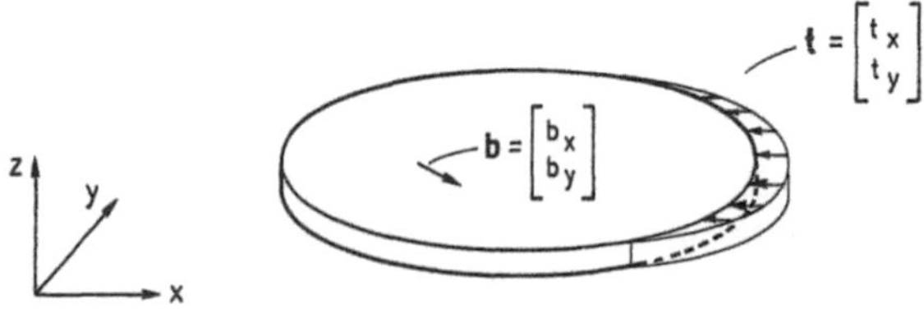

Figure 12.5 Thin disk loaded in plane stress. Traction vector $\mathbf{t}$ and body force vector $\mathbf{b}$ are located in the xy-plane

located in the xy-plane. Thin disks loaded by forces in the plane are examples of structures which may be analyzed using an assumption of plane stress (cf. Figure 12.5). In addition to the assumption above, it is also assumed that nothing depends on the z-coordinate, i.e. plane stress constitutes a two-dimensional problem.

12.2 Strains

Having obtained a description of the stresses in the body, we next consider its deformation. This deformation manifests itself as a change of distance between two neighbouring material points and as a change of angle between two intersecting lines. We will now proceed to quantify these phenomena.

Before deformation, a point in the body is described by the coordinates (x, y, z). After deformation, this point has moved so that it now has the coordinates $(x + u_x, y + u_y, z + u_z)$. The changes u_x, u_y, u_z caused by the deformation are termed *displacement components* and we collect these components in the *displacement vector* **u** given by

$$\mathbf{u} = \begin{bmatrix} u_x \\ u_y \\ u_z \end{bmatrix} \tag{12.29}$$

The displacements of a point (x, y, z) are given by $\mathbf{u} = \mathbf{u}(x, y, z)$. For a neighbouring point $(x + dx, y + dy, z + dz)$, the displacements become $\mathbf{u} + d\mathbf{u}$. Using the chain rule, $d\mathbf{u}$ is written as

$$du_x = \frac{\partial u_x}{\partial x}\,dx + \frac{\partial u_x}{\partial y}\,dy + \frac{\partial u_x}{\partial z}\,dz$$

$$du_y = \frac{\partial u_y}{\partial x}\,dx + \frac{\partial u_y}{\partial y}\,dy + \frac{\partial u_y}{\partial z}\,dz \tag{12.30}$$

$$du_z = \frac{\partial u_z}{\partial x}\,dx + \frac{\partial u_z}{\partial y}\,dy + \frac{\partial u_z}{\partial z}\,dz$$

A derivative like $\partial u_x / \partial x$ is called a *displacement gradient*.

Before any deformations, consider next the line AB parallel to the x-axis (see Figure 12.6). After deformation this line takes the position A′B′. Referring to this figure, the length $|AB|$ of AB is

$$|AB| = dx \tag{12.31}$$

Likewise the length $|A'B'|$ of A′B′ becomes

$$|A'B'| = [(x + dx + u_x + du_x - x - u_x)^2 + (y + u_y + du_y - y - u_y)^2$$
$$+ (z + u_z + du_z - z - u_z)^2]^{1/2}$$

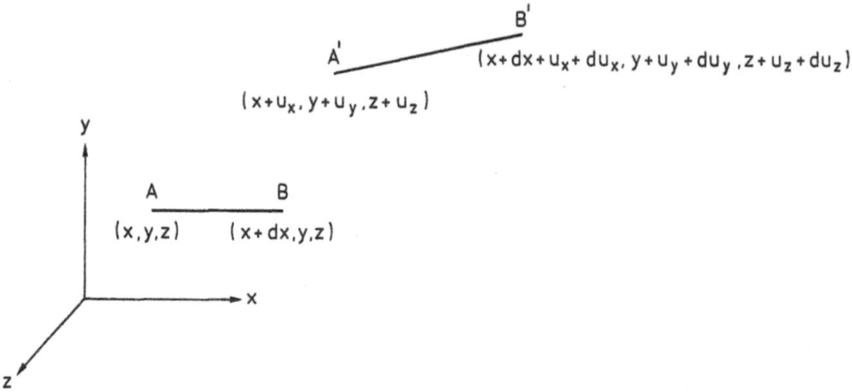

Figure 12.6   Deformation of line AB into line A'B'

i.e.

$$|A'B'| = [(dx + du_x)^2 + (du_y)^2 + (du_z)^2]^{1/2} \tag{12.32}$$

As the line AB is parallel to the x-axis, we have $dy = dz = 0$, i.e. (12.30) gives

$$du_x = \frac{\partial u_x}{\partial x}\,dx; \quad du_y = \frac{\partial u_y}{\partial x}\,dx; \quad du_z = \frac{\partial u_z}{\partial x}\,dx$$

Use of these expressions in (12.32) yields

$$|A'B'| = dx\left[\left(1 + \frac{\partial u_x}{\partial x}\right)^2 + \left(\frac{\partial u_y}{\partial x}\right)^2 + \left(\frac{\partial u_z}{\partial x}\right)^2\right]^{1/2} \tag{12.33}$$

This expression is exact. In most engineering applications, however, the displacement gradients are small when compared with unity, and, as examples, we have

$$\left|\frac{\partial u_x}{\partial x}\right| \ll 1; \quad \left|\frac{\partial u_y}{\partial x}\right| \ll 1; \quad \left|\frac{\partial u_z}{\partial x}\right| \ll 1 \tag{12.34}$$

The assumption of small displacement gradients implies that

$$\left(\frac{\partial u_y}{\partial x}\right)^2 + \left(\frac{\partial u_z}{\partial x}\right)^2 \ll \left(1 + \frac{\partial u_x}{\partial x}\right)^2$$

which simplifies (12.33) to

$$|A'B'| = dx\left(1 + \frac{\partial u_x}{\partial x}\right) \tag{12.35}$$

where $1 + \partial u_x/\partial x$ is positive (cf. (12.34)). We are now in a position to calculate the relative elongation of the infinitesimal line AB. From (12.31) and (12.35) it follows that

$$\frac{|A'B'| - |AB|}{|AB|} = \frac{\partial u_x}{\partial x} \tag{12.36}$$

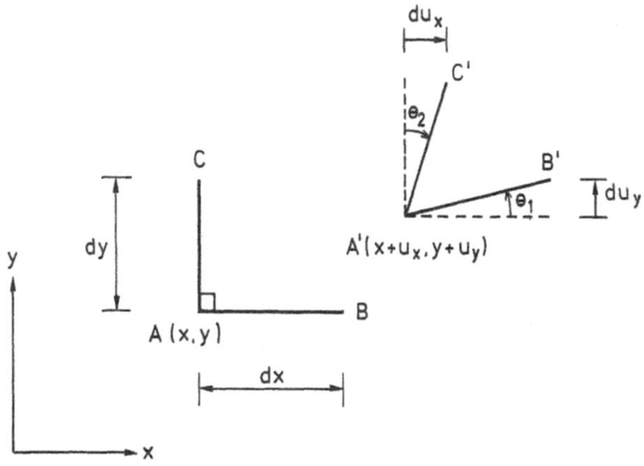

Figure 12.7 Distortion of the right angle CAB due to deformation

This relation was derived for a line parallel to the x-axis and in accordance with the usual terminology we denote this relative elongation as the *normal strain* ε_{xx} in the x-direction, i.e. $\varepsilon_{xx} = \partial u_x / \partial x$. Treating lines parallel to the y- and z-axes in the same manner, the following normal strains are derived:

$$\varepsilon_{xx} = \frac{\partial u_x}{\partial x}; \quad \varepsilon_{yy} = \frac{\partial u_y}{\partial y}; \quad \varepsilon_{zz} = \frac{\partial u_z}{\partial z} \tag{12.37}$$

Next let us evaluate changes of angles. Before the deformation, two orthogonal lines AB and AC parallel to the coordinate axes are considered (see Figure 12.7). Due to the deformation, points A, B and C move to A', B' and C', respectively, as shown in the figure.

As we are only interested in changes of angles, and as small displacement gradients are assumed, we can ignore changes in the length of AB and AC, i.e. $|A'B'| = |AB|$ and $|A'C'| = |AC|$. Therefore, Figure 12.7 provides that

$$\sin \theta_1 = \frac{du_y}{|A'B'|} = \frac{du_y}{dx}; \quad \sin \theta_2 = \frac{du_x}{|A'C'|} = \frac{du_x}{dy} \tag{12.38}$$

Along AB, $dy = dz = 0$ holds, i.e. (12.30) yields $du_y = (\partial u_y / \partial x)\, dx$. Likewise, along AC, $dx = dz = 0$ holds, which implies that $du_x = (\partial u_x / \partial y)\, dy$. Use of these expressions in (12.38) gives

$$\theta_1 = \frac{\partial u_y}{\partial x}; \quad \theta_2 = \frac{\partial u_x}{\partial y} \tag{12.39}$$

where $\sin \theta \simeq \theta$ for small angles. It follows that the orthogonal angle CAB has decreased due to the deformation by the amount

$$\theta_1 + \theta_2 = \frac{\partial u_y}{\partial x} + \frac{\partial u_x}{\partial y}$$

This amount is called the *shear strain* γ_{xy}, where the subscripts indicate that lines AB and AC are parallel to the x- and y-axes, respectively. Treating changes of angles between the other coordinate directions in the same manner, the following shear strains are derived:

$$\gamma_{xy} = \frac{\partial u_x}{\partial y} + \frac{\partial u_y}{\partial x}; \quad \gamma_{xz} = \frac{\partial u_x}{\partial z} + \frac{\partial u_z}{\partial x}; \quad \gamma_{yz} = \frac{\partial u_y}{\partial z} + \frac{\partial u_z}{\partial y} \tag{12.40}$$

It appears that the shear strains are symmetric; for instance, we have $\gamma_{xy} = \gamma_{yx}$.

With the normal strains (12.37) and the shear strains (12.40), a complete description of the deformation of the body has been achieved. As the strains are obtained as derivatives of the displacements, the strains are independent of *rigid-body motions* – as we would expect. However, we emphasize that the strains above were derived under the assumption of small displacement gradients, i.e. *small strains*. In practice, small strains mean strains smaller than, say, 3–5%. For large displacement gradients, proper strain measures become much more involved than (12.37) and (12.40) and we may refer to Fung (1965), Malvern (1969) and Spencer (1980) for a treatment of these matters.

Let us define the following matrices:

$$\boldsymbol{\varepsilon} = \begin{bmatrix} \varepsilon_{xx} \\ \varepsilon_{yy} \\ \varepsilon_{zz} \\ \gamma_{xy} \\ \gamma_{xz} \\ \gamma_{yz} \end{bmatrix}; \quad \tilde{\nabla} = \begin{bmatrix} \dfrac{\partial}{\partial x} & 0 & 0 \\[2mm] 0 & \dfrac{\partial}{\partial y} & 0 \\[2mm] 0 & 0 & \dfrac{\partial}{\partial z} \\[2mm] \dfrac{\partial}{\partial y} & \dfrac{\partial}{\partial x} & 0 \\[2mm] \dfrac{\partial}{\partial z} & 0 & \dfrac{\partial}{\partial x} \\[2mm] 0 & \dfrac{\partial}{\partial z} & \dfrac{\partial}{\partial y} \end{bmatrix} \tag{12.41}$$

where $\boldsymbol{\varepsilon}$ contains all the strains and $\tilde{\nabla}$ is a *matrix differential operator*. With these definitions (12.37) and (12.40) can be combined into the compact form

$$\boldsymbol{\varepsilon} = \tilde{\nabla}\mathbf{u} \tag{12.42}$$

It is of considerable interest that the matrix differential operator $\tilde{\nabla}$ enters both the

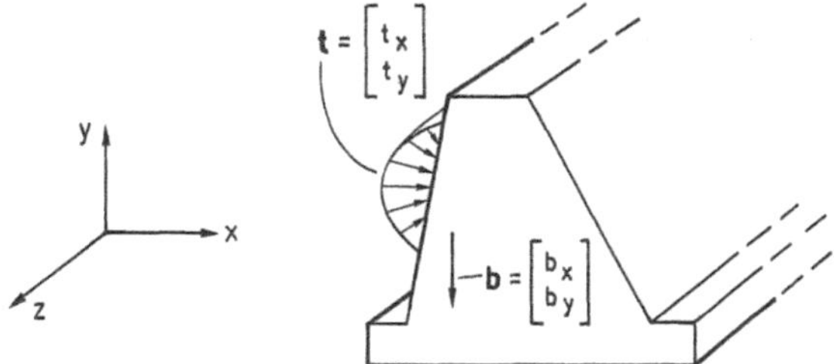

Figure 12.8 Long body loaded in plane strain. Traction vector **t** and body force vector **b** are located in the xy-plane and do not depend on the z-coordinate

equilibrium condition (12.21) and in (12.42). As (12.42) is based on purely kinematical considerations, it is also referred to as the *kinematic relation.*

12.2.1 *Plane strain*

A state of *plane strain* exists if the only non-zero strains are ε_{xx}, ε_{yy} and γ_{xy} and if nothing depends on the z-coordinate. This is equivalent to the following displacements:

$$u_x = u_x(x, y); \quad u_y = u_y(x, y); \quad u_z = 0 \tag{12.43}$$

As nothing depends on the z-coordinate, plane strain constitutes a two-dimensional problem. Let us define the matrices

$$\boldsymbol{\varepsilon} = \begin{bmatrix} \varepsilon_{xx} \\ \varepsilon_{yy} \\ \gamma_{xy} \end{bmatrix}; \quad \mathbf{u} = \begin{bmatrix} u_x \\ u_y \end{bmatrix}; \quad \tilde{\nabla} = \begin{bmatrix} \dfrac{\partial}{\partial x} & 0 \\ 0 & \dfrac{\partial}{\partial y} \\ \dfrac{\partial}{\partial y} & \dfrac{\partial}{\partial x} \end{bmatrix} \tag{12.44}$$

Then the kinematic relations (12.37) and (12.40) can again be written as

$$\boxed{\boldsymbol{\varepsilon} = \tilde{\nabla}\mathbf{u}} \tag{12.45}$$

We note that the matrix differential operator $\tilde{\nabla}$ defined above is the same as that entering the equilibrium differential equations (12.28) for plane stress.

Plane strain often occurs in practice when a long prismatic or cylindrical body is loaded by forces which are perpendicular to the longitudinal axis and which do not vary along this axis. In this case it can be assumed that all cross-sections are in the same state and if, moreover, the body is restricted from moving in the length direction, a state of plane strain exists. An example is a long retaining wall with lateral pressure (see Figure 12.8). We observe that plane strain and plane stress may be considered as two extreme and opposite conditions: whereas the first is applicable for long bodies, the latter may hold for thin bodies.

13

Linear elasticity

In the previous chapter, we established the concepts of stresses and strains. No reference was made to the material as such, and we emphasize that within the assumption of small strains the results of Chapter 12 hold for any material which may be treated as a continuum. It is obvious, however, that stresses and strains must be related in some way or another and the specific manner of this relation is controlled by the specific material in question. The relation between stresses and strains is called the *constitutive relation* and a variety of such relations has been established. Examples are elasticity, plasticity, viscoelasticity, viscoplasticity and creep. Here, we shall consider the simplest constitutive theory, namely *linear elasticity*.

In one dimension, linear elasticity is expressed by *Hooke's law* from 1676

$$\sigma = E\varepsilon \tag{13.1}$$

where the material constant E is Young's modulus. This expression is illustrated in Figure 13.1, and we note that loading as well as unloading follow the same path. This means that the material response is *path independent*; alternatively one says that the material response is *history independent*, as there exists a one-to-one relation between stress and strain. We also emphasize the linear response of the material.

These characteristics also hold for linear elasticity with several stress and strain components. In this general case the stresses are given by σ and the strains by ε (cf. (12.20) and (12.41)), and a direct generalization of (13.1), which maintains the concept

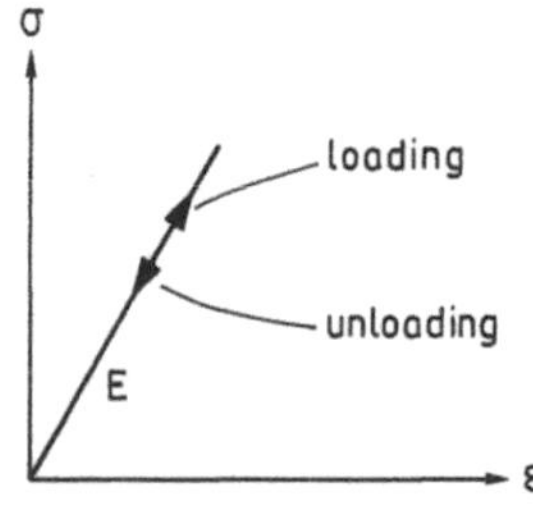

Figure 13.1   Linear elasticity

of linearity between stresses and strains, is given by

$$\boxed{\sigma = \mathbf{D}\varepsilon} \tag{13.2}$$

where

$$\sigma = \begin{bmatrix} \sigma_{xx} \\ \sigma_{yy} \\ \sigma_{zz} \\ \sigma_{xy} \\ \sigma_{xz} \\ \sigma_{yz} \end{bmatrix}; \quad \mathbf{D} = \begin{bmatrix} D_{11} & D_{12} & \cdots & D_{16} \\ D_{21} & D_{22} & \cdots & D_{26} \\ \vdots & \vdots & & \vdots \\ D_{61} & D_{62} & \cdots & D_{66} \end{bmatrix}; \quad \varepsilon = \begin{bmatrix} \varepsilon_{xx} \\ \varepsilon_{yy} \\ \varepsilon_{zz} \\ \gamma_{xy} \\ \gamma_{xz} \\ \gamma_{yz} \end{bmatrix} \tag{13.3}$$

and $\mathbf{D}$ is the *constitutive matrix*. As linearity is required, the matrix $\mathbf{D}$ is constant for a given position. Occasionally, (13.2) is referred to as *Hooke's generalized law*. This generalization is similar to that of Fourier's law from one to three dimensions (cf. (4.3) and (6.4)). The elasticity defined by (13.2), which possesses the properties of linearity and a one-to-one relation between stresses and strains, is also termed *Cauchy elasticity* (Cauchy, 1789–1857). If, in addition to these properties, we also require that the *strain energy* for a given strain state only depends on the strain state itself and not the manner in which this strain state was obtained, we obtain so-called *hyperelasticity* or *Green elasticity* from 1839. In this case, the constitutive matrix $\mathbf{D}$ is symmetric, i.e.

$$\boxed{\mathbf{D} = \mathbf{D}^{\mathrm{T}}} \tag{13.4}$$

as shown, for instance, by Malvern (1969) and Sokolnikoff (1946). The symmetry property (13.4) is used almost universally within linear elasticity, and it will also be adopted here. Due to this symmetry the 36 *elasticity coefficients* of $\mathbf{D}$ reduce to 21 independent coefficients.

Let us investigate the concept of *strain energy* W per unit volume of the body (see Figure 13.2), i.e. W has dimension $[\mathrm{Nm/m^3}]$. For a uniaxial stress state, the incremental strain energy is defined by

$$dW = \sigma \, d\varepsilon; \quad \text{i.e.} \quad W(\varepsilon) = \int_0^\varepsilon \sigma(\varepsilon) \, d\varepsilon \tag{13.5}$$

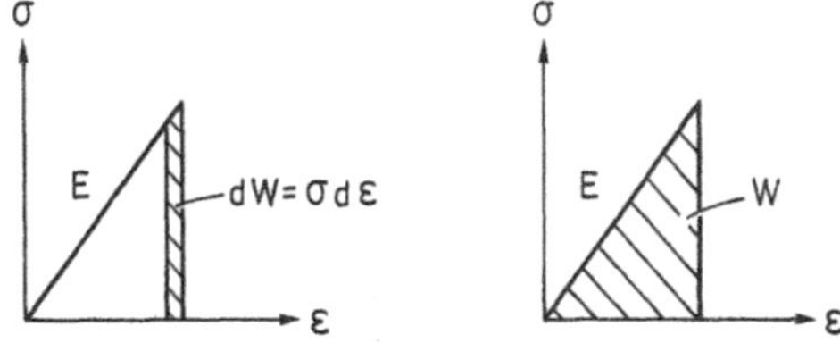

Figure 13.2 Incremental strain energy dW and strain energy W

Using Hooke's law (13.1) we obtain the familiar result

$$W = \tfrac{1}{2}E\varepsilon^2 \tag{13.6}$$

Adopting this approach in the general case gives

$$dW = \sigma^T \, d\varepsilon; \quad dW = d\varepsilon^T\sigma \tag{13.7}$$

where $dW = dW^T$. From Hooke's generalized law (13.2) it follows that

$$dW = \varepsilon^T\mathbf{D} \, d\varepsilon; \quad dW = d\varepsilon^T\mathbf{D}\varepsilon \tag{13.8}$$

where the symmetry property (13.4) was used. We may rewrite (13.8) in the form

$$dW = \tfrac{1}{2}(\varepsilon^T\mathbf{D} \, d\varepsilon + d\varepsilon^T\mathbf{D}\varepsilon) = \tfrac{1}{2}\,d(\varepsilon^T\mathbf{D}\varepsilon) \tag{13.9}$$

This gives

$$W(\varepsilon) = \int_0^\varepsilon dW = \int_0^\varepsilon \tfrac{1}{2}\,d(\varepsilon^T\mathbf{D}\varepsilon)$$

where the limits of integration indicates that the integration should be taken from zero to the current strain state. This integration yields

$$\boxed{W = \tfrac{1}{2}\varepsilon^T\mathbf{D}\varepsilon} \tag{13.10}$$

The strain energy W for a uniaxial state of stress given by (13.6) shows that this strain energy is positive (as we certainly expect Young's modulus E to be positive). It seems natural also to expect that the strain energy is positive for the general case, i.e. (13.10) gives

$$\boxed{\varepsilon^T\mathbf{D}\varepsilon > 0} \tag{13.11}$$

for all $\varepsilon \neq \mathbf{0}$, which shows that $\mathbf{D}$ is positive definite. We may refer to the similar result for heat flow (cf. (6.8)). As $\mathbf{D}$ is positive definite, we conclude from (2.67) that $\mathbf{D}$ is non-singular, i.e.

$$\boxed{\det \mathbf{D} \neq 0} \tag{13.12}$$

This means that (13.2) can be inverted to yield

$$\boxed{\varepsilon = \mathbf{C}\sigma; \quad \mathbf{C} = \mathbf{D}^{-1}} \tag{13.13}$$

where $\mathbf{C}$ is termed the *compliance* or *flexibility matrix*. After having discussed these general properties of the constitutive matrix $\mathbf{D}$, we shall now evaluate its components, i.e. the elasticity coefficients. We will review certain material types and refer to Lekhnitskii (1981), Love (1944), Malvern (1969) and Sokolnikoff (1946) for a detailed discussion. It is emphasized that the intention of this review is confined to provide the reader with some general information on different **D**-matrices that may be encountered in practice and not to present a detailed discussion.

13.1 Symmetry properties

It turns out that, in general, the **D**-matrix changes if another coordinate system is chosen. A *symmetry plane* is said to exist if two coordinate systems, which are mirror images of each other with respect to this plane, leave the **D**-matrix unchanged. If no symmetry planes exist the material is *anisotropic* (see Figure 13.3) and the symmetric **D**-matrix contains 21 independent coefficients; see (13.3) and (13.4). If one symmetry plane exists and if this plane is parallel to the xy-plane (cf. Figure 13.3) it can be shown that 13 independent coefficients exist as shown below:

$$
\mathbf{D} = \begin{bmatrix}
D_{11} & D_{12} & D_{13} & D_{14} & 0 & 0 \\
D_{21} & D_{22} & D_{23} & D_{24} & 0 & 0 \\
D_{31} & D_{32} & D_{33} & D_{34} & 0 & 0 \\
D_{41} & D_{42} & D_{43} & D_{44} & 0 & 0 \\
0 & 0 & 0 & 0 & D_{55} & D_{56} \\
0 & 0 & 0 & 0 & D_{65} & D_{66}
\end{bmatrix}
\tag{13.14}
$$

If three planes of symmetry exist and if the coordinate planes are parallel to these

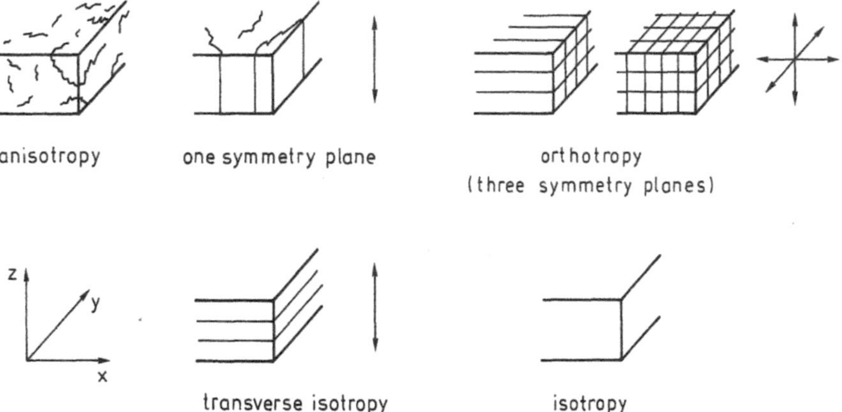

Figure 13.3 Illustration of increasing degree of symmetry

planes, the **D**-matrix turns out to be

$$\mathbf{D} = \begin{bmatrix} D_{11} & D_{12} & D_{13} & 0 & 0 & 0 \\ D_{21} & D_{22} & D_{23} & 0 & 0 & 0 \\ D_{31} & D_{32} & D_{33} & 0 & 0 & 0 \\ 0 & 0 & 0 & D_{44} & 0 & 0 \\ 0 & 0 & 0 & 0 & D_{55} & 0 \\ 0 & 0 & 0 & 0 & 0 & D_{66} \end{bmatrix} \tag{13.15}$$

i.e. we have nine independent coefficients. Such a material is said to be *orthotropic* and it has important practical applications, for instance for wood. It can be shown that two symmetry planes imply the existence of three symmetry planes (cf. Figure 13.3). *Isotropy* exists if the **D**-matrix is the same for all coordinate systems. If a plane of symmetry exists and if isotropy exists in this plane, one speaks of a *transversely isotropic* material. If the plane is taken as the xy-plane, the **D**-matrix is given by

$$\mathbf{D} = \begin{bmatrix} D_{11} & D_{12} & D_{13} & 0 & 0 & 0 \\ D_{21} & D_{22} & D_{13} & 0 & 0 & 0 \\ D_{31} & D_{31} & D_{33} & 0 & 0 & 0 \\ 0 & 0 & 0 & \tfrac{1}{2}(D_{11} - D_{12}) & 0 & 0 \\ 0 & 0 & 0 & 0 & D_{55} & 0 \\ 0 & 0 & 0 & 0 & 0 & D_{55} \end{bmatrix} \tag{13.16}$$

i.e. five independent coefficients. This type of material is illustrated in Figure 13.3 and is typical of stratified materials. Finally, if every plane is a symmetry plane, i.e. if the **D**-matrix takes the same form irrespective of the coordinate system, we have an *isotropic* material. Here

$$\mathbf{D} = \frac{E}{(1 + v)(1 - 2v)} \begin{bmatrix} 1 - v & v & v & 0 & 0 & 0 \\ v & 1 - v & v & 0 & 0 & 0 \\ v & v & 1 - v & 0 & 0 & 0 \\ 0 & 0 & 0 & \tfrac{1}{2}(1 - 2v) & 0 & 0 \\ 0 & 0 & 0 & 0 & \tfrac{1}{2}(1 - 2v) & 0 \\ 0 & 0 & 0 & 0 & 0 & \tfrac{1}{2}(1 - 2v) \end{bmatrix} \tag{13.17}$$

i.e. two independent coefficients. These coefficients are *Young's modulus E* and *Poisson's ratio v*. We recall that an isotropic material has the same properties in all directions (cf. Figure 13.3). This type of material is the one most often used in engineering applications. It is also recalled that the *shear modulus G* is given by

$$G = \frac{E}{2(1 + v)} \tag{13.18}$$

and with (13.2), (13.17) and (13.18) it follows that

$$\sigma_{xy} = G\gamma_{xy}; \quad \sigma_{xz} = G\gamma_{xz}; \quad \sigma_{yz} = G\gamma_{yz} \tag{13.19}$$

From (13.2) and (13.17), we observe that the shear stresses and shear strains are independent of (i.e. uncoupled from) the normal stresses and normal strains and vice versa. Previously, we required **D** to be positive definite (cf. (13.11)) and for an isotropic material it can be shown (e.g. Malvern, 1969, p. 501) that this is fulfilled if

$$E > 0; \quad -1 < v < \tfrac{1}{2} \tag{13.20}$$

It is straightforward to invert **D** for isotropic materials to obtain

$$\mathbf{D}^{-1} = \mathbf{C} = \frac{1}{E}
\begin{bmatrix}
1 & -v & -v & 0 & 0 & 0 \\
-v & 1 & -v & 0 & 0 & 0 \\
-v & -v & 1 & 0 & 0 & 0 \\
0 & 0 & 0 & 2(1+v) & 0 & 0 \\
0 & 0 & 0 & 0 & 2(1+v) & 0 \\
0 & 0 & 0 & 0 & 0 & 2(1+v)
\end{bmatrix} \tag{13.21}$$

We finally remark that if **D** depends on position, i.e. $\mathbf{D} = \mathbf{D}(x, y, z)$, then the material is referred to as *inhomogeneous*; otherwise it is termed *homogeneous*.

13.2 Initial strains

From (13.13) it appears that zero stresses imply zero strains. However, there exist situations with the presence of non-zero strains for zero stresses, i.e.

$$\boldsymbol{\varepsilon} = \mathbf{C}\boldsymbol{\sigma} + \boldsymbol{\varepsilon}_0 \tag{13.22}$$

where $\boldsymbol{\varepsilon}_0$, the so-called *initial strains*, are the strains for zero stresses. As $\mathbf{D} = \mathbf{C}^{-1}$, (13.22) implies that

$$\boldsymbol{\sigma} = \mathbf{D}(\boldsymbol{\varepsilon} - \boldsymbol{\varepsilon}_0) \tag{13.23}$$

An important example of such initial strains is *thermal strains*, i.e. strains caused by the thermal expansion of the material. Consider a specimen of isotropic material which is free to expand as a result of a change in temperature. As the material is free to expand, $\sigma = 0$ holds, and the thermal strains are then given by

$$\varepsilon_0 = \alpha \Delta T \begin{bmatrix} 1 \\ 1 \\ 1 \\ 0 \\ 0 \\ 0 \end{bmatrix} \tag{13.24}$$

where α is the *thermal expansion coefficient* $[1/^\circ\text{C}]$ and ΔT is the change in temperature. In accordance with the assumption of isotropy, the thermal normal strains are equal and no thermal shear strains exist. With (13.17) and (13.24), (13.23) can be written as

$$\boxed{\sigma = \mathbf{D}\varepsilon - \mathbf{D}\varepsilon_0} \tag{13.25}$$

where

$$\mathbf{D}\varepsilon_0 = \frac{\alpha E \Delta T}{1 - 2v} \begin{bmatrix} 1 \\ 1 \\ 1 \\ 0 \\ 0 \\ 0 \end{bmatrix} \tag{13.26}$$

When thermal strains are considered in Hooke's law (13.25) one speaks of *thermoelasticity*.

13.3 Plane stress

For plane stress the only non-zero stresses are σ_{xx}, σ_{yy} and σ_{xy} (cf. (12.22)). Using these conditions in (13.22) and assuming isotropy and that ε_0 are the thermal strains given by (13.24), we obtain from the first, second and fourth equations

$$\begin{bmatrix} \varepsilon_{xx} \\ \varepsilon_{yy} \\ \gamma_{xy} \end{bmatrix} = \frac{1}{E} \begin{bmatrix} 1 & -v & 0 \\ -v & 1 & 0 \\ 0 & 0 & 2(1+v) \end{bmatrix} \begin{bmatrix} \sigma_{xx} \\ \sigma_{yy} \\ \sigma_{xy} \end{bmatrix} + \alpha \Delta T \begin{bmatrix} 1 \\ 1 \\ 0 \end{bmatrix} \tag{13.27}$$

and from the third, fifth and sixth equations

$$\varepsilon_{zz} = -\frac{v}{E}(\sigma_{xx} + \sigma_{yy}) + \alpha \Delta T \tag{13.28}$$

$$\gamma_{xz} = \gamma_{yz} = 0 \tag{13.29}$$

Let us define the following matrices:

$$\boldsymbol{\varepsilon} = \begin{bmatrix} \varepsilon_{xx} \\ \varepsilon_{yy} \\ \gamma_{xy} \end{bmatrix}; \quad \mathbf{C} = \frac{1}{E} \begin{bmatrix} 1 & -v & 0 \\ -v & 1 & 0 \\ 0 & 0 & 2(1+v) \end{bmatrix}; \quad \boldsymbol{\sigma} = \begin{bmatrix} \sigma_{xx} \\ \sigma_{yy} \\ \sigma_{xy} \end{bmatrix}; \quad \boldsymbol{\varepsilon}_0 = \alpha \Delta T \begin{bmatrix} 1 \\ 1 \\ 0 \end{bmatrix} \tag{13.30}$$

Then (13.27) can be written in the compact form

$$\boldsymbol{\varepsilon} = \mathbf{C}\boldsymbol{\sigma} + \boldsymbol{\varepsilon}_0 \tag{13.31}$$

which leads to

$$\boxed{\boldsymbol{\sigma} = \mathbf{D}\boldsymbol{\varepsilon} - \mathbf{D}\boldsymbol{\varepsilon}_0} \tag{13.32}$$

where

$$\mathbf{D} = \mathbf{C}^{-1} = \frac{E}{1-v^2} \begin{bmatrix} 1 & v & 0 \\ v & 1 & 0 \\ 0 & 0 & \frac{1}{2}(1-v) \end{bmatrix} \tag{13.33}$$

and

$$\mathbf{D}\boldsymbol{\varepsilon}_0 = \frac{\alpha E \Delta T}{1-v} \begin{bmatrix} 1 \\ 1 \\ 0 \end{bmatrix} \tag{13.34}$$

It is emphasized that the *in-plane stresses* σ_{xx}, σ_{yy}, σ_{xy} directly determine the *in-plane strains* ε_{xx}, ε_{yy}, γ_{xy} and vice versa. Moreover, when the in-plane stresses are known, these stresses determine the remaining strain components (cf. (13.28) and (13.29)). We note in particular that the *out-of-plane strain* ε_{zz}, in general, is different from zero.

13.4 Plane strain

For plane strain the only non-zero strains are ε_{xx}, ε_{yy} and γ_{xy} (cf. (12.43)). Using these conditions in (13.25) and assuming isotropy we obtain from the first, second and fourth equations

$$\begin{bmatrix} \sigma_{xx} \\ \sigma_{yy} \\ \sigma_{xy} \end{bmatrix} = \frac{E}{(1+v)(1-2v)} \begin{bmatrix} 1-v & v & 0 \\ v & 1-v & 0 \\ 0 & 0 & \frac{1}{2}(1-2v) \end{bmatrix} \begin{bmatrix} \varepsilon_{xx} \\ \varepsilon_{yy} \\ \gamma_{xy} \end{bmatrix} - \frac{\alpha E \Delta T}{1-2v} \begin{bmatrix} 1 \\ 1 \\ 0 \end{bmatrix} \tag{13.35}$$

and from the third, fifth and sixth equations

$$\sigma_{zz} = \frac{Ev}{(1+v)(1-2v)}(\varepsilon_{xx} + \varepsilon_{yy}) - \frac{\alpha E \Delta T}{1-2v} \tag{13.36}$$

$$\sigma_{xz} = \sigma_{yz} = 0 \tag{13.37}$$

Let us define the following matrices:

$$\boldsymbol{\sigma} = \begin{bmatrix} \sigma_{xx} \\ \sigma_{yy} \\ \sigma_{xy} \end{bmatrix}; \quad \mathbf{D} = \frac{E}{(1+v)(1-2v)} \begin{bmatrix} 1-v & v & 0 \\ v & 1-v & 0 \\ 0 & 0 & \tfrac{1}{2}(1-2v) \end{bmatrix};$$

$$\boldsymbol{\varepsilon} = \begin{bmatrix} \varepsilon_{xx} \\ \varepsilon_{yy} \\ \gamma_{xy} \end{bmatrix}; \quad \boldsymbol{\varepsilon}_0 = (1+v)\alpha\Delta T \begin{bmatrix} 1 \\ 1 \\ 0 \end{bmatrix} \tag{13.38}$$

Then (13.35) can be written in the form

$$\boxed{\boldsymbol{\sigma} = \mathbf{D}\boldsymbol{\varepsilon} - \mathbf{D}\boldsymbol{\varepsilon}_0} \tag{13.39}$$

This relation leads to

$$\boldsymbol{\varepsilon} = \mathbf{C}\boldsymbol{\sigma} + \boldsymbol{\varepsilon}_0 \tag{13.40}$$

where

$$\mathbf{C} = \mathbf{D}^{-1} = \frac{1+v}{E} \begin{bmatrix} 1-v & -v & 0 \\ -v & 1-v & 0 \\ 0 & 0 & 2 \end{bmatrix} \tag{13.41}$$

With (13.40) it is possible to rewrite (13.36) as

$$\sigma_{zz} = v(\sigma_{xx} + \sigma_{yy}) - \alpha E \Delta T \tag{13.42}$$

As in the case of plane stress, we emphasize that the in-plane strains ε_{xx}, ε_{yy}, γ_{xy} determine directly the in-plane stresses σ_{xx}, σ_{yy}, σ_{xy} and vice versa. When the in-plane strains are known these strains determine the remaining stress components (cf. (13.36) and (13.37)). We note that the out-of-plane stress σ_{zz}, in general, is different from zero.

13.5 Summary of equations for solid mechanics

It may be advantageous to summarize the relations established so far for three-dimensional solid mechanics.

The stresses σ must fulfil the differential equations of equilibrium (12.21), i.e.

$$\boxed{\tilde{\nabla}^{\mathrm{T}}\sigma + \mathbf{b} = \mathbf{0}} \tag{13.43}$$

where $\mathbf{b}$ is the body force vector. The strains ε must fulfil the kinematic relation (12.42), i.e.

$$\boxed{\varepsilon = \tilde{\nabla}\mathbf{u}} \tag{13.44}$$

where $\mathbf{u}$ is the displacement vector. Within our assumption of small strains, these equations hold for arbitrary solid materials which can be considered as continua. For a specific kind of material, the relation between stresses and strains is given in terms of the constitutive relation. Here, we assume the constitutive relation in the form of linear elasticity; that is, Hooke's generalized law (13.25) states that

$$\boxed{\sigma = \mathbf{D}\varepsilon - \mathbf{D}\varepsilon_0} \tag{13.45}$$

which is formulated here so that it includes the effect of thermal strains. The expressions (13.43)–(13.45) are generally referred to as the *field equations* for thermoelasticity.

These field equations have to be solved given certain boundary conditions. These boundary conditions may be expressed in terms of prescribed traction vectors $\mathbf{t}$ or prescribed displacements $\mathbf{u}$. That is, for three-dimensional problems, we obtain with (12.13)

$$\boxed{\begin{aligned} \mathbf{t} = \mathbf{Sn} = \mathbf{h} \quad &\text{on } S_h \\ \mathbf{u} = \mathbf{g} \quad\quad &\text{on } S_g \end{aligned}} \tag{13.46}$$

where $\mathbf{h}$ and $\mathbf{g}$ are given vectors. S_h is that part of the boundary S on which the traction vector $\mathbf{t}$ is prescribed and S_g is that part of the boundary S on which the displacement vector $\mathbf{u}$ is given. The sum of S_h and S_g constitutes the entire boundary S. We note that it is not possible to prescribe the traction vector and the displacement vector at the same position.

With these observations, the equations for solid mechanics may be summarized as shown in Figure 13.4.

In order to solve a specific problem, the number of unknowns must be the same as the number of equations. Let us check whether (13.43)–(13.45) fulfil this requirement, noting that the body forces $\mathbf{b}$ and the thermal strains ε_0 are prescribed. For three dimensions (3-D), we recall that (13.43) comprises three equations, (13.44) six equations and (13.45) six equations. A survey of equations and unknowns is given in Table 13.1.

It is no surprise that exact analytical solutions of three-dimensional elasticity problems present a formidable task, and as a consequence they are scarce. Moreover, these solutions are confined to physically simple problems with uncomplicated

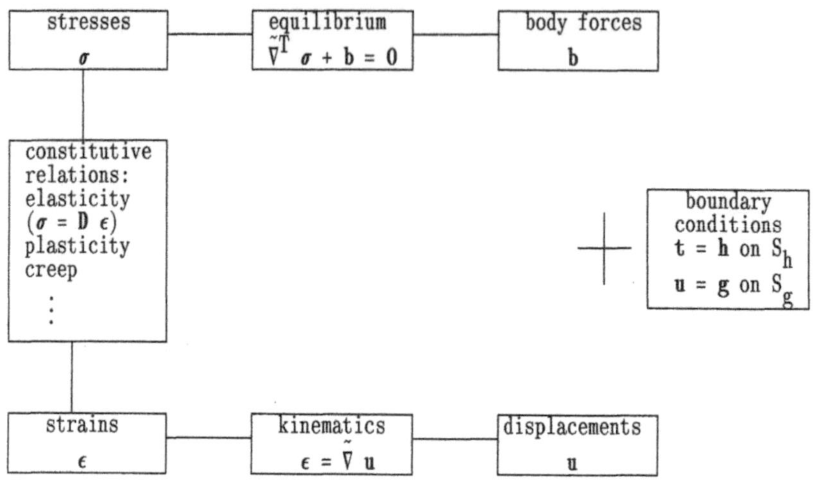

Figure 13.4   Illustration of fundamental equations of solid mechanics

Table 13.1   Summary of equations and unknowns

Type of equation		Number of equations		
		3-D	2-D	1-D
Equilibrium	$\tilde{\nabla}^T\sigma + \mathbf{b} = \mathbf{0}$	3	2	1
Constitutive relations	$\sigma = \mathbf{D}\varepsilon - \mathbf{D}\varepsilon_0$	6	3	1
Kinematics	$\varepsilon = \tilde{\nabla}\mathbf{u}$	6	3	1
Total		15	8	3

Type of unknown		Number of unknowns		
		3-D	2-D	1-D
Stresses	σ	6	3	1
Strains	ε	6	3	1
Displacements	$\mathbf{u}$	3	2	1
Total		15	8	3

geometry. The advantage of the FE method is that we can in principle solve any problem, irrespective of the geometry and even considering complex constitutive relations. Here we confine ourselves to thermoelastic problems, but much more involved constitutive models like, for instance, plasticity and creep can also be treated by the FE method (cf. for instance Zienkiewicz, 1977). The expense that we pay for this gain in generality is that the FE method provides an approximate solution, which, however, converges towards the exact solution for a decreasing element size.

13.5.1 *Comparison of the equations of heat flow and solid mechanics*

Having established the equations governing solid mechanics, it may be of interest to make a comparison with the equations of heat flow. The result is given in Table 13.2.

The variable T or $\mathbf{u}$ is often termed the 'state' variable as it comprises the unknown, i.e. the state itself. The distribution of the state variable gives rise to a certain physical quantity called the 'flux' vector ($\mathbf{q}$ in heat flow, $\boldsymbol{\sigma}$ in solid mechanics). At the boundary, this flux vector is determined by q_n for heat flow and by the traction vector $\mathbf{t}$ in solid mechanics. It is characteristic that the divergence always enters the balance principle, which takes the form of a conservation principle in heat flow and an equilibrium condition for solid mechanics. In order to relate the state variables T or $\mathbf{u}$ to the fluxes $\mathbf{q}$ or $\boldsymbol{\sigma}$, a 'kinematic' quantity is derived from the state variable. In heat flow, this 'kinematic' quantity is the temperature gradient ∇T whereas in solid mechanics it is the strain $\boldsymbol{\varepsilon}$. The flux and kinematic quantity are related by the constitutive law in terms of Fourier's or Hooke's law. Finally, the similarity of the boundary conditions is also apparent from the table.

We note that insertion of the constitutive law into the balance equation gives

$$\text{div}(\mathbf{D}\nabla T) + Q = 0 \tag{13.47}$$

for heat flow, whereas for elasticity we obtain

$$\boxed{\tilde{\nabla}^{\mathrm{T}}\mathbf{D}\tilde{\nabla}\mathbf{u} + \mathbf{b} = 0} \tag{13.48}$$

Table 13.2 Equations of heat flow and of solid mechanics

Heat flow (scalar field problem)	Solid mechanics (vector field problem)	Quantity	Remarks
T	$\mathbf{u}$	state variable	
$\mathbf{q}$	$\boldsymbol{\sigma}$	flux vector	
$q_n = \mathbf{q}^{\mathrm{T}}\mathbf{n}$	$\mathbf{t} = \mathbf{Sn}$	flux at boundary	
$\text{div }\mathbf{q} = Q$	$\left.\begin{array}{l}\text{div }\mathbf{s}_x + b_x = 0 \\ \text{div }\mathbf{s}_y + b_y = 0 \\ \text{div }\mathbf{s}_z + b_z = 0\end{array}\right\}\tilde{\nabla}^{\mathrm{T}}\boldsymbol{\sigma} + \mathbf{b} = 0$		balance or conservation principle
∇T	$\boldsymbol{\varepsilon} = \tilde{\nabla}\mathbf{u}$	'kinematic' quantity	
$\mathbf{q} = -\mathbf{D}\nabla T$	$\boldsymbol{\sigma} = \mathbf{D}\boldsymbol{\varepsilon}$		constitutive law – Fourier and Hooke
$T = g$ on S_g $q_n = h$ on S_h	$\mathbf{u} = \mathbf{g}$ on S_g $\mathbf{t} = \mathbf{h}$ on S_h		boundary conditions

This last set of equations is termed *Navier's equations* from 1821. For orthotropic and isotropic materials, all the terms appearing in (13.47) are given by (6.38) and (6.39). In a similar manner, the reader is invited to identify all the terms in Navier's equations (13.48) and it will turn out that, even for isotropic elasticity, the result is highly complicated. This supports the fact already stated that exact solutions of the field equations of elasticity are scarce.

14

FE formulation of torsion of non-circular shafts

We have emphasized the difficulties of obtaining exact solutions to the field equations of elasticity. Now we shall consider one of the exceptions for which we may obtain an advantageous reformulation of these equations that allows for exact solutions for some geometries and that provides the basis for an FE analysis for general geometries.

The problem we consider is that of torsion of non-circular elastic shafts, i.e. prismatic members. Whereas the solution of circular shafts is trivial, this is certainly not the case for non-circular shafts. The response of such shafts is of great technical importance, and the theory presented below is essentially indebted to Saint-Venant in 1855. This so-called *Saint-Venant torsion theory* is also of historical significance, since it provides the first exact solution to the field equations of elasticity theory. Moreover, it is of considerable interest that Saint-Venant torsion theory leads to equations that are similar to those of heat flow, and therefore we may take advantage of the results obtained previously.

14.1 Saint-Venant torsion theory

14.1.1 *Differential equation for Saint-Venant torsion theory*

The non-circular shaft shown in Figure 14.1 is subjected to a twisting moment (i.e. torque T) about the axis of the prismatic member. The cross-section is constant along the axis and the torque is applied at the ends of the shaft. A coordinate system is chosen so that the z-axis is in the axial direction of the shaft. The essential characteristic of torsion of non-circular shafts is that they *warp*, i.e. we have displacements in the z-direction. The theory below follows partly that given, for instance, by Fung (1965) and Timoshenko and Goodier (1970). The key point is that certain assumptions for the displacements are made and it is then checked that all field equations, in fact, are satisfied exactly.

A cross-section of the shaft perpendicular to the z-axis is shown in Figure 14.2. Apart from the z-axis being in the axial direction, the location of the coordinate

Figure 14.1   Torsion of non-circular shaft

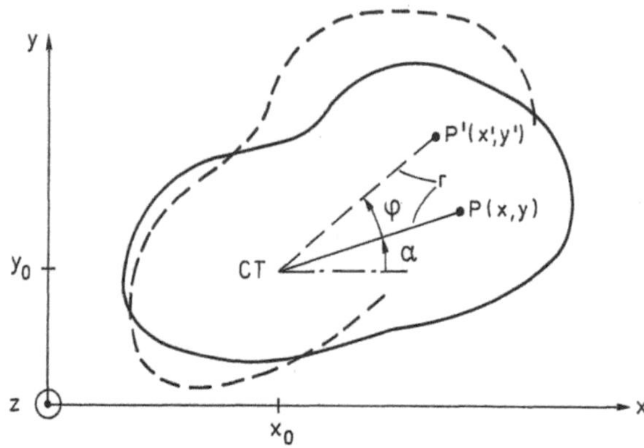

Figure 14.2   Cross-section of member and centre of twist (CT)

system is arbitrary. It is assumed that the shape of this cross-section remains unchanged during torsion; that is, the projection of the cross-section is assumed to rotate in the xy-plane as a rigid body about a centre, the so-called *centre of twist* (CT), with the position (x_0, y_0) (cf. Figure 14.2). Before any deformation, we consider the point P shown in Figure 14.2 with the coordinates

$$x = x_0 + r \cos \alpha; \quad y = y_0 + r \sin \alpha \tag{14.1}$$

where r is the distance between P and CT and α is the angle shown in the figure. Due to the deformation, the point P rotates an angle φ about CT to the position P'. As the projection of the cross-section in Figure 14.2 is assumed to rotate as a rigid body, the distance from P' to CT is r. The coordinates of P' are

$$x' = x_0 + r \cos(\alpha + \varphi); \quad y' = y_0 + r \sin(\alpha + \varphi) \tag{14.2}$$

As usual, we assume small strains; that is, the rotation angle φ is small and thus $\cos \varphi \approx 1$ and $\sin \varphi \approx \varphi$. Using standard trigonometric relations, (14.2) then becomes

$$x' = x_0 + r \cos \alpha - r\varphi \sin \alpha; \quad y' = y_0 + r \sin \alpha + r\varphi \cos \alpha$$

which, due to (14.1) and Figure 14.2, can be written as

$$x' = x - \varphi(y - y_0); \quad y' = y + \varphi(x - x_0) \tag{14.3}$$

With u_x and u_y being the displacements in the x- and y-directions, respectively, we then obtain

$$u_x = x' - x = -\varphi(y - y_0); \quad u_y = y' - y = \varphi(x - x_0) \tag{14.4}$$

Since it was assumed that the projection of the cross-section in Figure 14.2 rotates as a rigid body, the rotation angle φ can only depend on z, i.e.

$$\varphi = \varphi(z) \tag{14.5}$$

Having determined the displacements u_x and u_y, we shall now investigate the displacement u_z in the z-direction. The displacement u_z is called the *warping* and it is characteristic for non-circular shafts that they warp, i.e. $u_z \neq 0$. We consider a shaft for which the cross-section does not change with position along the axis and the shaft is loaded by opposite torques at the ends of the shaft. On this background, it seems reasonable to assume that the warping is independent of the z-coordinate, i.e.

$$u_z = u_z(x, y) \tag{14.6}$$

We emphasize that the displacements given by (14.4)–(14.6) are assumed quantities. However, it will appear that under certain circumstances, these displacements are, in fact, the exact ones. To see this, we shall in the following prove that all field equations and boundary conditions are fulfilled exactly.

From the displacements given by (14.4) and (14.6), the corresponding strains are obtained from the kinematic relations (12.37) and (12.40), i.e.

$$\varepsilon_{xx} = \varepsilon_{yy} = \varepsilon_{zz} = \gamma_{xy} = 0 \tag{14.7}$$

and

$$\gamma_{xz} = -\frac{d\varphi}{dz}(y - y_0) + \frac{\partial u_z}{\partial x}; \quad \gamma_{yz} = \frac{d\varphi}{dz}(x - x_0) + \frac{\partial u_z}{\partial y} \tag{14.8}$$

Having determined the strains, we next consider the constitutive model which is that of linear isotropic elasticity. From (13.2), (13.17) and (13.18) it follows that

$$\sigma_{xx} = \sigma_{yy} = \sigma_{zz} = \sigma_{xy} = 0 \tag{14.9}$$

and

$$\sigma_{xz} = G\left(-\frac{d\varphi}{dz}(y - y_0) + \frac{\partial u_z}{\partial x}\right); \quad \sigma_{yz} = G\left(\frac{d\varphi}{dz}(x - x_0) + \frac{\partial u_z}{\partial y}\right) \tag{14.10}$$

Finally, the last set of field equations to be invoked is the differential equations of equilibrium given by (12.19). With (14.9), and as we have no body forces, these differential equations take the form

$$\frac{\partial \sigma_{xz}}{\partial z} = 0; \quad \frac{\partial \sigma_{yz}}{\partial z} = 0 \tag{14.11}$$

and

$$\frac{\partial \sigma_{xz}}{\partial x} + \frac{\partial \sigma_{yz}}{\partial y} = 0 \tag{14.12}$$

We assume that the shear modulus G does not depend on z, i.e.

$$\boxed{G = G(x, y)} \tag{14.13}$$

Use of (14.10) in (14.11) then leads to

$$\frac{d^2 \varphi}{dz^2} = 0 \tag{14.14}$$

since u_z is independent of z (cf. (14.6)). From (14.14) we conclude that

$$\boxed{\theta = \frac{d\varphi}{dz} = \text{constant}} \tag{14.15}$$

i.e. the *rate of twist* θ is constant along the member. The dimension for θ is $[\text{rad/m}]$.

Let us, following Prandtl in 1903, define a so-called *stress function* $\phi(x, y)$ by

$$\boxed{\sigma_{xz} = \frac{\partial \phi}{\partial y}; \quad \sigma_{yz} = -\frac{\partial \phi}{\partial x}} \tag{14.16}$$

This definition has the important consequence that all the equilibrium differential equations (14.11) and (14.12) are identically satisfied. Moreover, it turns out that the problem of torsion can be expressed entirely in terms of the stress function $\phi(x, y)$. To see this, insert (14.16) in (14.10) and use (14.15) to obtain

$$\frac{1}{G} \frac{\partial \phi}{\partial y} = -\theta(y - y_0) + \frac{\partial u_z}{\partial x}; \quad \frac{1}{G} \frac{\partial \phi}{\partial x} = -\theta(x - x_0) - \frac{\partial u_z}{\partial y}$$

Differentiating the first of these equations with respect to y and the second equation with respect to x and adding the results, we obtain

$$\boxed{\frac{\partial}{\partial x}\left(\frac{1}{G} \frac{\partial \phi}{\partial x}\right) + \frac{\partial}{\partial y}\left(\frac{1}{G} \frac{\partial \phi}{\partial y}\right) + 2\theta = 0} \tag{14.17}$$

This is the quasi-harmonic differential equation that we have previously seen govern two-dimensional heat flow (cf. (6.26)). If G is constant, (14.17) reduces to Poisson's differential equation

$$\frac{\partial^2 \phi}{\partial x^2} + \frac{\partial^2 \phi}{\partial y^2} + 2G\theta = 0 \tag{14.18}$$

(cf. (6.27)). With the definition

$$\mathbf{D} = \begin{bmatrix} 1/G & 0 \\ 0 & 1/G \end{bmatrix} = \frac{1}{G}\mathbf{I} \tag{14.19}$$

(14.17) can be rewritten according to

$$\boxed{\operatorname{div}(\mathbf{D}\nabla\phi) + 2\theta = 0} \tag{14.20}$$

where $\nabla\phi$ is the gradient of ϕ (cf. (5.2)). We may also compare (14.20) with the differential equation (6.32) for two-dimensional heat flow.

At this point we have shown that a solution to (14.17) is in accordance with the exact fulfilment of all kinematic, constitutive and equilibrium equations.

14.1.2 *Boundary conditions*

To solve (14.17) we need to establish proper boundary conditions for the stress function $\phi(x, y)$. As ϕ through (14.16) determines the stresses, it is no surprise that the boundary conditions for ϕ emerge from the stress conditions along the boundary $\mathscr{L}$ of the cross-section; see Figure 14.3.

Let $\mathbf{n}$ denote the unit vector normal to the boundary and directed outwards from the body. Moreover, the unit vector $\mathbf{m}$ is tangential to the boundary and its direction is chosen so that $\mathbf{n}$ and $\mathbf{m}$, after a suitable rigid-body rotation, have the same directions

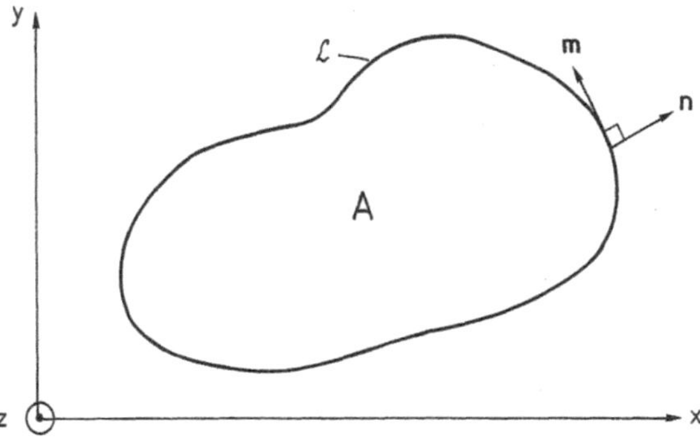

Figure 14.3   Establishment of boundary conditions

as the x- and y-axes, respectively. We have

$$\mathbf{n} = \begin{bmatrix} n_x \\ n_y \end{bmatrix}; \quad \mathbf{m} = \begin{bmatrix} m_x \\ m_y \end{bmatrix}; \quad |\mathbf{n}| = |\mathbf{m}| = 1 \tag{14.21}$$

The orthogonality of these two vectors is expressed through

$$\mathbf{n}^T\mathbf{m} = 0; \quad \text{i.e. } n_x m_x + n_y m_y = 0 \tag{14.22}$$

It appears that this relation is fulfilled for

$$\begin{bmatrix} m_x \\ m_y \end{bmatrix} = \begin{bmatrix} -n_y \\ n_x \end{bmatrix} \tag{14.23}$$

This expression is also easily seen to fulfil the previously mentioned orientation of the $\mathbf{m}$-vector.

After these preliminary remarks, we are able to establish the boundary conditions. We recall that the torque T is applied at the ends of the member, i.e. the traction vector $\mathbf{t}$ is zero along the boundary $\mathscr{L}$ of the cross-section. As the only non-zero stresses are σ_{xz} and σ_{yz}, and as $\mathbf{n}$ is located in the xy-plane, the boundary conditions (12.14) reduce to

$$\sigma_{xz} n_x + \sigma_{yz} n_y = 0$$

Inserting (14.16) and (14.23) yields

$$\frac{\partial \phi}{\partial x} m_x + \frac{\partial \phi}{\partial y} m_y = 0$$

or

$$(\nabla\phi)^T\mathbf{m} = 0 \tag{14.24}$$

i.e. the gradient $\nabla\phi$ is orthogonal to the tangential vector $\mathbf{m}$. We have previously shown in (5.8) that the slope $d\phi/dm$ of a quantity ϕ in the direction of the unit vector $\mathbf{m}$ is given by

$$\frac{d\phi}{dm} = (\nabla\phi)^T\mathbf{m} \tag{14.25}$$

and a comparison with (14.24) shows that $d\phi/dm = 0$, i.e. ϕ is constant along the boundary $\mathscr{L}$ of the cross-section. As the differential equation (14.17) only contains derivatives of ϕ, we may conveniently set

$$\boxed{\phi = 0 \quad \text{along boundary } \mathscr{L}} \tag{14.26}$$

This condition holds if the cross-section of the member is simply connected, i.e. if the member is solid. If the member is not solid, i.e. if the cross-section contains holes, then ϕ takes different constant values on each individual closed boundary. For simplicity, we shall consider here situations where the member is solid, i.e. (14.26) is valid.

The boundary condition (14.26) is then the one for which the differential equation (14.17) is to be solved. The objective of the analysis is primarily to determine the shear stresses σ_{xz} and σ_{yz}, and once the ϕ-function has been determined, these stresses are derived from relation (14.16). For rather simple geometries of the cross-section, like, for instance, ellipses and rectangles, exact closed-form solutions can be obtained for ϕ (cf. for instance Timoshenko and Goodier, 1970). For other more complicated geometries, it is obvious that we can obtain an FE formulation of (14.17) and (14.26) similar to that of two-dimensional heat flow, and we shall investigate this topic later in detail.

14.1.3 *Determination of torque*

In the previous section, we used the fact that the traction vector **t** is zero along the prismatic boundary of the shaft to determine the boundary conditions for Prandtl's stress function ϕ. Let us now consider the stress boundary conditions at the ends of the shaft.

From (14.9) we have that $\sigma_{zz} = 0$ in accordance with the fact that only a torque T is applied at the ends of the shaft. The remaining stress components σ_{xz} and σ_{yz} at the end of the shaft give rise to torque T and if these stress components are those that follow from the solution of (14.17) then all field equations and boundary conditions are exactly fulfilled by our approach.

One point needs some clarification. In the differential equation (14.17), the rate of twist θ enters. This rate of twist is most often unknown and instead the applied torque T is known. Let us therefore derive an expression for T. The torque T is measured as positive if it acts in the counter-clockwise direction (see Figure 14.4). Taking moments about an arbitrary point, for instance the origin of the coordinate

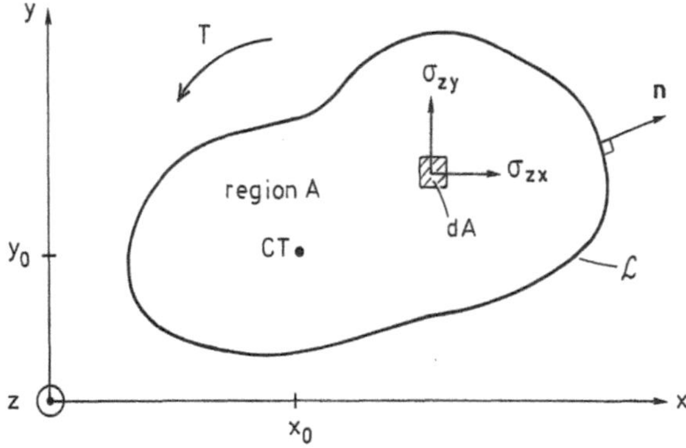

Figure 14.4   Establishment of torque T

system, we obtain from Figure 14.4

$$T = \int_A (x\sigma_{zy} - y\sigma_{zx})\, \mathrm{d}A$$

With (14.16) we obtain

$$T = -\int_A \left(x\frac{\partial\phi}{\partial x} + y\frac{\partial\phi}{\partial y} \right) \mathrm{d}A = -\int_A (\nabla\phi)^{\mathrm{T}} \mathbf{x}\, \mathrm{d}A \tag{14.27}$$

where

$$\mathbf{x} = \begin{bmatrix} x \\ y \end{bmatrix}$$

Use of the Green–Gauss theorem (5.26) in (14.27) results in

$$T = \int_A \phi \operatorname{div} \mathbf{x}\, \mathrm{d}A - \oint_{\mathscr{L}} \phi\mathbf{x}^{\mathrm{T}}\mathbf{n}\, \mathrm{d}\mathscr{L}$$

where the unit vector $\mathbf{n}$ is as shown in Figure 14.4. As we have chosen $\phi = 0$ along the boundary $\mathscr{L}$ and as $\operatorname{div} \mathbf{x} = 1 + 1 = 2$, it follows that

$$T = 2\int_A \phi\, \mathrm{d}A \tag{14.28}$$

That is, the torque is proportional to the volume beneath the surface described by ϕ.

The solution ϕ of (14.17) subjected to the boundary condition (14.26) is proportional to the rate of twist θ. Therefore, according to (14.28) the torque T is also proportional to θ and we may write

$$T = C\theta \tag{14.29}$$

where C is a certain constant for the specific member in question. This constant is termed the *torsional rigidity* or *torsional stiffness*. In most applications, the shear modulus $G(x, y)$ is a constant, and in this case the differential equation reduces to the Poisson equation (14.18). This implies that the solution ϕ becomes proportional to $G\theta$, i.e. instead of (14.29) we have

$$T = GK_v\theta \tag{14.30}$$

where the proportionality constant K_v only depends on the geometry of the cross-section. For a variety of different cross-sections, exact expressions for K_v may be found in the literature, for instance Roark (1975) and Timoshenko and Goodier (1970).

We mentioned previously that in the differential equation (14.17) or (14.20), the loading emerges only via the specification of the value of the rate of twist θ. If, instead, the torque T is known, we first solve (14.17) for an arbitrary value θ_1 of the rate of twist. From the resulting ϕ-solution, the corresponding torque T_1 is calculated according to (14.28), and following (14.29) we then have

$$\theta = \frac{\theta_1}{T_1} T \tag{14.31}$$

Finally, as the solution ϕ is proportional to the rate of twist θ, the same holds for the shear stresses σ_{xz} and σ_{yz}.

14.1.4 *Soap film analogy*

Let us combine the stress components σ_{xz} and σ_{yz} into the matrix $\boldsymbol{\sigma}$ given by

$$\boldsymbol{\sigma} = \begin{bmatrix} \sigma_{xz} \\ \sigma_{yz} \end{bmatrix} \tag{14.32}$$

Referring to Figure 14.4, it is obvious that $\boldsymbol{\sigma}$ may be considered as a vector and the length of this vector is

$$|\boldsymbol{\sigma}| = (\sigma_{xz}^2 + \sigma_{yz}^2)^{1/2} \tag{14.33}$$

It appears that $|\boldsymbol{\sigma}|$ is the magnitude of the maximal shear stress occurring at a particular position. Use of (14.16) in (14.33) gives

$$|\boldsymbol{\sigma}| = \left[\left(\frac{\partial \phi}{\partial x} \right)^2 + \left(\frac{\partial \phi}{\partial y} \right)^2 \right]^{1/2} = |\nabla \phi| \tag{14.34}$$

i.e. large shear stresses occur in regions where the gradient $\nabla \phi$ is large.

Equation (14.24) shows that $\nabla \phi$ is orthogonal to the boundary. With $\mathbf{n}$ being the outer unit vector normal to the boundary, we therefore have

$$\frac{\mathrm{d}\phi}{\mathrm{d}n} = (\nabla \phi)^{\mathrm{T}} \mathbf{n} = |\nabla \phi| |\mathbf{n}| \cos \alpha = \pm |\nabla \phi|$$

since $|\mathbf{n}| = 1$ and $\alpha = 0°$ or $180°$. From (14.34) we then conclude that

$$\boxed{\frac{\mathrm{d}\phi}{\mathrm{d}n} = \pm |\boldsymbol{\sigma}|} \tag{14.35}$$

holds along the boundary of the member.

The result (14.34) opens the way for the appealing interpretation given by *Prandtl's soap film* or *membrane analogy*. For a thin film of liquid, such as that of a soap bubble, the predominant force is the surface tension force S per unit length, which may be considered as constant. Following Timoshenko and Goodier (1970, p. 304), the deflection w of a soap film loaded by a lateral pressure p turns out to be

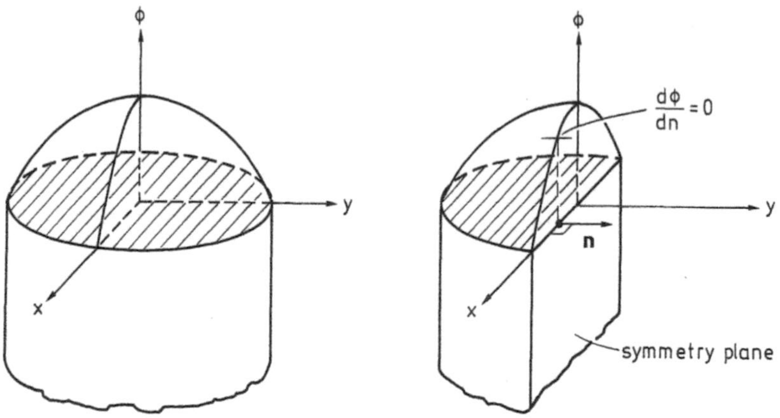

Figure 14.5   Illustration of ϕ-function

governed by the differential equation

$$\frac{\partial^2 w}{\partial x^2} + \frac{\partial^2 w}{\partial y^2} + \frac{p}{S} = 0 \tag{14.36}$$

This expression may be appreciated by referring to its one-dimensional counterpart – the flexible string (4.28). However, the important point is that (14.36) is completely similar to (14.18). In addition, the boundary condition for (14.36) is clearly that $w = 0$ along the edges of the soap film. This boundary condition is in perfect agreement with (14.26). Therefore, if a soap film is placed on a section with a similar geometry as that of the cross-section of the twisted member and then loaded by a pressure, the deflection w is proportional to the function ϕ. By investigating this soap film, it becomes possible to identify very easily where large shear stresses occur, since these regions – according to (14.34) – coincide with the regions where the slope of the soap film is large. Moreover, even in FE analysis, the imagination of such a soap film may be helpful for a qualitative evaluation of the character of the ϕ-function. An illustration of the ϕ-function is given in Figure 14.5 and it is seen that along a symmetry plane we have $d\phi/dn = 0$.

14.1.5 *Discussion of Saint-Venant torsion theory*

We have previously discussed that if the torque T is applied at the end surfaces of the member in accordance with the shear stress distribution following from the solution ϕ then an analytical solution of (14.17) subjected to (14.26) is the exact solution to the elasticity problem considered and not just an engineering approximation. That it is the exact solution follows from the fact that all equations

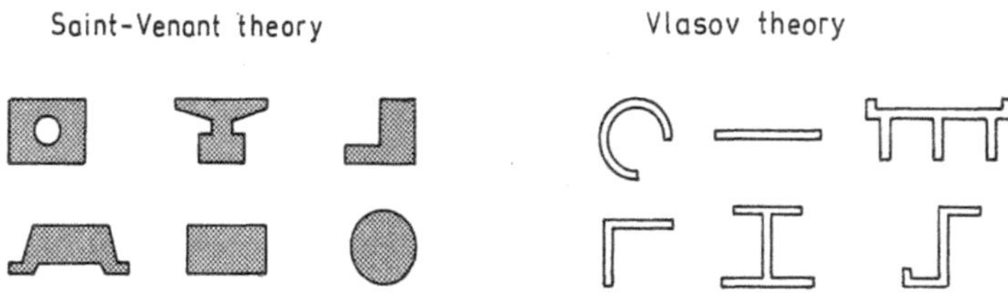

Figure 14.6 Applicability regions for Saint-Venant and Vlasov theory

of interest (differential equations of equilibrium, kinematic relations, constitutive equations and boundary conditions) are satisfied exactly. We observe that, initially, we did not try to solve all these equations directly, but rather guessed one part of the solution, namely certain assumptions for the displacements and then determined the remaining part of the solution rationally so that all equations of interest were satisfied. In general, such an approach is called a *semi-inverse method* and in the present case one speaks of Saint-Venant's semi-inverse method.

The Saint–Venant theory is exact, but it should be noted that we have assumed that the warping u_z given by (14.6), i.e. the displacement in the z-direction, can occur without any constraints, since we have that $\sigma_{zz} = 0$ (cf. (14.9)). Often the warping is restrained, for instance at the end surfaces of the member, and then deviations from the Saint-Venant theory will occur. Generally speaking, it the warping is small, as it usually is for solid or closed thin-walled members, Saint-Venant theory provides close predictions. For open thin-walled members, however, the warping is frequently so large that Saint-Venant theory becomes questionable and in that case the more complicated so-called *Vlasov theory* must be applied (see for instance Kollbrunner and Hajdin, 1972). However, the ramifications of this theory are beyond the scope of this text. Following this discussion we may summarize the applicability regions for Saint-Venant and Vlasov theory as shown in Figure 14.6.

14.2 FE formulation

From (14.20) and (14.26) the strong form of the torsion problem is given by

$$
\begin{aligned}
\mathrm{div}(\mathbf{D}\nabla\phi) + 2\theta &= 0 \quad \text{in region } A &\quad (14.37)\\
\phi &= 0 \quad \text{on boundary } \mathscr{L} &\quad (14.38)
\end{aligned}
$$

where

$$
\mathbf{D} = \begin{bmatrix} 1/G & 0 \\ 0 & 1/G \end{bmatrix} = \frac{1}{G}\mathbf{I}; \quad \nabla\phi = \begin{bmatrix} \dfrac{\partial\phi}{\partial x} \\[2mm] \dfrac{\partial\phi}{\partial y} \end{bmatrix} \qquad (14.39)
$$

The analogy with two-dimensional heat flow is striking (cf. (6.32)–(6.34)). Therefore, we may obtain the FE formulation of the torsion problem by making suitable changes in the FE formulation for two-dimensional heat flow. However, it is more instructive to obtain the FE formulation for torsion directly from (14.37).

Therefore, we first derive the weak form of (14.37). To achieve this form, multiply (14.37) by the arbitrary weight function v and integrate over the region A, i.e.

$$\int_A [v \operatorname{div}(\mathbf{D}\nabla\phi) + 2v\theta]\, \mathrm{d}A = 0 \tag{14.40}$$

The first term is integrated by parts using the Green–Gauss theorem (5.26), i.e.

$$\int_A v \operatorname{div}(\mathbf{D}\nabla\phi)\, \mathrm{d}A = \oint_{\mathscr{L}} v(\mathbf{D}\nabla\phi)^{\mathrm{T}}\mathbf{n}\, \mathrm{d}\mathscr{L} - \int_A (\nabla v)^{\mathrm{T}}\mathbf{D}\nabla\phi\, \mathrm{d}A$$

Use of this expression in (14.40) provides the following weak form:

$$\int_A (\nabla v)^{\mathrm{T}}\mathbf{D}\nabla\phi\, \mathrm{d}A = \oint_{\mathscr{L}} v(\mathbf{D}\nabla\phi)^{\mathrm{T}}\mathbf{n}\, \mathrm{d}\mathscr{L} + 2\int_A v\theta\, \mathrm{d}A \tag{14.41}$$

Let us evaluate the boundary term in more detail. From (14.39) it follows that

$$\mathbf{D}\nabla\phi = \frac{1}{G}\nabla\phi$$

i.e.

$$(\mathbf{D}\nabla\phi)^{\mathrm{T}}\mathbf{n} = \frac{1}{G}(\nabla\phi)^{\mathrm{T}}\mathbf{n} \tag{14.42}$$

From (5.8) we have that

$$\frac{\mathrm{d}\phi}{\mathrm{d}n} = (\nabla\phi)^{\mathrm{T}}\mathbf{n} \tag{14.43}$$

where $\mathrm{d}\phi/\mathrm{d}n$ is the slope of ϕ in the direction of $\mathbf{n}$. Combining (14.42) and (14.43) yields

$$(\mathbf{D}\nabla\phi)^{\mathrm{T}}\mathbf{n} = \frac{1}{G}\frac{\mathrm{d}\phi}{\mathrm{d}n} \tag{14.44}$$

Using (14.44) the weak form (14.41) may be written as

$$\int_A (\nabla v)^{\mathrm{T}}\mathbf{D}\nabla\phi\, \mathrm{d}A = \int_{\mathscr{L}_h} v\frac{1}{G}\frac{\mathrm{d}\phi}{\mathrm{d}n}\, \mathrm{d}\mathscr{L} + \int_{\mathscr{L}_g} v\frac{1}{G}\frac{\mathrm{d}\phi}{\mathrm{d}n}\, \mathrm{d}\mathscr{L} + 2\int_A v\theta\, \mathrm{d}A \tag{14.45}$$

In this expression we have split the boundary term into the two parts

$$\boxed{\begin{aligned} &\frac{1}{G}\frac{\mathrm{d}\phi}{\mathrm{d}n} \quad \text{known along } \mathscr{L}_h \\[2mm] &\phi = 0 \quad \text{known along } \mathscr{L}_g \end{aligned}} \tag{14.46}$$

It may be somewhat surprising that we make this division of the boundary $\mathscr{L}$, since we emphasized that $\phi = 0$ holds all along the boundary (cf. (14.38)). However, when considering symmetry of a region, we do not know the ϕ-value along the symmetry lines, but instead we know the value of $\mathrm{d}\phi/\mathrm{d}n$, namely that $\mathrm{d}\phi/\mathrm{d}n = 0$ (cf. Figure 14.5), and this is the reason for (14.46). In the example considered later, we shall see an application of this kind. However, along the boundary of the member itself, we recall from (14.35) that $\mathrm{d}\phi/\mathrm{d}n$ determines the maximum shear stress along this boundary.

Having obtained the weak form we now introduce the approximation for ϕ. As usual we write

$$\phi = \mathbf{N}\mathbf{a} \tag{14.47}$$

where $\mathbf{N}$ is the global shape function matrix and $\mathbf{a}$ contains the ϕ-value at the nodal points in the entire cross-section. That is,

$$\mathbf{N} = [N_1 \quad N_2 \quad \cdots \quad N_n]; \quad \mathbf{a} = \begin{bmatrix} \phi_1 \\ \phi_2 \\ \vdots \\ \phi_n \end{bmatrix} \tag{14.48}$$

where n is the number of nodal points for the entire cross-section and a component N_i depends on x and y, i.e. $N_i = N_i(x, y)$. From (14.47) we obtain

$$\nabla\phi = \mathbf{B}\mathbf{a} \quad \text{where } \mathbf{B} = \nabla\mathbf{N} \tag{14.49}$$

which implies that

$$\mathbf{B} = \begin{bmatrix} \dfrac{\partial N_1}{\partial x} & \dfrac{\partial N_2}{\partial x} & \cdots & \dfrac{\partial N_n}{\partial x} \\[2mm] \dfrac{\partial N_1}{\partial y} & \dfrac{\partial N_2}{\partial y} & \cdots & \dfrac{\partial N_n}{\partial y} \end{bmatrix} \tag{14.50}$$

Inserting (14.49) into (14.45) gives

$$\left(\int_A (\nabla v)^{\mathrm{T}} \mathbf{D}\mathbf{B}\, \mathrm{d}A \right)\mathbf{a} = \int_{\mathscr{L}_h} v\, \frac{1}{G}\frac{\mathrm{d}\phi}{\mathrm{d}n}\, \mathrm{d}\mathscr{L} + \int_{\mathscr{L}_g} v\, \frac{1}{G}\frac{\mathrm{d}\phi}{\mathrm{d}n}\, \mathrm{d}\mathscr{L} + 2\int_A v\theta\, \mathrm{d}A \tag{14.51}$$

The final step is to choose the arbitrary weight function v in accordance with the Galerkin method. That is, we set

$$v = \mathbf{N}\mathbf{c} \tag{14.52}$$

Since v is arbitrary, the matrix $\mathbf{c}$ is arbitrary. From (14.52) it follows that

$$\nabla v = \mathbf{B}\mathbf{c} \tag{14.53}$$

As $v = v^T$, (14.52) may also be written as

$$v = \mathbf{c}^T \mathbf{N}^T \tag{14.54}$$

Inserting (14.53) and (14.54) into (14.51) and using the fact that $\mathbf{c}$ is independent of position result in

$$\mathbf{c}^T\left[\left(\int_A \mathbf{B}^T\mathbf{D}\mathbf{B}\,dA\right)\mathbf{a} - \int_{\mathscr{L}_h} \mathbf{N}^T\frac{1}{G}\frac{d\phi}{dn}\,d\mathscr{L} - \int_{\mathscr{L}_g} \mathbf{N}^T\frac{1}{G}\frac{d\phi}{dn}\,d\mathscr{L}\right.$$
$$\left.- 2\int_A \mathbf{N}^T\theta\,dA\right] = 0$$

As this expression should hold for arbitrary $\mathbf{c}^T$-matrices, we conclude that

$$\left(\int_A \mathbf{B}^T\mathbf{D}\mathbf{B}\,dA\right)\mathbf{a} = \int_{\mathscr{L}_h} \mathbf{N}^T\frac{1}{G}\frac{d\phi}{dn}\,d\mathscr{L} + \int_{\mathscr{L}_g} \mathbf{N}^T\frac{1}{G}\frac{d\phi}{dn}\,d\mathscr{L} + 2\int_A \mathbf{N}^T\theta\,dA \tag{14.55}$$

which is the FE formulation for the torsion problem. The reader may notice that the derivations starting from the strong form (14.37) and ending with the FE formulation (14.55) contain a summary of much of what has been presented in previous chapters.

As usual we define the stiffness matrix $\mathbf{K}$, the boundary vector $\mathbf{f}_b$ and the load vector $\mathbf{f}_l$ by

$$\boxed{\begin{aligned}
\mathbf{K} &= \int_A \mathbf{B}^T\mathbf{D}\mathbf{B}\,dA \quad \text{where } \mathbf{D} = \frac{1}{G}\mathbf{I} \\[2ex]
\mathbf{f}_b &= \int_{\mathscr{L}_h} \mathbf{N}^T\frac{1}{G}\frac{d\phi}{dn}\,d\mathscr{L} + \int_{\mathscr{L}_g} \mathbf{N}^T\frac{1}{G}\frac{d\phi}{dn}\,d\mathscr{L} \\[2ex]
\mathbf{f}_l &= 2\theta\int_A \mathbf{N}^T\,dA
\end{aligned}} \tag{14.56}$$

where it was used that θ is constant (cf. (14.15)). We may then write (14.55) as

$$\boxed{\mathbf{Ka} = \mathbf{f}_b + \mathbf{f}_l} \tag{14.57}$$

The force vector $\mathbf{f}$ is defined by

$$\boxed{\mathbf{f} = \mathbf{f}_b + \mathbf{f}_l} \tag{14.58}$$

which leads to the standard form

$$\boxed{\mathbf{Ka} = \mathbf{f}} \tag{14.59}$$

As in the heat flow problem, we may obtain the global stiffness matrix $\mathbf{K}$ and the load vector $\mathbf{f}_l$ as a summation of all expanded element stiffness matrices and expanded element load vectors, respectively (cf. (10.34) and (10.35)). Moreover, we can derive the FE formulation of one element using only those degrees of freedom which belong to this element. In the usual notation we obtain

$$\mathbf{K}^e\mathbf{a}^e = \mathbf{f}^e \tag{14.60}$$

where

$$\mathbf{f}^e = \mathbf{f}^e_b + \mathbf{f}^e_l \tag{14.61}$$

and

$$
\mathbf{K}^e = \int_{A_\alpha} \mathbf{B}^{eT}\mathbf{D}\mathbf{B}^e \, \mathrm{d}A \quad \text{where } \mathbf{D} = \frac{1}{G}\mathbf{I}
$$

$$
\mathbf{f}^e_b = \int_{\mathscr{L}_{h\alpha}} \mathbf{N}^{eT}\frac{1}{G}\frac{\mathrm{d}\phi}{\mathrm{d}n}\,\mathrm{d}\mathscr{L} + \int_{\mathscr{L}_{g\alpha}} \mathbf{N}^{eT}\frac{1}{G}\frac{\mathrm{d}\phi}{\mathrm{d}n}\,\mathrm{d}\mathscr{L} \tag{14.62}
$$

$$
\mathbf{f}^e_l = 2\theta \int_{A_\alpha} \mathbf{N}^{eT}\,\mathrm{d}A
$$

Having obtained the ϕ-solution for a given rate of twist θ, the torque T is calculated from (14.28), i.e.

$$T = 2\int_A \phi \, \mathrm{d}A \tag{14.63}$$

Using that $\phi = \mathbf{N}\mathbf{a}$ we may calculate the torque according to

$$T = 2\left(\int_A \mathbf{N}\,\mathrm{d}A\right)\mathbf{a} \tag{14.64}$$

However, in this case it is also advantageous to calculate T from the contributions of each element. In (14.63) we can write the area integration as the sum of integrations over all element regions A_α, i.e.

$$T = 2\sum_{\alpha=1}^{n_{el}} \int_{A_\alpha} \phi \, \mathrm{d}A \tag{14.65}$$

where n_{el} is the total number of elements. In each element, we have $\phi = \mathbf{N}^e\mathbf{a}^e$, i.e.

$$T = 2\sum_{\alpha=1}^{n_{el}}\left[\left(\int_{A_\alpha} \mathbf{N}^e\,\mathrm{d}A\right)\mathbf{a}^e\right] \tag{14.66}$$

We finally mention that when the ϕ-solution has been obtained, then within each element we have

$$\nabla\phi = \mathbf{B}^e\mathbf{a}^e; \quad \nabla\phi = \begin{bmatrix} \dfrac{\partial\phi}{\partial x} \\[2mm] \dfrac{\partial\phi}{\partial y} \end{bmatrix} \tag{14.67}$$

From this value of the gradient $\nabla\phi$ the shear stresses are determined according to (14.16), i.e.

$$\sigma_{xz} = \frac{\partial\phi}{\partial y}; \quad \sigma_{yz} = -\frac{\partial\phi}{\partial x} \tag{14.68}$$

14.2.1 *Example*

Consider the square shaft shown in Figure 14.7(a). As indicated we have four lines of symmetry, implying that only one-eighth of the total cross-section needs to be analyzed. Along these symmetry lines we have the boundary condition $d\phi/dn = 0$ as indicated in the figure. The fractional part of the cross-section is divided into two three-node triangles as shown in Figure 14.7(b). This extremely simple mesh is chosen in order to obtain moderate hand calculations. The dimensions of the cross-section are given in Figure 14.7(b), and in addition we have

$$G = \tfrac{4}{5} \times 10^{11}\,\mathrm{N/m^2}; \quad \theta = 10^{-4}\,\mathrm{rad/m} \tag{14.69}$$

where the value for the rate of twist θ is chosen arbitrarily (cf. the discussion of

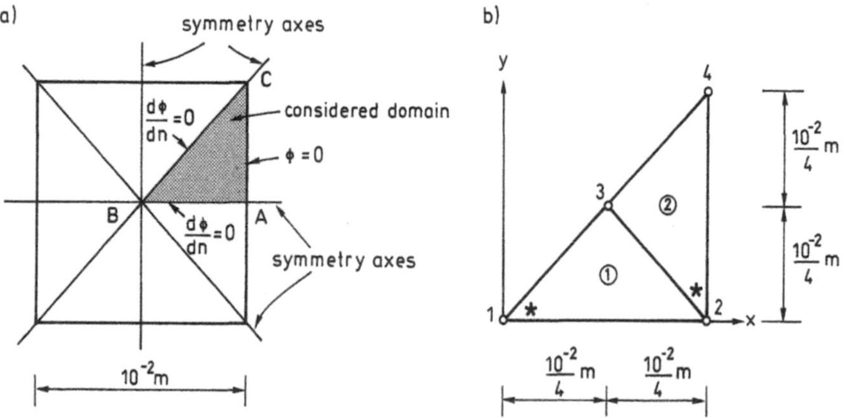

Figure 14.7   (a) Square shaft: use of symmetry; (b) element mesh and global nodal numbers

(14.29)–(14.31)). The local nodal point numbers i, j, k of the three-node triangles start with i located at the asterisk shown in Figure 14.7(b) and, as usual, the sequence i, j, k is taken in the counter-clockwise direction.

First let us evaluate the element stiffness matrices. As the shear modulus G in the present case is constant, (14.62) yields

$$\mathbf{K}^e = \frac{1}{G} \int_{A_\alpha} \mathbf{B}^{eT} \mathbf{B}^e \, dA$$

For the three-node triangle, the $\mathbf{B}^e$-matrix is constant (cf. (7.99)), i.e. we obtain

$$\mathbf{K}^e = \frac{A_\alpha}{G} \mathbf{B}^{eT} \mathbf{B}^e \tag{14.70}$$

Referring to Figure 14.7(b), the element areas A_α are equal and given by

$$A_\alpha = \frac{1}{2} \frac{10^{-2}}{2} \frac{10^{-2}}{4} = \frac{10^{-4}}{16} \tag{14.71}$$

Inserting (14.69) and (14.71) into (14.70) gives

$$\mathbf{K}^e = \tfrac{5}{64} 10^{-15} \mathbf{B}^{eT} \mathbf{B}^e \tag{14.72}$$

Next let us evaluate the element load vectors. From (14.62), we get

$$\mathbf{f}_1^e = 2\theta \int_{A_\alpha} \mathbf{N}^{eT} \, dA$$

As the element shape functions for the three-node triangle vary in a manner similar to Figure 10.5, we obtain with (10.47) that

$$\mathbf{f}_1^e = \frac{2}{3} \theta A_\alpha \begin{bmatrix} 1 \\ 1 \\ 1 \end{bmatrix}$$

From (14.69) and (14.71) we then conclude that

$$\mathbf{f}_1^e = \frac{10^{-8}}{24} \begin{bmatrix} 1 \\ 1 \\ 1 \end{bmatrix} \tag{14.73}$$

It is observed that this expression for the element load vector holds for all the elements.

After these preliminary calculations, *element 1* is considered. From Figure 14.7(b) we have

$$x_i = y_i = 0; \quad x_j = \frac{10^{-2}}{2}, \, y_j = 0; \quad x_k = y_k = \frac{10^{-2}}{4}$$

That is, $\mathbf{B}^e$ given by (7.99) becomes

$$\mathbf{B}^e = 200 \begin{bmatrix} -1 & 1 & 0 \\ -1 & -1 & 2 \end{bmatrix} \tag{14.74}$$

From (14.72) and (14.74) the element stiffness matrix becomes

$$\mathbf{K}^e = \frac{5}{8} \times 10^{-11} \begin{bmatrix} 1 & 0 & -1 \\ 0 & 1 & -1 \\ -1 & -1 & 2 \end{bmatrix} \tag{14.75}$$

Referring to Figure 14.7(b), the local nodal numbers i, j, k correspond to the global nodal numbers 1, 2, 3. That is, after the contributions from the first element given by (14.75) and (14.73), the global stiffness matrix $\mathbf{K}$ and the global load vector $\mathbf{f}_l$ become

$$\mathbf{K} = \mathbf{K}_1^{ee} = \frac{5}{8} \times 10^{-11} \begin{bmatrix} 1 & 0 & -1 & 0 \\ 0 & 1 & -1 & 0 \\ -1 & -1 & 2 & 0 \\ 0 & 0 & 0 & 0 \end{bmatrix}; \quad \mathbf{f}_l = \mathbf{f}_{11}^{ee} = \frac{10^{-8}}{24} \begin{bmatrix} 1 \\ 1 \\ 1 \\ 0 \end{bmatrix} \tag{14.76}$$

Next consider *element 2*. From Figure 14.7(b) we have

$$x_i = \frac{10^{-2}}{2}, \; y_i = 0; \quad x_j = y_j = \frac{10^{-2}}{2}; \quad x_k = y_k = \frac{10^{-2}}{4}$$

That is, $\mathbf{B}^e$ given by (7.99) becomes

$$\mathbf{B}^e = 200 \begin{bmatrix} 1 & 1 & -2 \\ -1 & 1 & 0 \end{bmatrix} \tag{14.77}$$

From (14.72) and (14.77) the element stiffness matrix becomes

$$\mathbf{K}^e = \frac{5}{8} \times 10^{-11} \begin{bmatrix} 1 & 0 & -1 \\ 0 & 1 & -1 \\ -1 & -1 & 2 \end{bmatrix}; \quad \mathbf{K}_2^{ee} = \frac{5}{8} \times 10^{-11} \begin{bmatrix} 0 & 0 & 0 & 0 \\ 0 & 1 & -1 & 0 \\ 0 & -1 & 2 & -1 \\ 0 & 0 & -1 & 1 \end{bmatrix} \tag{14.78}$$

Referring to Figure 14.7(b), the local nodal numbers i, j, k correspond to the global nodal numbers 2, 4, 3. For clarity, the expanded element stiffness matrix $\mathbf{K}_2^{ee}$ is also given in (14.78). Therefore, adding the element stiffness components of (14.78) and the element load components (14.73) to the proper components of $\mathbf{K}$ and $\mathbf{f}_l$ given by

(14.76), we arrive at

$$\mathbf{K} = \frac{5}{8} \times 10^{-11} \begin{bmatrix} 1 & 0 & -1 & 0 \\ 0 & 2 & -2 & 0 \\ -1 & -2 & 4 & -1 \\ 0 & 0 & -1 & 1 \end{bmatrix}; \quad \mathbf{f}_1 = \frac{10^{-8}}{24} \begin{bmatrix} 1 \\ 2 \\ 2 \\ 1 \end{bmatrix} \tag{14.79}$$

These expressions are the final ones for the global stiffness matrix $\mathbf{K}$ and the global load vector $\mathbf{f}_1$.

Now let us evaluate the boundary vector $\mathbf{f}_b$. From (14.56) and Figure 14.7(a) we get

$$\mathbf{f}_b = \int_{\mathscr{L}_{ABC}} \mathbf{N}^T \frac{1}{G} \frac{d\phi}{dn} d\mathscr{L} + \int_{\mathscr{L}_{AC}} \mathbf{N}^T \frac{1}{G} \frac{d\phi}{dn} d\mathscr{L}$$

and as $d\phi/dn = 0$ along ABC, it follows that

$$\mathbf{f}_b = \int_{\mathscr{L}_{AC}} \mathbf{N}^T \frac{1}{G} \frac{d\phi}{dn} d\mathscr{L}$$

Along the boundary AC the only non-zero global shape functions are N_2 and N_4, i.e.

$$\mathbf{f}_b = \begin{bmatrix} 0 \\ R_2 \\ 0 \\ R_4 \end{bmatrix} \tag{14.80}$$

where the notations

$$R_2 = \int_{\mathscr{L}_{AC}} N_2 \frac{1}{G} \frac{d\phi}{dn} d\mathscr{L}; \quad R_4 = \int_{\mathscr{L}_{AC}} N_4 \frac{1}{G} \frac{d\phi}{dn} d\mathscr{L} \tag{14.81}$$

have been introduced for brevity.

From (14.79) and (14.80), the final FE system of equations is

$$\frac{5}{8} \times 10^{-11} \begin{bmatrix} 1 & 0 & -1 & 0 \\ 0 & 2 & -2 & 0 \\ -1 & -2 & 4 & -1 \\ 0 & 0 & -1 & 1 \end{bmatrix} \begin{bmatrix} \phi_1 \\ 0 \\ \phi_3 \\ 0 \end{bmatrix} = \begin{bmatrix} 0 \\ R_2 \\ 0 \\ R_4 \end{bmatrix} + \frac{10^{-8}}{24} \begin{bmatrix} 1 \\ 2 \\ 2 \\ 1 \end{bmatrix} \tag{14.82}$$

where the boundary condition $\phi = 0$ along AC has been used, i.e.

$$\phi_2 = \phi_4 = 0 \tag{14.83}$$

From the first and third equations of (14.82), we obtain

$$\frac{5}{8} \times 10^{-11} \begin{bmatrix} 1 & -1 \\ -1 & 4 \end{bmatrix} \begin{bmatrix} \phi_1 \\ \phi_3 \end{bmatrix} = \frac{10^{-8}}{24} \begin{bmatrix} 1 \\ 2 \end{bmatrix}$$

with the solution

$$\phi_1 = \frac{2}{15} \times 10^3 = 133.3 \text{ N/m}; \quad \phi_3 = \frac{10^3}{15} = 66.7 \text{ N/m} \tag{14.84}$$

In order to calculate the torque T and the stresses, it is advantageous to establish the vector $\mathbf{a}^e$ for the two elements. With reference to Figure 14.7(b) it follows that

$$\text{element 1: } \mathbf{a}^e = \begin{bmatrix} \phi_1 \\ \phi_2 \\ \phi_3 \end{bmatrix}; \quad \text{element 2: } \mathbf{a}^e = \begin{bmatrix} \phi_2 \\ \phi_4 \\ \phi_3 \end{bmatrix} \tag{14.85}$$

To determine the torque T, we first evaluate the term

$$\int_{A_\alpha} \mathbf{N}^e \, dA = \int_{A_\alpha} [N_i^e \quad N_j^e \quad N_k^e] \, dA$$

The element shape functions vary in the manner shown in Figure 10.5, i.e. (10.47) implies that

$$\int_{A_\alpha} \mathbf{N}^e \, dA = \frac{A_\alpha}{3} [1 \quad 1 \quad 1] \tag{14.86}$$

From (14.66), (14.85) and (14.86) we obtain

$$T = 8 \times 2 \left(\frac{A_1}{3} [1 \quad 1 \quad 1] \begin{bmatrix} \phi_1 \\ \phi_2 \\ \phi_3 \end{bmatrix} + \frac{A_2}{3} [1 \quad 1 \quad 1] \begin{bmatrix} \phi_2 \\ \phi_4 \\ \phi_3 \end{bmatrix} \right)$$

where the factor 8 emerges because we have only considered one-eighth of the total cross-section. As $A_1 = A_2 = A_\alpha$, where A_α is given by (14.71), and with the solution (14.83) and (14.84), we arrive at

$$T = \tfrac{4}{45} \times 10^{-1} \text{ N m} \tag{14.87}$$

This is the torque required to produce the assumed rate of twist $\theta = 10^{-4} [\text{rad/m}]$ (cf. (14.69)). The torsional rigidity C (cf. (14.29)) is

$$C = \frac{T}{\theta} = \frac{4}{45} 10^3 = 88.9 \text{ N m}^2 \tag{14.88}$$

From Timoshenko and Goodier (1970, p. 313), the exact torsional rigidity can be calculated as

$$C_{\text{exact}} = 112.5 \text{ N m}^2 \tag{14.89}$$

and, recognizing the extremely simple finite mesh adopted, the solution (14.88) is surprisingly close.

Finally let us determine the shear stresses given by (14.68). For element 1, it follows from (14.67), (14.74) and (14.85) that

$$\begin{bmatrix} \dfrac{\partial \phi}{\partial x} \\[2mm] \dfrac{\partial \phi}{\partial y} \end{bmatrix} = 200 \begin{bmatrix} -1 & 1 & 0 \\ -1 & -1 & 2 \end{bmatrix} \begin{bmatrix} \phi_1 \\ \phi_2 \\ \phi_3 \end{bmatrix}$$

and with the solution (14.83) and (14.84), we get

$$\begin{bmatrix} \dfrac{\partial \phi}{\partial x} \\[2mm] \dfrac{\partial \phi}{\partial y} \end{bmatrix} = -\frac{80}{3} \times 10^3 \begin{bmatrix} 1 \\ 0 \end{bmatrix}$$

From (14.68), the shear stresses in element 1 then become

$$\sigma_{xz} = 0; \quad \sigma_{yz} = 26.7 \times 10^3 \ \text{N}/\text{m}^2 \tag{14.90}$$

We emphasize that, due to the adoption of the three-node triangle, which possesses a constant gradient $\nabla \phi$, the stresses are also constant within the element. Proceeding in the same manner for element 2, we find that

$$\sigma_{xz} = 0; \quad \sigma_{yz} = 26.7 \times 10^3 \ \text{N}/\text{m}^2 \tag{14.91}$$

15

Approximating functions for the FE method – vector problems

Up until now, the FE approach has been formulated for problems having one unknown function. When dealing with solid mechanics we have, in general, more unknown functions, namely the displacements in the x-, y- and z-directions. These displacements define the displacement vector $\mathbf{u}$ and we shall now discuss the manner in which we obtain approximate expressions for this vector. It will turn out that such approximations are easily established when use is made of our previous results from Chapter 7. However, let us start with the convergence criteria.

15.1 Convergence criteria

We consider a problem for which the unknowns are given in terms of the displacement vector $\mathbf{u}$. For decreasing element size, it is required that the approximation for $\mathbf{u}$ should converge towards the exact displacement vector. For an infinitely small region, the exact displacement components may either be constant or possess constant derivatives with respect to the coordinates, i.e. they exhibit constant gradients. In three dimensions, the approximation for the displacement components must therefore be of the type

$$
\begin{aligned}
u_x &= \alpha_1 + \alpha_2 x + \alpha_3 y + \alpha_4 z + \text{possibly other terms} \\
u_y &= \beta_1 + \beta_2 x + \beta_3 y + \beta_4 z + \text{possibly other terms} \\
u_z &= \gamma_1 + \gamma_2 x + \gamma_3 y + \gamma_4 z + \text{possibly other terms}
\end{aligned}
\tag{15.1}
$$

where $\alpha_1 \ldots \alpha_4$, $\beta_1 \ldots \beta_4$ and $\gamma_1 \ldots \gamma_4$ are constant parameters within the region – or element – considered. With (15.1), we can determine the strains according to (12.37) and (12.40) to obtain

$$
\begin{aligned}
\varepsilon_{xx} &= \alpha_2; & \varepsilon_{yy} &= \beta_3; & \varepsilon_{zz} &= \gamma_4 \\
\gamma_{xy} &= \alpha_3 + \beta_2; & \gamma_{xz} &= \alpha_4 + \gamma_2; & \gamma_{yz} &= \beta_4 + \gamma_3
\end{aligned}
\tag{15.2}
$$

where only the linear terms in (15.1) have been considered. As all the parameters $\alpha_1 \ldots \alpha_4$, $\beta_1 \ldots \beta_4$ and $\gamma_1 \ldots \gamma_4$ are independent, it appears from (15.2) that the approximation (15.1) for the displacement components implies that an arbitrary constant strain state can be approximated within a region. Moreover, as the terms α_1, β_1 and γ_1 correspond to an arbitrary rigid-body translation, we are led to the following *completeness requirements*:

Completeness

- The approximation of the displacement vector **u** must be able to represent an arbitrary constant rigid-body motion.
- The approximation of the displacement vector **u** must be able to represent an arbitrary constant strain state.

These completeness requirements are fulfilled for the approximations given by (15.1).

In addition, we require fulfilment of the *compatibility* or *conforming requirement* stating that:

Compatibility or conforming requirement

- The approximation of the displacement vector **u** must vary in a continuous manner over element boundaries.

If both the completeness and compatibility requirements are fulfilled, we have ensured convergence in the sense that, for a decreasing element size, the approximation for the displacement vector **u** approaches the exact one, i.e.

---

convergence criteria = completeness + compatibility requirements

---

We observe that it *may* be allowable to relax the compatibility requirement provided that the discontinuity across the element boundaries approaches zero for decreasing element size. We may refer to Hughes (1987), Strang and Fix (1973) and Zienkiewicz and Taylor (1989) for a more stringent treatment. Such non-conforming elements are used especially in plate analysis, discussed in Chapter 18.

The continuity requirement for the displacement vector **u** clearly means that the displacement components are also required to vary continuously over the element boundaries. With this observation we may compare (15.1) with (7.2) and it is concluded that each displacement component can be approximated in exactly the same manner as that used for temperature approximations. We may therefore refer to Chapter 7 for a discussion of the rate of convergence. Likewise, more information about non-conforming elements, especially the patch test that such elements must pass, is given on page 93. It is emphasized that the displacement components are approximated independently of each other.

15.2 General approximation techniques

With the observation that every displacement component can be approximated in exactly the same manner as that adopted for temperature approximations and independently of each other, we can directly establish the approximation technique for the displacement vector **u**.

First consider two-dimensional solid mechanics, where only the displacement components u_x and u_y are considered. Assume that the element contains n_e nodal points. For each nodal point i we have a displacement u_{xi} in the x-direction and a displacement u_{yi} in the y-direction, as illustrated in Figure 15.1. From Chapter 7 it follows directly that if the same *type* of approximation is used for u_x and u_y, we get

$$u_x = \begin{bmatrix} N_1^e & N_2^e & \dots & N_{n_e}^e \end{bmatrix} \begin{bmatrix} u_{x1} \\ u_{x2} \\ \vdots \\ u_{xn_e} \end{bmatrix}; \quad u_y = \begin{bmatrix} N_1^e & N_2^e & \dots & N_{n_e}^e \end{bmatrix} \begin{bmatrix} u_{y1} \\ u_{y2} \\ \vdots \\ u_{yn_e} \end{bmatrix} \quad (15.3)$$

It is emphasized that the element shape function N_i^e belonging to nodal point i is exactly the same as that discussed in Chapter 7. Moreover, as the same *type* of approximation is used for u_x and u_y it follows that N_i^e in the expression for u_x is the same as N_i^e in the expression for u_y. Instead of the formulation (15.3), we may write

$$u_x = \begin{bmatrix} N_1^e & 0 & N_2^e & 0 & \dots & N_{n_e}^e & 0 \end{bmatrix} \begin{bmatrix} u_{x1} \\ u_{y1} \\ u_{x2} \\ u_{y2} \\ \vdots \\ u_{xn_e} \\ u_{yn_e} \end{bmatrix};$$

$$u_y = \begin{bmatrix} 0 & N_1^e & 0 & N_2^e & \dots & 0 & N_{n_e}^e \end{bmatrix} \begin{bmatrix} u_{x1} \\ u_{y1} \\ u_{x2} \\ u_{y2} \\ \vdots \\ u_{xn_e} \\ u_{yn_e} \end{bmatrix} \quad (15.4)$$

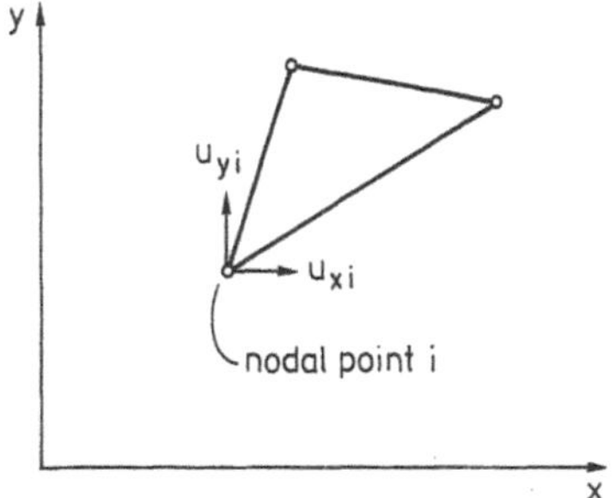

Figure 15.1   Nodal displacements u_{xi} and u_{yi} for nodal point i

These two expressions can be combined into

$$\mathbf{u} = \mathbf{N}^e \mathbf{a}^e \tag{15.5}$$

where

$$\mathbf{u} = \begin{bmatrix} u_x \\ u_y \end{bmatrix}; \quad \mathbf{N}^e = \begin{bmatrix} N_1^e & 0 & N_2^e & 0 & \cdots & N_{n_e}^e & 0 \\ 0 & N_1^e & 0 & N_2^e & \cdots & 0 & N_{n_e}^e \end{bmatrix}; \quad \mathbf{a}^e = \begin{bmatrix} u_{x1} \\ u_{y1} \\ u_{x2} \\ u_{y2} \\ \vdots \\ u_{xn_e} \\ u_{yn_e} \end{bmatrix} \tag{15.6}$$

In two-dimensional solid mechanics, we only consider the strains

$$\varepsilon_{xx} = \frac{\partial u_x}{\partial x}; \quad \varepsilon_{yy} = \frac{\partial u_y}{\partial y}; \quad \gamma_{xy} = \frac{\partial u_x}{\partial y} + \frac{\partial u_y}{\partial x} \tag{15.7}$$

According to (12.44) and (12.45), these expressions can be written as

$$\boldsymbol{\varepsilon} = \tilde{\nabla} \mathbf{u} \tag{15.8}$$

where

$$\boldsymbol{\varepsilon} = \begin{bmatrix} \varepsilon_{xx} \\ \varepsilon_{yy} \\ \gamma_{xy} \end{bmatrix}; \quad \tilde{\nabla} = \begin{bmatrix} \dfrac{\partial}{\partial x} & 0 \\ 0 & \dfrac{\partial}{\partial y} \\ \dfrac{\partial}{\partial y} & \dfrac{\partial}{\partial x} \end{bmatrix} \tag{15.9}$$

Combining (15.5) and (15.8) yields

$$\boxed{\boldsymbol{\varepsilon} = \mathbf{B}^e \mathbf{a}^e} \quad \text{where} \quad \boxed{\mathbf{B}^e = \tilde{\nabla} \mathbf{N}^e} \tag{15.10}$$

With (15.6) and (15.10) we get

$$\mathbf{B}^e = \begin{bmatrix} \dfrac{\partial N_1^e}{\partial x} & 0 & \dfrac{\partial N_2^e}{\partial x} & 0 & \cdots & \dfrac{\partial N_{n_e}^e}{\partial x} & 0 \\[2mm] 0 & \dfrac{\partial N_1^e}{\partial y} & 0 & \dfrac{\partial N_2^e}{\partial y} & \cdots & 0 & \dfrac{\partial N_{n_e}^e}{\partial y} \\[2mm] \dfrac{\partial N_1^e}{\partial y} & \dfrac{\partial N_1^e}{\partial x} & \dfrac{\partial N_2^e}{\partial y} & \dfrac{\partial N_2^e}{\partial x} & \cdots & \dfrac{\partial N_{n_e}^e}{\partial y} & \dfrac{\partial N_{n_e}^e}{\partial x} \end{bmatrix} \tag{15.11}$$

Defining a global shape function N_i belonging to nodal point i in exactly the same manner as in Chapter 7 (cf. (7.135) and (7.136)), we can write the approximation of the displacement vector $\mathbf{u}$ over the entire body as

$$\boxed{\mathbf{u} = \mathbf{N}\mathbf{a}} \tag{15.12}$$

where

$$\mathbf{N} = \begin{bmatrix} N_1 & 0 & N_2 & 0 & \cdots & N_n & 0 \\ 0 & N_1 & 0 & N_2 & \cdots & 0 & N_n \end{bmatrix}; \quad \mathbf{a} = \begin{bmatrix} u_{x1} \\ u_{y1} \\ u_{x2} \\ u_{y2} \\ \vdots \\ u_{xn} \\ u_{yn} \end{bmatrix} \tag{15.13}$$

and where n is the number of nodal points for the entire body. It follows that $\mathbf{N}$ has the dimension $2 \times 2n$ and $\mathbf{a}$ the dimension $2n \times 1$. From (15.12) and with (15.9) we obtain

$$\boxed{\boldsymbol{\varepsilon} = \mathbf{B}\mathbf{a}} \quad \text{where} \quad \boxed{\mathbf{B} = \tilde{\nabla} \mathbf{N}} \tag{15.14}$$

where $\mathbf{B}$ has the dimension $3 \times 2n$.

The generalization to three dimensions is straightforward and we obtain for the approximation over the entire body that

$$\boxed{\mathbf{u} = \mathbf{N}\mathbf{a}} \tag{15.15}$$

where

$$\mathbf{u} = \begin{bmatrix} u_x \\ u_y \\ u_z \end{bmatrix}; \quad \mathbf{N} = \begin{bmatrix} N_1 & 0 & 0 & N_2 & 0 & 0 & \cdots & N_n & 0 & 0 \\ 0 & N_1 & 0 & 0 & N_2 & 0 & \cdots & 0 & N_n & 0 \\ 0 & 0 & N_1 & 0 & 0 & N_2 & \cdots & 0 & 0 & N_n \end{bmatrix};$$

$$\mathbf{a} = \begin{bmatrix} u_{x1} \\ u_{y1} \\ u_{z1} \\ u_{x2} \\ u_{y2} \\ u_{z2} \\ \vdots \\ u_{xn} \\ u_{yn} \\ u_{zn} \end{bmatrix} \tag{15.16}$$

and where n is the number of nodal points in the entire body. That is, $\mathbf{N}$ has the dimension $3 \times 3n$ and $\mathbf{a}$ has the dimension $3n \times 1$. From (15.15) and (12.42) we get

$$\boxed{\boldsymbol{\varepsilon} = \mathbf{Ba}} \quad \text{where} \quad \boxed{\mathbf{B} = \tilde{\nabla}\mathbf{N}} \tag{15.17}$$

and

$$\boldsymbol{\varepsilon} = \begin{bmatrix} \varepsilon_{xx} \\ \varepsilon_{yy} \\ \varepsilon_{zz} \\ \gamma_{xy} \\ \gamma_{xz} \\ \gamma_{yz} \end{bmatrix}; \quad \tilde{\nabla} = \begin{bmatrix} \dfrac{\partial}{\partial x} & 0 & 0 \\[2mm] 0 & \dfrac{\partial}{\partial y} & 0 \\[2mm] 0 & 0 & \dfrac{\partial}{\partial z} \\[2mm] \dfrac{\partial}{\partial y} & \dfrac{\partial}{\partial x} & 0 \\[2mm] \dfrac{\partial}{\partial z} & 0 & \dfrac{\partial}{\partial x} \\[2mm] 0 & \dfrac{\partial}{\partial z} & \dfrac{\partial}{\partial y} \end{bmatrix} \tag{15.18}$$

The dimension for **B** is $6 \times 3n$. The approximation over one element follows the procedure above and we need not write the expressions explicitly.

From the discussion above it is evident that, by using the results of Chapter 7, we may formulate a sequence of finite elements applicable to solid mechanics. As the establishment of these elements is straightforward, we confine ourselves to an illustration of the two simplest two-dimensional elements: the three-node triangle and the four-node rectangle.

15.2.1 *Three-node element – constant strain element*

In order to illustrate the approach outlined above, we consider the simplest possible element applicable for two-dimensional solid mechanics: the three-node triangle shown in Figure 15.2. We recall that two unknowns are related to each nodal point: a displacement in the x-direction and a displacement in the y-direction. As usual, nodal numbers i, j, k are taken in a counter-clockwise direction. From (15.5) and (15.6) we have

$$\mathbf{u} = \mathbf{N}^e \mathbf{a}^e \tag{15.19}$$

where

$$\mathbf{u} = \begin{bmatrix} u_x \\ u_y \end{bmatrix}; \quad \mathbf{N}^e = \begin{bmatrix} N_i^e & 0 & N_j^e & 0 & N_k^e & 0 \\ 0 & N_i^e & 0 & N_j^e & 0 & N_k^e \end{bmatrix}; \quad \mathbf{a}^e = \begin{bmatrix} u_{xi} \\ u_{yi} \\ u_{xj} \\ u_{yj} \\ u_{xk} \\ u_{yk} \end{bmatrix} \tag{15.20}$$

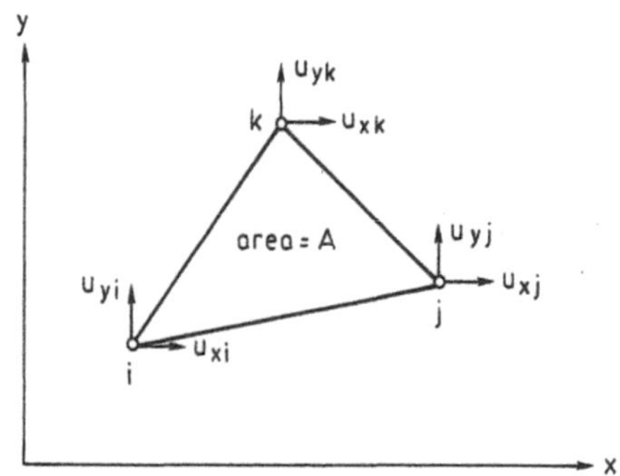

Figure 15.2   Constant strain element

According to (7.92) the element shape functions are given by

$$N_i^e = \frac{1}{2A}\left[x_j y_k - x_k y_j + (y_j - y_k)x + (x_k - x_j)y\right]$$

$$N_j^e = \frac{1}{2A}\left[x_k y_i - x_i y_k + (y_k - y_i)x + (x_i - x_k)y\right] \qquad (15.21)$$

$$N_k^e = \frac{1}{2A}\left[x_i y_j - x_j y_i + (y_i - y_j)x + (x_j - x_i)y\right]$$

where A is the area of the triangle. From (15.10) and (15.11) it follows that

$$\boldsymbol{\varepsilon} = \mathbf{B}^e \mathbf{a}^e \qquad (15.22)$$

where

$$\mathbf{B}^e = \frac{1}{2A}\begin{bmatrix} y_j - y_k & 0 & y_k - y_i & 0 & y_i - y_j & 0 \\ 0 & x_k - x_j & 0 & x_i - x_k & 0 & x_j - x_i \\ x_k - x_j & y_j - y_k & x_i - x_k & y_k - y_i & x_j - x_i & y_i - y_j \end{bmatrix} \qquad (15.23)$$

It appears that $\mathbf{B}^e$ is a constant matrix, i.e. the strains $\boldsymbol{\varepsilon}$ are constant within the element. For this reason the element is often termed the *constant strain element* (CST element).

15.2.2 *Four-node rectangle – Melosh element*

As the next example, consider the rectangular four-node element shown in Figure 15.3 – the so-called *Melosh element*; cf. Chapter 7. From (15.5) and (15.6) we have

$$\mathbf{u} = \mathbf{N}^e \mathbf{a}^e \qquad (15.24)$$

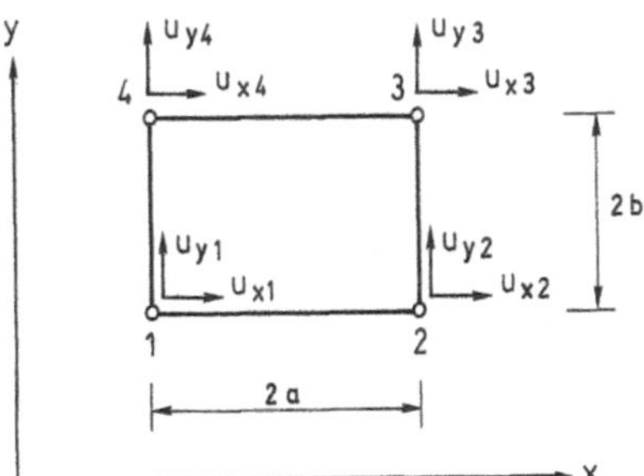

Figure 15.3   Melosh element

where

$$\mathbf{u} = \begin{bmatrix} u_x \\ u_y \end{bmatrix}; \quad \mathbf{N}^e = \begin{bmatrix} N_1^e & 0 & N_2^e & 0 & N_3^e & 0 & N_4^e & 0 \\ 0 & N_1^e & 0 & N_2^e & 0 & N_3^e & 0 & N_4^e \end{bmatrix}; \quad \mathbf{a}^e = \begin{bmatrix} u_{x1} \\ u_{y1} \\ u_{x2} \\ u_{y2} \\ u_{x3} \\ u_{y3} \\ u_{x4} \\ u_{y4} \end{bmatrix}$$

$$(15.25)$$

The element shape functions are given by (7.110), i.e.

$$
\begin{aligned}
N_1^e &= \frac{1}{4ab}(x - x_2)(y - y_4) \\[2mm]
N_2^e &= -\frac{1}{4ab}(x - x_1)(y - y_3) \\[2mm]
N_3^e &= \frac{1}{4ab}(x - x_4)(y - y_2) \\[2mm]
N_4^e &= -\frac{1}{4ab}(x - x_3)(y - y_1)
\end{aligned}
$$

$$(15.26)$$

With (15.10) and (15.11) it is concluded that

$$\boldsymbol{\varepsilon} = \mathbf{B}^e \mathbf{a}^e \tag{15.27}$$

where

$$\mathbf{B}^e = \frac{1}{4ab}\begin{bmatrix} y - y_4 & 0 & y_3 - y & 0 & y - y_2 & 0 & y_1 - y & 0 \\ 0 & x - x_2 & 0 & x_1 - x & 0 & x - x_4 & 0 & x_3 - x \\ x - x_2 & y - y_4 & x_1 - x & y_3 - y & x - x_4 & y - y_2 & x_3 - x & y_1 - y \end{bmatrix}$$

$$(15.28)$$

We observe that along a line parallel to the x-axis, i.e. $y = $ constant, the first row of

$\mathbf{B}^e$ consists of constant components, i.e. the strain ε_{xx} is constant. A similar observation holds for a line parallel to the y-axis.

As in the discussion in Chapter 7, we recall that in order to fulfil the compatibility requirement, the boundaries of the four-node element must be parallel to the coordinate axes. We also mention that this restriction can be relaxed by means of the isoparametric formulation presented in Chapter 19.

16

FE formulation of three- and two-dimensional elasticity

One of the most important applications of the FE method is in three- and two-dimensional elasticity. The approximations used for the displacements are those treated in the previous chapter. In Chapter 13 it was emphasized that the expressions obtained for the stresses and strains are general and – within the assumption of small strains – they apply for any solid material. This implies that the only topic which differentiates elasticity theory from, for instance, plasticity theory is the form of the constitutive relation. It is therefore of considerable importance to keep the formulation as general as possible and only introduce the particular constitutive relation at the latest possible stage. With such a viewpoint it becomes possible to formulate the FE method easily, not only for linear elasticity, but also for more involved constitutive theories.

Since the differential equations of equilibrium hold irrespective of the constitutive relation, we can achieve the objective discussed above by first making a weak formulation of these differential equations.

16.1 Weak form of equilibrium equations – three-dimensional case

For three-dimensional problems, the differential equations of equilibrium are given by (12.21), i.e.

$$\boxed{\tilde{\nabla}^{\mathrm{T}}\sigma + \mathbf{b} = \mathbf{0}} \tag{16.1}$$

where

$$
\tilde{\nabla}^{\mathrm{T}} =
\begin{bmatrix}
\dfrac{\partial}{\partial x} & 0 & 0 & \dfrac{\partial}{\partial y} & \dfrac{\partial}{\partial z} & 0 \\[2ex]
0 & \dfrac{\partial}{\partial y} & 0 & \dfrac{\partial}{\partial x} & 0 & \dfrac{\partial}{\partial z} \\[2ex]
0 & 0 & \dfrac{\partial}{\partial z} & 0 & \dfrac{\partial}{\partial x} & \dfrac{\partial}{\partial y}
\end{bmatrix} ; \quad
\sigma =
\begin{bmatrix}
\sigma_{xx} \\ \sigma_{yy} \\ \sigma_{zz} \\ \sigma_{xy} \\ \sigma_{xz} \\ \sigma_{yz}
\end{bmatrix} ;
\tag{16.2}
$$

$$
\mathbf{b} =
\begin{bmatrix}
b_x \\ b_y \\ b_z
\end{bmatrix}
$$

Carrying out the matrix multiplications of (16.1) gives (as $\sigma_{xy} = \sigma_{yx}$, etc.)

$$
\frac{\partial \sigma_{xx}}{\partial x} + \frac{\partial \sigma_{xy}}{\partial y} + \frac{\partial \sigma_{xz}}{\partial z} + b_x = 0
$$

$$
\frac{\partial \sigma_{yx}}{\partial x} + \frac{\partial \sigma_{yy}}{\partial y} + \frac{\partial \sigma_{yz}}{\partial z} + b_y = 0
\tag{16.3}
$$

$$
\frac{\partial \sigma_{zx}}{\partial x} + \frac{\partial \sigma_{zy}}{\partial y} + \frac{\partial \sigma_{zz}}{\partial z} + b_z = 0
$$

in accordance with (12.19). On the boundary of the body, the traction vector $\mathbf{t}$ with the components

$$
\mathbf{t} =
\begin{bmatrix}
t_x \\ t_y \\ t_z
\end{bmatrix}
\tag{16.4}
$$

acts. Moreover, this traction vector must fulfil the boundary condition (12.14), i.e.

$$
\boxed{
\begin{aligned}
t_x &= \sigma_{xx} n_x + \sigma_{xy} n_y + \sigma_{xz} n_z \\
t_y &= \sigma_{yx} n_x + \sigma_{yy} n_y + \sigma_{yz} n_z \\
t_z &= \sigma_{zx} n_x + \sigma_{zy} n_y + \sigma_{zz} n_z
\end{aligned}
}
\tag{16.5}
$$

The objective is to determine the weak form of the differential equations of equilibrium. Before doing so, we shall first derive a preliminary result.

Consider the arbitrary vector $\mathbf{v}$

$$
\mathbf{v} =
\begin{bmatrix}
v_x \\ v_y \\ v_z
\end{bmatrix}
\tag{16.6}
$$

In accordance with the kinematic relation (12.42), we have

$$
\tilde{\nabla}\mathbf{v} =
\begin{bmatrix}
\dfrac{\partial v_x}{\partial x} \\[2ex]
\dfrac{\partial v_y}{\partial y} \\[2ex]
\dfrac{\partial v_z}{\partial z} \\[2ex]
\dfrac{\partial v_x}{\partial y} + \dfrac{\partial v_y}{\partial x} \\[2ex]
\dfrac{\partial v_x}{\partial z} + \dfrac{\partial v_z}{\partial x} \\[2ex]
\dfrac{\partial v_y}{\partial z} + \dfrac{\partial v_z}{\partial y}
\end{bmatrix}
\tag{16.7}
$$

i.e.

$$
\begin{aligned}
(\tilde{\nabla}\mathbf{v})^{\mathrm{T}}\boldsymbol{\sigma} &= \frac{\partial v_x}{\partial x}\sigma_{xx} + \frac{\partial v_y}{\partial y}\sigma_{yy} + \frac{\partial v_z}{\partial z}\sigma_{zz} + \left(\frac{\partial v_x}{\partial y} + \frac{\partial v_y}{\partial x}\right)\sigma_{xy} \\[2ex]
&\quad + \left(\frac{\partial v_x}{\partial z} + \frac{\partial v_z}{\partial x}\right)\sigma_{xz} + \left(\frac{\partial v_y}{\partial z} + \frac{\partial v_z}{\partial y}\right)\sigma_{yz}
\end{aligned}
\tag{16.8}
$$

We are now in a position to derive the weak form of (16.3). Multiply the first equation of (16.3) by the arbitrary function v_x and integrate over the volume V to obtain

$$
\int_V v_x \frac{\partial \sigma_{xx}}{\partial x}\,\mathrm{d}V + \int_V v_x \frac{\partial \sigma_{xy}}{\partial y}\,\mathrm{d}V + \int_V v_x \frac{\partial \sigma_{xz}}{\partial z}\,\mathrm{d}V + \int_V v_x b_x\,\mathrm{d}V = 0
$$

An integration by parts using the Green–Gauss theorem is now performed. Using this theorem in the form given by (5.34), the result becomes

$$
\begin{aligned}
&\int_S v_x \sigma_{xx} n_x\,\mathrm{d}S - \int_V \frac{\partial v_x}{\partial x}\sigma_{xx}\,\mathrm{d}V + \int_S v_x \sigma_{xy} n_y\,\mathrm{d}S - \int_V \frac{\partial v_x}{\partial y}\sigma_{xy}\,\mathrm{d}V \\[2ex]
&\quad + \int_S v_x \sigma_{xz} n_z\,\mathrm{d}S - \int_V \frac{\partial v_x}{\partial z}\sigma_{xz}\,\mathrm{d}V + \int_V v_x b_x\,\mathrm{d}V = 0
\end{aligned}
$$

The component t_x of the traction vector $\mathbf{t}$ is given by (16.5), i.e. the expression above reduces to

$$
\int_S v_x t_x\,\mathrm{d}S - \int_V \left(\frac{\partial v_x}{\partial x}\sigma_{xx} + \frac{\partial v_x}{\partial y}\sigma_{xy} + \frac{\partial v_x}{\partial z}\sigma_{xz}\right)\mathrm{d}V + \int_V v_x b_x\,\mathrm{d}V = 0
\tag{16.9}
$$

In a similar way, the second and third equations of (16.3) are multiplied by the arbitrary functions v_y and v_z respectively, and we find that

$$\int_S v_y t_y \, \mathrm{d}S - \int_V \left(\frac{\partial v_y}{\partial x} \sigma_{yx} + \frac{\partial v_y}{\partial y} \sigma_{yy} + \frac{\partial v_y}{\partial z} \sigma_{yz} \right) \mathrm{d}V + \int_V v_y b_y \, \mathrm{d}V = 0 \qquad (16.10)$$

$$\int_S v_z t_z \, \mathrm{d}S - \int_V \left(\frac{\partial v_z}{\partial x} \sigma_{zx} + \frac{\partial v_z}{\partial y} \sigma_{zy} + \frac{\partial v_z}{\partial z} \sigma_{zz} \right) \mathrm{d}V + \int_V v_z b_z \, \mathrm{d}V = 0 \qquad (16.11)$$

Addition of (16.9)–(16.11) yields

$$\int_S (v_x t_x + v_y t_y + v_z t_z) \, \mathrm{d}S + \int_V (v_x b_x + v_y b_y + v_z b_z) \, \mathrm{d}V$$

$$- \int_V \left[\frac{\partial v_x}{\partial x} \sigma_{xx} + \frac{\partial v_y}{\partial y} \sigma_{yy} + \frac{\partial v_z}{\partial z} \sigma_{zz} + \left(\frac{\partial v_x}{\partial y} + \frac{\partial v_y}{\partial x} \right) \sigma_{xy} + \left(\frac{\partial v_x}{\partial z} + \frac{\partial v_z}{\partial x} \right) \sigma_{xz} \right.$$

$$\left. + \left(\frac{\partial v_y}{\partial z} + \frac{\partial v_z}{\partial y} \right) \sigma_{yz} \right] \mathrm{d}V = 0$$

Finally, use of (16.8) provides

$$\boxed{\int_V (\tilde{\nabla} \mathbf{v})^{\mathrm{T}} \boldsymbol{\sigma} \, \mathrm{d}V = \int_S \mathbf{v}^{\mathrm{T}} \mathbf{t} \, \mathrm{d}S + \int_V \mathbf{v}^{\mathrm{T}} \mathbf{b} \, \mathrm{d}V} \qquad (16.12)$$

This is the weak form of the differential equations of equilibrium (16.1) subjected to the boundary conditions (16.5), and we stress that the *weight vector* $\mathbf{v}$ is arbitrary and that (16.12) holds for any constitutive relation. This weak form is of fundamental importance in solid mechanics where it is often termed the *virtual work equation* or *virtual work principle*. It is emphasized that the arbitrary weight vector $\mathbf{v}$ and the stresses $\boldsymbol{\sigma}$ are completely unrelated, and even though $\mathbf{v}$ may be interpreted as a displacement vector, $\mathbf{v}$ has nothing to do with the real displacements of the body. Clearly, $\mathbf{v}$ may be chosen as the real displacements $\mathbf{u}$, but in general this is not the case. When the weak form (16.12) is derived, the traction vector $\mathbf{t}$ emerges in the boundary term. In accordance with the discussion of heat flow, the traction vector $\mathbf{t}$ is therefore a *natural boundary condition*.

The celebrated virtual work principle was derived from the differential equations of equilibrium and the boundary conditions for the traction vector. Just as for the heat flow problem, it can be shown that as $\mathbf{v}$ in the weak form (16.12) denotes *any* vector, the weak form implies (16.1) and (16.5). The reader is referred to Hughes (1987) for details.

16.2 FE formulation of three-dimensional elasticity

With the weak form (16.12) of the equilibrium equations, it is straightforward to derive the FE equations for three-dimensional elasticity. As mentioned previously, we shall introduce the assumption of elasticity at the latest possible stage. To achieve this purpose, we already know that the displacement vector $\mathbf{u}$ will be approximated by

$$\mathbf{u} = \mathbf{N}\mathbf{a} \tag{16.13}$$

(cf. (15.15)). The Galerkin method means that the weight vector $\mathbf{v}$ is chosen in accordance with

$$\mathbf{v} = \mathbf{N}\mathbf{c} \tag{16.14}$$

As $\mathbf{v}$ is arbitrary, the matrix $\mathbf{c}$ is arbitrary. From (16.14) it follows that

$$\tilde{\nabla}\mathbf{v} = \mathbf{B}\mathbf{c} \quad \text{where} \quad \mathbf{B} = \tilde{\nabla}\mathbf{N} \tag{16.15}$$

(cf. (15.17)). Inserting (16.14) and (16.15) in the weak form (16.12), and noting that $\mathbf{c}$ is independent of the coordinates, yields

$$\mathbf{c}^{\mathrm{T}}\left(\int_V \mathbf{B}^{\mathrm{T}}\boldsymbol{\sigma}\,\mathrm{d}V - \int_S \mathbf{N}^{\mathrm{T}}\mathbf{t}\,\mathrm{d}S - \int_V \mathbf{N}^{\mathrm{T}}\mathbf{b}\,\mathrm{d}V \right) = 0$$

As the $\mathbf{c}$-matrix is arbitrary, we conclude that

$$\boxed{\int_V \mathbf{B}^{\mathrm{T}}\boldsymbol{\sigma}\,\mathrm{d}V = \int_S \mathbf{N}^{\mathrm{T}}\mathbf{t}\,\mathrm{d}S + \int_V \mathbf{N}^{\mathrm{T}}\mathbf{b}\,\mathrm{d}V} \tag{16.16}$$

As this expression was derived from the weak form of the equilibrium conditions that hold for arbitrary weight vectors $\mathbf{v}$, (16.16) holds exactly and expresses the balance principle for the body. It is emphasized that (16.16) applies irrespective of the constitutive model.

The global shape function matrix $\mathbf{N}$ has the dimension $3 \times 3n$, where n is the number of nodal points for the entire body (cf. (15.16)), and as $\mathbf{b}$ has the dimension 3×1, $\mathbf{N}^{\mathrm{T}}\mathbf{b}$ has the dimension $3n \times 1$. This means that (16.16) constitutes a system of equations with $3n$ equations. As in the discussion of heat flow (cf. Chapter 10) the right-hand side of (16.16) can be viewed as forces acting at the nodal points. These nodal forces have components in the x-, y- and z-directions and, as in the discussion in Chapter 10, it can be shown that the nodal forces fulfil exactly the conditions for global equilibrium of the entire body. This is certainly not surprising since (16.16) was derived from the equilibrium conditions for the entire body.

At this stage we introduce the constitutive model, and we assume that the material responds thermoelastically. Consequently, the constitutive model

$$\boxed{\boldsymbol{\sigma} = \mathbf{D}\boldsymbol{\varepsilon} - \mathbf{D}\boldsymbol{\varepsilon}_0} \tag{16.17}$$

is adopted (cf. (13.23), where $\mathbf{D}$ is the constitutive matrix and $\boldsymbol{\varepsilon}_0$ contains the initial strains. These initial strains are assumed to be known. In most cases, $\boldsymbol{\varepsilon}_0$ is due to thermal strains, in which case it is determined by an independent temperature calculation, for instance in terms of a heat flow FE calculation. The kinematic relation

(12.42) states that

$$\varepsilon = \tilde{\nabla}\mathbf{u} \tag{16.18}$$

From (16.13) and (16.15) we get

$$\varepsilon = \mathbf{Ba} \tag{16.19}$$

i.e. (16.17) becomes

$$\boldsymbol{\sigma} = \mathbf{DBa} - \mathbf{D}\varepsilon_0 \tag{16.20}$$

With (16.20), (16.16) takes the form

$$\left(\int_V \mathbf{B}^\mathrm{T}\mathbf{DB}\,dV\right)\mathbf{a} = \int_S \mathbf{N}^\mathrm{T}\mathbf{t}\,dS + \int_V \mathbf{N}^\mathrm{T}\mathbf{b}\,dV + \int_V \mathbf{B}^\mathrm{T}\mathbf{D}\varepsilon_0\,dV \tag{16.21}$$

Let us now consider the boundary conditions, which are expressed either in terms of a prescribed traction vector $\mathbf{t}$ – the *natural boundary condition* – or a prescribed displacement vector $\mathbf{u}$ – the *essential boundary condition*. Instead of the traction vector given by (16.5), we may use the formulation (12.13) and the boundary conditions become

$$\boxed{\begin{aligned} \mathbf{t} = \mathbf{Sn} = \mathbf{h} \quad &\text{on } S_h \\ \mathbf{u} = \mathbf{g} \quad &\text{on } S_g \end{aligned}} \tag{16.22}$$

where $\mathbf{h}$ and $\mathbf{g}$ are known vectors. This means that the traction vector $\mathbf{t}$ is known along the boundary S_h and the displacement $\mathbf{u}$ is known along the boundary S_g. The total boundary S consists of S_h and S_g (note the difference between the total boundary S and the stress tensor $\mathbf{S}$). With these formulations (16.21) takes the form

$$\left(\int_V \mathbf{B}^\mathrm{T}\mathbf{DB}\,dV\right)\mathbf{a} = \int_{S_h} \mathbf{N}^\mathrm{T}\mathbf{h}\,dS + \int_{S_g} \mathbf{N}^\mathrm{T}\mathbf{t}\,dS + \int_V \mathbf{N}^\mathrm{T}\mathbf{b}\,dV + \int_V \mathbf{B}^\mathrm{T}\mathbf{D}\varepsilon_0\,dV \tag{16.23}$$

which is the FE formulation sought.

In order to write this formulation in a compact fashion, the following matrices are defined:

$$\boxed{\begin{aligned} \mathbf{K} &= \int_V \mathbf{B}^\mathrm{T}\mathbf{DB}\,dV \\[2mm] \mathbf{f}_b &= \int_{S_h} \mathbf{N}^\mathrm{T}\mathbf{h}\,dS + \int_{S_g} \mathbf{N}^\mathrm{T}\mathbf{t}\,dS \\[2mm] \mathbf{f}_l &= \int_V \mathbf{N}^\mathrm{T}\mathbf{b}\,dV \\[2mm] \mathbf{f}_0 &= \int_V \mathbf{B}^\mathrm{T}\mathbf{D}\varepsilon_0\,dV \end{aligned}} \tag{16.24}$$

where $\mathbf{K}$ is the stiffness matrix, $\mathbf{f}_b$ the boundary vector, $\mathbf{f}_l$ the load vector and $\mathbf{f}_0$ the *initial strain vector*. Using (16.24), (16.23) can be written as

$$\mathbf{Ka} = \mathbf{f}_b + \mathbf{f}_l + \mathbf{f}_0 \qquad (16.25)$$

If n is the number of nodal points for the entire body, it can easily be seen that $\mathbf{K}$ has the dimension $3n \times 3n$, $\mathbf{a}$ has the dimension $3n \times 1$ and the right-hand side of (16.25) has the dimension $3n \times 1$. Moreover, defining the force vector $\mathbf{f}$ by

$$\mathbf{f} = \mathbf{f}_b + \mathbf{f}_l + \mathbf{f}_0 \qquad (16.26)$$

we obtain the standard FE formulation

$$\mathbf{Ka} = \mathbf{f} \qquad (16.27)$$

where $\mathbf{f}$ has the dimension of force, i.e. [N]. We observe that the inclusion of initial strains $\boldsymbol{\varepsilon}_0$ in the analysis presents no difficulties in the FE formulation, and it is characteristic for the FE method that various effects may easily be accounted for. We note that in order to determine the initial strain vector $\mathbf{f}_0$, it is required that the initial strains $\boldsymbol{\varepsilon}_0$ are known. If these initial strains are thermal strains, then an FE analysis of the corresponding heat flow problem is first carried out. With the known temperatures, the initial strain vector $\boldsymbol{\varepsilon}_0$ is determined from (13.24), and with (16.25) we obtain the elastic response of the body exposed to combined mechanical and thermal loading.

The contribution of the load vector $\mathbf{f}_l$ from concentrated forces can be dealt with by means of Dirac's delta function in complete analogy with the treatment of point sources in heat flow problems.

According to (13.4), $\mathbf{D}$ is symmetric; that is, it follows from (16.24) that the stiffness matrix $\mathbf{K}$ is also symmetric, i.e.

$$\mathbf{K} = \mathbf{K}^T \qquad (16.28)$$

We observe that if $\mathbf{a}$ corresponds to a rigid-body motion, this creates no strains in the body, i.e. $\boldsymbol{\varepsilon} = \mathbf{Ba} = \mathbf{0}$. As in the discussion of (9.38)–(9.40) it follows that $\mathbf{K}$ is singular, i.e.

$$\det \mathbf{K} = 0 \qquad (16.29)$$

Moreover, as $\mathbf{D}$ is positive definite (cf. (13.11)), it follows from (16.24) that $\mathbf{K}$ is positive semi-definite, i.e.

$$\mathbf{a}^T \mathbf{Ka} \geq 0 \qquad (16.30)$$

for all $\mathbf{a}$; only if $\mathbf{a}$ corresponds to a rigid-body motion is the quadratic form above equal to zero. These observations are in complete analogy with the heat flow problems, and it follows that if sufficient boundary conditions for the displacements are prescribed so that rigid-body motions are eliminated, then the corresponding submatrix $\underset{\sim}{\mathbf{K}}$ of $\mathbf{K}$ is non-singular. In analogy with the heat flow problem, the prescription of these boundary conditions allows for a modification of the system of equations (16.27) so that a unique solution can be obtained.

As before, we may obtain the FE equations for one element, and with obvious notation we get

$$\boxed{\mathbf{K}^e \mathbf{a}^e = \mathbf{f}^e} \tag{16.31}$$

where

$$\boxed{\mathbf{f}^e = \mathbf{f}^e_b + \mathbf{f}^e_l + \mathbf{f}^e_0} \tag{16.32}$$

and

$$\boxed{\begin{aligned}
\mathbf{K}^e &= \int_{V_\alpha} \mathbf{B}^{eT} \mathbf{D} \mathbf{B}^e \, dV \\
\mathbf{f}^e_b &= \int_{S_{h\alpha}} \mathbf{N}^{eT} \mathbf{h} \, dS + \int_{S_{g\alpha}} \mathbf{N}^{eT} \mathbf{t} \, dS \\
\mathbf{f}^e_l &= \int_{V_\alpha} \mathbf{N}^{eT} \mathbf{b} \, dV \\
\mathbf{f}^e_0 &= \int_{V_\alpha} \mathbf{B}^{eT} \mathbf{D} \boldsymbol{\varepsilon}_0 \, dV
\end{aligned}} \tag{16.33}$$

In these expressions, V_α and S_α are the volume and surface of the element, respectively. In general, the boundary conditions along the element surface are unknown.

16.3 Weak form of equilibrium equations – two-dimensional case

In the present text, two-dimensional solid mechanics is defined according to:

$$\boxed{\text{two-dimensional problem} = \text{plane stress or plane strain conditions}} \tag{16.34}$$

In general, however, *axisymmetric* problems can also be treated as a two-dimensional problem and even though the FE formulation of such problems can be dealt with in a fashion closely related to the exposition given below, we shall confine ourselves here to the definition given by (16.34).

For plane stress conditions, we may derive the weak form of the equilibrium differential equations (12.26) in a manner completely analogous to the derivation of the weak form (16.12) applicable to three-dimensional solid mechanics. However, we want to obtain the weak form of the equilibrium differential equations not only for plane stress, but also for plane strain conditions, and we shall perform this derivation in a unified manner here.

The weak form (16.12) of the equilibrium equations holds for the general three-dimensional case, so it also applies for two-dimensional problems. In (16.12), the weight vector $\mathbf{v}$ is arbitrary and it is allowable to choose its components according to

$$v_x = v_x(x, y); \quad v_y = v_y(x, y); \quad v_z = 0 \tag{16.35}$$

This choice implies that (16.8) reduces to

$$(\tilde{\nabla}\mathbf{v})^{\mathrm{T}}\boldsymbol{\sigma} = \frac{\partial v_x}{\partial x}\,\sigma_{xx} + \frac{\partial v_y}{\partial y}\,\sigma_{yy} + \left(\frac{\partial v_x}{\partial y} + \frac{\partial v_y}{\partial x}\right)\sigma_{xy} \tag{16.36}$$

From the following redefinition of matrices

$$\tilde{\nabla} = \begin{bmatrix} \dfrac{\partial}{\partial x} & 0 \\[2ex] 0 & \dfrac{\partial}{\partial y} \\[2ex] \dfrac{\partial}{\partial y} & \dfrac{\partial}{\partial x} \end{bmatrix}; \quad \mathbf{v} = \begin{bmatrix} v_x \\ v_y \end{bmatrix}; \quad \boldsymbol{\sigma} = \begin{bmatrix} \sigma_{xx} \\ \sigma_{yy} \\ \sigma_{xy} \end{bmatrix} \tag{16.37}$$

it follows that (16.36) is fulfilled. We note that this definition of the operator $\tilde{\nabla}$ occurs also for plane stress (cf. (12.27)) as well as for plane strain (cf. (12.44)). We also introduce the definitions

$$\mathbf{t} = \begin{bmatrix} t_x \\ t_y \end{bmatrix}; \quad \mathbf{b} = \begin{bmatrix} b_x \\ b_y \end{bmatrix} \tag{16.38}$$

With the definitions (16.37) and (16.38), the weak form (16.12) becomes

$$\int_V (\tilde{\nabla}\mathbf{v})^{\mathrm{T}}\boldsymbol{\sigma}\,\mathrm{d}V = \int_S \mathbf{v}^{\mathrm{T}}\mathbf{t}\,\mathrm{d}S + \int_V \mathbf{v}^{\mathrm{T}}\mathbf{b}\,\mathrm{d}V \tag{16.39}$$

This formulation looks like (16.12) but it is different since $\mathbf{v}$, $\tilde{\nabla}$, $\boldsymbol{\sigma}$, $\mathbf{t}$ and $\mathbf{b}$ are now defined by (16.37) and (16.38).

Let us now observe a characteristic feature of two-dimensional problems, namely:

> For two-dimensional problems, the displacements, strains, stresses, tractions and body forces do not depend on the z-coordinate.

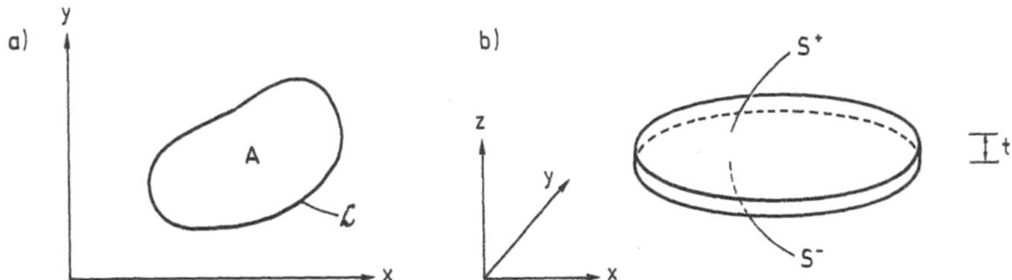

Figure 16.1 (a) Region A and boundary $\mathcal{L}$ for two-dimensional problems; (b) illustration of upper surface S^+, lower surface S^- and thickness t

For the two-dimensional problem described by the x- and y-coordinates, we denote the region in the xy-plane by A and the boundary of this region by $\mathcal{L}$; see Figure 16.1(a). The thickness t in the z-direction is shown in Figure 16.1(b) (note the difference between the thickness t and the traction vector $\mathbf{t}$). As all integrands in (16.39) are independent on the z-coordinate, we obtain

$$\int_V (\tilde{\nabla}\mathbf{v})^{\mathrm{T}}\boldsymbol{\sigma}\,dV = \int_A \left[\int_z (\tilde{\nabla}\mathbf{v})^{\mathrm{T}}\boldsymbol{\sigma}\,dz \right] dA = \int_A (\tilde{\nabla}\mathbf{v})^{\mathrm{T}}\boldsymbol{\sigma}t\,dA \tag{16.40}$$

$$\int_V \mathbf{v}^{\mathrm{T}}\mathbf{b}\,dV = \int_A \left[\int_z \mathbf{v}^{\mathrm{T}}\mathbf{b}\,dz \right] dA = \int_A \mathbf{v}^{\mathrm{T}}\mathbf{b}t\,dA$$

Consider now the surface integral in (16.39). Referring to Figure 16.1(b), let S^+ and S^- denote the upper and lower surface of the body, respectively. We then get that

$$\int_S \mathbf{v}^{\mathrm{T}}\mathbf{t}\,dS = \int_{S+} \mathbf{v}^{\mathrm{T}}\mathbf{t}\,dS + \int_{S-} \mathbf{v}^{\mathrm{T}}\mathbf{t}\,dS + \oint_{\mathcal{L}} \left[\int_z \mathbf{v}^{\mathrm{T}}\mathbf{t}\,dz \right] d\mathcal{L} \tag{16.41}$$

Referring to (16.38), the traction vector $\mathbf{t}$ is given by the components t_x and t_y. Along S^+ and S^- we have $n_x = n_y = 0$, i.e. (16.5) gives

$$\mathbf{t} = \begin{bmatrix} t_x \\ t_y \end{bmatrix} = \begin{bmatrix} \sigma_{xz}n_z \\ \sigma_{yz}n_z \end{bmatrix}$$

along S^+ and S^-. For plane stress conditions, we have by definition that $\sigma_{xz} = \sigma_{yz} = 0$; that is, (16.41) reduces to

$$\int_S \mathbf{v}^{\mathrm{T}}\mathbf{t}\,dS = \oint_{\mathcal{L}} \mathbf{v}^{\mathrm{T}}\mathbf{t}t\,d\mathcal{L} \tag{16.42}$$

For plane strain conditions, the components $\sigma_{xz} = \sigma_{xz}(x, y)$ and $\sigma_{yz} = \sigma_{yz}(x, y)$ may, in general, be different from zero. However, along the upper surface S^+ we have $n_z = 1$, whereas $n_z = -1$ holds along the lower surface S^-. With obvious notation, this implies that $\mathbf{t}^+ = -\mathbf{t}^-$ and as $\mathbf{v}$ is independent of the z-coordinate, (16.41) reduces to (16.42) even for plane strain conditions.

Therefore, insertion of (16.40) and (16.42) in (16.39) yields

$$\int_A (\tilde{\nabla}\mathbf{v})^{\mathrm{T}}\boldsymbol{\sigma} t \, \mathrm{d}A = \oint_{\mathscr{L}} \mathbf{v}^{\mathrm{T}}\mathbf{t}t \, \mathrm{d}\mathscr{L} + \int_A \mathbf{v}^{\mathrm{T}}\mathbf{b}t \, \mathrm{d}A \qquad (16.43)$$

which is the weak form – or the virtual work principle – for two-dimensional problems. Again we recall the difference between the thickness t and the traction vector $\mathbf{t}$. In (16.43), the thickness t may vary, but in order to approximate the requirements for the existence of plane stress (or plane strain), such a variation must, in general, be small. It is emphasized that (16.43) holds irrespective of the constitutive model.

If the thickness t of the body is constant, the virtual work principle may be simplified to

$$\int_A (\tilde{\nabla}\mathbf{v})^{\mathrm{T}}\boldsymbol{\sigma} \, \mathrm{d}A = \oint_{\mathscr{L}} \mathbf{v}^{\mathrm{T}}\mathbf{t} \, \mathrm{d}\mathscr{L} + \int_A \mathbf{v}^{\mathrm{T}}\mathbf{b} \, \mathrm{d}A \qquad (16.44)$$

16.4 FE formulation of two-dimensional elasticity

The derivation of the FE formulation from (16.43) closely follows that of three-dimensional elasticity. As before, we introduce the specific assumption for the constitutive relation at the latest possible stage. We already know that the displacement vector

$$\mathbf{u} = \begin{bmatrix} u_x \\ u_y \end{bmatrix} \qquad (16.45)$$

is to be approximated by

$$\mathbf{u} = \mathbf{N}\mathbf{a} \qquad (16.46)$$

The reason that $\mathbf{u}$ only comprises u_x and u_y and not u_z will be revealed in a moment. According to the Galerkin method we choose the arbitrary weight vector $\mathbf{v}$ as

$$\mathbf{v} = \mathbf{N}\mathbf{c} \qquad (16.47)$$

As $\mathbf{v}$ is arbitrary, $\mathbf{c}$ is also arbitrary. From (16.47) it follows that

$$\tilde{\nabla}\mathbf{v} = \mathbf{B}\mathbf{c} \quad \text{where} \quad \mathbf{B} = \tilde{\nabla}\mathbf{N} \qquad (16.48)$$

(cf. (15.14)). Inserting (16.48) in the weak form (16.43) gives

$$\mathbf{c}^{\mathrm{T}}\left(\int_A \mathbf{B}^{\mathrm{T}}\boldsymbol{\sigma} t \, \mathrm{d}A - \oint_{\mathscr{L}} \mathbf{N}^{\mathrm{T}}\mathbf{t}t \, \mathrm{d}\mathscr{L} - \int_A \mathbf{N}^{\mathrm{T}}\mathbf{b}t \, \mathrm{d}A \right) = 0$$

and as the c-matrix is arbitrary, we conclude that

$$\int_A \mathbf{B}^T \boldsymbol{\sigma} t \, dA = \oint_{\mathscr{L}} \mathbf{N}^T \mathbf{t} t \, d\mathscr{L} + \int_A \mathbf{N}^T \mathbf{b} t \, dA \qquad (16.49)$$

The similarity with (16.16) should be noted. As (16.49) is only based on the equilibrium equations it holds for arbitrary constitutive models.

We now introduce our assumption for the material behaviour, i.e. the constitutive relation. In the present case we assume thermoelasticity and we note from (13.32), applicable for plane stress, and from (13.39), applicable for plane strain, that both cases can be formulated as

$$\boldsymbol{\sigma} = \mathbf{D}\boldsymbol{\varepsilon} - \mathbf{D}\boldsymbol{\varepsilon}_0 \qquad (16.50)$$

where

$$\boldsymbol{\sigma} = \begin{bmatrix} \sigma_{xx} \\ \sigma_{yy} \\ \sigma_{xy} \end{bmatrix}; \quad \boldsymbol{\varepsilon} = \begin{bmatrix} \varepsilon_{xx} \\ \varepsilon_{yy} \\ \gamma_{xy} \end{bmatrix} \qquad (16.51)$$

It is emphasized, however, that $\mathbf{D}$ and $\boldsymbol{\varepsilon}_0$ take different forms depending on whether plane stress or plane strain is assumed: cf. (13.33) and (13.30) with (13.38). Just as for three-dimensional elasticity, the initial strains $\boldsymbol{\varepsilon}_0$ are assumed to be known. For thermoelasticity, $\boldsymbol{\varepsilon}_0$ is due to temperature changes which are determined by an independent temperature calculation, for instance in terms of a heat flow FE analysis.

From (16.50) and (16.51) it follows that the stresses given by $\boldsymbol{\sigma}$ only depend on the in-plane strain components ε_{xx}, ε_{yy} and γ_{xy}, and this is the reason why the displacement vector $\mathbf{u}$ given by (16.45) only comprises the components u_x and u_y even though u_z for plane stress conditions, in general, is different from zero.

Irrespective of whether plane stress or plane strain is assumed, the strains $\boldsymbol{\varepsilon}$ defined by (16.51) are derived from the displacements (16.45) using the kinematic relation

$$\boldsymbol{\varepsilon} = \tilde{\nabla}\mathbf{u} \qquad (16.52)$$

From (16.46) and (16.48) we obtain

$$\boldsymbol{\varepsilon} = \mathbf{B}\mathbf{a} \qquad (16.53)$$

i.e. (16.50) takes the form

$$\boldsymbol{\sigma} = \mathbf{D}\mathbf{B}\mathbf{a} - \mathbf{D}\boldsymbol{\varepsilon}_0 \qquad (16.54)$$

Inserting (16.54) into (16.49) results in

$$\left(\int_A \mathbf{B}^T \mathbf{D}\mathbf{B} t \, dA \right) \mathbf{a} = \oint_{\mathscr{L}} \mathbf{N}^T \mathbf{t} t \, d\mathscr{L} + \int_A \mathbf{N}^T \mathbf{b} t \, dA + \int_A \mathbf{B}^T \mathbf{D}\boldsymbol{\varepsilon}_0 t \, dA \qquad (16.55)$$

The unit vector $\mathbf{n}$ normal to the boundary $\mathscr{L}$ is located in the xy-plane and the traction vector $\mathbf{t}$ along $\mathscr{L}$ is given by (16.38). That is, the traction boundary conditions (16.5) can be written as

$$\mathbf{t} = \mathbf{Sn} \tag{16.56}$$

where

$$\mathbf{S} = \begin{bmatrix} \sigma_{xx} & \sigma_{xy} \\ \sigma_{yx} & \sigma_{yy} \end{bmatrix}; \quad \mathbf{n} = \begin{bmatrix} n_x \\ n_y \end{bmatrix} \tag{16.57}$$

All boundary conditions can therefore be given in the form

$$\boxed{\begin{aligned} \mathbf{t} = \mathbf{Sn} = \mathbf{h} &\quad \text{on } \mathscr{L}_h \\ \mathbf{u} = \mathbf{g} &\quad \text{on } \mathscr{L}_g \end{aligned}} \tag{16.58}$$

where $\mathbf{h}$ and $\mathbf{g}$ are known vectors. This means that the traction vector $\mathbf{t}$ – the *natural boundary condition* – is known along the boundary $\mathscr{L}_h$ and the displacement vector $\mathbf{u}$ – the *essential boundary condition* – is known along the boundary $\mathscr{L}_g$. The total boundary $\mathscr{L}$ consists of $\mathscr{L}_h$ and $\mathscr{L}_g$; see Figure 16.2. Using (16.58) we may reformulate (16.55) according to

$$\left(\int_A \mathbf{B}^{\mathrm{T}}\mathbf{DB}t \, \mathrm{d}A \right)\mathbf{a} = \int_{\mathscr{L}_h} \mathbf{N}^{\mathrm{T}}\mathbf{h}t \, \mathrm{d}\mathscr{L} + \int_{\mathscr{L}_g} \mathbf{N}^{\mathrm{T}}\mathbf{t}t \, \mathrm{d}\mathscr{L} + \int_A \mathbf{N}^{\mathrm{T}}\mathbf{b}t \, \mathrm{d}A$$
$$+ \int_A \mathbf{B}^{\mathrm{T}}\mathbf{D}\boldsymbol{\varepsilon}_0 t \, \mathrm{d}A \tag{16.59}$$

which is the FE formulation for two-dimensional elasticity.

In order to write this formulation in a more compact fashion the following matrices are defined:

$$\boxed{\begin{aligned} \mathbf{K} &= \int_A \mathbf{B}^{\mathrm{T}}\mathbf{DB}t \, \mathrm{d}A \\[2mm] \mathbf{f}_b &= \int_{\mathscr{L}_h} \mathbf{N}^{\mathrm{T}}\mathbf{h}t \, \mathrm{d}\mathscr{L} + \int_{\mathscr{L}_g} \mathbf{N}^{\mathrm{T}}\mathbf{t}t \, \mathrm{d}\mathscr{L} \\[2mm] \mathbf{f}_l &= \int_A \mathbf{N}^{\mathrm{T}}\mathbf{b}t \, \mathrm{d}A \\[2mm] \mathbf{f}_0 &= \int_A \mathbf{B}^{\mathrm{T}}\mathbf{D}\boldsymbol{\varepsilon}_0 t \, \mathrm{d}A \end{aligned}} \tag{16.60}$$

where $\mathbf{K}$ is the stiffness matrix, $\mathbf{f}_b$ the boundary vector, $\mathbf{f}_l$ the load vector and $\mathbf{f}_0$ the

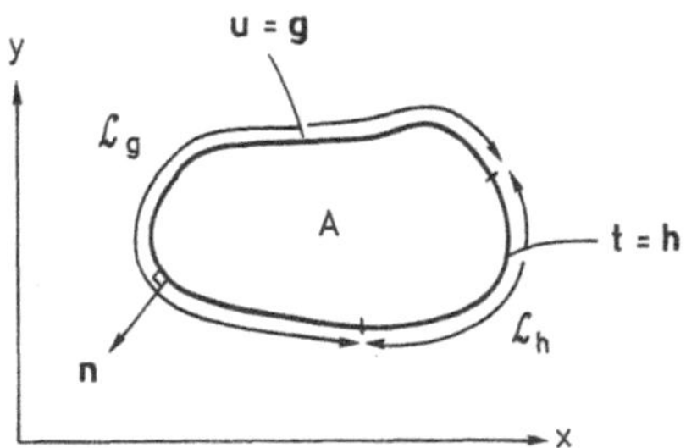

Figure 16.2   Two-dimensional region with boundary $\mathcal{L} = \mathcal{L}_h + \mathcal{L}_g$

initial strain vector. Using (16.60), (16.59) can be written as

$$\mathbf{Ka} = \mathbf{f}_b + \mathbf{f}_l + \mathbf{f}_0 \tag{16.61}$$

If n is the number of nodal points for the entire body, it is easily shown that $\mathbf{K}$ has the dimension $2n \times 2n$, $\mathbf{a}$ has the dimension $2n \times 1$ and the right-hand side of (16.61) has the dimension $2n \times 1$. As usual, the force vector $\mathbf{f}$ is defined by

$$\mathbf{f} = \mathbf{f}_b + \mathbf{f}_l + \mathbf{f}_0 \tag{16.62}$$

That is, (16.61) takes the standard form

$$\mathbf{Ka} = \mathbf{f} \tag{16.63}$$

where it can easily be seen that $\mathbf{f}$ has the dimension of force (i.e. [N]) just as for the three-dimensional case (cf. (16.27)). Indeed, the close analogy with the formulation for three-dimensional elasticity is distinct. Even the analogy with one-dimensional elasticity, i.e. the response of an axially loaded elastic bar (cf. (9.133) and (9.131)), should be noted. It is recalled that in the formulation presented for two-dimensional problems, the thickness t may vary, but in order to approximate the requirements for the existence of plane stress or plane strain, such a thickness variation must, in general, be small.

As before, the contribution of the load vector $\mathbf{f}_l$ from concentrated forces can be dealt with by means of Dirac's delta function in complete analogy with the treatment of point sources in heat flow problems.

We have seen that plane stress and plane strain problems result in FE formulations which, in principle, are identical. As already mentioned, they differ only in the expressions for the constitutive matrix $\mathbf{D}$ and the initial strains $\boldsymbol{\varepsilon}_0$. The reader's attention is also drawn to the fact that even *axisymmetric problems* can be dealt with in a fashion very similar to that of plane stress and strain; see e.g. Zienkiewicz and Taylor (1989).

We also mention that for both two- and three-dimensional elasticity, the $\mathbf{B}$-matrix contains first order derivatives of the global shape functions. With reference to the discussion in Chapter 9, we conclude that C^0-continuity is required for the global shape functions. This C^0-continuity requirement implies fulfilment of the compatibility requirement discussed in the previous chapter. For more details of the FE approach for two- and three-dimensional solid mechanics problems in elasticity, plasticity and creep we may refer, for instance, to Bathe (1982), Cook *et al.* (1989), Hughes (1987), Owen and Hinton (1980) and Zienkiewicz and Taylor (1989).

Finally, let us obtain the FE formulation for one element. With evident notation, the result is

$$\boxed{\mathbf{K}^e\mathbf{a}^e = \mathbf{f}^e} \tag{16.64}$$

where

$$\boxed{\mathbf{f}^e = \mathbf{f}_b^e + \mathbf{f}_l^e + \mathbf{f}_0^e} \tag{16.65}$$

and

$$\boxed{\begin{aligned}
\mathbf{K}^e &= \int_{A_\alpha} \mathbf{B}^{e\mathrm{T}}\mathbf{D}\mathbf{B}^e t\, \mathrm{d}A \\[2ex]
\mathbf{f}_b^e &= \int_{\mathscr{L}_{h\alpha}} \mathbf{N}^{e\mathrm{T}}\mathbf{h}t\, \mathrm{d}\mathscr{L} + \int_{\mathscr{L}_{g\alpha}} \mathbf{N}^{e\mathrm{T}}\mathbf{t}t\, \mathrm{d}\mathscr{L} \\[2ex]
\mathbf{f}_l^e &= \int_{A_\alpha} \mathbf{N}^{e\mathrm{T}}\mathbf{b}t\, \mathrm{d}A \\[2ex]
\mathbf{f}_0^e &= \int_{A_\alpha} \mathbf{B}^{e\mathrm{T}}\mathbf{D}\boldsymbol{\varepsilon}_0 t\, \mathrm{d}A
\end{aligned}} \tag{16.66}$$

16.4.1 *Example 1*

As the FE formulation for two- and three-dimensional elasticity contains two and three unknowns per nodal point, respectively, it becomes difficult to illustrate the FE solution process by moderate hand calculations. However, in order to provide some kind of illustration, the element load vector $\mathbf{f}_l^e$ will be considered in this example and the boundary vector $\mathbf{f}_b$ in the next example.

The four-node rectangle shown in Figure 16.3 is adopted and the problem is to determine the element load vector $\mathbf{f}_l^e$ due to the body forces $\mathbf{b}$. In the present example, these body forces are due to gravity and they act in the negative direction of the

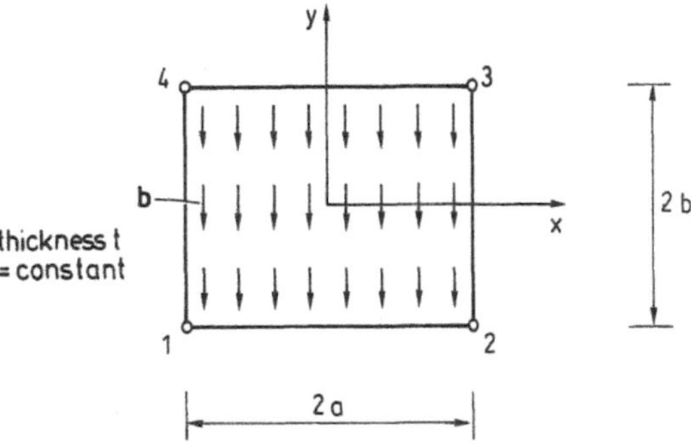

Figure 16.3   Four-node rectangle – Melosh element

y-axis, i.e.

$$\mathbf{b} = \begin{bmatrix} 0 \\ -\rho g \end{bmatrix}$$ (16.67)

where ρ is the mass density and g the acceleration due to gravity. From (16.66) it follows that

$$\mathbf{f}_1^e = \int_{A_\alpha} \mathbf{N}^{eT} \mathbf{b} t \, dA = -t \int_{A_\alpha} \begin{bmatrix} N_1^e & 0 \\ 0 & N_1^e \\ N_2^e & 0 \\ 0 & N_2^e \\ N_3^e & 0 \\ 0 & N_3^e \\ N_4^e & 0 \\ 0 & N_4^e \end{bmatrix} \begin{bmatrix} 0 \\ \rho g \end{bmatrix} dA = -\rho g t \int_{A_\alpha} \begin{bmatrix} 0 \\ N_1^e \\ 0 \\ N_2^e \\ 0 \\ N_3^e \\ 0 \\ N_4^e \end{bmatrix} dA$$ (16.68)

where the assumption of a constant thickness t was utilized. From (15.26) and Figure 16.3, we have

$$\int_{A_\alpha} N_1^e \, dA = \frac{1}{4ab} \int_{-b}^{b} \left(\int_{-a}^{a} (x - a)(y - b) \, dx \right) dy = ab$$ (16.69)

and with the geometrical interpretation of the element shape functions in mind, it is evident that the same value is obtained for the other integrations. Using (16.69) in

(16.68) we obtain

$$\mathbf{f}_1^e = -\rho gabt \begin{bmatrix} 0 \\ 1 \\ 0 \\ 1 \\ 0 \\ 1 \\ 0 \\ 1 \end{bmatrix} \tag{16.70}$$

It is evident that $\rho g4abt$ is the total vertical force acting on the element. The result (16.70) shows that one-fourth of this total force is distributed to each nodal point.

16.4.2 *Example 2*

Assume that a rectangular disk is loaded by a constant traction vector along the boundary AE; see Figure 16.4. The forces q_1 and q_2 per unit area act in the directions shown, i.e. the traction vector $\mathbf{t}$ along AE is given by

$$\mathbf{t} = \mathbf{h} = \begin{bmatrix} h_x \\ h_y \end{bmatrix} = \begin{bmatrix} q_1 \\ -q_2 \end{bmatrix} \tag{16.71}$$

In accordance with (16.58) the notation $\mathbf{h}$ is used for the traction vector, since it is a known quantity. As in the heat flow problem (cf. the discussion of (10.29)–(10.33)) it is evident that the tractions along AE are equivalent to some point forces that act at the nodal points. Our objective is to determine these equivalent nodal forces. Referring to (16.60) for the boundary vector $\mathbf{f}_b$, the contribution from the known

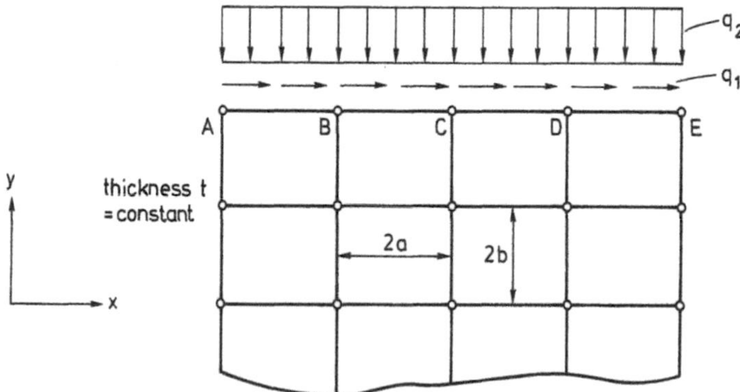

Figure 16.4   Boundary with constant traction vector; four-node rectangles of the same size

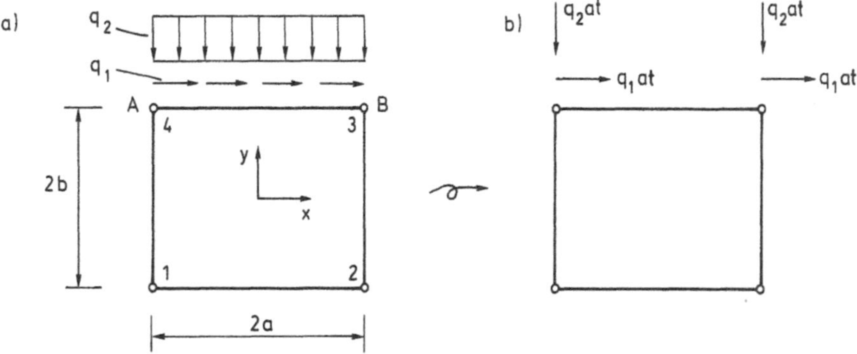

Figure 16.5   (a) Local coordinate system and uniform traction; (b) equivalent nodal forces

tractions to the nodal forces is given by

$$\int_{\mathscr{L}_{AE}} \mathbf{N}^T \mathbf{h} t \, d\mathscr{L} = \int_{\mathscr{L}_{AB}} \mathbf{N}^T \mathbf{h} t \, d\mathscr{L} + \int_{\mathscr{L}_{BC}} \mathbf{N}^T \mathbf{h} t \, d\mathscr{L} + \int_{\mathscr{L}_{CD}} \mathbf{N}^T \mathbf{h} t \, d\mathscr{L}$$

$$+ \int_{\mathscr{L}_{DE}} \mathbf{N}^T \mathbf{h} t \, d\mathscr{L} \tag{16.72}$$

As in the assembling process, we may evaluate these contributions using the element shape functions and then – using the topology data – determine the proper locations in the boundary vector for these contributions. Along **AB**, we therefore consider the term

$$\int_{\mathscr{L}_{AB}} \mathbf{N}^{eT} \mathbf{h} t \, d\mathscr{L}$$

To evaluate this expression we may actually use any convenient local coordinate system and we choose the one shown in Figure 16.5(a).

Along AB the only non-zero element shape functions are N_3^e and N_4^e and from (15.25) and (16.71) we obtain for a constant thickness t

$$\int_{\mathscr{L}_{AB}} \mathbf{N}^{eT} \mathbf{h} t \, d\mathscr{L} = t \int_{\mathscr{L}_{AB}} \begin{bmatrix} N_1^e & 0 \\ 0 & N_1^e \\ N_2^e & 0 \\ 0 & N_2^e \\ N_3^e & 0 \\ 0 & N_3^e \\ N_4^e & 0 \\ 0 & N_4^e \end{bmatrix} \begin{bmatrix} q_1 \\ -q_2 \end{bmatrix} d\mathscr{L} = t \int_{\mathscr{L}_{AB}} \begin{bmatrix} 0 \\ 0 \\ 0 \\ 0 \\ q_1 N_3^e \\ -q_2 N_3^e \\ q_1 N_4^e \\ -q_2 N_4^e \end{bmatrix} d\mathscr{L}$$

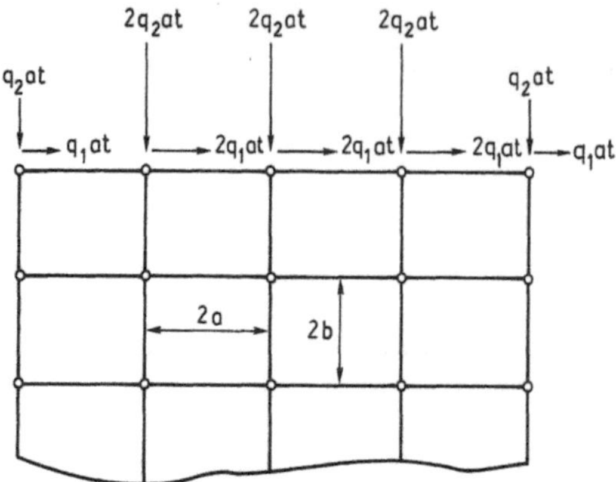

Figure 16.6   Equivalent nodal forces for the loading shown in Figure 16.4

As both N_3^e and N_4^e vary linearly along AB, we derive, for instance, that

$$t \int_{\mathscr{L}_{AB}} q_1 N_3^e \, d\mathscr{L} = tq_1 \int_{\mathscr{L}_{AB}} N_3^e \, d\mathscr{L} = tq_1 \times \tfrac{1}{2}2a = q_1 at$$

i.e.

$$\int_{\mathscr{L}_{AB}} \mathbf{N}^{eT} \mathbf{h} t \, d\mathscr{L} = \begin{bmatrix} 0 \\ 0 \\ 0 \\ 0 \\ q_1 at \\ -q_2 at \\ q_1 at \\ -q_2 at \end{bmatrix} \tag{16.73}$$

That is, the uniformly distributed traction shown in Figure 16.5(a) is equivalent to the nodal forces shown in Figure 16.5(b). It is obvious that the same results are obtained when considering the other elements and this implies that the uniform traction along A, B, C, D and E of Figure 16.4 is equivalent to the nodal forces shown in Figure 16.6.

17

FE formulation of beams

The FE formulation just presented for two- and three-dimensional elasticity theory was based on the exact field equations (equilibrium, kinematics and constitutive relation) and the corresponding boundary conditions. In many cases, however, the body is of such a geometry that it becomes possible to introduce a sequence of assumptions which significantly simplify the problem formulation. Two of the most outstanding examples of such an approach are the theories for beam and plate bending. In principle, such structures are three dimensional, but as a beam is dominated by its extension in the axial direction and a plate by its extension in the plane, it becomes possible to make certain assumptions about the structural deformation (i.e. about the kinematic relations), which significantly simplify the problem. However, one should be aware of the fact that, apart from very special situations, these theories are engineering approximations that violate some of the field equations.

A beam with cross-sectional area $A(x)$ is shown in Figure 17.1. The x-axis is directed in the axial direction and the transverse loading $q(x)$ is measured as positive in the z-direction. This loading q is the force per unit length, i.e. q has the dimension [N/m]. The material and cross-section of the beam are assumed to be symmetric about the xz-plane, as also indicated in Figure 17.1. Moreover, only loadings normal to the xy-plane and located symmetrically about the xz-plane are considered. This implies that the deflection w of the beam occurs in the xz-plane and w is measured as positive in the z-direction. For beam theories considering non-symmetric

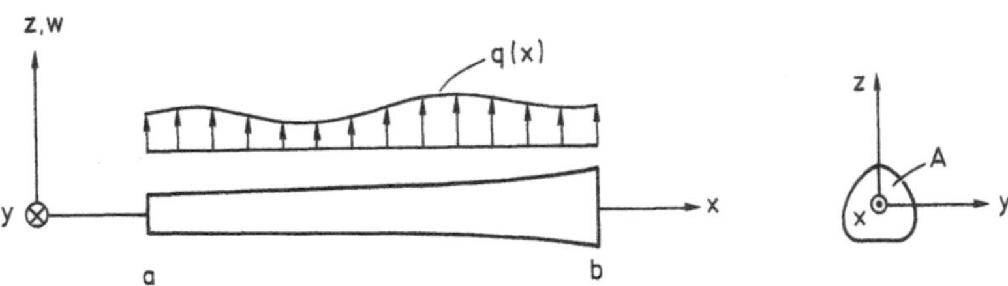

Figure 17.1 Configuration and loading of beam

cross-sections and more general loadings, we may refer to Timoshenko and Gere (1972).

17.1 Beam theory

17.1.1 *Equilibrium conditions*

For a section normal to the x-axis, we have the stress components σ_{xx}, σ_{xy} and σ_{xz}. Due to the symmetry in geometry and loading about the xz-plane, we can restrict ourselves to consider the effect of the components σ_{xx} and σ_{xz} for which we have

$$\sigma_{xx} \neq 0; \quad \sigma_{xz} \neq 0 \tag{17.1}$$

These stress components give rise to a *bending moment M* and a vertical *shear force V* defined by

$$M = \int_A z\sigma_{xx}\, \mathrm{d}A; \quad V = \int_A \sigma_{xz}\, \mathrm{d}A \tag{17.2}$$

where M is the moment about the y-axis. In accordance with the positive directions for σ_{xx} and σ_{xz} (cf. Figure 12.2), the positive directions for M and V are shown in Figure 17.2.

Let us now establish the equilibrium conditions for the beam. First it is noted that the stress component σ_{xx} also results in a *normal force N* in the x-direction given by

$$N = \int_A \sigma_{xx}\, \mathrm{d}A \tag{17.3}$$

As no resulting forces act in the x-direction, horizontal equilibrium implies that

$$N = 0 \tag{17.4}$$

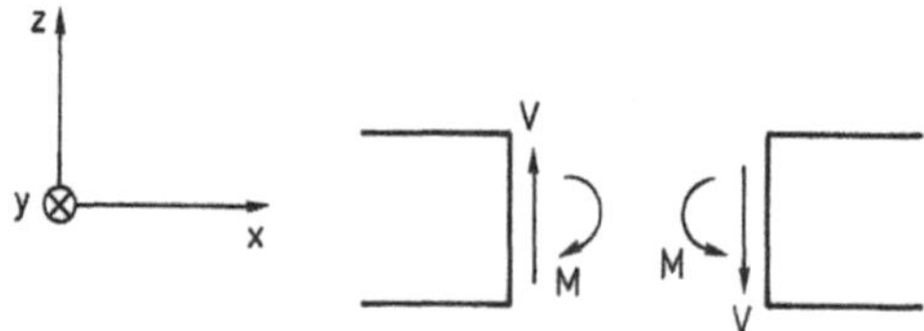

Figure 17.2   Positive directions for bending moment M and shear force V

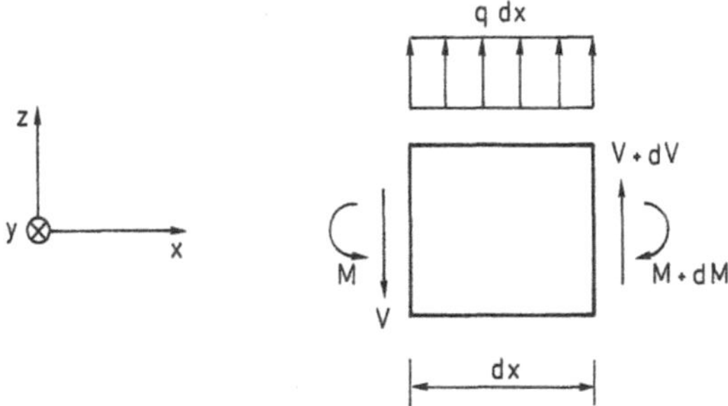

Figure 17.3 Infinitely small part of beam

For an infinitely small part of the beam, the remaining forces and moments caused by σ_{xx} and σ_{xz} are shown in Figure 17.3. Vertical equilibrium requires that

$$q\,dx - V + (V + dV) = 0$$

i.e.

$$\boxed{\frac{dV}{dx} = -q} \tag{17.5}$$

Evaluating the moment equilibrium, for instance, about the left end of the infinitely small part shown in Figure 17.3, we obtain

$$M + q\,dx\frac{dx}{2} + (V + dV)\,dx - (M + dM) = 0$$

i.e.

$$\frac{1}{2}q\,dx + V + dV - \frac{dM}{dx} = 0$$

As $q\,dx$ and dV are infinitesimal quantities, we conclude that

$$\boxed{\frac{dM}{dx} = V} \tag{17.6}$$

17.1.2 *Kinematic relations*

The essential feature of beam theory is that certain kinematic assumptions are made. The fundamental kinematic assumption for engineering beam theory, which dates

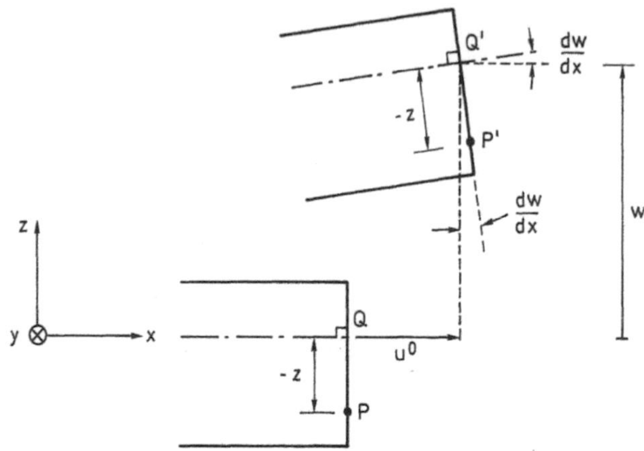

Figure 17.4   Deformation of Bernoulli beam

back to Jacob Bernoulli (1654–1705), is expressed by:

> Plane sections normal to the beam axis remain plane and
> normal to the beam axis during the deformation.

To investigate the consequences of *Bernoulli's assumption*, a plane section normal to
the x-axis before the deformation is considered (see Figure 17.4). The two points P
and Q are located on this plane and point Q is located on the x-axis. Due to the
deformation, the points P and Q move to the positions P′ and Q′ and, following
Bernoulli's assumption, the plane defined by P′Q′ is normal to the deformed beam
axis. Since P is located below the x-axis, the distance (a positive quantity) between
P and Q and between P′ and Q′ is given by $-z$. From Figure 17.4 and assuming
that the slope $\mathrm{d}w/\mathrm{d}x$ is small, the displacements u_x and u_z in the x- and z-directions
of point P are given by

$$u_x = u^0 - z\frac{\mathrm{d}w}{\mathrm{d}x}; \quad u_z = w \tag{17.7}$$

where u^0 is the displacement in the x-direction of the point Q. Moreover, it is assumed
that

$$u_y = 0 \tag{17.8}$$

and that u^0 and the deflection w only depend on x, i.e.

$$u^0 = u^0(x); \quad w = w(x) \tag{17.9}$$

With these kinematic assumptions, the strains given by (12.37) and (12.40) become

$$\varepsilon_{xx} = \frac{\mathrm{d}u^0}{\mathrm{d}x} - z\frac{\mathrm{d}^2 w}{\mathrm{d}x^2} \tag{17.10}$$

$$\varepsilon_{yy} = \varepsilon_{zz} = \gamma_{xy} = \gamma_{yz} = \gamma_{xz} = 0 \tag{17.11}$$

i.e. the only non-zero strain component is ε_{xx}.

17.1.3 *Constitutive relation*

Linear elasticity in terms of Hooke's law for isotropic materials is assumed. According to (17.10) and (17.11) the only non-zero strain is ε_{xx}, i.e. use of (13.2) and (13.17) results in

$$\begin{bmatrix} \sigma_{xx} \\ \sigma_{yy} \\ \sigma_{zz} \end{bmatrix} = \frac{E\varepsilon_{xx}}{(1+v)(1-2v)} \begin{bmatrix} 1-v \\ v \\ v \end{bmatrix} \tag{17.12}$$

and

$$\sigma_{xy} = \sigma_{xz} = \sigma_{yz} = 0 \tag{17.13}$$

A comparison of (17.13) with (17.1) reveals evident contradictions, since (17.13) predicts the shear stress σ_{xz} to be zero even though in reality it must be non-zero in order to have a non-zero shear force V. Stated explicitly, we have $\sigma_{xz} \neq 0$ and $\gamma_{xz} = 0$ and the constitutive relation $\sigma_{xz} = G\gamma_{xz}$. This kind of inconsistency is typical of simplifying engineering theories which try to reformulate problems that are actually three dimensional into a simpler form.

Instead of the equations above, a uniaxial state of stress is assumed; that is, we assume that the relation

$$\boxed{\sigma_{xx} = E\varepsilon_{xx}} \tag{17.14}$$

holds and we accept the contradiction $\sigma_{xz} \neq 0$ and $\gamma_{xz} = 0$. We may even include the effect of initial strains (e.g. thermal strains) as

$$\sigma_{xx} = E\varepsilon_{xx} - \alpha E \Delta T \tag{17.15}$$

but for simplicity we shall not pursue this topic in what follows.

17.1.4 *Choice of position of x-axis*

Insertion of (17.10) into (17.14) yields

$$\sigma_{xx} = E\frac{du^0}{dx} - Ez\frac{d^2w}{dx^2} \tag{17.16}$$

and using (17.2) the bending moment M is determined from

$$M = \frac{du^0}{dx} \int_A Ez \, dA - \frac{d^2w}{dx^2} \int_A Ez^2 \, dA \tag{17.17}$$

where u^0 and w only depend on the coordinate x (cf. (17.9)). Up until now we have not specified the location of the x-axis – we have just said that the x-axis should be in the axial direction of the beam. To obtain the simplest possible formulation, we now choose the vertical position of the x-axis so that

$$\int_A Ez \, dA = 0 \tag{17.18}$$

If E is constant within the cross-section, this requirement implies that

$$\int_A z \, dA = 0$$

i.e. the x-axis should be placed at the centroid of the cross-section. Equation (17.18) can therefore be interpreted so that the x-axis is located along the centroid of the cross-section weighted with the material parameter E. Using (17.18), (17.17) becomes

$$M = -\frac{d^2 w}{dx^2} \int_A Ez^2 \, dA \tag{17.19}$$

If E is constant over the cross-section, (17.19) reduces to the well-known formula

$$M = -EI \frac{d^2 w}{dx^2}; \quad I = \int_A z^2 \, dA \tag{17.20}$$

where I is the *moment of inertia* about the y-axis. The term EI is called the *bending stiffness* and when E varies over the cross-section we may for convenience introduce the notation

$$E^* I^* = \int_A Ez^2 \, dA \tag{17.21}$$

where $E^* I^*$ is again the bending stiffness. This notation means that (17.19) may be written in the compact form

$$M = -E^* I^* \frac{d^2 w}{dx^2} \tag{17.22}$$

i.e. it is possible to express the moment M in terms of the deflection w.

The normal force N is given by (17.3) which with (17.16) and (17.18) becomes

$$N = \frac{du^0}{dx} \int_A E \, dA \tag{17.23}$$

In the present situation we have assumed that $N = 0$ (cf. (17.4)). Equation (17.23) then gives

$$\frac{\mathrm{d}u^0}{\mathrm{d}x} = 0 \tag{17.24}$$

i.e. there is no elongation of the x-axis. However, even in the situation where the normal force $N \neq 0$, we observe from (17.22) and (17.23) that, whereas the bending moment M is controlled by the term $\mathrm{d}^2w/\mathrm{d}x^2$, the normal force is determined by the quantity $\mathrm{d}u^0/\mathrm{d}x$. It is therefore concluded that the phenomena of bending and elongation of the beam are *uncoupled* phenomena, i.e. they can be treated separately. In turn, this means that the FE formulation in the present chapter can be combined directly with the FE formulation of the elastic bar treated in Chapter 9 if elongation of the beam occurs. It is emphasized that this advantageous uncoupling is a result of our choice of the vertical position of the x-axis as given by (17.18).

17.1.5 *Differential equations for Bernoulli's beam theory*

In the present situation where the normal force N is zero, (17.24) holds. Thus (17.10) and (17.16) reduce to

$$\boxed{\varepsilon_{xx} = -z\frac{\mathrm{d}^2w}{\mathrm{d}x^2}} \qquad \boxed{\sigma_{xx} = -zE\frac{\mathrm{d}^2w}{\mathrm{d}x^2}} \tag{17.25}$$

i.e. the axial strain and stress are zero along the x-axis. The *neutral axis* is in general defined as that axis along which no axial strain occurs. With the location of the x-axis defined by (17.18) and with the normal force $N = 0$, the x-axis becomes the neutral axis.

It is of interest to evaluate the term $\mathrm{d}^2w/\mathrm{d}x^2$ present in (17.22) and (17.25). From mathematical textbooks on differential geometry we find that the *curvature* κ of the curve $w(x)$ is defined by

$$\kappa = \frac{\mathrm{d}^2w/\mathrm{d}x^2}{[1 + (\mathrm{d}w/\mathrm{d}x)^2]^{3/2}}$$

As the slope $\mathrm{d}w/\mathrm{d}x$ is assumed to be small, we obtain approximately that

$$\kappa = \frac{\mathrm{d}^2w}{\mathrm{d}x^2} \tag{17.26}$$

i.e. we may write (17.22) as

$$M = -E^*I^*\kappa \tag{17.27}$$

We observe from (17.22) that it is possible to express the bending moment M in terms of the deflection w. However, the shear stress σ_{xz} cannot be determined using the constitutive relation since the shear strain γ_{xz} is assumed to be zero. Consequently, the shear force V cannot be expressed by kinematic quantities. Our present objective is to obtain a differential equation for the beam behaviour expressed in terms of the deflection w. For this purpose we recognize the equilibrium conditions given by (17.5) and (17.6). As only the moment M can be related to the deflection w, we eliminate the shear force V from (17.5) and (17.6) to obtain

$$\boxed{\frac{\mathrm{d}^2 M}{\mathrm{d}x^2} + q = 0} \qquad (17.28)$$

It is emphasized that this equilibrium condition holds irrespective of the constitutive and kinematic assumptions. Use of (17.22) in (17.28) yields the differential equation sought:

$$\boxed{\frac{\mathrm{d}^2}{\mathrm{d}x^2}\left(E^*I^*\frac{\mathrm{d}^2 w}{\mathrm{d}x^2}\right) - q = 0} \qquad (17.29)$$

It appears that when the deflection w has been determined from this differential equation, all quantities of interest can be derived.

Although Bernoulli's beam theory presented above involves quite a number of assumptions, it provides close predictions for a variety of problems of practical importance. The most critical point in these approximations is the contradiction between the existence of a shear stress σ_{xz} – necessary to maintain equilibrium – and a zero shear strain γ_{xz}. For long slender beams where the ratio $L/h > 5$–10, where L is the beam length and h the beam height, the real shear strain γ_{xz} is generally small and Bernoulli's theory is usually satisfactory. For higher beams, more refined beam theories can be used which include the effect of a non-zero shear strain γ_{xz}, for instance *Timoshenko beam theory* (Przemieniecki, 1968; Timoshenko and Gere, 1972) or *Mindlin–Reissner beam theory* (Bathe, 1982; Hughes, 1987). For very high beams, one has to resort to the general plane stress elasticity theory described in Chapter 16.

17.2 Weak form of equilibrium equation

As in the previous chapter, it is advantageous to obtain the weak form of the equilibrium condition before the constitutive assumption is introduced. The pertinent differential equation is given by (17.28), and multiplying by an arbitrary weight function $v(x)$ and integrating over the region (see Figure 17.1) we obtain

$$\int_a^b v\frac{\mathrm{d}^2 M}{\mathrm{d}x^2}\,\mathrm{d}x + \int_a^b vq\,\mathrm{d}x = 0$$

The first term is integrated by parts and this results in

$$\int_a^b \frac{dv}{dx}\frac{dM}{dx}\,dx = \left[v\frac{dM}{dx}\right]_a^b + \int_a^b vq\,dx$$

Use of (17.6) yields

$$\int_a^b \frac{dv}{dx}\frac{dM}{dx}\,dx = [vV]_a^b + \int_a^b vq\,dx$$

The left-hand side of this expression may even be integrated by parts once more to obtain

$$\boxed{\int_a^b \frac{d^2v}{dx^2}M\,dx = \left[\frac{dv}{dx}M\right]_a^b - [vV]_a^b - \int_a^b vq\,dx}\tag{17.30}$$

This weak form of the equilibrium equation holds irrespective of constitutive and kinematic assumptions. As the moment M and the shear force V emerge in the boundary terms of this weak form, they are the *natural boundary conditions*.

17.3 FE formulation

From the weak form of the equilibrium condition (17.30) we shall now derive the corresponding FE formulation. As the deflection w is the unknown function, we can write the approximation for w quite generally as

$$\boxed{w = \mathbf{Na}}\tag{17.31}$$

where

$$\mathbf{N} = [N_1 \quad N_2 \quad \cdots \quad N_n]; \quad \mathbf{a} = \begin{bmatrix} u_1 \\ u_2 \\ \vdots \\ u_n \end{bmatrix}\tag{17.32}$$

and n is the number of unknowns for the entire beam. Note that n now denotes the number of unknowns and not, as in previous chapters, the number of nodal points. At this point we shall leave the meaning of the shape functions N_i and the nodal values u_i unspecified, and merely observe that whereas N_i depends on x, i.e. $N_i = N_i(x)$, the u_i components are parameters. From (17.31) it follows that

$$\boxed{\frac{d^2w}{dx^2} = \mathbf{Ba}} \quad \text{where} \quad \boxed{\mathbf{B} = \frac{d^2\mathbf{N}}{dx^2}}\tag{17.33}$$

i.e.

$$\mathbf{B} = \left[\frac{d^2 N_1}{dx^2} \quad \frac{d^2 N_2}{dx^2} \quad \cdots \quad \frac{d^2 N_n}{dx^2} \right] \tag{17.34}$$

Adopting the Galerkin method, we use the following expression for the arbitrary weight function v:

$$v = \mathbf{N}\mathbf{c} \tag{17.35}$$

With the weight function v being arbitrary, the parameters given by $\mathbf{c}$ are also arbitrary. As $v = v^{\mathrm{T}}$, (17.35) can be rewritten as

$$v = \mathbf{c}^{\mathrm{T}}\mathbf{N}^{\mathrm{T}} \tag{17.36}$$

which leads to

$$\frac{dv}{dx} = \mathbf{c}^{\mathrm{T}}\frac{d\mathbf{N}^{\mathrm{T}}}{dx}; \quad \frac{d^2 v}{dx^2} = \mathbf{c}^{\mathrm{T}}\mathbf{B}^{\mathrm{T}} \tag{17.37}$$

Inserting (17.36) and (17.37) into (17.30) and noting that the $\mathbf{c}$-matrix is independent of position, we get

$$\mathbf{c}^{\mathrm{T}}\left(\int_a^b \mathbf{B}^{\mathrm{T}}M \, dx - \left[\frac{d\mathbf{N}^{\mathrm{T}}}{dx}M \right]_a^b + [\mathbf{N}^{\mathrm{T}}V]_a^b + \int_a^b \mathbf{N}^{\mathrm{T}}q \, dx \right) = 0$$

As $\mathbf{c}$ is arbitrary, it is concluded that

$$\boxed{\int_a^b \mathbf{B}^{\mathrm{T}}M \, dx = \left[\frac{d\mathbf{N}^{\mathrm{T}}}{dx}M \right]_a^b - [\mathbf{N}^{\mathrm{T}}V]_a^b - \int_a^b \mathbf{N}^{\mathrm{T}}q \, dx} \tag{17.38}$$

The similarity with the corresponding forms for two- and three-dimensional elasticity may be noted. We recall that (17.38) is an expression for the equilibrium conditions for the entire beam and that it holds irrespective of the constitutive model adopted and irrespective of the kinematic assumptions for the beam response.

At this point it is indeed very easy to derive the FE formulation. From (17.22) and (17.33) we get

$$\boxed{M = -E^*I^*\mathbf{B}\mathbf{a}} \tag{17.39}$$

and insertion into (17.38) results in

$$\left(\int_a^b \mathbf{B}^{\mathrm{T}}E^*I^*\mathbf{B} \, dx \right)\mathbf{a} = [\mathbf{N}^{\mathrm{T}}V]_a^b - \left[\frac{d\mathbf{N}^{\mathrm{T}}}{dx}M \right]_a^b + \int_a^b \mathbf{N}^{\mathrm{T}}q \, dx \tag{17.40}$$

which is the FE formulation sought. To write this result in a more compact fashion, the following matrices are defined:

$$
\begin{aligned}
\mathbf{K} &= \int_a^b \mathbf{B}^\mathrm{T} E * I * \mathbf{B}\, \mathrm{d}x \\[2mm]
\mathbf{f}_b &= [\mathbf{N}^\mathrm{T} V]_a^b - \left[\frac{\mathrm{d}\mathbf{N}^\mathrm{T}}{\mathrm{d}x} M\right]_a^b \\[2mm]
\mathbf{f}_l &= \int_a^b \mathbf{N}^\mathrm{T} q\, \mathrm{d}x
\end{aligned}
\tag{17.41}
$$

where $\mathbf{K}$ is the stiffness matrix, $\mathbf{f}_b$ the boundary vector and $\mathbf{f}_l$ the load vector. With (17.41), (17.40) may be written in the standard form

$$
\mathbf{Ka} = \mathbf{f}_b + \mathbf{f}_l
\tag{17.42}
$$

As usual, we define the force vector $\mathbf{f}$ by

$$
\mathbf{f} = \mathbf{f}_b + \mathbf{f}_l
\tag{17.43}
$$

i.e.

$$
\mathbf{Ka} = \mathbf{f}
\tag{17.44}
$$

The boundary conditions for this system of equations are described either by *static conditions*, i.e. the moment M and shear force V (the natural boundary conditions), or by *kinematic conditions*, i.e. the deflection w and slope $\mathrm{d}w/\mathrm{d}x$ (the essential boundary conditions). We therefore have

$$
\begin{aligned}
w;\, \mathrm{d}w/\mathrm{d}x &\quad \text{kinematic boundary conditions} \\
M;\, V &\quad \text{static boundary conditions}
\end{aligned}
$$

Examples of such boundary conditions are shown in Figure 17.5 and we observe that static and kinematic boundary conditions may be prescribed at the same boundary.

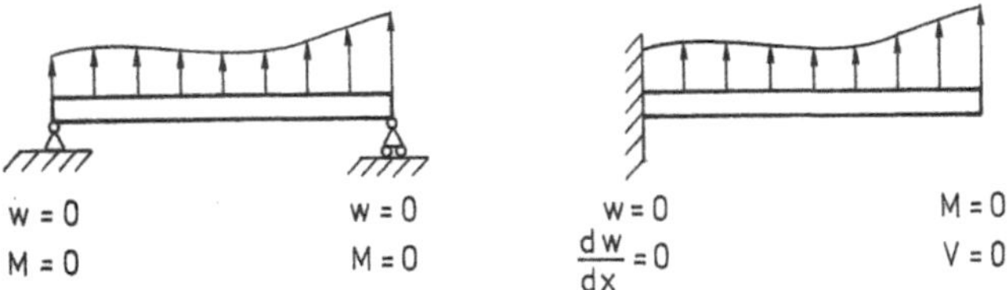

Figure 17.5 Examples of boundary conditions

Let us now consider the FE formulation of one element. With the usual notation, the approximation of the deflection w for one element is written as

$$\boxed{w = \mathbf{N}^e \mathbf{a}^e} \tag{17.45}$$

where

$$\mathbf{N}^e = [N_1^e \quad N_2^e \quad \cdots \quad N_{n_e}^e]; \quad \mathbf{a}^e = \begin{bmatrix} u_1^e \\ u_2^e \\ \vdots \\ u_{n_e}^e \end{bmatrix} \tag{17.46}$$

where n_e is the number of unknowns for one element. Again we deviate from the notation in the previous chapters and emphasize that n_e is the number of unknowns and not the number of nodal points. We define

$$\boxed{\mathbf{B}^e = \frac{d^2 \mathbf{N}^e}{dx^2}} \tag{17.47}$$

It is obvious that

$$\boxed{\mathbf{K}^e \mathbf{a}^e = \mathbf{f}^e} \tag{17.48}$$

where

$$\boxed{\mathbf{f}^e = \mathbf{f}_b^e + \mathbf{f}_l^e} \tag{17.49}$$

and

$$\boxed{\begin{aligned} \mathbf{K}^e &= \int_{L_\alpha} \mathbf{B}^{eT} E * I * \mathbf{B}^e \, dx \\ \mathbf{f}_b^e &= [\mathbf{N}^{eT} V]_{L_\alpha} - \left[\frac{d\mathbf{N}^{eT}}{dx} M \right]_{L_\alpha} \\ \mathbf{f}_l^e &= \int_{L_\alpha} \mathbf{N}^{eT} q \, dx \end{aligned}} \tag{17.50}$$

where L_α is the region of the element and the notation defined by (9.79) has been used in the expression for $\mathbf{f}_b^e$. We recall that the boundary conditions for an element are unknown except when an element boundary coincides with the boundary of the beam.

17.3.1 *Completeness and compatibility requirements*

Up until now we have not specified how the approximation (17.31) or (17.45) was made, and the meaning of the shape function components N_i and the nodal value components u_i have been left undefined. In order to be more specific, we shall investigate the convergence requirements that are fulfilled if the completeness and compatibility requirements are met.

For infinitely small beam elements, the deflection within each element will, in the limit, be given by an arbitrary rigid-body motion superposed by an arbitrary constant curvature. We note that a constant curvature is the simplest deflection state that creates strains (cf. (17.25)). We are then led to the following completeness requirements:

Completeness

- The approximation for the deflection w must be able to represent an arbitrary rigid-body motion.
- The approximation for the deflection w must be able to represent an arbitrary constant curvature.

An arbitrary rigid-body motion consists of a translation in the z-direction as well as a rotation in the xz-plane (cf. Figure 17.1). That is, the approximation for the deflection w must at least include the expression $\alpha_1 + \alpha_2 x$ where α_1 and α_2 are arbitrary parameters. The curvature κ is given by (17.26) and we therefore conclude that any approximation for the deflection w must take the form

$$w = \alpha_1 + \alpha_2 x + \alpha_3 x^2 + \text{possibly other terms} \tag{17.51}$$

To derive the compatibility criterion, we shall take advantage of the concept of C^n-continuity discussed in Chapter 9 (cf. page 202). In all our previous applications of the FE method, the **B**-matrix in the stiffness matrix **K** contained only first derivatives of the shape functions. Therefore, C^0-continuity was required and, in turn, this implied the compatibility condition that the approximation of the variable must vary in a continuous manner over the element boundaries. According to (17.41) and (17.33), the stiffness matrix now involves second derivatives of the shape functions. We immediately conclude that C^1-*continuity* is required for the shape functions and thus also for the deflection w. This compatibility requirement can be stated as follows:

Compatibility

- The approximation for the deflection w must vary continuously and with continuous slopes over the element boundaries.

(17.52)

Previously, we claimed that non-conforming elements are acceptable (cf. p. 93). However, we note that Bernoulli beam elements must always be compatible since a possible discontinuity does not disappear for infinitely small elements.

17.3.2 *Simplest possible beam element*

We shall now establish the simplest possible compatible beam element. Based on (17.51), we choose the approximation for the deflection w as

$$w = \alpha_1 + \alpha_2 x + \alpha_3 x^2 + \alpha_4 x^3 \qquad (17.53)$$

where α_1, α_2, α_3 and α_4 are certain parameters. This means that the completeness criterion is fulfilled. To determine the four parameters in (17.53), the beam element must possess four unknowns and these unknowns are taken as the quantities u_1, u_2, u_3 and u_4 shown in Figure 17.6. Here, u_1 and u_3 are the nodal displacements in the z-direction, whereas u_2 and u_4 are the slopes at the nodal points measured as positive in the counter-clockwise direction. It appears that two unknowns are related to each nodal point: one for the deflection w and another for the slope dw/dx.

As values for w and dw/dx are connected to each nodal point and as such a nodal point is common for two neighbouring elements, not only the deflection w but also its derivative dw/dx will vary in a continuous manner over the element boundaries, i.e. the simple beam element of Figure 17.6 fulfils the compatibility requirement given by (17.52).

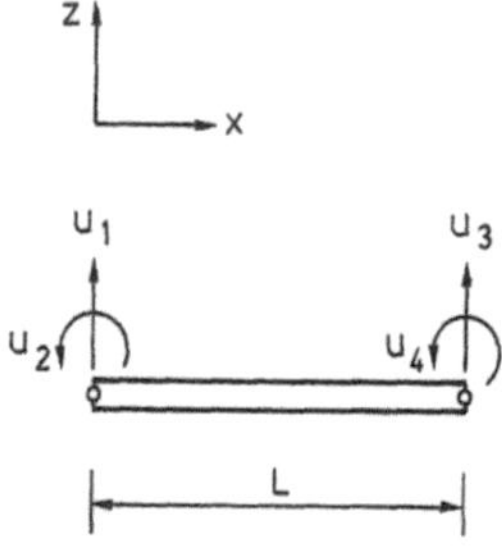

Figure 17.6   Simplest possible beam element

In order to determine the parameters α_1, α_2, α_3 and α_4 in (17.53), the **C**-matrix method of Chapter 7 is applied; cf. for instance (7.45)–(7.52). Therefore, (17.53) is written as

$$w = \bar{\mathbf{N}}\boldsymbol{\alpha} \tag{17.54}$$

where

$$\bar{\mathbf{N}} = [1 \quad x \quad x^2 \quad x^3]; \quad \boldsymbol{\alpha} = \begin{bmatrix} \alpha_1 \\ \alpha_2 \\ \alpha_3 \\ \alpha_4 \end{bmatrix} \tag{17.55}$$

We next express the conditions for w at the nodal points, i.e.

$$w_{x=0} = u_1; \quad \left(\frac{dw}{dx}\right)_{x=0} = u_2; \quad w_{x=L} = u_3; \quad \left(\frac{dw}{dx}\right)_{x=L} = u_4 \tag{17.56}$$

Use of these conditions in (17.54) results in

$$\mathbf{a}^e = \mathbf{C}\boldsymbol{\alpha} \tag{17.57}$$

where

$$\mathbf{a}^e = \begin{bmatrix} u_1 \\ u_2 \\ u_3 \\ u_4 \end{bmatrix}; \quad \mathbf{C} = \begin{bmatrix} 1 & 0 & 0 & 0 \\ 0 & 1 & 0 & 0 \\ 0 & L & L^2 & L^3 \\ 0 & 1 & 2L & 3L^2 \end{bmatrix} \tag{17.58}$$

From (17.57) it follows that

$$\boldsymbol{\alpha} = \mathbf{C}^{-1}\mathbf{a}^e \quad \text{where} \quad \mathbf{C}^{-1} = \begin{bmatrix} 1 & 0 & 0 & 0 \\ 0 & 1 & 0 & 0 \\ -3/L^2 & -2/L & 3/L^2 & -1/L \\ 2/L^3 & 1/L^2 & -2/L^3 & 1/L^2 \end{bmatrix} \tag{17.59}$$

and use of this result in (17.54) gives

$$\boxed{w = \mathbf{N}^e \mathbf{a}^e} \tag{17.60}$$

where

$$\mathbf{N}^e = \bar{\mathbf{N}}\mathbf{C}^{-1} = [N_1^e \quad N_2^e \quad N_3^e \quad N_4^e] \tag{17.61}$$

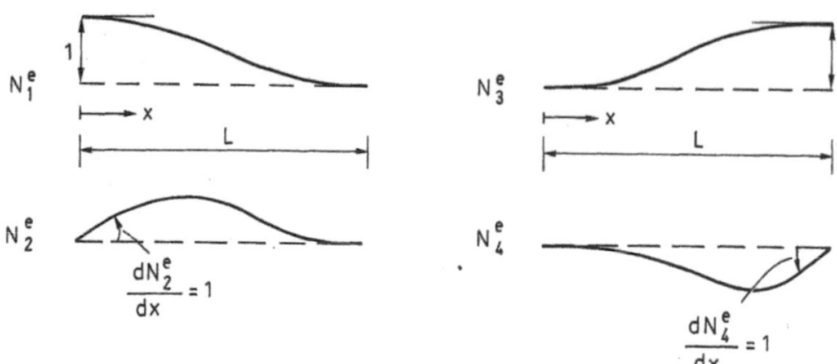

Figure 17.7   Element shape functions

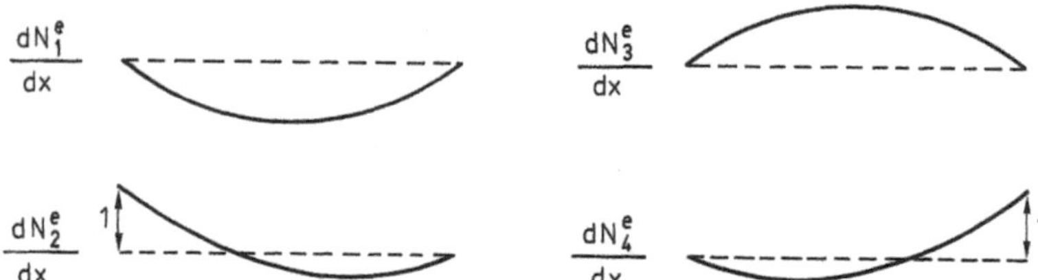

Figure 17.8   First derivatives of element shape functions

From this expression we derive

$$N_1^e = 1 - 3\frac{x^2}{L^2} + 2\frac{x^3}{L^3}; \qquad N_3^e = \frac{x^2}{L^2}\left(3 - 2\frac{x}{L}\right)$$
$$N_2^e = x\left(1 - 2\frac{x}{L} + \frac{x^2}{L^2}\right); \qquad N_4^e = \frac{x^2}{L}\left(\frac{x}{L} - 1\right) \tag{17.62}$$

and these element shape functions are illustrated in Figure 17.7. From (17.60) it follows that

$$w = N_1^e u_1 + N_2^e u_2 + N_3^e u_3 + N_4^e u_4 \tag{17.63}$$

As with previous FE formulations, it appears that the deflection w within the element is composed of contributions from the element shape functions N_1^e, N_2^e, N_3^e and N_4^e weighted with the nodal quantities u_1, u_2, u_3 and u_4. The first and second derivatives of the element shape functions are shown in Figures 17.8 and 17.9, respectively. From (17.39) and Figure 17.9 it follows that the simple beam element is able to model an arbitrary linear moment variation within the element.

It appears that, whereas the element shape functions N_1^e and N_3^e related to the displacement degrees of freedom possess the traditional properties, the element shape functions N_2^e and N_4^e related to the rotational degrees of freedom possess new

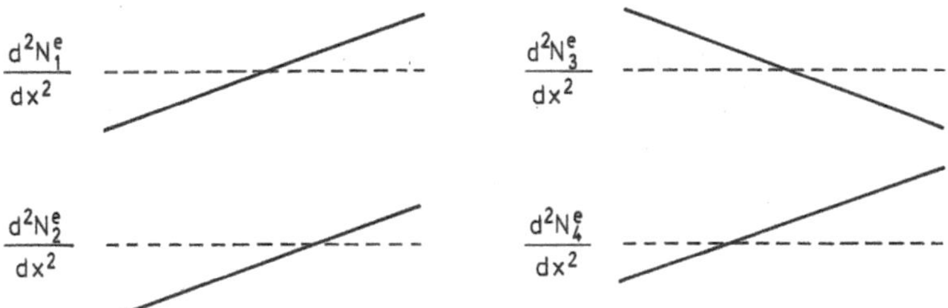

Figure 17.9 Second derivatives of element shape functions

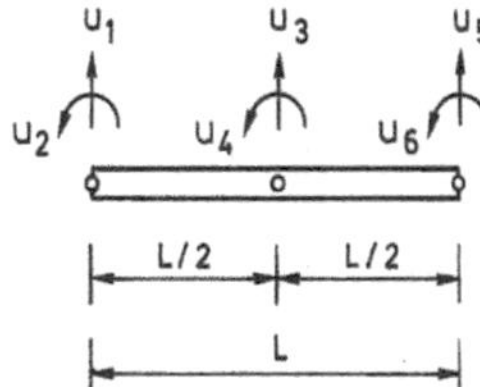

Figure 17.10 Three-node beam element

characteristics. The derivatives dN_2^e/dx and dN_4^e/dx, however, possess the features of the element shape functions dealt with in the previous chapters; for instance, we have that $(dN_2^e/dx)_{x=0} = 1$ and $(dN_2^e/dx)_{x=L} = 0$. The properties of the element shape functions shown in Figure 17.7 and 17.8 are clearly necessary so that the conditions (17.56) can be fulfilled.

We may consider beam elements of higher order, and Figure 17.10 shows an example with three nodal points and a total of six degrees of freedom. The approximation for the displacement w is given by

$$w = \alpha_1 + \alpha_2 x + \alpha_3 x^2 + \alpha_4 x^3 + \alpha_5 x^4 + \alpha_6 x^5 \tag{17.64}$$

Irrespective of the beam element considered, it follows directly that the global shape functions can be established from the element shape functions in the same manner as discussed in Chapter 7. For more details about beam elements, we refer to Przemieniecki (1968) and Bathe (1982), which even provide the FE formulation of the Timoshenko beam theory discussed previously.

Let us return to the topic discussed above, namely that the **B**-matrix now contains second derivatives of the shape functions, whereas in the previous chapters all **B**-matrices only involved first derivatives of the shape functions. The reason is that the differential equation for the beam (cf. (17.29)) is of fourth order, whereas all the previous physical problems were governed by differential equations of second order. In accordance with this, the weak form of the equilibrium conditions for the beam (cf. (17.30)) was obtained by invoking integration by parts twice, which implied that the order of differentiation of the weight function became two. As a consequence, the approximation for the deflection w was required to be C^1-continuous in contrast to the previous chapters where only C^0-continuity was demanded. The C^1-continuity means that the deflection w varies continuously and with continuous first derivatives – i.e. slopes – over the element boundaries. Approximations of functions which fulfil not only the function value itself at some specific points but also its slopes at these points are called *Hermite interpolations* and a simple example was given by (17.62) and (17.63).

17.3.3 *Evaluation of element stiffness matrix*

From (17.50) we obtain for the simple beam element given by Figure 17.6

$$\mathbf{K}^e = \int_0^L E^*I^* \begin{bmatrix} B_1^e B_1^e & B_1^e B_2^e & B_1^e B_3^e & B_1^e B_4^e \\ B_2^e B_1^e & B_2^e B_2^e & B_2^e B_3^e & B_2^e B_4^e \\ B_3^e B_1^e & B_3^e B_2^e & B_3^e B_3^e & B_3^e B_4^e \\ B_4^e B_1^e & B_4^e B_2^e & B_4^e B_3^e & B_4^e B_4^e \end{bmatrix} dx \tag{17.65}$$

We now assume that the bending stiffness E^*I^* does not vary along the beam axis; then, for example, the component K_{32}^e becomes

$$K_{32}^e = \int_0^L B_3^e E^*I^* B_2^e \, dx = E^*I^* \int_0^L B_3^e B_2^e \, dx = E^*I^* \int_0^L \frac{d^2 N_3^e}{dx^2} \frac{d^2 N_2^e}{dx^2} \, dx$$

From (17.62) it follows that

$$\mathbf{K}_{32}^e = E^*I^* \int_0^L \left(\frac{6}{L^2} - \frac{12x}{L^3} \right) \left(-\frac{4}{L} + \frac{6x}{L^2} \right) dx = -\frac{6E^*I^*}{L^2}$$

Evaluating all the components of (17.65) in this manner we arrive at

$$\mathbf{K}^e = \begin{bmatrix} \dfrac{12E^*I^*}{L^3} & \dfrac{6E^*I^*}{L^2} & -\dfrac{12E^*I^*}{L^3} & \dfrac{6E^*I^*}{L^2} \\[2.5ex] \dfrac{6E^*I^*}{L^2} & \dfrac{4E^*I^*}{L} & -\dfrac{6E^*I^*}{L^2} & \dfrac{2E^*I^*}{L} \\[2.5ex] -\dfrac{12E^*I^*}{L^3} & -\dfrac{6E^*I^*}{L^2} & \dfrac{12E^*I^*}{L^3} & -\dfrac{6E^*I^*}{L^2} \\[2.5ex] \dfrac{6E^*I^*}{L^2} & \dfrac{2E^*I^*}{L} & -\dfrac{6E^*I^*}{L^2} & \dfrac{4E^*I^*}{L} \end{bmatrix} \tag{17.66}$$

17.3.4 *Evaluation of element boundary vector*

Consider the element boundary vector $\mathbf{f}_b^e$ given by (17.50). Evaluating this vector for the simple beam element shown in Figure 17.6, we obtain

$$\mathbf{f}_b^e = [\mathbf{N}^{eT} V]_0^L - \left[\frac{d\mathbf{N}^{eT}}{dx} M\right]_0^L$$

i.e.

$$\mathbf{f}_b^e = \begin{bmatrix} (N_1^e V)_{x=L} - (N_1^e V)_{x=0} - \left(\dfrac{dN_1^e}{dx} M\right)_{x=L} + \left(\dfrac{dN_1^e}{dx} M\right)_{x=0} \\[2ex] (N_2^e V)_{x=L} - (N_2^e V)_{x=0} - \left(\dfrac{dN_2^e}{dx} M\right)_{x=L} + \left(\dfrac{dN_2^e}{dx} M\right)_{x=0} \\[2ex] (N_3^e V)_{x=L} - (N_3^e V)_{x=0} - \left(\dfrac{dN_3^e}{dx} M\right)_{x=L} + \left(\dfrac{dN_3^e}{dx} M\right)_{x=0} \\[2ex] (N_4^e V)_{x=L} - (N_4^e V)_{x=0} - \left(\dfrac{dN_4^e}{dx} M\right)_{x=L} + \left(\dfrac{dN_4^e}{dx} M\right)_{x=0} \end{bmatrix}$$

Considering the properties of the element shape functions given by Figures 17.7 and 17.8, it follows that

$$\mathbf{f}_b^e = \begin{bmatrix} -V_{x=0} \\ M_{x=0} \\ V_{x=L} \\ -M_{x=L} \end{bmatrix} \tag{17.67}$$

The positive directions of V and M are given in Figure 17.2. Therefore, when all the components of the element boundary vector $\mathbf{f}_b^e$ are positive, these components take the form shown in Figure 17.11. It appears that the directions shown in Figure 17.11 are in accordance with the directions of the kinematic quantities shown in Figure 17.6.

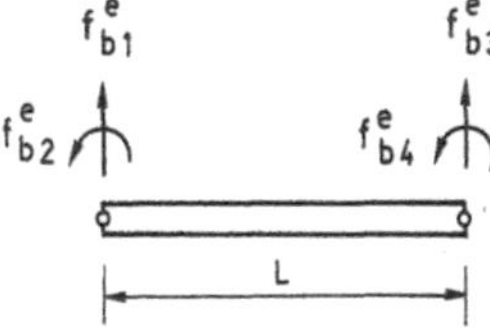

Figure 17.11 Illustration of element boundary vector $\mathbf{f}_b^e$

17.3.5 *Beam element with no distributed loading*

Consider the simple beam element with no distributed loading, i.e. $q = 0$. This implies that $\mathbf{f}_i^e = 0$ and that (17.48) reduces to

$$\mathbf{K}^e\mathbf{a}^e = \mathbf{f}_b^e \tag{17.68}$$

where $\mathbf{f}_b^e$ is given by (17.67) and the positive directions of the $\mathbf{f}_b^e$-components are shown in Figure 17.11. Assume now that the bending stiffness E^*I^* does not vary along the beam axis, i.e. $\mathbf{K}^e$ is given by (17.66). From (17.68) we may obtain some interesting results. Assume that

$$\mathbf{a}^e = \begin{bmatrix} 1 \\ 0 \\ 0 \\ 0 \end{bmatrix} \tag{17.69}$$

With the $\mathbf{a}^e$-components given by Figure 17.6, the nodal values of (17.69) correspond to the deflection of the beam element as shown in Figure 17.12(a). Using (17.69) in

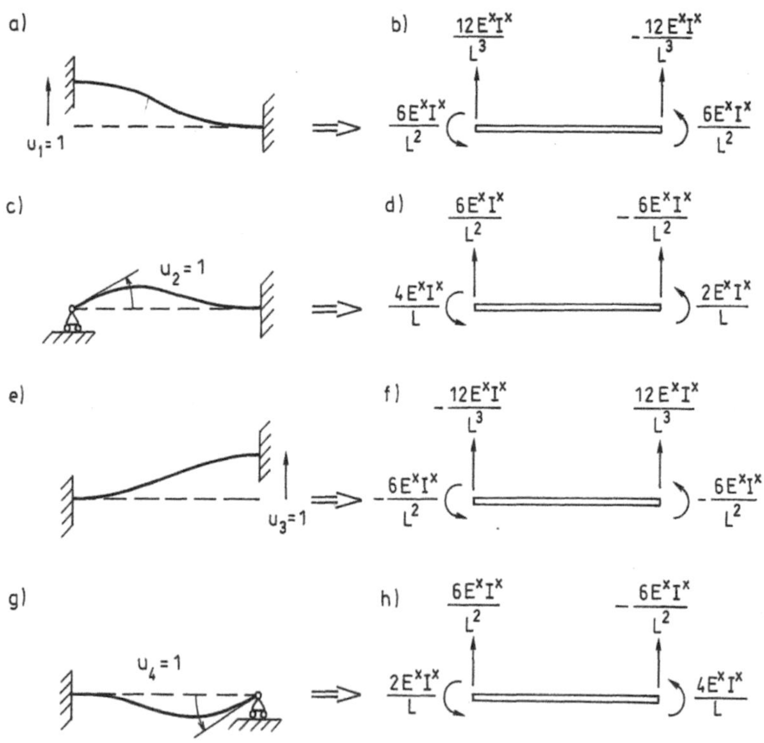

Figure 17.12 Deformations and corresponding forces and moments

(17.68) with $\mathbf{K}^e$ given by (17.66) we obtain

$$
\mathbf{f}_b^e =
\begin{bmatrix}
\dfrac{12E^*I^*}{L^3} \\[2ex]
\dfrac{6E^*I^*}{L^2} \\[2ex]
-\dfrac{12E^*I^*}{L^3} \\[2ex]
\dfrac{6E^*I^*}{L^2}
\end{bmatrix}
\tag{17.70}
$$

These components are illustrated in Figure 17.12(b). Choosing next $u_2 = 1$ and $u_1 = u_3 = u_4 = 0$, we derive the results shown in Figure 17.12(c) and (d). Proceeding in this way, we arrive at all the results shown in Figure 17.12.

In comparison with standard results in textbooks on beam theory, it appears that the results of Figure 17.12 are precisely those derived directly from beam theory. The reason why our FE solution provides the exact solutions and not just approximate ones is easily explained. With the bending stiffness E^*I^* independent of x and with no distributed load, i.e. $q = 0$, the differential equation (17.29) becomes

$$
E^*I^*\frac{d^4 w}{dx^4} = 0
$$

with the solution

$$
w = C_1 + C_2 x + C_3 x^2 + C_4 x^3
$$

where $C_1 \ldots C_4$ are arbitrary constants. However, this expression for the deflection is exactly the same as the one our FE solution was based on (cf. (17.53)). We conclude that if $q = 0$ and if E^*I^* is independent of x, the simple beam element provides exact results and not just approximations. In general, however, use of this simple element will give results which are approximations to the beam theory.

17.3.6 *Evaluation of element load vector – uniform load*

Assume that the simple beam element of Figure 17.6 is loaded by a uniform load q; cf. Figure 17.13. The element load vector $\mathbf{f}_l^e$ is to be calculated. From (17.50) we obtain

$$
\mathbf{f}_l^e = \int_0^L \mathbf{N}^{e\mathrm{T}} q \, dx = q \int_0^L
\begin{bmatrix}
N_1^e \\[1ex]
N_2^e \\[1ex]
N_3^e \\[1ex]
N_4^e
\end{bmatrix}
dx
$$

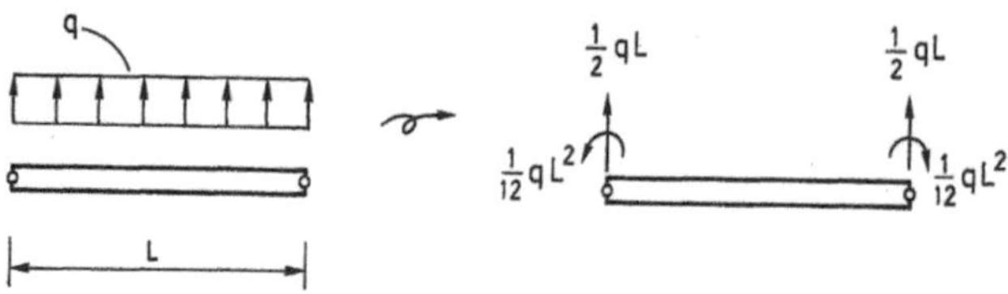

Figure 17.13 Uniform load q and equivalent nodal forces and nodal moments

and use of (17.62) implies that

$$
\mathbf{f}_1^e =
\begin{bmatrix}
\tfrac{1}{2}qL \\[4pt]
\tfrac{1}{12}qL^2 \\[4pt]
\tfrac{1}{2}qL \\[4pt]
-\tfrac{1}{12}qL^2
\end{bmatrix}
\tag{17.71}
$$

These equivalent nodal forces and nodal moments are illustrated in Figure 17.13.

17.3.7 *Evaluation of element load vector – concentrated force and moment*

Let us evaluate the element load vector $\mathbf{f}_1^e$ for an arbitrary type of beam element when the load q takes the form of a concentrated force P; see Figure 17.14. The dimension of P is [N] and P is measured as positive in the z-direction. As before, we use Dirac's delta function and write q as

$$
q = P\delta(x - a); \quad \text{i.e.} \int_0^L q\,\mathrm{d}x = \int_0^L P\delta(x - a)\,\mathrm{d}x = P
$$

This implies that $\mathbf{f}_1^e$ given by (17.50) becomes

$$
\mathbf{f}_1^e = \int_0^L \mathbf{N}^{eT} q\,\mathrm{d}x = \int_0^L \mathbf{N}^{eT} P\delta(x - a)\,\mathrm{d}x = P \int_0^L \mathbf{N}^{eT}\delta(x - a)\,\mathrm{d}x
$$

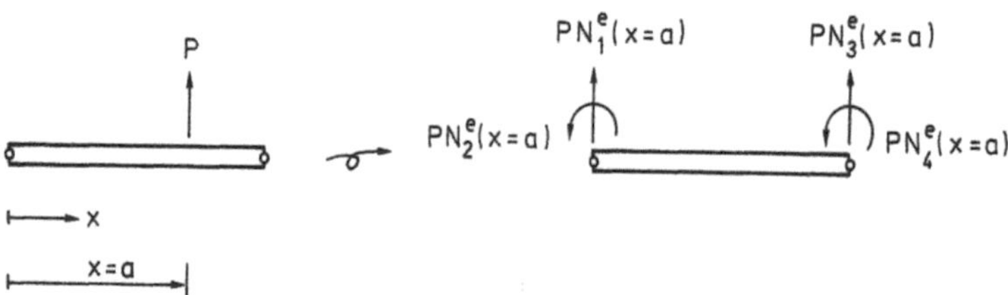

Figure 17.14 Concentrated force P and equivalent nodal forces and nodal moments

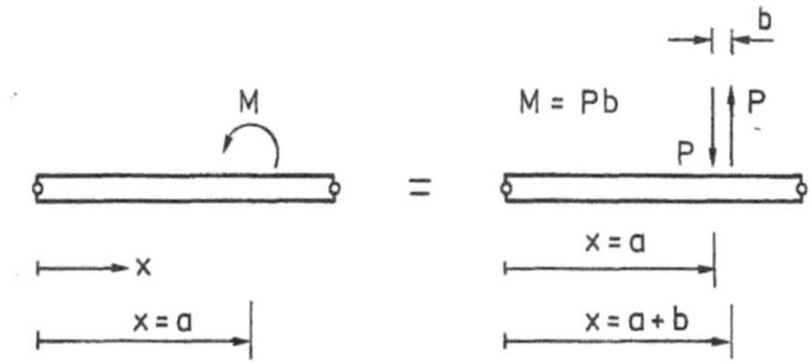

Figure 17.15 Equivalence between moment M and force couple

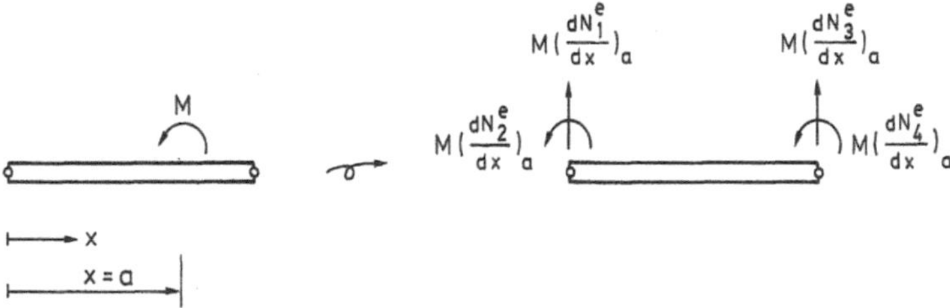

Figure 17.16 Moment M and equivalent nodal forces and nodal moments for simple beam element

i.e.

$$\boxed{\mathbf{f}_l^e = P\mathbf{N}_{x=a}^{eT}} \tag{17.72}$$

If the simple beam element is considered, the components of $\mathbf{f}_l^e$ are as illustrated in Figure 17.14. In general, the force P gives rise to both nodal forces and nodal moments. As easily shown by means of (17.72) and Figure 17.7, the only situation where the force P creates nodal forces but zero nodal moments is when the force is located at one of the end points of the simple beam element.

Another situation which often occurs in practice is the application of a moment M. For an arbitrary beam element, let us evaluate the element load vector $\mathbf{f}_l^e$ for such a loading. The moment M with the dimension $[\mathrm{N\,m}]$ is measured as positive in the counter-clockwise direction (see Figure 17.15). It is obvious that the effect of this moment is equivalent to the force couple shown in Figure 17.15. Here the distance b between the two opposite directed forces is taken to be infinitely small.

Referring to (17.72), the element load vector $\mathbf{f}_l^e$ for the force couple of Figure 17.15 is

$$\mathbf{f}_l^e = -P\mathbf{N}_{x=a}^{eT} + P\mathbf{N}_{x=a+b}^{eT} \tag{17.73}$$

As b denotes an infinitesimal distance, a Taylor expansion of $\mathbf{N}^{eT}$ about the point $x = a$ gives

$$\mathbf{N}_{x=a+b}^{eT} \simeq \mathbf{N}_{x=a}^{eT} + \left(\frac{d\mathbf{N}^{eT}}{dx}\right)_{x=a} b$$

Use of this expression in (17.73) yields

$$\mathbf{f}_i^e = M\left(\frac{d\mathbf{N}^{eT}}{dx}\right)_{x=a} \tag{17.74}$$

where $M = Pb$ (cf. Figure 17.15).

If the simple beam element is considered, the components of $\mathbf{f}_i^e$ are as illustrated in Figure 17.16. It appears that the moment M, in general, gives rise to both nodal forces and nodal moments. As easily shown by means of (17.74) and Figure 17.8, the only situation where the moment M creates nodal moments but zero nodal forces is when the moment is applied at one of the end points of the simple beam element.

18

FE formulation of plates

Following the treatment of the engineering theory of beams it is natural to consider the engineering theory of plates. Again we will discover that the engineering theory of plates is not a theory that fulfils all the field equations. This means that the theory of plates is an engineering approximation which reduces the original three-dimensional problem to a simpler two-dimensional problem. For many applications, however, this engineering theory provides realistic solutions, especially if the plate is thin.

Generally speaking, a plate is a structure with a thickness t that is small compared with all other dimensions of the plate; see Figure 18.1. Moreover, a plate is loaded by forces normal to the plane of the plate. To be more specific, we introduce a coordinate system and assume that the configuration of the plate is symmetric about the xy-plane, as shown in Figure 18.1. This means that the xy-plane is located in the mid-plane of the plate, and even though the plate thickness t in the z-direction may vary, it varies symmetrically about the xy-plane. The plate is loaded by a transverse loading q measured as positive in the z-direction. This loading q is force per unit area, i.e. q has the dimension $[\text{N/m}^2]$. The deflection w of the plate is measured as positive in the z-direction.

18.1 Plate theory

18.1.1 *Equilibrium conditions*

For sections normal to the x- and y-axes, the stress components σ_{xx}, σ_{xy}, σ_{xz} and σ_{yx}, σ_{yy}, σ_{yz} exist. These stress components give rise to the following forces and moments:

$$
\boxed{
\begin{aligned}
V_{xz} &= \int_{-t/2}^{t/2} \sigma_{xz}\, \mathrm{d}z \\
V_{yz} &= \int_{-t/2}^{t/2} \sigma_{yz}\, \mathrm{d}z
\end{aligned}
}
\tag{18.1}
$$

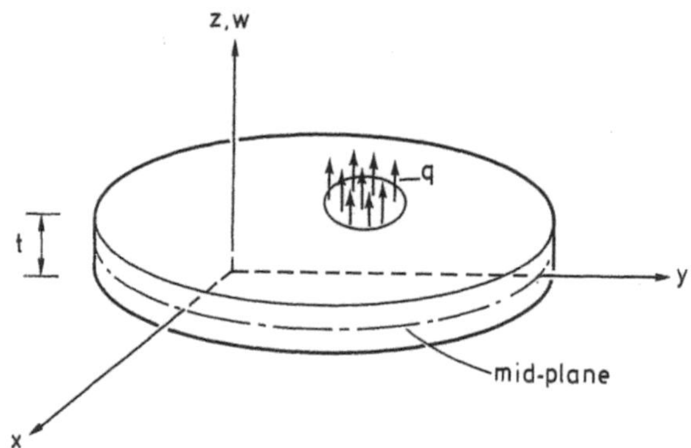

Figure 18.1   Configuration and loading of plate

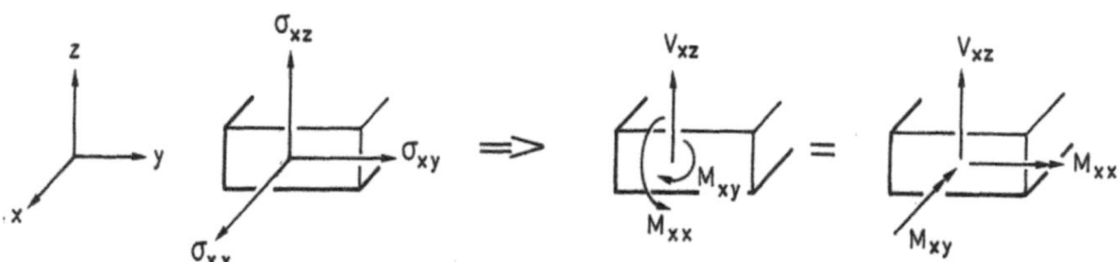

Figure 18.2   Illustration of M_{xx}, M_{xy} and V_{xz}

and

$$
\begin{aligned}
M_{xx} &= \int_{-t/2}^{t/2} z\sigma_{xx}\, \mathrm{d}z \\[2mm]
M_{yy} &= \int_{-t/2}^{t/2} z\sigma_{yy}\, \mathrm{d}z \\[2mm]
M_{xy} &= M_{yx} = \int_{-t/2}^{t/2} z\sigma_{xy}\, \mathrm{d}z
\end{aligned}
\tag{18.2}
$$

To illustrate the quantities defined by (18.1) and (18.2), consider first a plane cut in the plate and normal to the x-axis; see Figure 18.2. The stress components σ_{xx}, σ_{xy} and σ_{xz} acting on this plane create the quantities M_{xx}, M_{xy} and V_{xz} defined by (18.1) and (18.2) and shown in Figure 18.2. It appears that M_{xx} is the *bending moment* per unit length, M_{xy} is the *twisting moment* per unit length and V_{xz} is the *vertical shear force* per unit length. The positive directions for M_{xx}, M_{xy} and V_{xz} are also shown in Figure 18.2, and to illustrate the moments in a convenient manner, the standard conventional double-headed arrows has been adopted.

Considering next a plane cut in the plate and normal to the y-axis, the stress

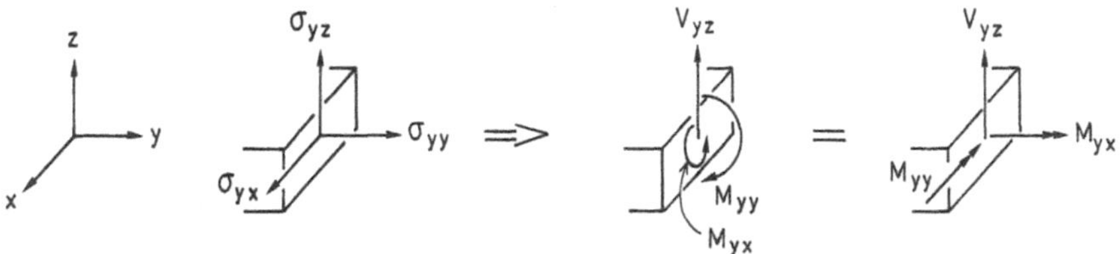

Figure 18.3 Illustration of M_{yy}, M_{yx} and V_{yz}

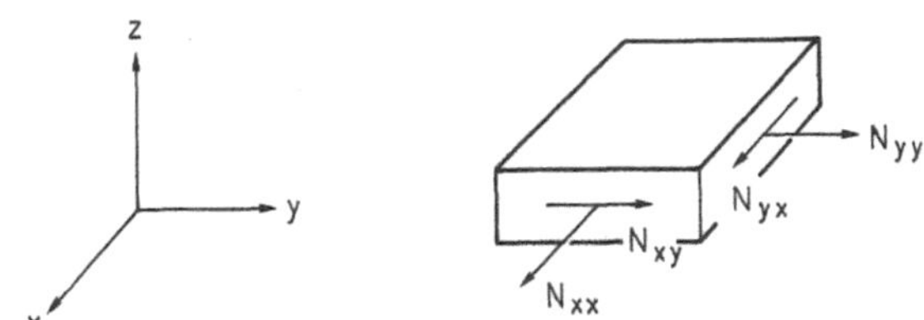

Figure 18.4 Illustration of horizontal forces N_{xx}, N_{yy} and N_{xy}

components σ_{yx}, σ_{yy} and σ_{yz} acting on this plane create the twisting moment M_{yx} per unit length, the bending moment M_{yy} per unit length and the vertical shear force V_{yz} per unit length. The positive directions of these quantities are shown in Figure 18.3.

In addition to these moments and vertical shear forces, we note that the stress components σ_{xx}, σ_{yy} and σ_{xy} also result in the following *horizontal forces*:

$$
\begin{aligned}
N_{xx} &= \int_{-t/2}^{t/2} \sigma_{xx}\, dz \\[2mm]
N_{yy} &= \int_{-t/2}^{t/2} \sigma_{yy}\, dz \\[2mm]
N_{xy} &= N_{xy} = \int_{-t/2}^{t/2} \sigma_{xy}\, dz
\end{aligned}
\tag{18.3}
$$

These forces act in the xy-plane as shown in Figure 18.4. It appears that N_{xx} is the *normal force* per unit length in the x-direction, N_{yy} is the *normal force* per unit length in the y-direction and N_{xy} is the *horizontal shear force* per unit length.

Let us now establish the equilibrium conditions for the plate. The plate is assumed to be loaded by transverse forces only. Therefore, as no resulting forces (or restraints) act in the xy-plane, horizontal equilibrium requires that

$$
N_{xx} = N_{yy} = N_{xy} = 0
\tag{18.4}
$$

Consider now an infinitesimally small part of the plate, Figure 18.5. As no resulting forces act in the xy-plane (cf. (18.4)), all forces and moments acting on this small

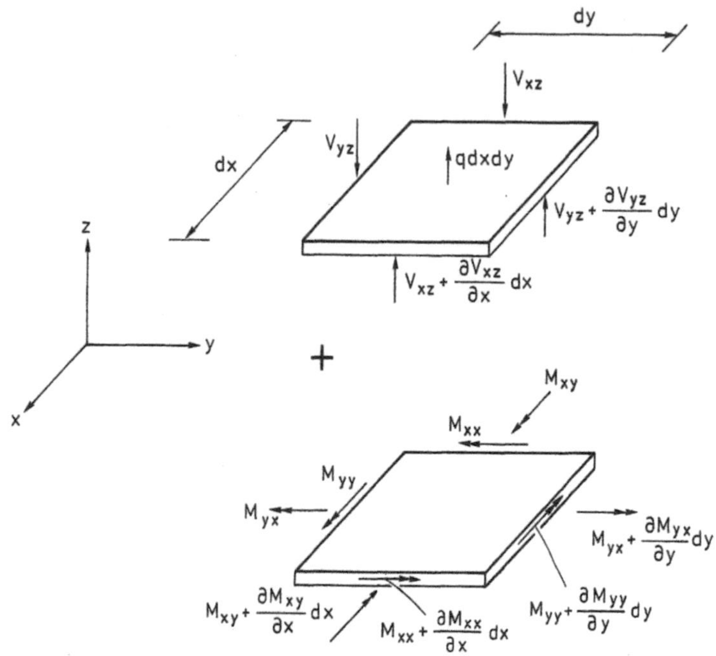

Figure 18.5   Vertical shear forces and moments acting on an infinitesimally small part of the plate

part of the plate appear from Figure 18.5. Vertical equilibrium requires that

$$q\,dx\,dy - V_{yz}\,dx + \left(V_{xz} + \frac{\partial V_{xz}}{\partial x}\,dx \right)dy + \left(V_{yz} + \frac{\partial V_{yz}}{\partial y}\,dy \right)dx - V_{xz}\,dy = 0$$

i.e.

$$\frac{\partial V_{xz}}{\partial x} + \frac{\partial V_{yz}}{\partial y} + q = 0 \tag{18.5}$$

Considering next the moment equilibrium about the right side of the small part of the plate, which is parallel to the x-axis, we obtain

$$-q\,dx\,dy\,\tfrac{1}{2}\,dy + V_{yz}\,dx\,dy - \left(V_{xz} + \frac{\partial V_{xz}}{\partial x}\,dx \right)dy\,\tfrac{1}{2}\,dy + V_{xz}\,dy\,\tfrac{1}{2}\,dy$$

$$+ M_{yy}\,dx - \left(M_{xy} + \frac{\partial M_{xy}}{\partial x}\,dx \right)dy - \left(M_{yy} + \frac{\partial M_{yy}}{\partial y}\,dy \right)dx + M_{xy}\,dy = 0$$

i.e.

$$-q\tfrac{1}{2}\,dy + V_{yz} - \frac{\partial V_{xz}}{\partial x}\tfrac{1}{2}\,dy - \frac{\partial M_{xy}}{\partial x} - \frac{\partial M_{yy}}{\partial y} = 0$$

As dy is an infinitely small quantity we get

$$\boxed{\frac{\partial M_{xy}}{\partial x} + \frac{\partial M_{yy}}{\partial y} = V_{yz}}$$

(18.6)

Considering the moment equilibrium about one of the sides parallel to the y-axis, we derive in a similar manner from Figure 18.5 that

$$\boxed{\frac{\partial M_{xx}}{\partial x} + \frac{\partial M_{xy}}{\partial y} = V_{xz}}$$

(18.7)

18.1.2 *Kinematic relations*

The plate is assumed to deform in accordance with Bernoulli's assumption for beam behaviour, i.e. that plane sections normal to the mid-plane remain plane and normal to the mid-plane during deformation. In complete analogy with Figure 17.4 and (17.7), we obtain the following displacements in the x-, y- and z-plane directions:

$$u_x = u^0 - z\frac{\partial w}{\partial x}$$

$$u_y = v^0 - z\frac{\partial w}{\partial y}$$

(18.8)

$$u_z = w$$

where u^0 and v^0 are the displacements of the mid-plane in the x- and y-directions, respectively. It follows that

$$u^0 = u^0(x, y); \quad v^0 = v^0(x, y) \quad \text{and} \quad w = w(x, y)$$

(18.9)

where we have introduced the assumption that the deflection w is independent of z, i.e. $w = w(x, y)$. From (18.8) and (18.9), the strains as given by (12.37) and (12.40) become

$$\varepsilon_{xx} = \frac{\partial u^0}{\partial x} - z\frac{\partial^2 w}{\partial x^2}$$

$$\varepsilon_{yy} = \frac{\partial v^0}{\partial y} - z\frac{\partial^2 w}{\partial y^2}$$

(18.10)

$$\gamma_{xy} = \frac{\partial u^0}{\partial y} + \frac{\partial v^0}{\partial x} - 2z\frac{\partial^2 w}{\partial x\partial y}$$

and

$$\varepsilon_{zz} = \gamma_{xz} = \gamma_{yz} = 0 \tag{18.11}$$

We note that the kinematic assumptions imply that the shear strains γ_{xz} and γ_{yz} are zero.

18.1.3 *Constitutive relation*

Hooke's law is assumed to be applicable. As in the discussion of beam behaviour given in relation to (17.12)–(17.14), it is not possible to obtain a correspondence between the non-zero shear stresses σ_{xz} and σ_{yz} necessary to maintain equilibrium (cf. (18.5)) and the zero shear strains γ_{xz} and γ_{yz}. However, we assume the plate to be thin, i.e. the largest stresses will be σ_{xx}, σ_{yy} and σ_{xy} that are illustrated in Figure 18.6.

This observation suggests that the assumption of plane stress is most applicable, i.e.

$$\boxed{\sigma = \mathbf{D}\varepsilon} \tag{18.12}$$

where

$$\sigma = \begin{bmatrix} \sigma_{xx} \\ \sigma_{yy} \\ \sigma_{xy} \end{bmatrix}; \quad \varepsilon = \begin{bmatrix} \varepsilon_{xx} \\ \varepsilon_{yy} \\ \gamma_{xy} \end{bmatrix} \tag{18.13}$$

For isotropic elasticity, the plane stress constitutive matrix $\mathbf{D}$ is given by (13.33), i.e.

$$\mathbf{D} = \frac{E}{1 - v^2} \begin{bmatrix} 1 & v & 0 \\ v & 1 & 0 \\ 0 & 0 & \tfrac{1}{2}(1 - v) \end{bmatrix} \tag{18.14}$$

In the following, however, the $\mathbf{D}$-matrix may take any form. We also observe that it is even possible to consider initial strains, ε_0 for instance, in terms of thermal strains. In this case (18.12) is replaced by

$$\sigma = \mathbf{D}\varepsilon - \mathbf{D}\varepsilon_0 \tag{18.15}$$

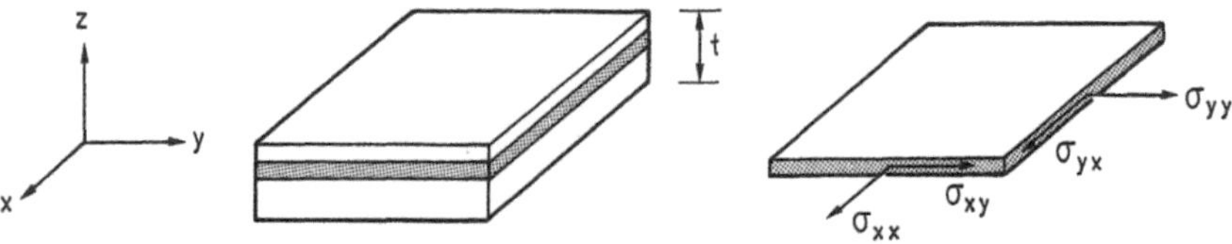

Figure 18.6   Illustration of stresses σ_{xx}, σ_{yy} and σ_{xy}

but we shall for simplicity consider the constitutive relation in the form given by (18.12).

18.1.4 *Further derivations*

The strains (18.10) may be written in the form

$$\boldsymbol{\varepsilon} = \boldsymbol{\varepsilon}^0 - z\boldsymbol{\kappa} \tag{18.16}$$

where

$$\boldsymbol{\varepsilon}^0 = \begin{bmatrix} \dfrac{\partial u^0}{\partial x} \\[2mm] \dfrac{\partial v^0}{\partial y} \\[2mm] \dfrac{\partial u^0}{\partial y} + \dfrac{\partial v^0}{\partial x} \end{bmatrix}; \qquad \boldsymbol{\kappa} = \begin{bmatrix} \dfrac{\partial^2 w}{\partial x^2} \\[2mm] \dfrac{\partial^2 w}{\partial y^2} \\[2mm] 2\dfrac{\partial^2 w}{\partial x \partial y} \end{bmatrix} \tag{18.17}$$

and $\boldsymbol{\kappa}$ is denoted the *curvature matrix*; cf. the discussion of (17.26). We emphasize that $\boldsymbol{\varepsilon}^0$ expresses the straining of the mid-plane and should not be confused with the initial strains given by $\boldsymbol{\varepsilon}_0$. Using (18.16) and (18.12) we get

$$\boldsymbol{\sigma} = \mathbf{D}\boldsymbol{\varepsilon}^0 - z\mathbf{D}\boldsymbol{\kappa} \tag{18.18}$$

We define the matrix $\mathbf{M}$ by

$$\mathbf{M} = \begin{bmatrix} M_{xx} \\ M_{yy} \\ M_{xy} \end{bmatrix} \tag{18.19}$$

Then the expression for the moments given by (18.2) can be written in the compact form

$$\mathbf{M} = \int_{-t/2}^{t/2} \boldsymbol{\sigma} z \, dz \tag{18.20}$$

Inserting (18.18) into (18.20) and recalling that neither $\boldsymbol{\varepsilon}^0$ nor $\boldsymbol{\kappa}$ depends on the coordinate z yield

$$\mathbf{M} = \mathbf{D}\boldsymbol{\varepsilon}^0 \int_{-t/2}^{t/2} z \, dz - \mathbf{D}\boldsymbol{\kappa} \int_{-t/2}^{t/2} z^2 \, dz \tag{18.21}$$

where it was assumed that $\mathbf{D}$ is independent of z, i.e.

$$\mathbf{D} = \mathbf{D}(x, y) \tag{18.22}$$

We have

$$\int_{-t/2}^{t/2} z \, dz = 0; \qquad \int_{-t/2}^{t/2} z^2 \, dz = \frac{t^3}{12}$$

That is, irrespective of the value of ε^0-strains, (18.21) reduces to

$$\boxed{\mathbf{M} = -\tilde{\mathbf{D}}\boldsymbol{\kappa}} \quad \text{where} \quad \boxed{\tilde{\mathbf{D}} = \frac{t^3}{12}\mathbf{D}} \tag{18.23}$$

With $\boldsymbol{\sigma}$ defined by (18.13), the horizontal forces given by (18.3) can be expressed as

$$\begin{bmatrix} N_{xx} \\ N_{yy} \\ N_{xy} \end{bmatrix} = \int_{-t/2}^{t/2} \boldsymbol{\sigma} \, dz$$

Use of (18.18) yields

$$\begin{bmatrix} N_{xx} \\ N_{yy} \\ N_{xy} \end{bmatrix} = \mathbf{D}\varepsilon^0 t \tag{18.24}$$

In the present case, no resulting horizontal forces act in the mid-plane (cf. (18.4)) and we conclude that

$$\varepsilon^0 = \mathbf{0} \tag{18.25}$$

In accordance with expectations, this shows that there is no straining of the mid-plane. However, even in the situation where the horizontal forces are different from zero, we observe from (18.23) and (18.24) that, whereas the moments $\mathbf{M}$ are controlled by the curvature matrix $\boldsymbol{\kappa}$, the horizontal forces are determined by the in-plane strains ε^0. This implies that the phenomena of bending and straining of the mid-plane are *uncoupled* phenomena, that can be treated separately if (18.22) holds. If the mid-plane is deformed, i.e. $\varepsilon^0 \neq \mathbf{0}$, we have so-called *membrane action*. It is concluded that, if membrane action is present, the FE formulation of the present chapter can be combined directly with the FE formulation of plane stress elasticity treated in Chapter 16. This uncoupled response is similar to that observed for beam theory, and it implies that shell structures in which membrane action is of prominent importance may be analyzed by a *shell element* which comprises a combined plate and plane stress element.

18.1.5 *Differential equations for plate theory*

In the present situation where no resulting horizontal forces act in the mid-plane, (18.25) holds, i.e. (18.16) and (18.18) reduce to

$$\boxed{\varepsilon = -z\boldsymbol{\kappa}} \quad \text{and} \quad \boxed{\boldsymbol{\sigma} = -z\mathbf{D}\boldsymbol{\kappa}} \tag{18.26}$$

The equilibrium conditions were given by (18.5)–(18.7). As only the moments can be expressed in terms of kinematic quantities, we eliminate the shear forces V_{xz} and V_{yz} from (18.5)–(18.7). Therefore, we differentiate (18.6) with respect to y and (18.7) with respect to x; we add the results and use (18.5) to obtain

$$\boxed{\frac{\partial^2 M_{xx}}{\partial x^2} + 2\frac{\partial^2 M_{xy}}{\partial x \partial y} + \frac{\partial^2 M_{yy}}{\partial y^2} + q = 0} \tag{18.27}$$

This equilibrium condition holds irrespective of the constitutive assumption. Let us now introduce the matrix differential operator $\overset{*}{\nabla}$ defined by

$$\overset{*}{\nabla} = \begin{bmatrix} \dfrac{\partial^2}{\partial x^2} \\[2mm] \dfrac{\partial^2}{\partial y^2} \\[2mm] 2\dfrac{\partial^2}{\partial x \partial y} \end{bmatrix} \tag{18.28}$$

It appears that the equilibrium condition (18.27) may then be written as

$$\boxed{\overset{*}{\nabla}{}^{\mathrm{T}}\mathbf{M} + q = 0} \tag{18.29}$$

Moreover, we may rewrite (18.17) as

$$\boldsymbol{\kappa} = \overset{*}{\nabla} w \tag{18.30}$$

i.e. (18.23) takes the form

$$\boxed{\mathbf{M} = -\tilde{\mathbf{D}}\overset{*}{\nabla} w} \tag{18.31}$$

Introducing this expression into (18.29) yields the following differential equation for plate theory:

$$\boxed{\overset{*}{\nabla}{}^{\mathrm{T}}\tilde{\mathbf{D}}\overset{*}{\nabla} w = q} \tag{18.32}$$

When the deflection w has been determined from this differential equation, all quantities of interest can be derived. Assume now that the thickness t is constant and that isotropic elasticity is considered, i.e. $\mathbf{D}$ is given by (18.14). If, in addition, $\mathbf{D}$ does not depend on x and y, (18.32) takes the form

$$\boxed{\frac{\partial^4 w}{\partial x^4} + 2\frac{\partial^4 w}{\partial x^2 \partial y^2} + \frac{\partial^4 w}{\partial y^4} = \frac{12(1-v^2)}{Et^3}q} \tag{18.33}$$

This differential equation is called the *biharmonic equation* and it was derived in 1811 by Lagrange, even though his plate theory was not entirely satisfactory. The first convincing plate theory was established by Kirchhoff (1850) and what was presented above is essentially due to him. This plate theory is therefore also termed *Kirchhoff plate theory* and we may refer, for instance, to the textbooks by Boresi *et al.* (1978) and Timoshenko and Woinowsky-Krieger (1959) for more details. The most critical point in the Kirchhoff theory is the contradiction between the existence of the shear stresses σ_{xz} and σ_{yz} – necessary to maintain equilibrium – and the zero shear strains γ_{xz} and γ_{yz}. In analogy with the Bernoulli beam theory, the Kirchhoff plate theory works well for thin plates where the real shear strains γ_{xz} and γ_{yz} are small. For thicker plates, the Kirchhoff theory becomes questionable, and more refined plate theories must be invoked which allow for the existence of non-zero shear strains. Among such refined plate theories, the one proposed by Mindlin (1951) and Reissner (1945) is the most prominent. This *Mindlin–Reissner plate theory* is also interesting from a finite element point of view since only C^0-continuity is required, whereas – as we shall see – the Kirchhoff theory requires C^1-continuity. Here we shall not treat the FE formulation of Mindlin–Reissner plates, but the interested reader may consult, for instance, Hughes (1987) and Cook *et al.* (1989).

18.1.6 *Moments and shear forces acting on an arbitrary plane*

Consider a section in the plate normal to the mid-plane and defined by the unit normal vector **n**, which is located in the xy-plane (see Figure 18.7). A unit vector **m** located in the xy-plane is also defined and taken to be tangential to the section, i.e. orthogonal to **n**. Moreover, the vector **m** is chosen so that **n** and **m**, after a suitable rigid-body rotation, follow the same directions as the x- and y-axes, respectively. We

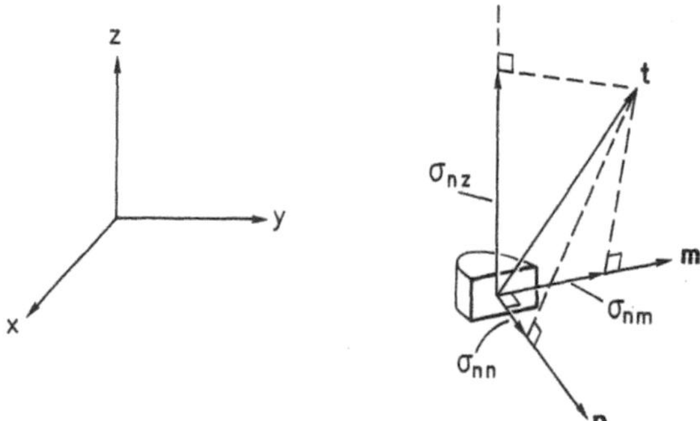

Figure 18.7   Illustration of stress components σ_{nn}, σ_{nm} and σ_{nz} in a plane defined by the unit normal vector **n**

have

$$\mathbf{n} = \begin{bmatrix} n_x \\ n_y \\ n_z \end{bmatrix} = \begin{bmatrix} n_x \\ n_y \\ 0 \end{bmatrix}; \quad \mathbf{m} = \begin{bmatrix} m_x \\ m_y \\ m_z \end{bmatrix} = \begin{bmatrix} m_x \\ m_y \\ 0 \end{bmatrix} \tag{18.34}$$

and

$$|\mathbf{n}| = |\mathbf{m}| = 1; \quad \mathbf{n}^T \mathbf{m} = 0 \tag{18.35}$$

The traction vector $\mathbf{t}$ acting on the section defined by $\mathbf{n}$ is given by (12.14), and as $n_z = 0$, we get

$$\mathbf{t} = \begin{bmatrix} t_x \\ t_y \\ t_z \end{bmatrix} = \begin{bmatrix} \sigma_{xx} n_x + \sigma_{xy} n_y \\ \sigma_{yx} n_x + \sigma_{yy} n_y \\ \sigma_{zx} n_x + \sigma_{zy} n_y \end{bmatrix} \tag{18.36}$$

Previously, when defining stress components, we considered, for instance, a plane normal to the x-axis, and the stress components σ_{xx}, σ_{xy} and σ_{xz} were defined as the components of the traction vector $\mathbf{t}$ in the x-, y- and z-directions, respectively; cf. Figure 12.2 and expression (12.2). In a similar fashion, for the plane identified by the unit normal vector $\mathbf{n}$ we now define the stress components σ_{nn}, σ_{nm} and σ_{nz} as the components of the traction vector in the $\mathbf{n}$-, $\mathbf{m}$- and z-directions, respectively. These stress components are illustrated in Figure 18.7 and we obtain

$$\sigma_{nn} = \mathbf{n}^T \mathbf{t}$$
$$\sigma_{nm} = \mathbf{m}^T \mathbf{t}$$
$$\sigma_{nz} = \begin{bmatrix} 0 & 0 & 1 \end{bmatrix} \mathbf{t}$$

With $\mathbf{n}$ and $\mathbf{m}$ defined by (18.34) and the traction vector $\mathbf{t}$ by (18.36) this result becomes

$$\sigma_{nn} = n_x^2 \sigma_{xx} + n_y^2 \sigma_{yy} + 2 n_x n_y \sigma_{xy}$$
$$\sigma_{nm} = n_x m_x \sigma_{xx} + n_y m_y \sigma_{yy} + (n_y m_x + n_x m_y) \sigma_{xy} \tag{18.37}$$
$$\sigma_{nz} = n_x \sigma_{xz} + n_y \sigma_{yz}$$

We may note that the first two equations of (18.37) correspond in fact to the results obtained by the well-known *Mohr's circles of stress*.

As in the definitions (18.1) and (18.2), we define the bending moment M_{nn} per unit length, twisting moment M_{nm} per unit length and vertical shear force V_{nz} per unit length acting on the section defined by the unit vector **n** according to

$$
\begin{aligned}
M_{nn} &= \int_{-t/2}^{t/2} z\sigma_{nn}\,\mathrm{d}z \\[2mm]
M_{nm} &= \int_{-t/2}^{t/2} z\sigma_{nm}\,\mathrm{d}z \\[2mm]
V_{nz} &= \int_{-t/2}^{t/2} \sigma_{nz}\,\mathrm{d}z
\end{aligned}
\tag{18.38}
$$

Inserting (18.37) into (18.38) and making use of the definitions (18.1) and (18.2) provide

$$
\begin{aligned}
M_{nn} &= n_x^2 M_{xx} + n_y^2 M_{yy} + 2n_x n_y M_{xy} \\[1mm]
M_{nm} &= n_x m_x M_{xx} + n_y m_y M_{yy} + (n_y m_x + n_x m_y)M_{xy} \\[1mm]
V_{nz} &= n_x V_{xz} + n_y V_{yz}
\end{aligned}
\tag{18.39}
$$

It may be observed that these expressions are completely similar to (18.37). Therefore, just as it is possible to identify the *principle stresses*, i.e. sections for which the shear stress $\sigma_{nm} = 0$, it is possible to determine the *principle moments*, i.e. sections for which the twisting moment $M_{nm} = 0$. We shall not pursue this subject, but refer to standard textbooks for details.

For future purposes, some useful relations will be derived from (18.39). From the three-dimensional vectors **n** and **m** defined by (18.34) and (18.35), we define the two-dimensional orthogonal vectors **n** and **m** by

$$
\mathbf{n} = \begin{bmatrix} n_x \\ n_y \end{bmatrix}; \quad
\mathbf{m} = \begin{bmatrix} m_x \\ m_y \end{bmatrix}; \quad
|\mathbf{n}| = |\mathbf{m}| = 1; \quad \mathbf{n}^{\mathrm{T}}\mathbf{m} = 0
\tag{18.40}
$$

These two unit vectors are located in the xy-plane. By carrying out the matrix multiplications it follows that

$$
M_{nn} = \mathbf{n}^{\mathrm{T}}\mathbf{A}\mathbf{n}; \quad M_{nm} = \mathbf{m}^{\mathrm{T}}\mathbf{A}\mathbf{n}
\tag{18.41}
$$

where

$$
\mathbf{A} = \begin{bmatrix} M_{xx} & M_{xy} \\ M_{xy} & M_{yy} \end{bmatrix}
\tag{18.42}
$$

Using (18.40)–(18.42) the following expression is derived:

$$
\mathbf{n}M_{nn} + \mathbf{m}M_{nm} = (\mathbf{n}\mathbf{n}^{\mathrm{T}} + \mathbf{m}\mathbf{m}^{\mathrm{T}})\mathbf{A}\mathbf{n}
\tag{18.43}
$$

Let us investigate the square matrix **R** defined by

$$
\mathbf{R} = \mathbf{n}\mathbf{n}^{\mathrm{T}} + \mathbf{m}\mathbf{m}^{\mathrm{T}}
$$

As **n** and **m** are orthogonal, any vector **r** in the xy-plane may be expressed as

$$\mathbf{r} = \alpha\mathbf{n} + \beta\mathbf{m}$$

where α and β are some parameters. Since **n** and **m** are orthogonal unit vectors, it appears that

$$\mathbf{Rr} = \mathbf{r}; \quad \text{i.e. } (\mathbf{R} - \mathbf{I})\mathbf{r} = \mathbf{0}$$

As this relation holds for arbitrary vectors **r** we conclude that $\mathbf{R} = \mathbf{I}$, i.e.

$$\mathbf{nn}^{\mathrm{T}} + \mathbf{mm}^{\mathrm{T}} = \mathbf{I} \tag{18.44}$$

where **I** is the unit square matrix. That is (18.43) reduces to

$$\mathbf{n}M_{nn} + \mathbf{m}M_{nm} = \mathbf{An}$$

Multiplication by the quantity $(\nabla v)^{\mathrm{T}}$, where $v = v(x, y)$ is an arbitrary function and ∇v denotes the gradient of v, yields

$$(\nabla v)^{\mathrm{T}}\mathbf{n}M_{nn} + (\nabla v)^{\mathrm{T}}\mathbf{m}M_{nm} = (\nabla v)^{\mathrm{T}}\mathbf{An}$$

Use of (5.8) and (18.42) gives the following relation, which will later be used in the weak formulation:

$$\frac{\mathrm{d}v}{\mathrm{d}n}M_{nn} + \frac{\mathrm{d}v}{\mathrm{d}m}M_{nm} = \begin{bmatrix} \dfrac{\partial v}{\partial x} & \dfrac{\partial v}{\partial y} \end{bmatrix} \begin{bmatrix} M_{xx}n_x + M_{xy}n_y \\ M_{xy}n_x + M_{yy}n_y \end{bmatrix} \tag{18.45}$$

18.1.7 *Discussion of proper boundary conditions*

The boundary conditions for Kirchhoff plate theory turn out to be non-trivial. In reality, the identification of proper boundary conditions had already been established by Kirchhoff himself in 1850, but even today a profound understanding of these boundary conditions is something which most students do not fully achieve. For this reason, we shall first try to illuminate the problem and then, in the next section, provide the solution.

From previous results we have arrived at a two-dimensional problem, where everything depends on the coordinates x and y in the mid-plane of the plate. The region spanned by this mid-plane is denoted by A and its boundary by $\mathscr{L}$; see Figure 18.8.

The assumption for the kinematics of the plate is such that a straight line normal to the mid-plane remains straight during deformation. Along the boundary $\mathscr{L}$, the movement of such a straight line is given by its slope θ_n normal to the boundary and its slope θ_m tangential to the boundary; see Figure 18.9. Therefore, along the boundary we can identify the following three kinematic quantities: the deflection w and the slopes θ_n and θ_m. In Kirchhoff plate theory, it is assumed that a straight line

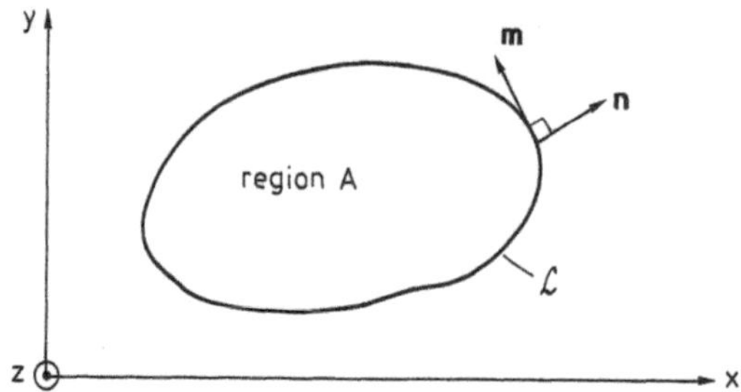

Figure 18.8   Two-dimensional problem formulation of plate theory

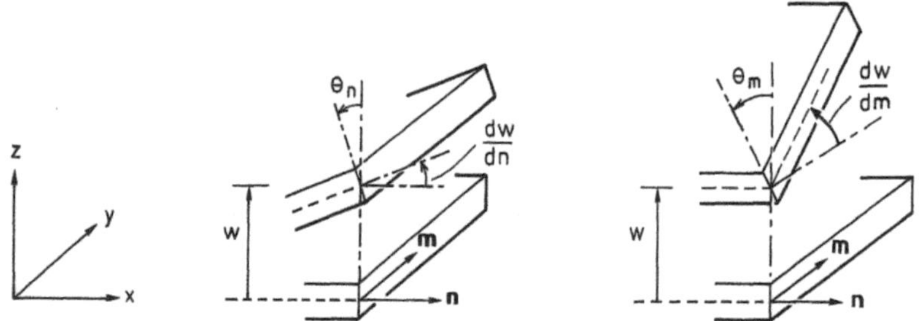

Figure 18.9   Kinematic quantities w, θ_n and θ_m along the boundary

normal to the mid-plane remains straight *and* normal to the mid-plane during deformation, i.e. we have $\theta_n = dw/dn$ and $\theta_m = dw/dm$. That is, the three kinematic quantities that we can identify are w, dw/dn, dw/dm. Suppose now that the deflection w is known along part of the boundary. We may write this as $w = w(m)$; cf. Figure 18.8. However, if $w(m)$ is known along the boundary, then the slope dw/dm is also known. This means that we only have two independent kinematic quantities that we may prescribe, namely w and dw/dn. These may be termed the *kinematic boundary conditions*, i.e.

$$\boxed{\quad w; \quad \frac{dw}{dn} \quad} \tag{18.46}$$

are the independent kinematic boundary conditions.

Consider next the boundary conditions given in terms of moments and shear forces, which may be denoted as the *static boundary conditions*. Along the boundary and referring to (18.38), we are able to prescribe the bending moment M_{nn}, the twisting moment M_{nm} and the shear force V_{nz}, i.e.

$$M_{nn}; \quad M_{nm}; \quad V_{nz} \tag{18.47}$$

are the static boundary conditions. It is somewhat surprising that, whereas we have two independent kinematic boundary conditions given by (18.46), we *apparently* have three independent static boundary conditions (18.47).

To investigate this problem, consider the differential equation in w given by (18.32). Without loss of generality we may for simplicity consider the biharmonic differential equation (18.33) valid for isotropic elasticity. It appears that this differential equation is of fourth order and we may recall that the differential equation for beam theory is also of fourth order. Assuming for convenience that the boundary $\mathcal{L}$ consists of two regions, one part where the kinematic boundary conditions are prescribed and another part where the static boundary conditions are prescribed, we are led to the conclusion that, along each of these parts, the solution of the fourth-order differential equation requires the specification of two independent boundary conditions. This requirement is fulfilled for the kinematic boundary conditions (18.46), and it implies that the three static quantities M_{nn}, M_{nm} and V_{nz} must combine in such a fashion that they appear in the form of two independent terms. Therefore, we must establish a way in which we can identify these two independent static boundary conditions. We have previously seen that the weak form by itself gives the natural boundary conditions, so it is tempting to obtain a weak form of the plate problem and in this fashion identify the proper static boundary conditions. This will be outlined in the next section.

Before deriving the weak form, we mention that the reason why we have 'lost' one static boundary condition is due to the reformulation of the equilibrium equations. Originally, we had the three first-order differential equations (18.5)–(18.7) corresponding to three independent static boundary conditions. However, when eliminating the shear forces we arrived at the second-order differential equation (18.27), which only requires two static boundary conditions.

We also mention that in higher-order plate theories like the Mindlin–Reissner theory, the formulation is based on (18.5)–(18.7) and not on (18.27), i.e. we have three independent static boundary conditions given by M_{nn}, M_{nm} and V_{nz}. Moreover, as shear strains are now included, we also have three independent kinematic boundary conditions given by the deflection w and the slopes θ_n and θ_m where it now holds that $\theta_n \neq dw/dn$ and $\theta_m \neq dw/dm$.

18.2 Weak form of equilibrium equation – proper static boundary conditions

We shall now establish the weak form of the equilibrium equation given by (18.27). The reason is two-fold: it is the weak form on which our FE solution is based; and from the natural boundary conditions we will be able to decide on the proper static boundary conditions.

The equilibrium condition (18.27) can be written as

$$\frac{\partial}{\partial x}\left(\frac{\partial M_{xx}}{\partial x}\right) + \frac{\partial}{\partial x}\left(\frac{\partial M_{xy}}{\partial y}\right) + \frac{\partial}{\partial y}\left(\frac{\partial M_{xy}}{\partial x}\right) + \frac{\partial}{\partial y}\left(\frac{\partial M_{yy}}{\partial y}\right) + q = 0 \qquad (18.48)$$

To obtain the weak form of this differential equation, we multiply by the arbitrary weight function $v(x, y)$ and integrate over the region A, i.e.

$$\int_A v \frac{\partial}{\partial x}\left(\frac{\partial M_{xx}}{\partial x}\right) dA + \int_A v \frac{\partial}{\partial x}\left(\frac{\partial M_{xy}}{\partial y}\right) dA + \int_A v \frac{\partial}{\partial y}\left(\frac{\partial M_{xy}}{\partial x}\right) dA$$

$$+ \int_A v \frac{\partial}{\partial y}\left(\frac{\partial M_{yy}}{\partial y}\right) dA + \int_A vq \, dA = 0$$

To perform an integration by parts, the Green–Gauss theorem given in the form of (5.27) and (5.28) is used. This results in

$$\oint_{\mathscr{L}} v \frac{\partial M_{xx}}{\partial x} n_x \, d\mathscr{L} - \int_A \frac{\partial v}{\partial x}\frac{\partial M_{xx}}{\partial x} dA + \oint_{\mathscr{L}} v \frac{\partial M_{xy}}{\partial y} n_x \, d\mathscr{L} - \int_A \frac{\partial v}{\partial x}\frac{\partial M_{xy}}{\partial y} dA$$

$$+ \oint_{\mathscr{L}} v \frac{\partial M_{xy}}{\partial x} n_y \, d\mathscr{L} - \int_A \frac{\partial v}{\partial y}\frac{\partial M_{xy}}{\partial x} dA + \oint_{\mathscr{L}} v \frac{\partial M_{yy}}{\partial y} n_y \, d\mathscr{L} - \int_A \frac{\partial v}{\partial y}\frac{\partial M_{yy}}{\partial y} dA$$

$$+ \int_A vq \, dA = 0$$

Using (18.6) and (18.7) in the boundary integrals gives

$$\oint_{\mathscr{L}} v V_{xz} n_x \, d\mathscr{L} + \oint_{\mathscr{L}} v V_{yz} n_y \, d\mathscr{L} - \int_A \frac{\partial v}{\partial x}\frac{\partial M_{xx}}{\partial x} dA - \int_A \frac{\partial v}{\partial x}\frac{\partial M_{xy}}{\partial y} dA$$

$$- \int_A \frac{\partial v}{\partial y}\frac{\partial M_{xy}}{\partial x} dA - \int_A \frac{\partial v}{\partial y}\frac{\partial M_{yy}}{\partial y} + \int_A vq \, dA = 0$$

In the boundary integrals, advantage is taken of (18.39) and the result becomes

$$- \int_A \frac{\partial v}{\partial x}\frac{\partial M_{xx}}{\partial x} dA - \int_A \frac{\partial v}{\partial x}\frac{\partial M_{xy}}{\partial y} dA - \int_A \frac{\partial v}{\partial y}\frac{\partial M_{xy}}{\partial x} dA - \int_A \frac{\partial v}{\partial y}\frac{\partial M_{yy}}{\partial y} dA$$

$$+ \oint_{\mathscr{L}} v V_{nz} \, d\mathscr{L} + \int_A vq \, dA = 0$$

Repeated use of the Green–Gauss theorem (5.27) and (5.28) yields

$$- \oint_{\mathscr{L}} \frac{\partial v}{\partial x} M_{xx} n_x \, d\mathscr{L} + \int_A \frac{\partial^2 v}{\partial x^2} M_{xx} \, dA - \oint_{\mathscr{L}} \frac{\partial v}{\partial x} M_{xy} n_y \, d\mathscr{L} + \int_A \frac{\partial^2 v}{\partial y \partial x} M_{xy} \, dA$$

$$- \oint_{\mathscr{L}} \frac{\partial v}{\partial y} M_{xy} n_x \, d\mathscr{L} + \int_A \frac{\partial^2 v}{\partial x \partial y} M_{xy} \, dA - \oint_{\mathscr{L}} \frac{\partial v}{\partial y} M_{yy} n_y \, d\mathscr{L}$$

$$+ \int_A \frac{\partial^2 v}{\partial y^2} M_{yy} \, dA + \oint_{\mathscr{L}} v V_{nz} \, d\mathscr{L} + \int_A vq \, dA = 0$$

which can be written as

$$\int_A \left(\frac{\partial^2 v}{\partial x^2} M_{xx} + \frac{\partial^2 v}{\partial y^2} M_{yy} + 2 \frac{\partial^2 v}{\partial x \partial y} M_{xy} \right) dA$$

$$- \oint_{\mathscr{L}} \left(\frac{\partial v}{\partial x}(M_{xx}n_x + M_{xy}n_y) + \frac{\partial v}{\partial y}(M_{xy}n_x + M_{yy}n_y) \right) d\mathscr{L}$$

$$+ \oint_{\mathscr{L}} v V_{nz} \, d\mathscr{L} + \int_A vq \, dA = 0 \tag{18.49}$$

With the definitions (18.19) and (18.28) we have

$$(\overset{*}{\nabla} v)^{\mathrm{T}} \mathbf{M} = \frac{\partial^2 v}{\partial x^2} M_{xx} + \frac{\partial^2 v}{\partial y^2} M_{yy} + 2 \frac{\partial^2 v}{\partial x \partial y} M_{xy} \tag{18.50}$$

Moreover, in (18.45) it was shown that

$$\frac{dv}{dn} M_{nn} + \frac{dv}{dm} M_{nm} = \frac{\partial v}{\partial x}(M_{xx}n_x + M_{xy}n_y) + \frac{\partial v}{\partial y}(M_{xy}n_x + M_{yy}n_y) \tag{18.51}$$

where the unit vector $\mathbf{m}$ is orthogonal to $\mathbf{n}$, i.e. $\mathbf{m}$ is tangential to the boundary. With (18.50) and (18.51), (18.49) reduces to

$$\int_A (\overset{*}{\nabla} v)^{\mathrm{T}} \mathbf{M} \, dA = \oint_{\mathscr{L}} \left(\frac{dv}{dn} M_{nn} + \frac{dv}{dm} M_{nm} \right) d\mathscr{L} - \oint_{\mathscr{L}} v V_{nz} \, d\mathscr{L} - \int_A vq \, dA \tag{18.52}$$

Now, as $\mathbf{m}$ is tangential to the boundary, we may take $d\mathscr{L} = dm$, i.e.

$$\oint_{\mathscr{L}} \frac{dv}{dm} M_{nm} \, d\mathscr{L} = \oint_{\mathscr{L}} \frac{d}{dm}(v M_{nm}) \, d\mathscr{L} - \oint_{\mathscr{L}} v \frac{dM_{nm}}{dm} \, d\mathscr{L}$$

$$= \oint_{\mathscr{L}} d(v M_{nm}) - \oint_{\mathscr{L}} v \frac{dM_{nm}}{dm} \, d\mathscr{L} = - \oint_{\mathscr{L}} v \frac{dM_{nm}}{dm} \, d\mathscr{L}$$

Hence (18.52) takes the form

$$\boxed{\int_A (\overset{*}{\nabla} v)^{\mathrm{T}} \mathbf{M} \, dA = \oint_{\mathscr{L}} \frac{dv}{dn} M_{nn} \, d\mathscr{L} - \oint_{\mathscr{L}} v \left(V_{nz} + \frac{dM_{nm}}{dm} \right) d\mathscr{L} - \int_A vq \, dA} \tag{18.53}$$

which is the weak form of the equilibrium equation sought. We immediately conclude that the natural boundary conditions are given by

$$M_{nn}; \quad V_{nz} + \frac{dM_{nm}}{dm} \qquad (18.54)$$

Therefore, it is not the quantities M_{nn}, M_{nm} and V_{nz} that determine the static boundary conditions. Rather, it is M_{nn} and the term $V_{nz} + dM_{nm}/dm$ which matter, and in this term the value of V_{nz} or of dM_{nm}/dm is irrelevant; it is only the sum $V_{nz} + dM_{nm}/dm$ that counts. The proper static boundary conditions (18.54) are called *Kirchhoff's boundary conditions* and the term $V_{nz} + dM_{nm}/dm$ is often called the *effective shear force*. As M_{nm} is the twisting moment per unit length, dM_{nm}/dm has the dimension of force per unit length, i.e. [N/m], just like the shear force V_{nz}. Following a proposal by Kelvin and Tait in 1870, it is possible to obtain a physical interpretation of the effective shear force, and this explanation is the one which most often appears in textbooks; cf. Timoshenko and Woinowsky-Krieger (1959).

18.3 Concentrated shear forces at corners

In addition to the vertical shear force per unit length V_{nz} caused by the vertical shear stress σ_{nz}, we have seen the existence of the vertical shear force per unit length dM_{nm}/dm caused by the horizontal shear stress σ_{nm}. The shear force distribution dM_{nm}/dm, however, possesses some special properties that are of essential importance when evaluating a plate response. Even experienced engineers may be hesitant about these properties and we shall therefore present a detailed discussion of these matters.

The first point of interest is that the distribution of the shear force per unit length dM_{nm}/dm along the boundary of the plate comprises a self-equilibrating system of forces. To see this we integrate dM_{nm}/dm along the entire boundary, and as we may take $d\mathcal{L} = dm$ (cf. Figure 18.8) we obtain

$$\oint_{\mathcal{L}} \frac{dM_{nm}}{dm} \, d\mathcal{L} = \oint_{\mathcal{L}} \frac{dM_{nm}}{dm} \, dm = \oint_{\mathcal{L}} dM_{nm} = 0 \qquad (18.55)$$

Vertical equilibrium of the entire plate then requires that

$$\oint_{\mathcal{L}} V_{nz} \, d\mathcal{L} = \int_{A} q \, dA \qquad (18.56)$$

i.e. the total vertical load is in equilibrium with the total shear force due to V_{nz}.

The second point of interest is that the distribution of the shear force per unit length dM_{nm}/dm exhibits a discontinuity at sharp corners of the boundary. That is, at sharp corners of the boundary we have a concentrated shear force. To illustrate this remarkable feature, we consider Figure 18.10(a). At point P on one side of the corner we have the unit normal vector $\mathbf{n}^P$ and the unit tangential vector $\mathbf{m}^P$. At point Q on the other side of the corner, we have the unit normal vector $\mathbf{n}^Q$ and the

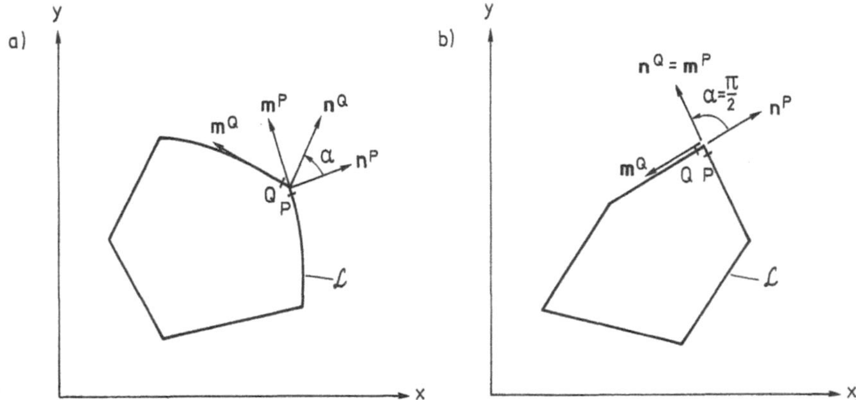

Figure 18.10   Conditions at sharp corners

unit tangential vector $\mathbf{m}^Q$. The tangential vectors are chosen so that the vectors $\mathbf{n}$ and $\mathbf{m}$, after a suitable rigid-body rotation, have the same directions as the x- and y-axes, respectively. The sharp corner is characterized by the angle α between the normal vectors $\mathbf{n}^P$ and $\mathbf{n}^Q$. This angle is measured as positive in the counter-clockwise direction.

Considering (18.53), let us now evaluate the contribution from the shear force per unit length $\mathrm{d}M_{nm}/\mathrm{d}m$, when we pass from point P to point Q. Choosing $\mathrm{d}\mathscr{L} = \mathrm{d}m$ we get

$$\int_P^Q \frac{\mathrm{d}M_{nm}}{\mathrm{d}m}\,\mathrm{d}\mathscr{L} = \int_P^Q \mathrm{d}M_{nm} = M_{nm}^Q - M_{nm}^P \tag{18.57}$$

The exact plate deflection w is a continuous function and so are arbitrary orders of differentiation of the deflection w and due to (18.31) we conclude that also the moments vary in a continuous manner. As points P and Q are located infinitely close, we therefore have

$$M_{xx}^P = M_{xx}^Q = M_{xx}; \quad M_{yy}^P = M_{yy}^Q = M_{yy}; \quad M_{xy}^P = M_{xy}^Q = M_{xy} \tag{18.58}$$

Using (18.58), (18.39) provides

$$M_{nm}^Q = n_x^Q m_x^Q M_{xx} + n_y^Q m_y^Q M_{yy} + (n_y^Q m_x^Q + n_x^Q m_y^Q) M_{xy}$$

$$M_{nm}^P = n_x^P m_x^P M_{xx} + n_y^P m_y^P M_{yy} + (n_y^P m_x^P + n_x^P m_y^P) M_{xy} \tag{18.59}$$

Inserting (18.59) into (18.57) yields

$$\int_P^Q \frac{\mathrm{d}M_{nm}}{\mathrm{d}m}\,\mathrm{d}\mathscr{L} = M_{nm}^Q - M_{nm}^P = V_*$$

$$= (n_x^Q m_x^Q - n_x^P m_x^P) M_{xx} + (n_y^Q m_y^Q - n_y^P m_y^P) M_{yy}$$

$$+ (n_y^Q m_x^Q - n_y^P m_x^P + n_x^Q m_y^Q - n_x^P m_y^P) M_{xy} \tag{18.60}$$

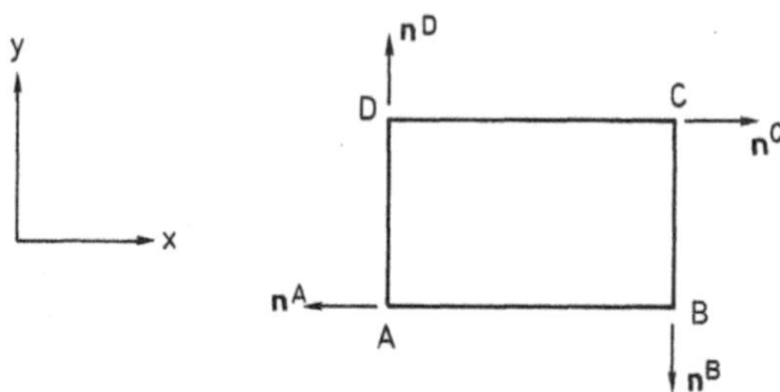

Figure 18.11 Rectangular plate

where the definition $V_* = M_{nm}^Q - M_{nm}^P$ has been introduced. From (18.60) it follows, as expected, that if the corner is smooth, i.e. $\mathbf{n}^P = \mathbf{n}^Q$ and $\mathbf{m}^P = \mathbf{m}^Q$, the distribution of the shear force per unit length $\mathrm{d}M_{nm}/\mathrm{d}m$ is continuous, i.e. $M_{nm}^Q = M_{nm}^P$ and $V_* = 0$. However, at sharp corners of the boundary we have $\mathbf{n}^P \neq \mathbf{n}^Q$ and $\mathbf{m}^P \neq \mathbf{m}^Q$. That is, (18.60) shows that the distribution of the shear force per unit length $\mathrm{d}M_{nm}/\mathrm{d}m$ exhibits a discontinuity since, in general, we will have $M_{nm}^Q \neq M_{nm}^P$, i.e. $V_* \neq 0$. Recalling that M_{nm} has the dimension of moment per unit length $[\mathrm{N\,m/m}]$, the difference $M_{nm}^Q - M_{nm}^P = V_*$ has the dimension of force $[\mathrm{N}]$, i.e. the discontinuity of the distribution of $\mathrm{d}M_{nm}/\mathrm{d}m$ at sharp corners emerges in the form of a concentrated vertical force, which is here termed V_*.

To evaluate this concentrated shear force for the case of most practical interest, we consider a corner in the form of a right angle, i.e. $\alpha = \pi/2$; cf. Figure 18.10(b). From this figure we conclude that

$$\mathbf{n}^P = \begin{bmatrix} n_x^P \\ n_y^P \end{bmatrix}; \quad \mathbf{m}^P = \mathbf{n}^Q = \begin{bmatrix} -n_y^P \\ n_x^P \end{bmatrix}; \quad \mathbf{m}^Q = -\begin{bmatrix} n_x^P \\ n_y^P \end{bmatrix} \tag{18.61}$$

Using (18.61) in (18.60) yields

$$V_* = M_{nm}^Q - M_{nm}^P = 2n_x^P n_y^P (M_{xx} - M_{yy}) + 2[(n_y^P)^2 - (n_x^P)^2] M_{xy} \tag{18.62}$$

As an illustration, consider the rectangular plate shown in Figure 18.11. At corner C, we have $(\mathbf{n}^P)^T = \begin{bmatrix} 1 & 0 \end{bmatrix}$, i.e. (18.62) gives $V_*^C = -2M_{xy}^C$. Proceeding in the same way for the other corners gives

$$\boxed{V_*^A = -2M_{xy}^A; \quad V_*^B = 2M_{xy}^B; \quad V_*^C = -2M_{xy}^C; \quad V_*^D = 2M_{xy}^D} \tag{18.63}$$

To evaluate these results further, we consider the particular case of a simply supported rectangular plate loaded by a distributed load q per unit area, which varies sinusoidally over the plate, i.e.

$$q = -q_0 \sin\frac{\pi x}{a} \sin\frac{\pi y}{b} \tag{18.64}$$

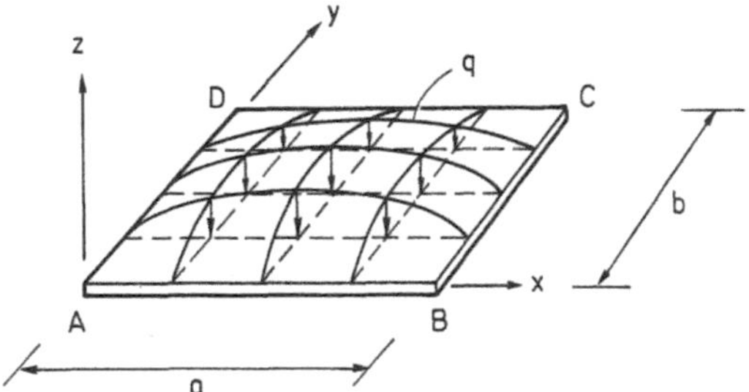

Figure 18.12 Simply supported rectangular plate exposed to a sinusoidal load q

see Figure 18.12. We take q_0 as a positive quantity; that is, the load q acts in the negative direction of the z-axis, or downwards.

The material is assumed to be isotropic and it is straightforward to check that the solution to (18.33) subjected to the boundary conditions $w = 0$ and $M_{nn} = 0$ is given by

$$w = -\frac{q_0}{\alpha} \sin\frac{\pi x}{a} \sin\frac{\pi y}{b} \quad \text{where} \quad \alpha = \frac{\pi^4 E t^3}{12(1 - v^2)}\left(\frac{1}{a^2} + \frac{1}{b^2}\right)^2 \tag{18.65}$$

The reader may also check that trivial but somewhat involved calculations based on (18.63) and (18.65) show that

$$V_*^A = V_*^B = V_*^C = V_*^D = -2\beta \tag{18.66}$$

where the positive quantity β is given by

$$\beta = \frac{q_0(1 - v)}{\pi^2 ab(1/a^2 + 1/b^2)^2} \tag{18.67}$$

Therefore, as the concentrated shear forces given by (18.66) are negative, they act in the negative direction of the z-axis, i.e. downwards. Moreover, along the boundaries AB and DC we get the following shear forces per unit length:

$$V_{nz} = \frac{\beta a \pi}{1 - v}\left(\frac{1}{a^2} + \frac{1}{b^2}\right)\sin\frac{\pi x}{a}; \quad \frac{\mathrm{d}M_{nm}}{\mathrm{d}m} = \beta\frac{\pi}{a}\sin\frac{\pi x}{a} \tag{18.68}$$

and along the boundaries BC and AD we obtain

$$V_{nz} = \frac{\beta b \pi}{1 - v}\left(\frac{1}{a^2} + \frac{1}{b^2}\right)\sin\frac{\pi y}{a}; \quad \frac{\mathrm{d}M_{nm}}{\mathrm{d}m} = \beta\frac{\pi}{b}\sin\frac{\pi y}{b} \tag{18.69}$$

The shear forces per unit length V_{nz} and $\mathrm{d}M_{nm}/\mathrm{d}m$ given by (18.68) and (18.69) are positive, i.e. they act in the positive direction of the z-axis, or upwards. It is recalled that the forces given by (18.66), (18.68) and (18.69) are forces which act on the plate.

As a specific example, assume that $a = 3$, $b = 2$ and $v = 0.3$. The concentrated shear forces V_* and the shear forces per unit length $\mathrm{d}M_{nm}/\mathrm{d}m$ are then as shown in

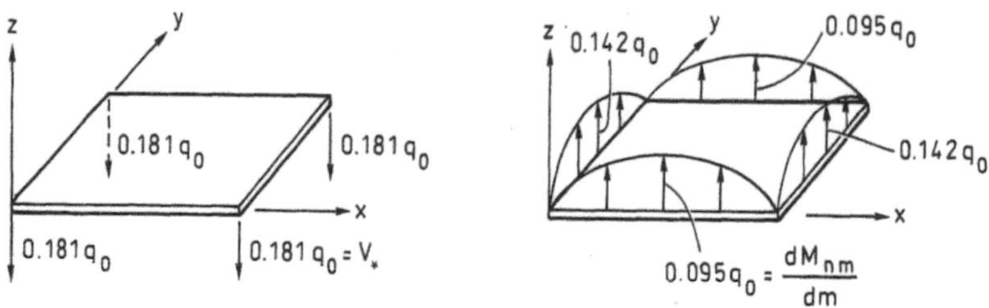

Figure 18.13   Self-equilibrating system of forces consisting of V_* and dM_{nm}/dm

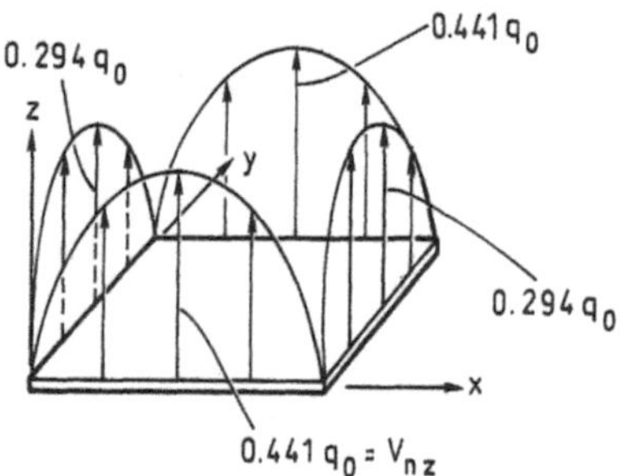

Figure 18.14   Distribution of V_{nz}

Figure 18.13, whereas the shear forces per unit length V_{nz} are as shown in Figure 18.14. In accordance with (18.55) the system of forces consisting of the concentrated forces V_* at the corners and the continuous distribution of dM_{nm}/dm away from the corners is self-equilibrating, as is also apparent from Figure 18.13. In accordance with (18.56) the shear forces per unit length V_{nz} shown in Figure 18.14 balance the external load due to the load q; cf. Figure 18.12. It is recalled that the forces V_*, dM_{nm}/dm and V_{nz} are forces acting on the plate. Therefore, Figure 18.13 shows that if the plate is not anchored at the corners, the corners will tend to rise, and this tendency is in accordance with observations found in practice.

18.4 FE formulation

Following this detailed discussion of various aspects of plate theory, it is timely to establish the FE formulation. As the deflection w is the unknown function in the plate theory (cf. (18.32) and (18.33)) we already know that at some stage we will write the approximation for w quite generally as

$$\boxed{w = \mathbf{N}\mathbf{a}} \tag{18.70}$$

where

$$\mathbf{N} = [N_1 \quad N_2 \quad \cdots \quad N_n]; \quad \mathbf{a} = \begin{bmatrix} u_1 \\ u_2 \\ \vdots \\ u_n \end{bmatrix} \tag{18.71}$$

and n is the number of unknowns for the entire plate. At this point we shall leave the meaning of the shape functions N_i and the nodal values u_i unspecified and just mention that, whereas N_i depends on x and y, i.e. $N_i = N_i(x, y)$, the u_i components are parameters. From (18.70) and (18.28) we obtain

$$\boxed{\overset{*}{\nabla} w = \mathbf{B}\mathbf{a}} \quad \text{where} \quad \boxed{\mathbf{B} = \overset{*}{\nabla} \mathbf{N}} \tag{18.72}$$

i.e.

$$\mathbf{B} = \begin{bmatrix} \dfrac{\partial^2 \mathbf{N}}{\partial x^2} \\[2ex] \dfrac{\partial^2 \mathbf{N}}{\partial y^2} \\[2ex] 2\dfrac{\partial^2 \mathbf{N}}{\partial x \partial y} \end{bmatrix} = \begin{bmatrix} \dfrac{\partial^2 N_1}{\partial x^2} & \dfrac{\partial^2 N_2}{\partial x^2} & \cdots & \dfrac{\partial^2 N_n}{\partial x^2} \\[2ex] \dfrac{\partial^2 N_1}{\partial y^2} & \dfrac{\partial^2 N_2}{\partial y^2} & \cdots & \dfrac{\partial^2 N_n}{\partial y^2} \\[2ex] 2\dfrac{\partial^2 N_1}{\partial x \partial y} & 2\dfrac{\partial^2 N_2}{\partial x \partial y} & \cdots & 2\dfrac{\partial^2 N_n}{\partial x \partial y} \end{bmatrix} \tag{18.73}$$

With the Galerkin method the following expression is used for the arbitrary weight function v:

$$v = \mathbf{N}\mathbf{c} \tag{18.74}$$

As v is arbitrary, the parameters $\mathbf{c}$ are also arbitrary. It appears that

$$\overset{*}{\nabla} v = \mathbf{B}\mathbf{c} \tag{18.75}$$

and (5.8) provides

$$\frac{\mathrm{d}v}{\mathrm{d}n} = (\nabla v)^\mathrm{T} \mathbf{n} = \mathbf{c}^\mathrm{T} (\nabla \mathbf{N})^\mathrm{T} \mathbf{n} \tag{18.76}$$

where

$$\nabla \mathbf{N} = \begin{bmatrix} \dfrac{\partial N_1}{\partial x} & \dfrac{\partial N_2}{\partial x} & \cdots & \dfrac{\partial N_n}{\partial x} \\[2ex] \dfrac{\partial N_1}{\partial y} & \dfrac{\partial N_2}{\partial y} & \cdots & \dfrac{\partial N_n}{\partial y} \end{bmatrix} \tag{18.77}$$

Moreover, since $v = v^\mathrm{T}$ (18.74) can be written as

$$v = \mathbf{c}^\mathrm{T} \mathbf{N}^\mathrm{T} \tag{18.78}$$

We are now in a position to derive the FE formulation. The weak formulation of the equilibrium condition is given by (18.53) which holds for arbitrary weight functions v. Inserting (18.75), (18.76) and (18.78) and noting that $\mathbf{c}^{\mathrm{T}}$ is independent of position yields

$$\mathbf{c}^{\mathrm{T}}\left[\int_A \mathbf{B}^{\mathrm{T}}\mathbf{M}\,\mathrm{d}A - \oint_{\mathscr{L}} (\nabla\mathbf{N})^{\mathrm{T}}\mathbf{n}M_{nn}\,\mathrm{d}\mathscr{L} + \oint_{\mathscr{L}} \mathbf{N}^{\mathrm{T}}\left(V_{nz} + \frac{\mathrm{d}M_{nm}}{\mathrm{d}m}\right)\mathrm{d}\mathscr{L} \right. $$
$$\left. + \int_A \mathbf{N}^{\mathrm{T}} q\,\mathrm{d}A\right] = 0$$

As $\mathbf{c}^{\mathrm{T}}$ is arbitrary, we conclude that

$$\int_A \mathbf{B}^{\mathrm{T}}\mathbf{M}\,\mathrm{d}A = \oint_{\mathscr{L}} (\nabla\mathbf{N})^{\mathrm{T}}\mathbf{n}M_{nn}\,\mathrm{d}\mathscr{L} - \oint_{\mathscr{L}} \mathbf{N}^{\mathrm{T}}\left(V_{nz} + \frac{\mathrm{d}M_{nm}}{\mathrm{d}m}\right)\mathrm{d}\mathscr{L}$$
$$- \int_A \mathbf{N}^{\mathrm{T}} q\,\mathrm{d}A \tag{18.79}$$

which is an expression for the equilibrium condition for the plate, i.e. (18.79) holds irrespective of the constitutive assumptions.

It is recalled that the essential, i.e. the kinematic, boundary conditions are given by

$$w; \quad \frac{\mathrm{d}w}{\mathrm{d}n} \tag{18.80}$$

and the natural, i.e. the static, boundary conditions are

$$M_{nn}; \quad V_{nz} + \frac{\mathrm{d}M_{nm}}{\mathrm{d}m} \tag{18.81}$$

Moreover, we emphasize that, similar to the situation for beams, it is not always possible to divide the boundary into parts where the kinematic boundary conditions are prescribed and parts where the static boundary conditions are prescribed. Indeed, it is entirely possible to have kinematic and static boundary conditions prescribed along the same part. For instance, along a simply supported boundary we have $w = 0$ and $M_{nn} = 0$. This example as well as other typical boundary conditions are illustrated in Figure 18.15.

At this point we introduce the constitutive and kinematic assumptions adopted. From (18.31) and (18.72) we get

$$\mathbf{M} = -\tilde{\mathbf{D}}\mathbf{Ba} \tag{18.82}$$

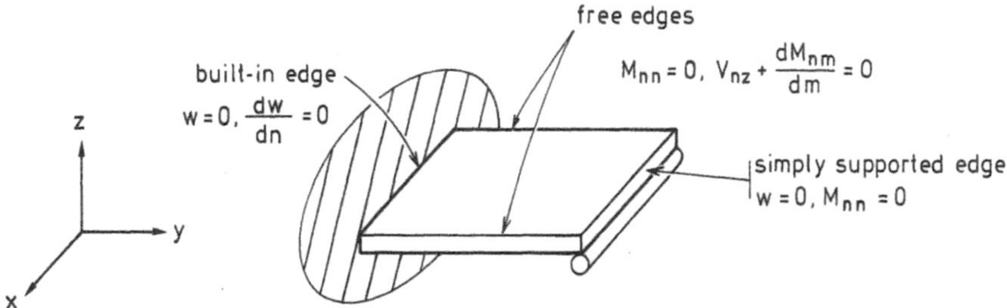

Figure 18.15   Examples of boundary conditions

and insertion into (18.79) yields

$$\left(\int_A \mathbf{B}^{\mathrm{T}} \tilde{\mathbf{D}} \mathbf{B} \, dA \right) \mathbf{a} = \oint_{\mathscr{L}} \mathbf{N}^{\mathrm{T}} \left(V_{nz} + \frac{dM_{nm}}{dm} \right) d\mathscr{L} - \oint_{\mathscr{L}} (\nabla \mathbf{N})^{\mathrm{T}} \mathbf{n} M_{nn} \, d\mathscr{L}$$

$$+ \int_A \mathbf{N}^{\mathrm{T}} q \, dA \qquad (18.83)$$

which is the FE formulation sought. To write this expression in a more compact fashion define the stiffness matrix $\mathbf{K}$, the boundary vector $\mathbf{f}_b$ and the load vector $\mathbf{f}_l$ by

$$\mathbf{K} = \int_A \mathbf{B}^{\mathrm{T}} \tilde{\mathbf{D}} \mathbf{B} \, dA$$

$$\mathbf{f}_b = \oint_{\mathscr{L}} \mathbf{N}^{\mathrm{T}} \left(V_{nz} + \frac{dM_{nm}}{dm} \right) d\mathscr{L} - \oint_{\mathscr{L}} (\nabla \mathbf{N})^{\mathrm{T}} \mathbf{n} M_{nn} \, d\mathscr{L} \qquad (18.84)$$

$$\mathbf{f}_l = \int_A \mathbf{N}^{\mathrm{T}} q \, dA$$

i.e. (18.83) can be written as

$$\mathbf{Ka} = \mathbf{f}_b + \mathbf{f}_l \qquad (18.85)$$

The force vector $\mathbf{f}$ is as usual defined by

$$\mathbf{f} = \mathbf{f}_b + \mathbf{f}_l \qquad (18.86)$$

i.e.

$$\mathbf{Ka} = \mathbf{f} \qquad (18.87)$$

The FE formulation for one element follows in the usual manner from the expressions above and need not be written explicitly.

18.4.1 *Completeness and compatibility requirements*

When we made the approximation for the deflection w given by (18.70), we did not specify the meaning of the shape functions N_i and the nodal values u_i. In order to be more specific, we shall investigate the convergence criteria that are fulfilled if the completeness and compatibility requirements are met.

For infinitely small plate elements, the deflection within each element will, in the limit, be characterized by an arbitrary rigid-body motion superposed by an arbitrary constant curvature. The curvature is determined by the curvature matrix κ defined by (18.17) and we note that a constant curvature is the simplest deflection state that create strains. It is concluded that the completeness requirements take the following form:

Completeness

- The approximation for the deflection w must be able to represent an arbitrary rigid-body motion.
- The approximation for the deflection w must be able to represent an arbitrary constant curvature matrix.

An arbitrary rigid-body motion consists of a translation in the z-direction as well as rotations about the x- and y-axes (cf. Figure 18.1). That is, the approximation for the deflection w must at least include the expression $\alpha_1 + \alpha_2 x + \alpha_3 y$. The curvature matrix κ is given by (18.17) and in order to fulfil the completeness criterion, it follows that any approximation for the deflection w must take the form

$$
\begin{aligned}
w = \alpha_1 &+ \alpha_2 x + \alpha_3 y + \alpha_4 x^2 + \alpha_5 xy + \alpha_6 y^2 \\
&+ \text{possibly other terms}
\end{aligned}
\tag{18.88}
$$

To derive the compatibility requirement, advantage is taken of the concept of C^n-continuity discussed in Chapter 9 (page 202). According to (18.84) and (18.72), the stiffness matrix contains second derivatives of the shape functions. This implies that C^1-*continuity* is required for the shape functions and thus also for the deflection w. We then arrive at the following compatibility requirement:

Compatibility

- The approximation for the deflection w must be continuous and possess continuous slopes over the element boundaries.

It is emphasized that, whereas the completeness requirement must be fulfilled, it may be possible to relax the compatibility requirement and to work with *non-conforming plate elements*.

The discussion above has much in common with the similar discussion for beam elements. However, whereas it is easy to construct conforming beam elements, it turns out to be impossible to establish conforming plate elements when use is made of a simple polynomial expression for the approximation of the deflection w (cf. Zienkiewicz, 1977). We conclude the following:

> Most plate elements are non-conforming.

As mentioned in Chapter 7 (page 93), the convergence criteria for non-conforming elements are more involved than for conforming elements. We shall not present an evaluation of these matters in this introductory text, but merely refer to Cook *et al.* (1989), Crisfield (1986), Hughes (1987), Zienkiewicz (1977) and Zienkiewicz and Taylor (1989) for a proper treatment of this topic.

18.4.2 *Two simple non-conforming plate elements*

For a presentation of plate elements we shall confiné ourselves to two simple plate elements used in practice. We shall also illustrate that both elements are non-conforming.

The simplest possible plate element having six degrees of freedom is shown in Figure 18.16. It was suggested by Morley (1971), and it consists of a triangle with the deflection degrees of freedom u_1, u_2, u_3 at the corners and the rotational degrees of freedom u_4, u_5, u_6 at the midpoint of the sides. The deflection w is approximated by

$$w = \alpha_1 + \alpha_2 x + \alpha_3 y + \alpha_4 x^2 + \alpha_5 xy + \alpha_6 y^2 \tag{18.89}$$

A comparison with (18.88) shows that this is the simplest possible approximation that fulfils the completeness requirement. It appears that the second derivatives of w are constants, i.e. the moments are constant within the element (cf. (18.23)).

To show that the element is non-conforming, consider the arbitrary element

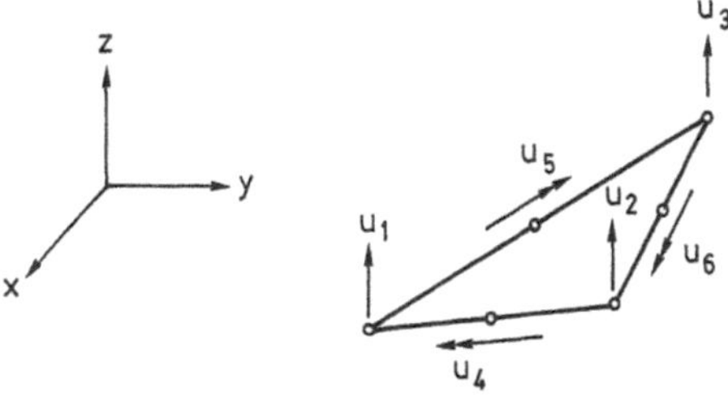

Figure 18.16　Simplest plate element

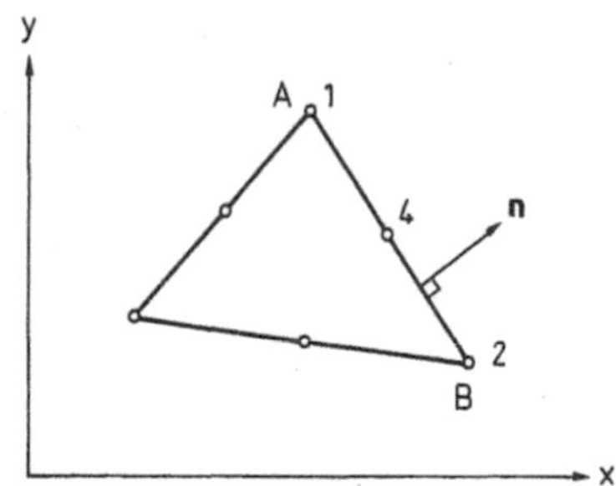

Figure 18.17 Triangular plate element with six degrees of freedom

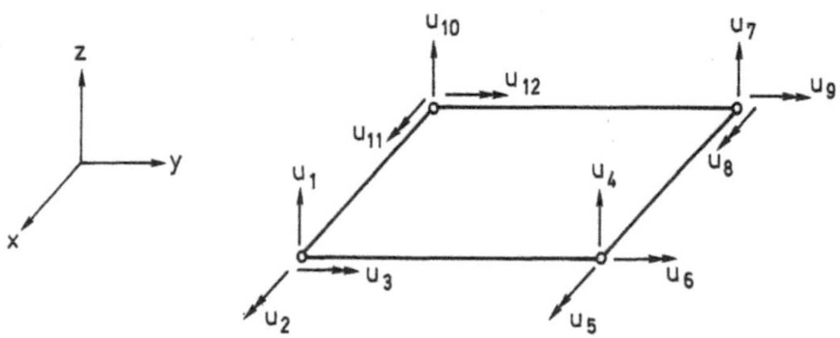

Figure 18.18 Rectangular plate element with 12 degrees of freedom

boundary AB with the nodal points 1, 4 and 2 in Figure 18.17. With w given by (18.89), we get from (5.8) that

$$\frac{dw}{dn} = (\nabla w)^{\mathrm{T}}\mathbf{n} = (\alpha_2 + 2\alpha_4 x + \alpha_5 y)n_x + (\alpha_3 + \alpha_5 x + 2\alpha_6 y)n_y$$

Along the boundary AB, we have $y = \beta_1 x + \beta_2$, i.e. (18.89) and the expression above provide

$$w = A_1 + A_2 x + A_3 x^2$$
$$\frac{dw}{dn} = B_1 + B_2 x$$

(18.90)

where A_1, A_2, A_3 and B_1, B_2 are certain constants. Along the element boundary AB, three quantities are known: the displacements at nodes 1 and 2 and the slope dw/dn at node 4. These three conditions are not sufficient to determine the five parameters A_1, A_2, A_3 and B_1, B_2 entering (18.90) in a unique manner. It is concluded that both the displacement w and the slope dw/dn, in general, vary discontinuously over common element boundaries, i.e. the triangular element of Figure 18.16 is non-conforming.

Consider next the rectangular plate element having 12 degrees of freedom as shown in Figure 18.18. It was suggested by Adini and Clough (1961) and Melosh (1963), and it has four deflection degrees of freedom u_1, u_4, u_7, u_{10} and eight rotational degrees of freedom $u_2, u_3, u_5, u_6, u_8, u_9, u_{11}, u_{12}$. The deflection is approximated by

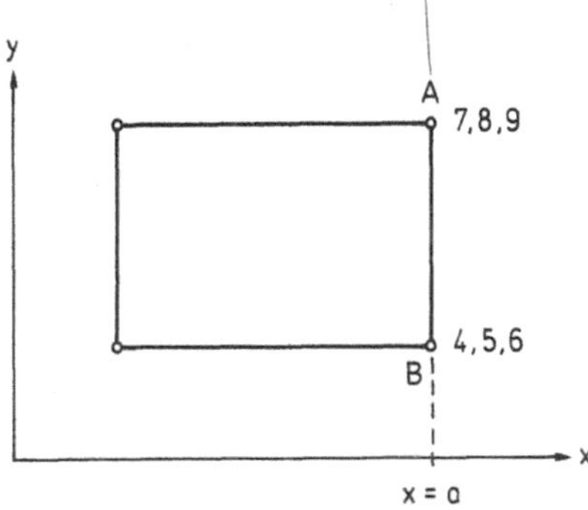

Figure 18.19   Rectangular plate element

$$w = \alpha_1 + \alpha_2 x + \alpha_3 y + \alpha_4 x^2 + \alpha_5 xy + \alpha_6 y^2$$
$$+ \alpha_7 x^3 + \alpha_8 x^2 y + \alpha_9 xy^2 + \alpha_{10} y^3 + \alpha_{11} x^3 y + \alpha_{12} xy^3 \tag{18.91}$$

It appears that this expression is able to model exactly an arbitrary linear variation of moments within the element.

To show that the element is non-conforming consider, for instance, the element boundary AB shown in Figure 18.19. Two nodal points are located on AB, each having three degrees of freedom. Along AB where $x = a$, (18.91) provides that

$$w = A_1 + A_2 y + A_3 y^2 + A_4 y^3$$
$$\frac{\partial w}{\partial y} = A_2 + 2A_3 y + 3A_4 y^2 \tag{18.92}$$

where A_1, A_2, A_3 and A_4 are certain constants. The deflection w and slope $\partial w/\partial y$ are known at both point A and point B. Insertion of these four conditions in (18.92) implies a unique determination of the four parameters A_1, A_2, A_3 and A_4. That is, the deflection w varies continuously over common element boundaries.

However, the slope $\partial w/\partial x$ does not vary continuously over the element boundary AB. To see this, we derive from (18.91) that

$$\frac{\partial w}{\partial x} = B_1 + B_2 y + B_3 y^2 + B_4 y^3$$

where B_1, B_2, B_3 and B_4 are certain parameters. A unique determination of these four parameters requires four conditions, but only two conditions are available: the value of $\partial w/\partial x$ at point A and its value at point B. That is, the slope $\partial w/\partial x$ in general varies discontinuously over the element boundary AB and we conclude that the rectangular element of Figure 18.18 is non-conforming.

Research activity on the FE modelling of plates is very intense, not only for Kirchhoff plate theory, but also for Mindlin–Reissner plate theory that only requires C^0-continuity and therefore facilitates the establishment of conforming elements. The reader is referred, for instance, to Cook *et al.* (1989), Crisfield (1986), Hughes (1987) and Zienkiewicz (1977) for an evaluation of these topics.

19
Isoparametric finite elements

It was emphasized in Chapter 7 that the sides of quadrilateral and brick elements must be parallel to the coordinate axes in order to behave in a compatible manner. This restriction has severe consequences when modelling bodies with arbitrary geometries and we shall now show how this obstacle can be avoided. It will turn out that it is possible to establish even compatible, i.e. conforming, finite elements that have curved boundaries. The concept we shall study is *isoparametric finite elements* originally introduced by Taig (1961) and in its general form by Irons (1966a,b).

Consider a *mapping* (i.e. a transformation) of one region into another region (Figure 19.1). A square region in the $\xi\eta$-coordinate system is bounded by the lines $\xi = \pm 1$ and $\eta = \pm 1$. This region is termed the *parent domain*. We want to map this region into another region defined in the xy-coordinate system. The region in the xy-plane is called the *global domain*. It follows that this mapping is described by

$$x = x(\xi, \eta); \quad y = y(\xi, \eta) \tag{19.1}$$

i.e. for every point given by its ξ, η-coordinates in the parent domain, there exists a corresponding point given by its x, y-coordinates in the global domain.

It is evident that if the general form of the region in the global domain could be used as a conforming element, we have achieved the objective mentioned above. It will turn out that this objective can be met using isoparametric elements. To show this, we need to consider the mapping technique as well as the manner in which the unknown function is approximated. Moreover, it will turn out that the establishment of the FE equations now requires that integrations are carried out in the parent domain.

19.1 Restrictions on the mapping

The mapping shown in Figure 19.1 creates a distorted quadrilateral in the global xy-plane, and it is natural to investigate whether there are limitations on the degree

364

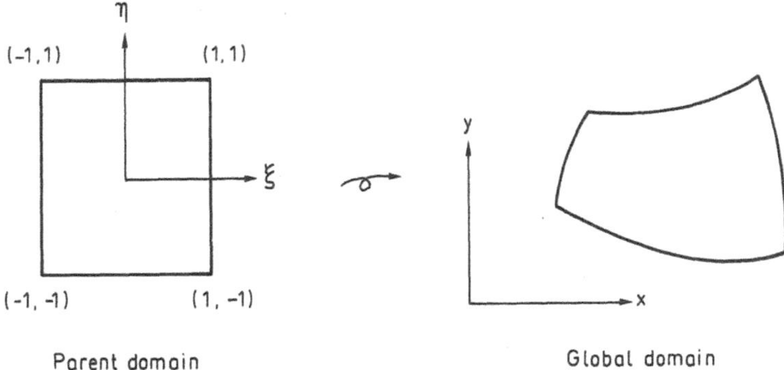

Figure 19.1 Mapping from parent domain to global domain

of distortion. Indeed, there are limitations, and they arise from the requirement that the mapping should be *one-to-one*, i.e. unique. Therefore, we require that one point in the parent domain should correspond to one point in the global domain and vice versa. To achieve this objective, we differentiate (19.1) using the chain rule to obtain

$$
\begin{bmatrix} dx \\ dy \end{bmatrix} = \begin{bmatrix} \dfrac{\partial x}{\partial \xi} & \dfrac{\partial x}{\partial \eta} \\[2mm] \dfrac{\partial y}{\partial \xi} & \dfrac{\partial y}{\partial \eta} \end{bmatrix} \begin{bmatrix} d\xi \\ d\eta \end{bmatrix}
\tag{19.2}
$$

The matrix **J** defined by

$$
\mathbf{J} = \begin{bmatrix} \dfrac{\partial x}{\partial \xi} & \dfrac{\partial x}{\partial \eta} \\[2mm] \dfrac{\partial y}{\partial \xi} & \dfrac{\partial y}{\partial \eta} \end{bmatrix}
\tag{19.3}
$$

is called the *Jacobian matrix* related to the mapping (19.1). Moreover, the determinant of **J**, i.e. det **J**, is called the *Jacobian*. For any given values of $d\xi$ and $d\eta$, (19.2) implies that dx and dy are uniquely determined. Suppose, however, that dx and dy are given and that we want to determine the corresponding $d\xi$- and $d\eta$-values. From (19.2) and (19.3) it follows that

$$
\begin{bmatrix} d\xi \\ d\eta \end{bmatrix} = \mathbf{J}^{-1} \begin{bmatrix} dx \\ dy \end{bmatrix}
\tag{19.4}
$$

and it is evident that we require det $\mathbf{J} \neq 0$. In the following discussion we shall assume that

$$
\det \mathbf{J} > 0
\tag{19.5}
$$

When (19.5) is fulfilled, a one-to-one relation exists between the values $d\xi$, $d\eta$ and dx, dy, i.e. the mapping is unique, at least locally. It is assumed that when (19.5) is fulfilled, the mapping (19.1) is unique even globally.

It is emphasized that, even when the mapping is unique, this does *not* necessarily imply that we are able to invert (19.1) and obtain explicit solutions in the form $\xi = \xi(x, y)$ and $\eta = \eta(x, y)$. For the mappings related to isoparametric finite elements such explicit forms are generally not obtainable.

19.2 Four-node isoparametric quadrilateral element

In order to introduce the concept of isoparametric finite elements, let us consider the four-node isoparametric finite element, which takes the form of an arbitrary quadrilateral.

The region in the parent domain is defined by the four corner points shown in Figure 19.2, and we want to map this region into the arbitrary quadrilateral shown in the global xy-plane. This quadrilateral is also defined by its four corner points. These corner points are identified by their global x, y-coordinates, which we assume are known quantities.

Each of the four corner points in the parent domain may be associated with an element shape function, i.e.

$$N_1^e = N_1^e(\xi, \eta); \quad N_2^e = N_2^e(\xi, \eta); \quad N_3^e = N_3^e(\xi, \eta); \quad N_4^e = N_e(\xi, \eta)$$

The explicit expressions for these element shape functions are taken as those of the four-node rectangular element treated in Chapter 7. Referring to Figure 7.27 and (7.110) we then have

$$\boxed{\begin{aligned} N_1^e &= \tfrac{1}{4}(\xi - 1)(\eta - 1); \quad N_2^e = -\tfrac{1}{4}(\xi + 1)(\eta - 1) \\ N_3^e &= \tfrac{1}{4}(\xi + 1)(\eta + 1); \quad N_4^e = -\tfrac{1}{4}(\xi - 1)(\eta + 1) \end{aligned}}$$

$$(19.6)$$

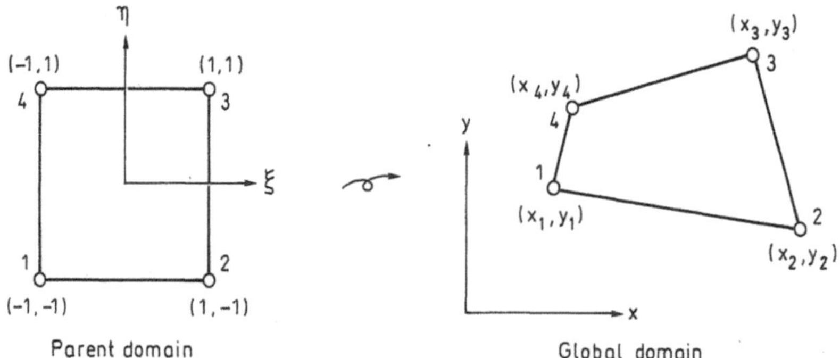

Figure 19.2   Mapping into four-node isoparametric quadrilateral element

These element shape functions possess the usual properties, and we have, for instance, that $N_1^e(-1, -1) = 1$ and $N_1^e(1, -1) = N_1^e(1, 1) = N_1^e(-1, 1) = 0$.

Expression (19.1) indicates the general form of a mapping from the parent domain to the global domain. It turns out that a very advantageous mapping is obtained if the element shape functions are used. Therefore, consider the mapping defined by

$$x = x(\xi, \eta) = N_1^e(\xi, \eta)x_1 + N_2^e(\xi, \eta)x_2 + N_3^e(\xi, \eta)x_3 + N_4^e(\xi, \eta)x_4$$
$$y = y(\xi, \eta) = N_1^e(\xi, \eta)y_1 + N_2^e(\xi, \eta)y_2 + N_3^e(\xi, \eta)y_3 + N_4^e(\xi, \eta)y_4 \tag{19.7}$$

where it is recalled that the x, y-coordinates of the corner points of the quadrilateral are known quantities (cf. Figure 19.2). If we define the following matrices:

$$\mathbf{N}^e(\xi, \eta) = [N_1^e \quad N_2^e \quad N_3^e \quad N_4^e]; \quad \mathbf{x}^e = \begin{bmatrix} x_1 \\ x_2 \\ x_3 \\ x_4 \end{bmatrix}; \quad \mathbf{y}^e = \begin{bmatrix} y_1 \\ y_2 \\ y_3 \\ y_4 \end{bmatrix} \tag{19.8}$$

the mapping (19.7) may be written in the compact form

$$\boxed{x = x(\xi, \eta) = \mathbf{N}^e(\xi, \eta)\mathbf{x}^e; \quad y = y(\xi, \eta) = \mathbf{N}^e(\xi, \eta)\mathbf{y}^e} \tag{19.9}$$

Bearing the properties of the element shape functions in mind, it follows from (19.7) that

$$x(-1, -1) = x_1; \quad x(1, -1) = x_2; \quad x(1, 1) = x_3; \quad x(-1, 1) = x_4$$
$$y(-1, -1) = y_1; \quad y(1, -1) = y_2; \quad y(1, 1) = y_3; \quad y(-1, 1) = y_4 \tag{19.10}$$

i.e. each corner point in the parent domain is mapped into a corner point of the quadrilateral. Consider next what happens when $\xi = -1$. With (19.6) and (19.7) we obtain

$$x(-1, \eta) = -\tfrac{1}{2}(\eta - 1)x_1 + \tfrac{1}{2}(\eta + 1)x_4$$
$$y(-1, \eta) = -\tfrac{1}{2}(\eta - 1)y_1 + \tfrac{1}{2}(\eta + 1)y_4 \tag{19.11}$$

i.e. x and y vary linearly with η between the corners 1 and 4 of the quadrilateral. Elimination of η in (19.11) provides a linear relation between x and y. Therefore, the straight line $\xi = -1$ between the corner points 1 and 4 in the parent domain is mapped into a straight line between the corner points 1 and 4 of the quadrilateral. In general, for $\xi = C$, where C is an arbitrary constant, we obtain a linear relation between x and y. Likewise, for $\eta = C$, where C is an arbitrary constant, a linear relation between x and y is also obtained. We have therefore proved that the mapping defined by (19.9) carries the square defined by its four corner points in the parent domain into the arbitrary quadrilateral that has straight boundaries and is defined by its four corner points.

In conclusion, we have the following: corner points in the parent domain are

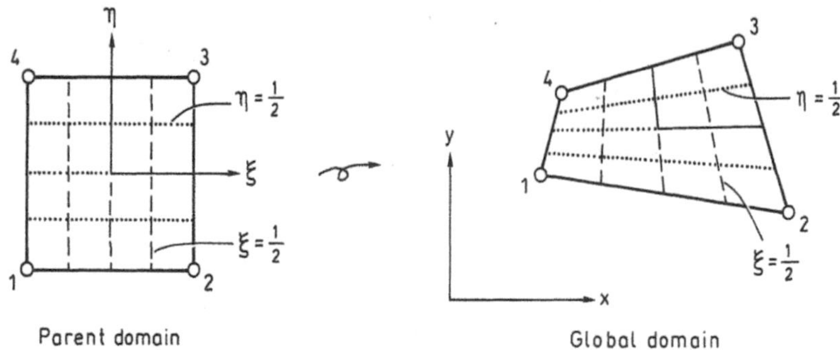

Figure 19.3   Mapping of straight lines given by $\xi = C$ or $\eta = C$ in the parent domain into straight lines in the global domain

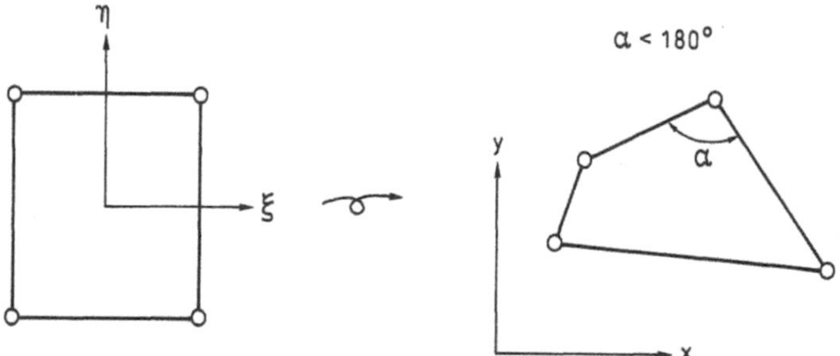

Figure 19.4   Geometrical restriction of the four-node isoparametric quadrilateral element

mapped into corner points in the global domain; the boundaries of the parent domain are mapped into the boundaries of the global domain; and any straight line $\xi = C$ or $\eta = C$ is mapped into a straight line in the global domain. These features are illustrated in Figure 19.3.

As already indicated in the figures, the corner points are the nodal points of the element. It is important that, to achieve the properties summarized above, the numbering of the nodal points should *not* be arbitrary. In fact, we must require that nodal numbers at one element boundary in the global domain should also be located at one element boundary in the parent domain. This objective is most easily fulfilled by labelling the nodal numbering in ascending order in the counter-clockwise direction. This convention is adopted here, as well as in other textbooks.

In order to achieve a one-to-one mapping, we previously derived the general requirement that det $\mathbf{J} > 0$ (cf. (19.5)). In the present case of a four-node quadrilateral, it can be shown that this requirement is fulfilled if all internal angles are less than $180°$; see Strang and Fix (1973, p. 158). Figure 19.4 illustrates this restriction of the otherwise arbitrary quadrilateral.

Having discussed the mapping technique using the element shape functions, we now turn to the approximation of the unknown function or functions. We have seen in Chapter 15 that for the approximation of the displacement vector $\mathbf{u}$ in solid mechanics, we adopted the technique of approximating each displacement component in the same manner as the one applied to the temperature. Here, we can therefore confine ourselves to considering the approximation of one unknown function such as the temperature.

For the four-node isoparametric quadrilateral element shown on the right in Figure 19.2, the temperature T within the element is approximated through

$$T = T(\xi, \eta) = N_1^e(\xi, \eta)T_1 + N_2^e(\xi, \eta)T_2 + N_3^e(\xi, \eta)T_3 + N_4^e(\xi, \eta)T_4 \quad (19.12)$$

where, as usual, $T_1 \ldots T_4$ are the temperatures at the nodal points of the quadrilateral. Moreover, the element shape functions are again given by (19.6). From (19.8), and with the standard notation $\mathbf{a}^{eT} = [\,T_1 \ \ T_2 \ \ T_3 \ \ T_4\,]$, (19.12) may be written in the compact form

$$\boxed{\,T = T(\xi, \eta) = \mathbf{N}^e(\xi, \eta)\mathbf{a}^e\,} \qquad (19.13)$$

Comparing (19.9) and (19.13), the characteristic feature of an isoparametric formulation emerges, namely that the same element shape functions are used in the mapping and in the approximation ('isoparametric' means 'the same parameters'). Moreover, these element shape functions are expressed in the *local* coordinates ξ and η.

We shall now see the advantage of expressing the element shape functions in terms of the ξ, η-coordinates. The objective of the isoparametric formulation of the four-node element is to obtain a conforming element for an arbitrary quadrilateral, and we shall now demonstrate that this is true. From (19.6) and (19.13) it follows that we may write the approximation of T as

$$T = T(\xi, \eta) = \alpha_1 + \alpha_2\xi + \alpha_3\eta + \alpha_4\xi\eta \qquad (19.14)$$

in accordance with (7.107). Along an element boundary of the quadrilateral, we have either $\xi = \pm 1$ or $\eta = \pm 1$. Assume, for instance, that $\xi = 1$, which means that (19.14) reduces to

$$T = (\alpha_1 + \alpha_2) + (\alpha_3 + \alpha_4)\eta$$

This expression, which is a linear relation in η, is uniquely determined by the two parameters $\alpha_1 + \alpha_2$ and $\alpha_3 + \alpha_4$. However, along the element boundary the temperatures at the two nodal points are known, and these two nodal temperatures uniquely determine the two parameters $\alpha_1 + \alpha_2$ and $\alpha_3 + \alpha_4$. Therefore, the temperature variation along an element boundary is uniquely determined by the two nodal temperatures at this element boundary. This implies that the four-node isoparametric element behaves in a conforming manner for an arbitrary quadrilateral.

Finally, we shall mention that if two of the nodal points of the quadrilateral are chosen to be one and the same point, it is possible to obtain exactly the three-node triangle treated in Chapter 7. The reader is referred to Hughes (1987, pp. 120–3) for details.

19.3 Eight-node isoparametric quadrilateral element

As the next example, we consider the isoparametric formulation of the eight-node serendipity element treated in Chapter 7. The mapping is illustrated in Figure 19.5, and as with (19.9) this mapping is given by

$$x = x(\xi, \eta) = \mathbf{N}^e(\xi, \eta)\mathbf{x}^e; \quad y = y(\xi, \eta) = \mathbf{N}^e(\xi, \eta)\mathbf{y}^e \tag{19.15}$$

where $\mathbf{x}^e$ and $\mathbf{y}^e$ contain the x- and y-coordinates of the nodes of the element in the xy-plane. The element shape functions are given by (7.120) and (7.121), i.e.

$$
\begin{aligned}
N_1^e &= -\tfrac{1}{4}(1 - \xi)(1 - \eta)(1 + \xi + \eta); & N_5^e &= \tfrac{1}{2}(1 - \xi^2)(1 - \eta) \\
N_2^e &= -\tfrac{1}{4}(1 + \xi)(1 - \eta)(1 - \zeta + \eta); & N_6^e &= \tfrac{1}{2}(1 + \xi)(1 - \eta^2) \\
N_3^e &= -\tfrac{1}{4}(1 + \xi)(1 + \eta)(1 - \xi - \eta); & N_7^e &= \tfrac{1}{2}(1 - \xi^2)(1 + \eta) \\
N_4^e &= -\tfrac{1}{4}(1 - \xi)(1 + \eta)(1 + \xi - \eta); & N_8^e &= \tfrac{1}{2}(1 - \xi)(1 - \eta^2)
\end{aligned}
\tag{19.16}
$$

From (19.15) and (19.16), it appears that $x(-1, -1) = x_1$, $y(-1, -1) = y_1$ and $x(0, -1) = x_5$, $y(0, -1) = y_5$, etc., and we therefore obtain the correspondence between nodal points in the parent and global domains indicated in Figure 19.5.

Let us now investigate how an element boundary in the parent domain maps into the global domain. As an example we assume that $\xi = -1$, and it follows from (19.16) that the only non-zero element shape functions are N_1^e, N_4^e and N_8^e. With (19.15) we then obtain

$$x(-1, \eta) = -\tfrac{1}{2}(1 - \eta)\eta x_1 + \tfrac{1}{2}(1 + \eta)\eta x_4 + (1 - \eta^2)x_8 \tag{19.17}$$

$$y(-1, \eta) = -\tfrac{1}{2}(1 - \eta)\eta y_1 + \tfrac{1}{2}(1 + \eta)\eta y_4 + (1 - \eta^2)y_8 \tag{19.18}$$

As both x and y, in general, vary quadratically with η, the straight boundary 1, 8, 4 in the parent domain is mapped into a curve through the global nodal points 1, 8, 4.

Now let us consider the special situation where the global nodal points 1, 8, 4

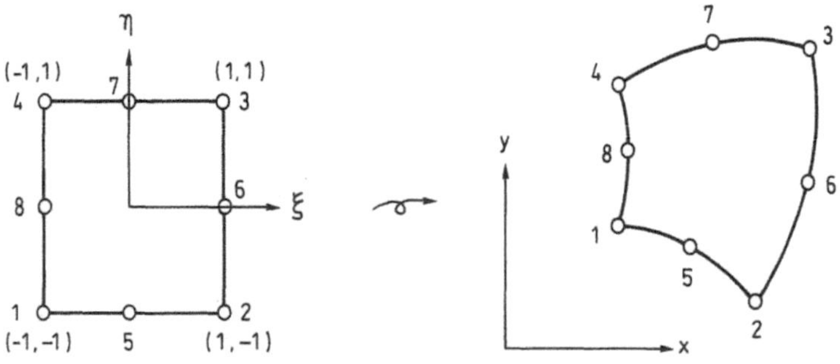

Figure 19.5 Mapping into eight-node isoparametric quadrilateral element

are located on the straight line $y = \alpha x + \beta$, and let us scrutinize the form of the *entire* global boundary 1, 8, 4. The coordinates of the three global nodal points are clearly related through

$$y_1 = \alpha x_1 + \beta; \quad y_4 = \alpha x_4 + \beta; \quad y_8 = \alpha x_8 + \beta$$

Inserting these values into (19.18) yields

$$y(-1, \eta) = -\tfrac{1}{2}(1 - \eta)\eta\alpha x_1 + \tfrac{1}{2}(1 + \eta)\eta\alpha x_4 + (1 - \eta^2)\alpha x_8 + \beta$$

and using (19.17) results in

$$y(-1, \eta) = \alpha x(-1, \eta) + \beta$$

That is, if the global nodal points 1, 8, 4 are located on a straight line, the *entire* global element boundary 1, 8, 4 is located on that line. In general, however, the global element boundary 1, 8, 4 will be curved. As in the derivation of (19.17) and (19.18), it follows that any straight line in the parent domain given by $\xi = C$ or $\eta = C$, where C is an arbitrary constant, maps into a curved line in the global domain. These features are illustrated in Figure 19.6.

As with the four-node isoparametric element, the numbering of the nodal points in the present case is *not* arbitrary. In order to achieve the mapping of element boundaries in the parent domain into element boundaries in the global domain, we must require that nodal points on one element boundary in the global domain are also located on one element boundary in the parent domain. Moreover, in the two domains, corner nodes must correspond to corner nodes and mid-side nodes must correspond to mid-side nodes. These requirements are fulfilled by the numbering shown in Figures 19.5 and 19.6.

In addition to these requirements, we also have the requirement of a one-to-one mapping, i.e. det $\mathbf{J} > 0$; cf. (19.5). For the eight-node element considered, it can be shown that this requirement is fulfilled if all internal angles are less than 180° and if the mid-side nodes are in the 'middle half' of the distance between adjacent corners, though the 'middle third' shown in Figure 19.7 is used in practice (see Zienkiewicz

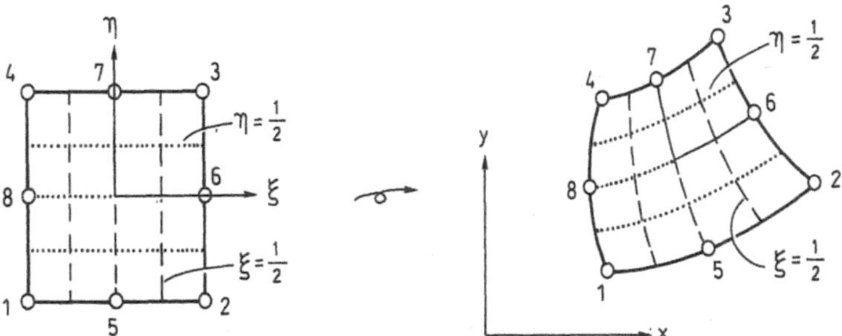

Figure 19.6 Mapping of straight lines given by $\xi = C$ or $\eta = C$ in the parent domain into curved lines in the global domain

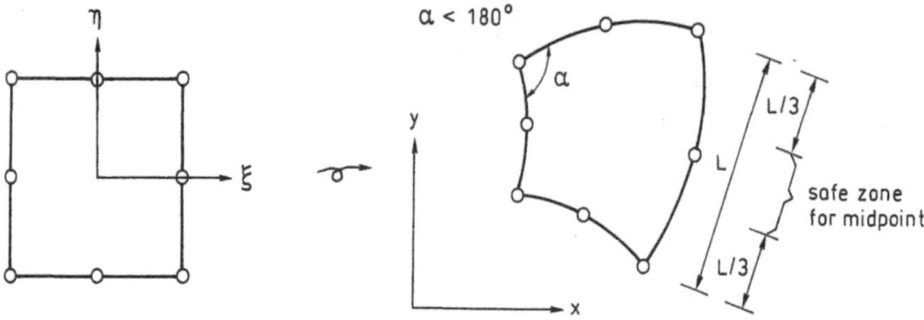

Figure 19.7   Geometrical restrictions on the eight-node isoparametric element

and Taylor, 1989, p. 158). After this discussion of various aspects of the mapping, we now turn to the approximation of the unknown function, for instance the temperature. As with the four-node element, the temperature T within the isoparametric element, shown on the right in Figure 19.5, is approximated through

$$T = T(\xi, \eta) = \mathbf{N}^e(\xi, \eta)\mathbf{a}^e \qquad (19.19)$$

where the components of $\mathbf{N}^e$ are given by (19.16) and $\mathbf{a}^e$ contains the temperatures at the nodal points. Again we observe the characteristic feature of the isoparametric concept, namely that the same element shape functions are used in the mapping and in the approximation (cf. (19.15)). Moreover, these element shape functions are expressed in the *local* coordinates ξ and η.

The objective of the isoparametric formulation of the eight-node element is to obtain a conforming element even for the complicated geometry illustrated in Figure 19.5, and we shall now demonstrate that this is true. From (19.16) and (19.19) it follows that we may write the approximation of T as

$$T = T(\xi, \eta) = \alpha_1 + \alpha_2 \xi + \alpha_3 \eta + \alpha_4 \xi^2 + \alpha_5 \xi\eta + \alpha_6 \eta^2 + \alpha_7 \xi^2\eta + \alpha_8 \xi\eta^2 \quad (19.20)$$

in accordance with (7.122). Along an element boundary we have either $\xi = \pm 1$ or $\eta = \pm 1$. Assume, for instance, that $\xi = 1$, which means that (19.20) reduces to

$$T = (\alpha_1 + \alpha_2 + \alpha_4) + (\alpha_3 + \alpha_5 + \alpha_7)\eta + (\alpha_6 + \alpha_8)\eta^2$$

This quadratic expression in η is uniquely determined once the three parameters $\alpha_1 + \alpha_2 + \alpha_4$, $\alpha_3 + \alpha_5 + \alpha_7$ and $\alpha_6 + \alpha_8$ are identified. Along the element boundary, the temperatures at three nodal points are known, and these three temperatures uniquely determine the three parameters $\alpha_1 + \alpha_2 + \alpha_4$, $\alpha_3 + \alpha_5 + \alpha_7$ and $\alpha_6 + \alpha_8$. Therefore, the temperature variation along an element boundary is uniquely determined by the three nodal temperatures at this boundary. This implies that the eight-node isoparametric element behaves in a conforming manner for the arbitrary geometry illustrated in Figure 19.5.

From this observation we may draw a further conclusion. As the temperature

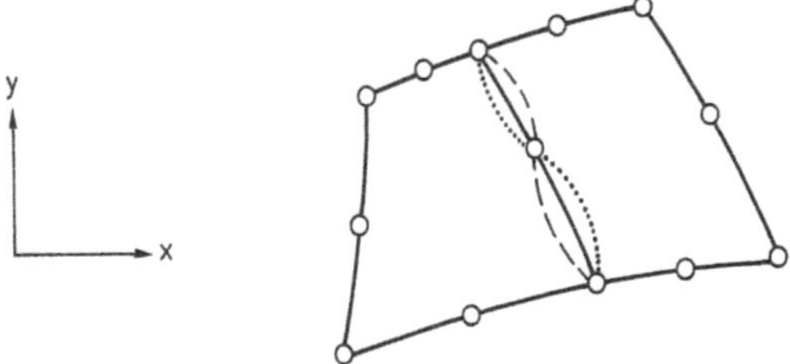

Figure 19.8   The mismatch illustrated is not possible

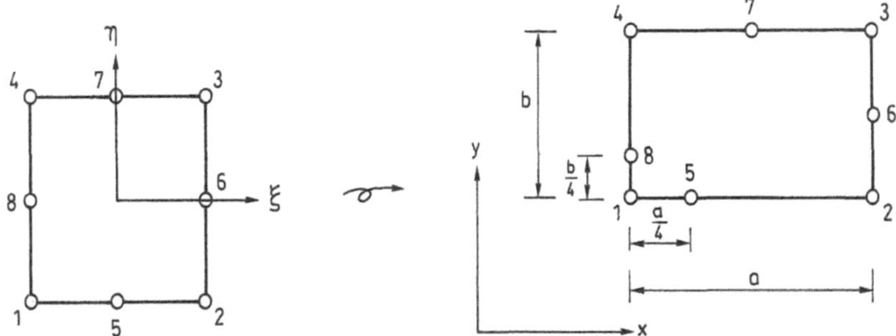

Figure 19.9   Modelling of singularity around crack tip by placing mid-nodes at quarter-points

and the geometry are modelled in the same way (cf. (19.15) and (19.19)), it follows that knowledge of the global nodal coordinates determines the global element boundaries in a unique manner. This, in turn, implies that two adjacent elements will join in a smooth way without any kind of gaps or overlapping. Consequently the mismatch between two adjacent elements illustrated in Figure 19.8 is not possible.

We have previously mentioned the restriction in selecting the position of the mid-side nodes (cf. Figure 19.7). It is of interest, however, that a violation of this restriction may result in some appealing new properties of the element. In fact, if two mid-side nodes are located at the quarter-points, as shown in Figure 19.9, and if linear elasticity is considered, it can be shown that a singularity arises in the displacement field and that this singularity is of the same type as that existing for problems in *linear elastic fracture mechanics*. As established by Henshell and Shaw (1975) as well as by Barsoum (1976), this interesting property can be used to model cracks in an efficient manner. In the case illustrated in Figure 19.9, the crack tip is located at nodal point 1.

19.4 Three-dimensional isoparametric elements

The two-dimensional four- and eight-node isoparametric elements have been discussed above, and by now it should be evident how to formulate other isoparametric elements.

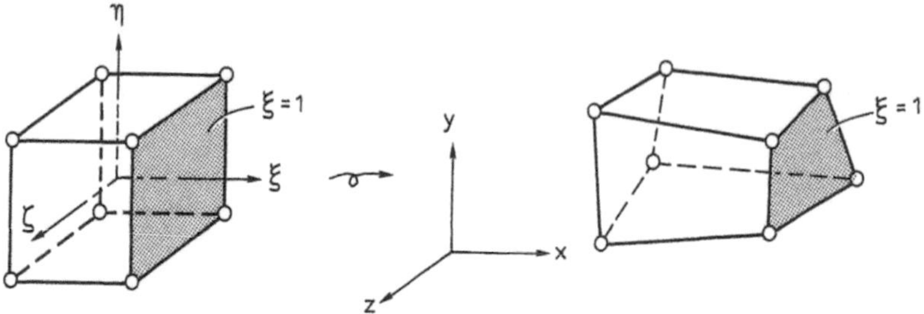

Figure 19.10   Eight-node three-dimensional isoparametric element

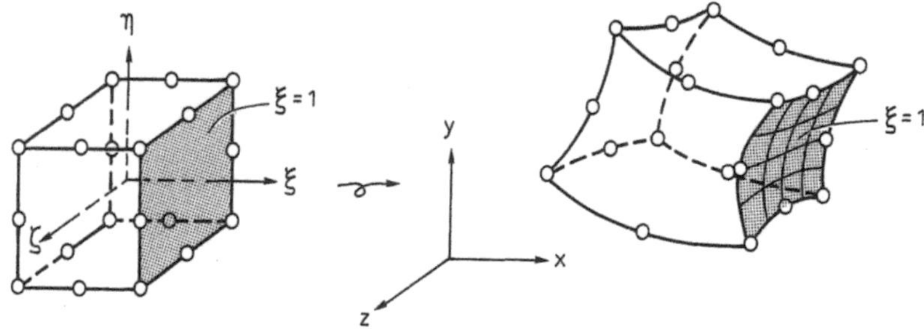

Figure 19.11   Twenty-node three-dimensional isoparametric element

Without going into any detail, the three-dimensional eight- and twenty-node isoparametric elements are illustrated in Figures 19.10 and 19.11, respectively.

19.5 General isoparametric formulation – convergence requirements

For two-dimensional problems, the mapping is obtained through

$$x = x(\xi, \eta) = \mathbf{N}^e(\xi, \eta)\mathbf{x}^e; \quad y = y(\xi, \eta) = \mathbf{N}^e(\xi, \eta)\mathbf{y}^e \tag{19.21}$$

where $\mathbf{N}^e(\xi, \eta)$ is the element shape function matrix, whereas $\mathbf{x}^e$ and $\mathbf{y}^e$ contain the x- and y-coordinates of the nodes of the element in the xy-plane.

The unknown function – the temperature T – is approximated through

$$T = T(\xi, \eta) = \mathbf{N}^e(\xi, \eta)\mathbf{a}^e \tag{19.22}$$

where $\mathbf{a}^e$ contains the temperatures at the global nodal points. Comparing (19.21) and (19.22), the fundamental issue of the isoparametric formulation is evident, as follows:

Isoparametric elements

The same functions – the element shape functions – are used to describe the geometry of the element and the approximation of the unknown function. Moreover, these shape functions are prescribed in terms of the ξ, η-coordinates.

This concept may be generalized directly to three dimensions, and with obvious notation we get

$$x = x(\xi, \eta, \zeta) = \mathbf{N}^e(\xi, \eta, \zeta)\mathbf{x}^e; \quad y = y(\xi, \eta, \zeta) = \mathbf{N}^e(\xi, \eta, \zeta)\mathbf{y}^e;$$
$$z = z(\xi, \eta, \zeta) = \mathbf{N}^e(\xi, \eta, \zeta)\mathbf{z}^e \tag{19.23}$$

and

$$T = T(\xi, \eta, \zeta) = \mathbf{N}^e(\xi, \eta, \zeta)\mathbf{a}^e \tag{19.24}$$

For one-dimensional problems we have

$$x = x(\xi) = \mathbf{N}^e(\xi)\mathbf{x}^e \tag{19.25}$$

and

$$T = T(\xi) = \mathbf{N}^e(\xi)\mathbf{a}^e \tag{19.26}$$

For the two-dimensional four- and eight-node isoparametric elements, we proved above that they fulfil the compatibility requirement and that no mismatch between adjacent elements occurs. These conclusions can easily be generalized to all the elements considered in Chapter 7 and we therefore arrive at the following general conclusions:

If an element behaves in a conforming, i.e. compatible, manner in the parent domain, its isoparametric version also behaves in a conforming way and no mismatch between adjacent elements exists.

As discussed in Chapter 7 (page 93), the convergence requirement is fulfilled if the compatibility and the completeness requirements are satisfied. It was shown above that the compatibility, or conforming, requirement is fulfilled, so it remains to prove that the completeness requirement is satisfied. It is recalled that the completeness requirement is met if an arbitrary constant temperature and an arbitrary constant temperature gradient can be modelled exactly by the approximation.

When a two-dimensional problem is considered, the temperature is approximated by (19.22). The global nodal temperatures given by $\mathbf{a}^e$ can be chosen arbitrarily and we may choose

$$\mathbf{a}^e = \alpha_1 \mathbf{e} + \alpha_2 \mathbf{x}^e + \alpha_3 \mathbf{y}^e \tag{19.27}$$

where $\mathbf{e}$ denotes a column matrix with all components equal to unity, i.e. $\mathbf{e}^T = [1\ 1\ \dots\ 1]$. Inserting (19.27) into (19.22) yields

$$T = \alpha_1 \mathbf{N}^e \mathbf{e} + \alpha_2 \mathbf{N}^e \mathbf{x}^e + \alpha_3 \mathbf{N}^e \mathbf{y}^e \tag{19.28}$$

We have

$$\mathbf{N}^e \mathbf{e} = [N_1^e \quad N_2^e \quad \dots \quad N_{n_e}^e] \begin{bmatrix} 1 \\ 1 \\ \vdots \\ 1 \end{bmatrix} = \sum_{i=1}^{n_e} N_i^e \tag{19.29}$$

where n_e denotes the number of nodal points in the element. From (7.130) it follows that

$$\boxed{\sum_{i=1}^{n} N_i^e = 1} \tag{19.30}$$

Using (19.29), (19.30) and (19.21) in (19.28) gives

$$T = \alpha_1 + \alpha_2 x + \alpha_3 y \tag{19.31}$$

However, this expression shows that an arbitrary constant temperature and an arbitrary constant temperature gradient can be modelled exactly by the approximation of the unknown function T. Hence the completeness criterion is fulfilled. It may be concluded that

> Isoparametric elements fulfil the completeness criterion if (19.30) is satisfied.

All the element shape functions considered in Chapter 7 satisfy (19.30), i.e. the corresponding isoparametric elements fulfil the completeness requirement.

19.6 Integral transformations

19.6.1 *Transformation of integrals over one-, two- and three-dimensional regions*

The determination of the stiffness matrix $\mathbf{K}$ and the load vector $\mathbf{f}_l$ requires that integrations should be performed over the region of interest (cf. for instance the FE formulation of two-dimensional heat flow given by (10.15) and (10.16)). The region of interest is the one given in the global domain. In order to perform these integrations, the pertinent integrands must be given in terms of the global variables, i.e. x, xy or xyz for one-, two- or three-dimensional problems, respectively. However, the shape function matrix $\mathbf{N}$ is given in terms of the local variables and, for example, we have for two-dimensional problems that $\mathbf{N} = \mathbf{N}(\xi, \eta)$. In this case the relations $x = x(\xi, \eta)$ and $y = y(\xi, \eta)$ are known, but explicit expressions are not known for the inverse relations $\xi = \xi(x, y)$ and $\eta = \eta(x, y)$.

Therefore, in order to perform the integrations necessary to obtain the stiffness matrix $\mathbf{K}$ and the load vector $\mathbf{f}_1$, we must transform the integrals from the global to the local variables. This transformation is discussed in the present section, whereas the next section deals with the corresponding transformation for boundary integrals present in the expression for the boundary vector $\mathbf{f}_b$.

We shall therefore first establish the relations between incremental volumes, areas and lengths in the parent and global domains. In order to do so, we shall draw on the classical results of vector algebra concerning the cross-product of two vectors (see for instance Sokolnikoff and Redheffer, 1958).

Consider in the xyz-coordinate system the two vectors $\mathbf{a}$ and $\mathbf{b}$ with the components

$$\mathbf{a} = \begin{bmatrix} a_x \\ a_y \\ a_z \end{bmatrix}; \quad \mathbf{b} = \begin{bmatrix} b_x \\ b_y \\ b_z \end{bmatrix} \tag{19.32}$$

The *cross-product* $\mathbf{a} \times \mathbf{b}$ of these two vectors is defined as the vector with the following components:

$$\mathbf{a} \times \mathbf{b} = \begin{bmatrix} a_y b_z - a_z b_y \\ a_z b_x - a_x b_z \\ a_x b_y - a_y b_x \end{bmatrix} \tag{19.33}$$

It is easy to check that the vector $\mathbf{a} \times \mathbf{b}$ is orthogonal to both $\mathbf{a}$ and $\mathbf{b}$, i.e. $\mathbf{a} \times \mathbf{b}$ is orthogonal to the plane defined by the vectors $\mathbf{a}$ and $\mathbf{b}$; see Figure 19.12. Moreover, from vector algebra it follows that the area A of the parallelogram spanned by $\mathbf{a}$ and $\mathbf{b}$ (cf. Figure 19.12) is given by the length of the vector $\mathbf{a} \times \mathbf{b}$, i.e.

$$A = |\mathbf{a} \times \mathbf{b}| \tag{19.34}$$

Let us now assume that the vectors $\mathbf{a}$ and $\mathbf{b}$ are located in the xy-plane. In this case, $a_z = b_z = 0$ and (19.33) and (19.34) reduce to

$$\mathbf{a} \times \mathbf{b} = \begin{bmatrix} 0 \\ 0 \\ a_x b_y - a_y b_x \end{bmatrix}; \quad \text{i.e. } A = |\mathbf{a} \times \mathbf{b}| = |a_x b_y - a_y b_x|$$

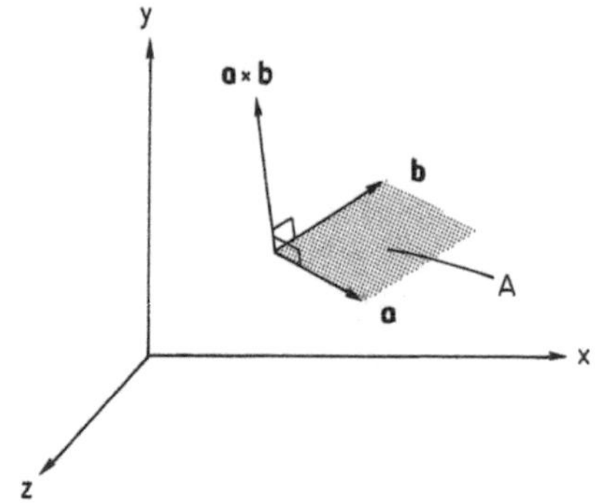

Figure 19.12   Cross-product of vectors **a** and **b**

and this expression for A may be written as

$$A = |\mathbf{a} \times \mathbf{b}| = \left| \det \begin{bmatrix} a_x & b_x \\ a_y & b_y \end{bmatrix} \right| \tag{19.35}$$

Considering two-dimensional elements, we have from (19.2) and (19.3) that

$$\begin{bmatrix} dx \\ dy \end{bmatrix} = \mathbf{J} \begin{bmatrix} d\xi \\ d\eta \end{bmatrix} \quad \text{where} \quad \mathbf{J} = \begin{bmatrix} \dfrac{\partial x}{\partial \xi} & \dfrac{\partial x}{\partial \eta} \\ \dfrac{\partial y}{\partial \xi} & \dfrac{\partial y}{\partial \eta} \end{bmatrix} \tag{19.36}$$

The components of the Jacobian matrix $\mathbf{J}$ are easily derived from (19.21), and we obtain

$$\boxed{\begin{aligned} \frac{\partial x}{\partial \xi} &= \frac{\partial \mathbf{N}^e}{\partial \xi} \mathbf{x}^e; & \frac{\partial x}{\partial \eta} &= \frac{\partial \mathbf{N}^e}{\partial \eta} \mathbf{x}^e \\[2mm] \frac{\partial y}{\partial \xi} &= \frac{\partial \mathbf{N}^e}{\partial \xi} \mathbf{y}^e; & \frac{\partial y}{\partial \eta} &= \frac{\partial \mathbf{N}^e}{\partial \eta} \mathbf{y}^e \end{aligned}} \tag{19.37}$$

We shall now determine the relation between incremental areas in the parent domain and the global domain. Consider two straight lines given by $\xi = C_1$ and $\eta = C_2$ in the parent domain where C_1 and C_2 are arbitrary constants (see Figure 19.13). According to the transformation $x = x(\xi, \eta)$ and $y = y(\xi, \eta)$, these two straight lines map into two lines in the global domain, and these lines will, in general, be curved; cf. Figure 19.13. At the point defined in the figure by the intersection of lines $\xi = C_1$ and $\eta = C_2$, we consider in the parent domain the increment $d\xi$ along $\eta = C_2$ and the increment $d\eta$ along $\xi = C_1$. The mapping (19.36) carries $d\xi$ into the vector **a** and $d\eta$ into the vector **b**. The vector **a** is obtained for $d\xi \neq 0$ and $d\eta = 0$, and the vector

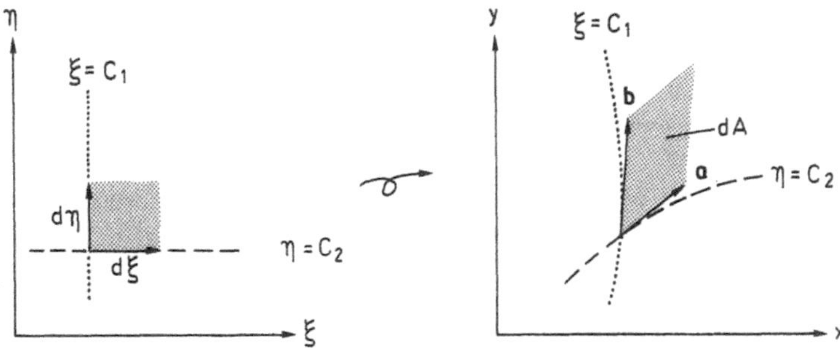

Figure 19.13 Transformation of incremental area in parent domain to global domain

b is obtained for $d\xi = 0$ and $d\eta \neq 0$, i.e. (19.36) yields

$$
\mathbf{a} = \begin{bmatrix} \dfrac{\partial x}{\partial \xi} \\[2mm] \dfrac{\partial y}{\partial \xi} \end{bmatrix} d\xi; \quad
\mathbf{b} = \begin{bmatrix} \dfrac{\partial x}{\partial \eta} \\[2mm] \dfrac{\partial y}{\partial \eta} \end{bmatrix} d\eta
$$

The incremental area dA spanned by the incremental vectors **a** and **b** is given by (19.35), i.e.

$$
dA = d\xi\, d\eta \left\| \begin{bmatrix} \dfrac{\partial x}{\partial \xi} \\[2mm] \dfrac{\partial y}{\partial \xi} \end{bmatrix} \times \begin{bmatrix} \dfrac{\partial x}{\partial \eta} \\[2mm] \dfrac{\partial y}{\partial \eta} \end{bmatrix} \right\| = d\xi\, d\eta \left| \det \begin{bmatrix} \dfrac{\partial x}{\partial \xi} & \dfrac{\partial x}{\partial \eta} \\[2mm] \dfrac{\partial y}{\partial \xi} & \dfrac{\partial y}{\partial \eta} \end{bmatrix} \right| = d\xi\, d\eta\, |\det \mathbf{J}|
$$

According to (19.5), we have that $\det \mathbf{J} > 0$, i.e. the following fundamental result is derived:

$$
\boxed{\; dA = d\xi\, d\eta\, \det \mathbf{J} \;} \tag{19.38}
$$

This result is easily generalized to three dimensions (cf. for instance Zienkiewicz and Taylor, 1989), and we obtain

$$
\boxed{\; dV = d\xi\, d\eta\, d\zeta\, \det \mathbf{J} \;} \tag{19.39}
$$

Here ξ, η and ζ are the coordinates in the parent domain and

$$
\begin{bmatrix} dx \\ dy \\ dz \end{bmatrix} = \mathbf{J} \begin{bmatrix} d\xi \\ d\eta \\ d\zeta \end{bmatrix}
\quad \text{where} \quad
\mathbf{J} = \begin{bmatrix} \dfrac{\partial x}{\partial \xi} & \dfrac{\partial x}{\partial \eta} & \dfrac{\partial x}{\partial \zeta} \\[2mm] \dfrac{\partial y}{\partial \xi} & \dfrac{\partial y}{\partial \eta} & \dfrac{\partial y}{\partial \zeta} \\[2mm] \dfrac{\partial z}{\partial \xi} & \dfrac{\partial z}{\partial \eta} & \dfrac{\partial z}{\partial \zeta} \end{bmatrix} \tag{19.40}
$$

In one dimension, we clearly get

$$\boxed{\mathrm{d}x = \mathrm{d}\xi \det \mathbf{J}}$$ (19.41)

where $\det \mathbf{J} = \mathrm{d}x/\mathrm{d}\xi$.

In the FE method, integrations over regions are in practice performed elementwise, and with the relations (19.38), (19.39) and (19.41), we obtain the following transformation of integrals over one-, two- and three-dimensional elements, respectively:

$$
\begin{aligned}
\int_{L_\alpha} f(x)\,\mathrm{d}x &= \int_{-1}^{1} f(x(\xi))\,\det \mathbf{J}\,\mathrm{d}\xi \\[2mm]
\int_{A_\alpha} f(x,y)\,\mathrm{d}A &= \int_{-1}^{1}\int_{-1}^{1} f(x(\xi,\eta),\,y(\xi,\eta))\,\det \mathbf{J}\,\mathrm{d}\xi\,\mathrm{d}\eta \\[2mm]
\int_{V_\alpha} f(x,y,z)\,\mathrm{d}V &= \int_{-1}^{1}\int_{-1}^{1}\int_{-1}^{1} f(x(\xi,\eta,\zeta),\,y(\xi,\eta,\zeta),\,z(\xi,\eta,\zeta)) \\
&\qquad \times \det \mathbf{J}\,\mathrm{d}\xi\,\mathrm{d}\eta\,\mathrm{d}\zeta
\end{aligned}
$$
   (19.42)

In these relations, f denotes an arbitrary function. Moreover, it is emphasized that L_α, A_α and V_α here refer to the regions occupied by one element in the one-, two- and three-dimensional case, respectively.

19.6.2 *Transformation of boundary integrals*

The transformation formulae given in (19.42) refer to integrations over the region that is occupied by one element, and as such they are applicable when evaluating the stiffness matrix $\mathbf{K}$ and the load vector $\mathbf{f}_l$. However, when determining the boundary vector $\mathbf{f}_b$, boundary integrals need to be evaluated (cf. for instance (10.15) and (10.16)).

For one-dimensional problems, the boundary integrals present in the expression for the boundary vector $\mathbf{f}_b$ reduce to boundary terms (cf. (9.13) and (9.14)). With $f(x)$ being an arbitrary function we clearly have

$$\boxed{\left[\,f(x)\,\right]_a^b = \left[\,f(x(\xi))\,\right]_{-1}^{1}}$$ (19.43)

where (a, b) refer to the region occupied by one element.

For two-dimensional problems (cf. for instance (10.15)), we need to express the incremental length $\mathrm{d}\mathscr{L}$ in terms of the local variables. From (5.17) it follows that

$$\mathrm{d}\mathscr{L} = (\mathrm{d}x^2 + \mathrm{d}y^2)^{1/2} \tag{19.44}$$

and it is recalled that $\mathrm{d}\mathscr{L}$, by definition, is a positive quantity. Evidently, the boundary of the two-dimensional body always coincides with some element boundaries. Along such an element boundary we have either $\mathrm{d}\xi = 0$ or $\mathrm{d}\eta = 0$ (cf. Figure 19.2). Assuming, for instance, that $\mathrm{d}\eta = 0$, (19.36) yields

$$\mathrm{d}x = \frac{\partial x}{\partial \xi}\,\mathrm{d}\xi; \quad \mathrm{d}y = \frac{\partial y}{\partial \xi}\,\mathrm{d}\xi \tag{19.45}$$

Inserting (19.45) into (19.44) gives

$$\mathrm{d}\mathscr{L} = |\mathrm{d}\xi|\left[\left(\frac{\partial x}{\partial \xi}\right)^2 + \left(\frac{\partial y}{\partial \xi}\right)^2\right]^{1/2} \tag{19.46}$$

where $\partial x/\partial \xi$ and $\partial y/\partial \xi$ are determined by (19.37). Moreover, as $\eta = 1$ or $\eta = -1$ along the element boundary considered, (19.46) should be evaluated accordingly. Therefore, letting $f = f(x, y)$ denote an arbitrary function and letting $\mathscr{L}_\alpha$ denote the pertinent boundary of one element, we obtain

$$\boxed{\int_{\mathscr{L}_\alpha} f(x, y)\,\mathrm{d}\mathscr{L} = \int_{-1}^{1} f(x(\xi, \eta), y(\xi, \eta))\left[\left(\frac{\partial x}{\partial \xi}\right)^2 + \left(\frac{\partial y}{\delta \xi}\right)^2\right]^{1/2}\,\mathrm{d}\xi} \tag{19.47}$$

where either $\eta = 1$ or $\eta = -1$. An analogous expression is derived if $\mathrm{d}\xi = 0$ holds along the element boundary considered.

For three-dimensional problems, the incremental area $\mathrm{d}S$ of the boundary surface has to be expressed in terms of the local variables. It is emphasized that, in contrast to $\mathrm{d}A$ given by (19.38), $\mathrm{d}S$ denotes the incremental area of the boundary surface that is located in the xyz-space. This boundary coincides with some element boundaries. Along an element boundary of a three-dimensional element we have $\mathrm{d}\xi = 0$ or $\mathrm{d}\eta = 0$ or $\mathrm{d}\zeta = 0$ (cf. Figure 19.11). Assume, for instance, that $\mathrm{d}\zeta = 0$. As in Figure 19.13, we then consider two straight lines given by $\xi = C_1$ and $\eta = C_2$ in the parent domain where C_1 and C_2 are arbitrary constants. Along $\eta = C_2$, we consider the incremental vector given by $\mathrm{d}\xi \neq 0$ and $\mathrm{d}\eta = 0$ (as well as $\mathrm{d}\zeta = 0$). With (19.40), this incremental vector transforms into the following vector $\mathbf{a}$ in the xyz-space:

$$\mathbf{a} = \begin{bmatrix} \mathrm{d}x \\ \mathrm{d}y \\ \mathrm{d}z \end{bmatrix} = \mathbf{a}_1\,\mathrm{d}\xi \quad \text{where} \quad \mathbf{a}_1 = \begin{bmatrix} \dfrac{\partial x}{\partial \xi} \\[6pt] \dfrac{\partial y}{\partial \xi} \\[6pt] \dfrac{\partial z}{\partial \xi} \end{bmatrix} \tag{19.48}$$

Likewise, the incremental vector given by $d\xi = 0$ and $d\eta \neq 0$ (as well as $d\zeta = 0$) is considered along $\xi = C_1$. Using (19.40), this incremental vector transforms into the following vector $\mathbf{b}$ in the xyz-space:

$$\mathbf{b} = \begin{bmatrix} dx \\ dy \\ dz \end{bmatrix} = \mathbf{b}_1 \, d\eta \quad \text{where} \quad \mathbf{b}_1 = \begin{bmatrix} \dfrac{\partial x}{\partial \eta} \\[2mm] \dfrac{\partial y}{\partial \eta} \\[2mm] \dfrac{\partial z}{\partial \eta} \end{bmatrix} \tag{19.49}$$

We emphasize that the element boundary was assumed to be given by $d\zeta = 0$. More specifically, we have either $\zeta = -1$ or $\zeta = 1$ (cf. Figure 19.11) and the vectors $\mathbf{a}_1$ and $\mathbf{b}_1$, given by (19.48) and (19.49) should be evaluated accordingly.

We are now in a position to derive the result sought. From (19.34), (19.48) and (19.49) it follows that

$$dS = |d\xi| \, |d\eta| \, |\mathbf{a}_1 \times \mathbf{b}_1| \tag{19.50}$$

With $f = f(x, y, z)$ being an arbitrary function, we can then conclude that

$$\boxed{\int_{S_z} f(x, y, z) \, dS = \int_{-1}^{1} \int_{-1}^{1} f(x(\xi, \eta, \zeta), y(\xi, \eta, \zeta), z(\xi, \eta, \zeta)) \\ \times |\mathbf{a}_1 \times \mathbf{b}_1| \, d\xi \, d\eta} \tag{19.51}$$

where either $\zeta = 1$ or $\zeta = -1$ and S refers to the pertinent boundary of one element. Analogous expressions are derived if $d\xi = 0$ or $d\eta = 0$ holds along the element boundary considered.

19.7 Evaluation of FE equations using isoparametric elements

The use of isoparametric elements implies that in order to evaluate the FE equations, some slight reformulations are needed. To see this, consider as an example the problem of two-dimensional heat flow. From (10.39)–(10.42) we have

$$\boxed{\mathbf{K}^e \mathbf{a}^e = \mathbf{f}_b^e + \mathbf{f}_l^e} \tag{19.52}$$

where

$$\mathbf{K}^e = \int_{A_\alpha} \mathbf{B}^{eT}\mathbf{D}\mathbf{B}^e t \, dA$$

$$\mathbf{f}_b^e = -\int_{\mathscr{L}_{h\alpha}} \mathbf{N}^{eT}ht \, d\mathscr{L} - \int_{\mathscr{L}_{g\alpha}} \mathbf{N}^{eT}q_n t \, d\mathscr{L} \qquad (19.53)$$

$$\mathbf{f}_l^e = -\int_{A_\alpha} \mathbf{N}^{eT}Qt \, dA$$

and

$$\mathbf{B}^e = \begin{bmatrix} \dfrac{\partial \mathbf{N}^e}{\partial x} \\[2ex] \dfrac{\partial \mathbf{N}^e}{\partial y} \end{bmatrix} \qquad (19.54)$$

The element boundary vector $\mathbf{f}_b^e$ and the element load vector $\mathbf{f}_l^e$ present no problems. The element shape function matrix $\mathbf{N}^e$ is given in terms of ξ and η, and the remaining quantities entering the integrands of $\mathbf{f}_b^e$ and $\mathbf{f}_l^e$ can also be expressed in the local variables ξ and η via the relations $x = x(\xi, \eta)$ and $y = y(\xi, \eta)$. When the entire integrands of $\mathbf{f}_b^e$ and $\mathbf{f}_l^e$ have been expressed in terms of ξ and η, the integration limits are changed in accordance with (19.47) and (19.42).

However, the evaluation of the element stiffness matrix $\mathbf{K}^e$ presents a new problem. The point is that the $\mathbf{B}^e$-matrix is obtained by differentiating $\mathbf{N}^e$ with respect to x and y (cf. (19.54)), but for an isoparametric formulation, the $\mathbf{N}^e$-matrix is given in terms of the ξ, η-coordinates and not the x, y-coordinates.

In order to determine the derivatives of $\mathbf{N}^e$ as given by (19.54), we first differentiate $\mathbf{N}^e$ with respect to ξ and η. It follows that

$$\begin{bmatrix} \dfrac{\partial \mathbf{N}^e}{\partial \xi} \\[2ex] \dfrac{\partial \mathbf{N}^e}{\partial \eta} \end{bmatrix} = \begin{bmatrix} \dfrac{\partial \mathbf{N}^e}{\partial x}\dfrac{\partial x}{\partial \xi} + \dfrac{\partial \mathbf{N}^e}{\partial y}\dfrac{\partial y}{\partial \xi} \\[2ex] \dfrac{\partial \mathbf{N}^e}{\partial x}\dfrac{\partial x}{\partial \eta} + \dfrac{\partial \mathbf{N}^e}{\partial y}\dfrac{\partial y}{\partial \eta} \end{bmatrix} = \begin{bmatrix} \dfrac{\partial x}{\partial \xi} & \dfrac{\partial y}{\partial \xi} \\[2ex] \dfrac{\partial x}{\partial \eta} & \dfrac{\partial y}{\partial \eta} \end{bmatrix} \begin{bmatrix} \dfrac{\partial \mathbf{N}^e}{\partial x} \\[2ex] \dfrac{\partial \mathbf{N}^e}{\partial y} \end{bmatrix}$$

Use of (19.3) gives

$$\begin{bmatrix} \dfrac{\partial \mathbf{N}^e}{\partial \xi} \\[2ex] \dfrac{\partial \mathbf{N}^e}{\partial \eta} \end{bmatrix} = \mathbf{J}^T \begin{bmatrix} \dfrac{\partial \mathbf{N}^e}{\partial x} \\[2ex] \dfrac{\partial \mathbf{N}^e}{\partial y} \end{bmatrix} \qquad (19.55)$$

i.e.

$$
\mathbf{B}^e = \begin{bmatrix} \dfrac{\partial \mathbf{N}^e}{\partial x} \\[2ex] \dfrac{\partial \mathbf{N}^e}{\partial y} \end{bmatrix} = (\mathbf{J}^{\mathrm{T}})^{-1} \begin{bmatrix} \dfrac{\partial \mathbf{N}^e}{\partial \xi} \\[2ex] \dfrac{\partial \mathbf{N}^e}{\partial y} \end{bmatrix}
\tag{19.56}
$$

Using this expression in (19.53) and changing the integration limits in accordance with (19.42) yield

$$
\mathbf{K}^e = \int_{-1}^{1} \int_{-1}^{1} \left[\dfrac{\partial \mathbf{N}^{e\mathrm{T}}}{\partial \xi} \quad \dfrac{\partial \mathbf{N}^{e\mathrm{T}}}{\partial \eta} \right] \mathbf{J}^{-1} \mathbf{D} (\mathbf{J}^{\mathrm{T}})^{-1} \begin{bmatrix} \dfrac{\partial \mathbf{N}^e}{\partial \xi} \\[2ex] \dfrac{\partial \mathbf{N}^e}{\partial \eta} \end{bmatrix} t \det \mathbf{J} \, d\xi \, d\eta
\tag{19.57}
$$

The constitutive matrix $\mathbf{D}$ and the thickness t are in general functions of x and y. Using $x = x(\xi, \eta)$ and $y = y(\xi, \eta)$, both $\mathbf{D}$ and t can be expressed directly in terms of ξ and η and the integration of (19.57) can then be carried out.

It is evident that if the Jacobian matrix $\mathbf{J}$ is not a constant matrix, the inverse matrix $\mathbf{J}^{-1}$ depends in a highly complicated manner on ξ and η. Consequently, an exact analytical integration of (19.57) is in general not possible and instead one has to resort to approximate integration techniques, the so-called *numerical integrations*. Such techniques are discussed in detail in the next chapter.

For elasticity problems, the expression for the $\mathbf{B}^e$-matrix is considerably more complicated than (19.56) and this implies also that the expression for the element stiffness matrix $\mathbf{K}^e$ becomes complex. To see this, consider for instance two-dimensional elasticity for which we have

$$
\mathbf{B}^e = \begin{bmatrix} \dfrac{\partial N_1^e}{\partial x} & 0 & \dfrac{\partial N_2^e}{\partial x} & 0 & \cdots & \dfrac{\partial N_{n_e}^e}{\partial x} & 0 \\[2ex] 0 & \dfrac{\partial N_1^e}{\partial y} & 0 & \dfrac{\partial N_2^e}{\partial y} & \cdots & 0 & \dfrac{\partial N_{n_e}^e}{\partial y} \\[2ex] \dfrac{\partial N_1^e}{\partial y} & \dfrac{\partial N_1^e}{\partial x} & \dfrac{\partial N_2^e}{\partial y} & \dfrac{\partial N_2^e}{\partial x} & \cdots & \dfrac{\partial N_{n_e}^e}{\partial y} & \dfrac{\partial N_{n_e}^e}{\partial x} \end{bmatrix}
\tag{19.58}
$$

where n_e is the number of nodal points for the element (cf. (15.11)). By analogy with (19.55), we find that

$$
\begin{bmatrix} \dfrac{\partial N_i^e}{\partial x} \\[2ex] \dfrac{\partial N_i^e}{\partial y} \end{bmatrix} = (\mathbf{J}^{\mathrm{T}})^{-1} \begin{bmatrix} \dfrac{\partial N_i^e}{\partial \xi} \\[2ex] \dfrac{\partial N_i^e}{\partial \eta} \end{bmatrix}
\tag{19.59}
$$

Use of these values for $\partial N_i^e/\partial x$ and $\partial N_i^e/\partial y$ in (19.58) means that all the components of the $\mathbf{B}^e$-matrix can be determined. However, contrary to the heat flow problem (cf. (19.56)), it is not possible to derive a simple expression for $\mathbf{B}^e$ and, consequently, a simple expression for the element stiffness matrix $\mathbf{K}^e$ – similar to (19.57) – cannot be established.

19.7.1 *Example*

In order to illustrate the mapping technique for isoparametric elements, we shall consider a special mapping by which the Jacobian matrix $\mathbf{J}$ becomes a constant matrix.

For this purpose we evaluate the four-node isoparametric element when it takes the form of an arbitrarily located parallelogram; see Figure 19.14. In this case we have the conditions

$$x_2 - x_1 = x_3 - x_4; \quad y_2 - y_1 = y_3 - y_4$$

Therefore, since $x_4 = x_1 - x_2 + x_3$ and $y_4 = y_1 - y_2 + y_3$, it follows from (19.6) and (19.9) that

$$x = -\tfrac{1}{2}(\xi - 1)x_1 + \tfrac{1}{2}(\xi - \eta)x_2 + \tfrac{1}{2}(\eta + 1)x_3$$

$$y = -\tfrac{1}{2}(\xi - 1)y_1 + \tfrac{1}{2}(\xi - \eta)y_2 + \tfrac{1}{2}(\eta + 1)y_3$$

The Jacobian matrix $\mathbf{J}$ is given by (19.3), and we obtain

$$\mathbf{J} = \begin{bmatrix} \tfrac{1}{2}(x_2 - x_1) & \tfrac{1}{2}(x_3 - x_2) \\ \tfrac{1}{2}(y_2 - y_1) & \tfrac{1}{2}(y_3 - y_2) \end{bmatrix} \tag{19.60}$$

It appears that $\mathbf{J}$ is a constant matrix, and it is easily shown that

$$\det \mathbf{J} = \frac{A}{4}$$

where A is the area of the parallelogram (cf. Figure 19.14). This result is in accordance

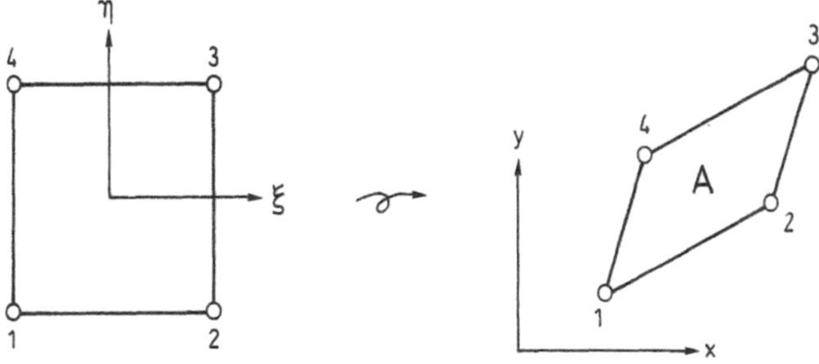

Figure 19.14   Four-node isoparametric element in the form of an arbitrary parallelogram

with (19.38), which, when det $\mathbf{J}$ is a constant, yields

$$A = \det \mathbf{J} \int_{-1}^{1} \int_{-1}^{1} \mathrm{d}\xi \, \mathrm{d}\eta = 4 \det \mathbf{J}$$

However, the important point is that when $\mathbf{J}$ is a constant matrix, the integrations necessary to obtain the FE equations for the four-node isoparametric element are precisely those integrations necessary to obtain the FE equations for the four-node element treated in Chapter 7 (cf. Figure 7.27). To show this, consider the element stiffness matrix $\mathbf{K}^e$ for two-dimensional heat flow. If the element formulation of Figure 7.27 is adopted, we obtain from (19.53)

$$\mathbf{K}^e = \int_{y_1}^{y_4} \int_{x_1}^{x_2} \left[\frac{\partial \mathbf{N}^{e\mathrm{T}}}{\partial x} \ \frac{\partial \mathbf{N}^{e\mathrm{T}}}{\partial y} \right] \mathbf{D} \begin{bmatrix} \dfrac{\partial \mathbf{N}^e}{\partial x} \\[2mm] \dfrac{\partial \mathbf{N}^e}{\partial y} \end{bmatrix} t \, \mathrm{d}x \, \mathrm{d}y \tag{19.61}$$

where the expression for $\mathbf{B}^e$ given by (19.54) has been used. In the present case where $\mathbf{J}$ is a constant matrix, a comparison of (19.61) with (19.57) reveals that the integrations of the two expressions are completely similar. This means that exact analytical integrations can be carried out also for the isoparametric formulation. However, when the four-node isoparametric element differs from the parallelogram shown in Figure 19.14, the Jacobian matrix $\mathbf{J}$ is not a constant matrix and the inverse matrix $\mathbf{J}^{-1}$ will therefore take a highly complex form. This implies that expression (19.57) for the element stiffness matrix $\mathbf{K}^e$ is not amenable to exact analytical integration.

It is observed that the arbitrarily located parallelogram treated above may take the simplified form of a rectangle.

19.8 Need for numerical integrations

The example just investigated provided the information that, apart from some simple geometrical configurations of isoparametric elements, the Jacobian matrix $\mathbf{J}$ and thus $\mathbf{J}^{-1}$ are functions of the local variables. This means that the integrations required to achieve the FE equations become so complex that exact analytical integration cannot be obtained. Therefore, it is necessary to perform this integration in an approximate manner, i.e. *numerical integration techniques* are called for. This is the subject of the next chapter.

Finally, we may refer the reader to the textbooks by Bathe (1982), Hughes (1987) and Zienkiewicz and Taylor (1989) for further information on isoparametric elements. In these books, the isoparametric formulation of triangular and tetrahedral elements is also treated.

20

Numerical integration

We have seen that the use of isoparametric elements forces us to perform the integrations required to obtain the FE equations in an approximate manner, i.e. by *numerical integration*. However, the existence of isoparametric elements is not the only motivation for the use of numerical integration techniques. Consider, as an example, an inhomogeneous material for which the constitutive matrix $\mathbf{D}$ depends on the coordinates, i.e. $\mathbf{D} = \mathbf{D}(x, y, z)$. If this expression is complex, then even when the elements of Chapter 7 are used, it may not be possible to perform the necessary integrations in an exact analytical manner. In addition, even though an exact analytical integration may be possible, it may be so complicated that it hampers the establishment of an efficient FE program. Surprisingly, we shall also see that, while an approximation is certainly related to numerical integration, this approximation *may*, in fact, improve the FE results.

In the following, we will first present some basic facts on numerical integration, and for a more comprehensive treatment, the reader is referred to Dahlquist and Björck (1974), Fröberg (1965) and Press *et al.* (1986). Then we will discuss various aspects of numerical integration in relation to isoparametric elements.

In order to derive suitable numerical integration techniques, it suffices to consider the problem

$$I = \int_{-1}^{1} f(\xi)\, d\xi \tag{20.1}$$

where $f(\xi)$ is an arbitrary function and I is the quantity to be determined. The problem of solving (20.1) can be viewed as determining the area below the $f(\xi)$-function, and methods of solving (20.1) are therefore also termed *quadrature formulae* ('quadrature' means 'area'); see Figure 20.1(a).

Quite generally, we may select some points ξ_i in the interval $-1 \leq \xi \leq 1$, and these points are termed *integration points*. We then have

$$I = \sum_{i=1}^{n} f(\xi_i) H_i + R \tag{20.2}$$

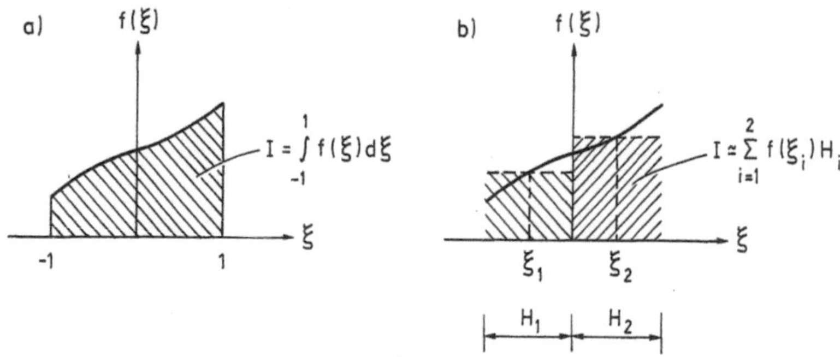

Figure 20.1   (a) Exact integration; (b) numerical integration

where n is the number of integration points, H_i denotes some parameters – or *weights* – related to each integration point and R is the so-called *remainder*. The objective is to make the remainder R as small as possible so that we may write, with close accuracy,

$$I \simeq \sum_{i=1}^{n} f(\xi_i) H_i \tag{20.3}$$

as illustrated in Figure 20.1(b).

The extension to multiple integrals is straightforward, and to illustrate this we consider

$$I = \int_{-1}^{1} \int_{-1}^{1} f(\xi, \eta)\, d\xi\, d\eta \tag{20.4}$$

It follows that

$$I \simeq \int_{-1}^{1} \left(\sum_{i=1}^{n} f(\xi_i, \eta) H_i \right) d\eta \tag{20.5}$$

where n is the number of integration points in the ζ-direction. Defining the function $g(\eta)$ by

$$g(\eta) = \sum_{i=1}^{n} f(\xi_i, \eta) H_i \tag{20.6}$$

(20.5) may be written as

$$I \simeq \int_{-1}^{1} g(\eta)\, d\eta$$

i.e.

$$I \simeq \sum_{j=1}^{m} g(\eta_j) H_j \tag{20.7}$$

where m is the number of integration points in the η-direction. It is emphasized that the number of integration points in the ξ- and η-directions may differ, i.e. we have in general that $n \neq m$. Combining (20.6) and (20.7) yields

$$I \simeq \sum_{j=1}^{m} \sum_{i=1}^{n} f(\xi_i, \eta_j) H_i H_j \tag{20.8}$$

This expression clearly demonstrates that we can focus our interest on the one-dimensional problem given by (20.3).

20.1 Newton–Cotes quadrature

The most obvious numerical integration is obtained by *a priori* selecting the position of the integration points. Moreover, in order to carry out the integration it is natural to approximate the function $f(\xi)$ by an easily integrated function that in practice takes the form of a polynomial.

As the function $f(\xi)$ is evaluated at the integration points ξ_i, the approximating polynomial is conveniently constructed using the Lagrange interpolation function $l_i^{n-1}(\xi)$; cf. (7.78) and Figure 7.18. With n denoting the number of integration points, the Lagrange polynomial $l_i^{n-1}(\xi)$ is given by

$$l_i^{n-1}(\xi) = \frac{(\xi - \xi_1)(\xi - \xi_2)\cdots(\xi - \xi_{i-1})(\xi - \xi_{i+1})\cdots(\xi - \xi_n)}{(\xi_i - \xi_1)(\xi_i - \xi_2)\cdots(\xi_i - \xi_{i-1})(\xi_i - \xi_{i+1})\cdots(\xi_i - \xi_n)} \tag{20.9}$$

and it appears that $l_i^{n-1}(\xi)$ is a polynomial of the order $n - 1$. In accordance with Figure 7.18, it is easily checked that $l_i^{n-1}(\xi_i) = 1$ and $l_i^{n-1}(\xi_j) = 0$ for $j \neq i$. Therefore, we approximate $f(\xi)$ by

$$f(\xi) \simeq \sum_{i=1}^{n} l_i^{n-1}(\xi) f(\xi_i) \tag{20.10}$$

which implies that the approximating polynomial takes the value $f(\xi_j)$ for $\xi = \xi_j$.

It follows that

$$I = \int_{-1}^{1} f(\xi)\, d\xi \simeq \int_{-1}^{1} \left(\sum_{i=1}^{n} l_i^{n-1}(\xi) f(\xi_i) \right) d\xi = \sum_{i=1}^{n} \left(f(\xi_i) \int_{-1}^{1} l_i^{n-1}(\xi)\, d\xi \right) \tag{20.11}$$

A comparison with (20.3) shows that the weights H_i are given by

$$H_i = \int_{-1}^{1} l_i^{n-1}(\xi)\, d\xi \tag{20.12}$$

We have now arrived at the *Newton–Cotes integration formula*. It is obvious that

> For n integration points, Newton–Cotes integration provides an exact integration of a polynomial of the order $n - 1$.

The weights determined by (20.12) are given in many textbooks; see for instance Fröberg (1965). Let us consider two simple examples in order to illustrate the procedure.

20.1.1 *Example 1*

Assume that two integration points ξ_1 and ξ_2 are chosen. We therefore have $n = 2$, and (20.9) provides

$$l_1^{n-1}(\xi) = l_1^1(\xi) = \frac{\xi - \xi_2}{\xi_1 - \xi_2}; \quad l_2^{n-1}(\xi) = l_2^1(\xi) = \frac{\xi - \xi_1}{\xi_2 - \xi_1}$$

Moreover, we choose the position of the integration points as

$$\xi_1 = -1; \quad \xi_2 = 1$$

From (20.12) we then obtain

$$H_1 = H_2 = 1$$

i.e.

$$\int_{-1}^{1} f(\xi)\,d\xi \simeq f(\xi_1)H_1 + f(\xi_2)H_2 = f(-1) \times 1 + f(1) \times 1$$

This result is illustrated in Figure 20.2(a). Moreover, the above result may be written as

$$\int_{-1}^{1} f(\xi)\,d\xi \simeq 2\tfrac{1}{2}[f(-1) + f(1)]$$

which shows that we have arrived at the well-known *trapezoidal rule*; cf. Figure 20.2(b).

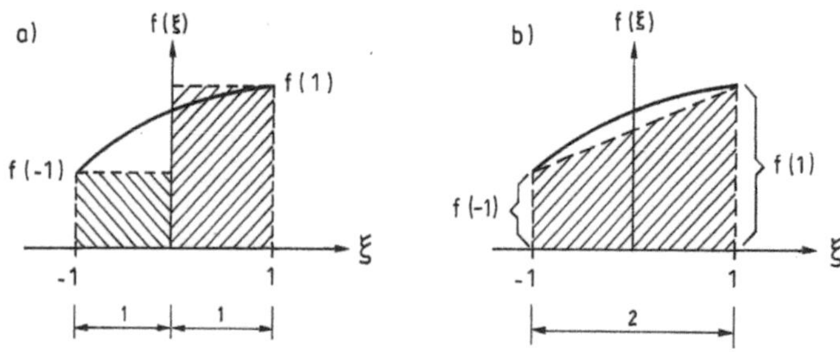

Figure 20.2   (a) Newton–Cotes integration with two integration points; (b) equivalence with trapezoidal rule

20.1.2 *Example 2*

Assume that three integration points ξ_1, ξ_2 and ξ_3 are chosen. We therefore have $n = 3$, and (20.9) provides

$$l_1^{n-1}(\xi) = l_1^2(\xi) = \frac{(\xi - \xi_2)(\xi - \xi_3)}{(\xi_1 - \xi_2)(\xi_1 - \xi_3)}$$

$$l_2^{n-1}(\xi) = l_2^2(\xi) = \frac{(\xi - \xi_1)(\xi - \xi_3)}{(\xi_2 - \xi_1)(\xi_2 - \xi_3)}$$

$$l_3^{n-1}(\xi) = l_3^2(\xi) = \frac{(\xi - \xi_1)(\xi - \xi_2)}{(\xi_3 - \xi_1)(\xi_3 - \xi_2)}$$

Choosing the position of the integration points as

$$\xi_1 = -1; \quad \xi_2 = 0; \quad \xi_3 = 1$$

we obtain from (20.12) that

$$H_1 = \tfrac{1}{3}; \quad H_2 = \tfrac{4}{3}; \quad H_3 = \tfrac{1}{3}$$

This implies that

$$\int_{-1}^{1} f(\xi)\,d\xi \simeq f(-1) \times \tfrac{1}{3} + f(0) \times \tfrac{4}{3} + f(1) \times \tfrac{1}{3}$$

which is the well-known *Simpson's formula*; cf. Figure 20.3.

20.2 Gauss integration

In Newton–Cotes integration, the position of the integration points was chosen *a priori*, and the corresponding weights were then determined so that a polynomial of the order $n - 1$ could be integrated exactly by using n integration points.

Another appealing strategy arises if not only the weights, but also the *positions*

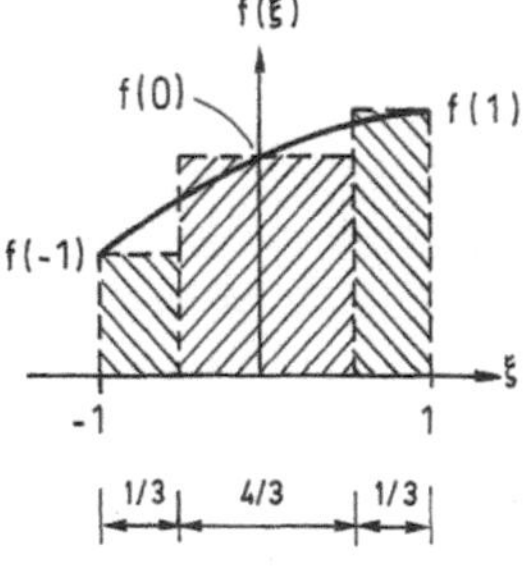

Figure 20.3   Newton–Cotes integration with three integration points; Simpson's formula

of the integration points are determined so that a given polynomial is integrated exactly. This approach leads to the *Gauss integration scheme* (occasionally termed *Gauss–Legendre integration*), and the positions of the integration points derived are termed *Gauss points*.

For this purpose consider the following polynomial of order $2n - 1$:

$$g(\xi) = \alpha_1 + \alpha_2 \xi + \alpha_3 \xi^2 + \cdots + \alpha_{2n-1} \xi^{2n-2} + \alpha_{2n} \xi^{2n-1} \tag{20.13}$$

We find that this polynomial contains $2n$ terms. Integration gives

$$I = \int_{-1}^{1} g(\xi)\,\mathrm{d}\xi = 2\alpha_1 + \frac{2}{3}\alpha_3 + \cdots + \frac{2}{2n-1}\alpha_{2n-1} \tag{20.14}$$

Assume that n integration points are adopted. The general integration scheme (20.3) then provides

$$I = \int_{-1}^{1} g(\xi)\,\mathrm{d}\xi = \sum_{i=1}^{n} g(\xi_i) H_i \tag{20.15}$$

With $g(\xi)$ given by (20.13), we obtain

$$I = \alpha_1 \sum_{i=1}^{n} H_i + \alpha_2 \sum_{i=1}^{n} \xi_i H_i + \alpha_3 \sum_{i=1}^{n} \xi_i^2 H_i + \cdots$$

$$+ \alpha_{2n-1} \sum_{i=1}^{n} \xi_i^{2n-2} H_i + \alpha_{2n} \sum_{i=1}^{n} \xi_i^{2n-1} H_i$$

A comparison with (20.14) implies the following conditions:

$$\boxed{\begin{array}{l} \displaystyle \sum_{i=1}^{n} H_i = 2; \quad \sum_{i=1}^{n} \xi_i H_i = 0; \quad \sum_{i=1}^{n} \xi_i^2 H_i = \frac{2}{3}; \quad \cdots; \\[2em] \displaystyle \sum_{i=1}^{n} \xi_i^{2n-2} H_i = \frac{2}{2n-1}; \quad \sum_{i=1}^{n} \xi_i^{2n-1} H_i = 0 \end{array}} \tag{20.16}$$

As (20.13) contains $2n$ terms, (20.16) provides $2n$ conditions. Moreover, in (20.15) we have $2n$ unknowns, namely n positions ξ_i and n weights H_i. That is, the conditions given by (20.16) provide sufficient information for the identification of the Gauss points ξ_i and the weights H_i.

We therefore conclude that

> For n integration points, Gauss integration provides an exact integration of a polynomial of the order $2n - 1$.

It follows that for a given number of integration points, Gauss integration provides the exact integration of a polynomial of a higher order than that obtained by Newton–Cotes integration. In practice, therefore, Gauss integration is used almost exclusively within isoparametric FE formulations.

Table 20.1 Positions of Gauss points ξ_i and corresponding weights H_i
$\int_{-1}^{1} f(\xi)\,\mathrm{d}\xi = \Sigma_{i=1}^{n} f(\xi_i)\,H_i$

ξ_i	H_i
	$n = 1$
0.000 000 000 000 000	2.000 000 000 000 000
	$n = 2$
$\pm 0.577\,350\,269\,189\,626$	1.000 000 000 000 000
	$n = 3$
0.000 000 000 000 000	0.888 888 888 888 889
$\pm 0.774\,596\,669\,241\,483$	0.555 555 555 555 556
	$n = 4$
$\pm 0.339\,981\,043\,584\,856$	0.652 145 154 862 546
$\pm 0.861\,136\,311\,594\,053$	0.347 854 845 137 454
	$n = 5$
0.000 000 000 000 000	0.568 888 888 888 889
$\pm 0.538\,469\,310\,105\,683$	0.478 628 670 499 366
$\pm 0.906\,179\,845\,938\,664$	0.236 926 885 056 189
	$n = 6$
$\pm 0.238\,619\,186\,083\,197$	0.467 913 934 572 691
$\pm 0.661\,209\,386\,466\,265$	0.360 761 573 048 139
$\pm 0.932\,469\,514\,203\,152$	0.171 324 492 379 170

The positions of the Gauss points and the corresponding weights are tabulated in many textbooks on numerical analysis and FE techniques. Some useful results taken from Zienkiewicz and Taylor (1989) are shown in Table 20.1.

The following two examples illustrate how the weights and the positions of the Gauss points can be determined.

20.2.1 *Example 3*

Assume that two Gauss points are adopted, i.e. $n = 2$. The conditions in (20.16) then reduce to

$$H_1 + H_2 = 2; \quad \xi_1 H_1 + \xi_2 H_2 = 0; \quad \xi_1^2 H_1 + \xi_2^2 H_2 = \tfrac{2}{3}; \quad \xi_1^3 H_1 + \xi_2^3 H_2 = 0$$

We could solve these equations directly, but there is no loss of generality in observing that symmetry reasons imply that $H_1 = H_2$ and $\xi_1 = -\xi_2$. With this observation the conditions above can easily be solved to obtain

$$H_1 = H_2 = 1; \quad \xi_1 = -1/\sqrt{3}; \quad \xi_2 = 1/\sqrt{3}$$

in accordance with Table 20.1.

20.2.2 *Example 4*

Consider next the integration scheme when three Gauss points are used, i.e. $n = 3$. From (20.16) it follows that

$$H_1 + H_2 + H_3 = 2; \quad \xi_1 H_1 + \xi_2 H_2 + \xi_3 H_3 = 0;$$

$$\xi_1^2 H_1 + \xi_2^2 H_2 + \xi_3^2 H_3 = \tfrac{2}{3}$$

$$\xi_1^3 H_1 + \xi_2^3 H_2 + \xi_3^3 H_3 = 0; \quad \xi_1^4 H_1 + \xi_2^4 H_2 + \xi_3^4 H_3 = \tfrac{2}{5};$$

$$\xi_1^5 H_1 + \xi_2^5 H_2 + \xi_3^5 H_3 = 0$$

For reasons of symmetry, we have $\xi_1 = -\xi_3, \xi_2 = 0$ and $H_1 = H_3$. It then follows that

$$\xi_1 = -\sqrt{\tfrac{3}{5}}; \quad \xi_2 = 0; \quad \xi_3 = \sqrt{\tfrac{3}{5}}; \quad \text{and} \quad H_1 = \tfrac{5}{9}; \quad H_2 = \tfrac{8}{9}; \quad H_3 = \tfrac{5}{9}$$

in accordance with Table 20.1.

20.3 Introductory remarks on the order of Gauss integration of isoparametric elements

With the choice of the Gauss method for integrating isoparametric finite elements, one question remains: namely, the number of Gauss points, i.e. the *order of Gauss integration*. The number of Gauss points is usually the same in the directions of the local, ξ-, η- and ζ-axes, but apart from that no precise answer can be given.

However, in order to provide some guidance on this delicate problem, we may ask whether situations exist where the numerical integration results in an exact integration. For this purpose consider Figure 19.14. In this figure, the four-node isoparametric element takes the form of an arbitrarily located parallelogram. As a consequence, it was shown that the Jacobian matrix $\mathbf{J}$ becomes a constant matrix (cf. (19.60)). According to (19.56) and (19.6) the components of the $\mathbf{B}^e$-matrix are then linear functions of ξ and η. Moreover, if the constitutive matrix $\mathbf{D}$ and the thickness t are assumed to be constants, it follows from (19.57) that the element stiffness matrix $\mathbf{K}^e$ is obtained as integrations of polynomials of the order of two. Consequently, when the four-node isoparametric element takes the form of an arbitrarily located parallelogram, a so-called *2 × 2 point integration* (i.e. two Gauss points in the ξ-direction and two Gauss points in the η-direction) implies that the numerical integration provides the exact result.

Considering the eight-node isoparametric element, it can easily be shown that when it takes the form of a rectangle, the Jacobian matrix $\mathbf{J}$ becomes constant. It follows that the $\mathbf{B}^e$-matrix contains quadratic polynomials in ξ and η, and in order to obtain the element stiffness matrix $\mathbf{K}^e$, integrations have to be performed of polynomials of the fourth order in ξ and η. Therefore, a 3 × 3 point integration suffices to provide the exact integration result.

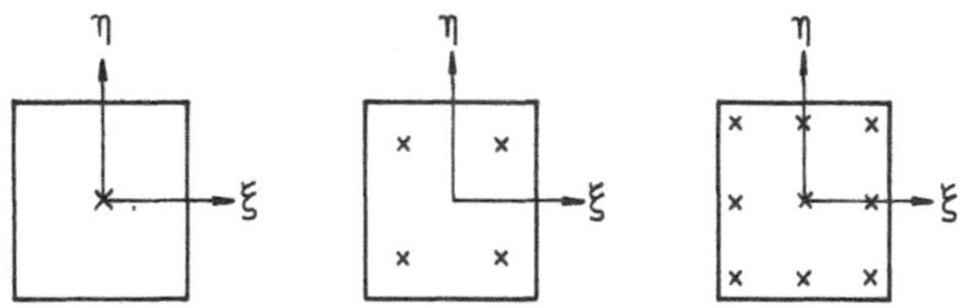

Figure 20.4   Locations of Gauss points for 1×1, 2×2 and 3×3 point integration in parent domain

Based on the results of Table 20.1 and with crosses indicating the positions of the Gauss points, the 1×1, 2×2 and 3×3 point integrations are illustrated in the parent domain in Figure 20.4. These locations apply whether the isoparametric element is distorted or not. However, it should be realized that the Jacobian matrix $\mathbf{J}$ is not constant when the isoparametric elements are distorted. In this situation, the establishment of the $\mathbf{K}^e$-matrix implies integrations of complex functions of ξ and η that are *not* integrated exactly when a 2×2 and a 3×3 point integration are adopted for the four-node and eight-node isoparametric elements, respectively. This suggests that when we use isoparametric elements, it is advantageous to adopt an FE mesh that distorts the elements as little as possible.

In general, it appears that numerical integration introduces an additional approximation into the FE method, and therefore we may expect a high order of integration to be preferable. In practice, this is not the case, and the reason is two-fold: (a) numerical integration requires a substantial computational effort that evidently increases with the order of integration; and (b) the approximation related to numerical integration may, in fact, *improve* the FE results, which suggests that a relatively low order of integration should be adopted.

This latter point is somewhat surprising, but it hinges on the fact that exact integration implies that the stiffness of the body is overestimated when compared with the true stiffness of the body, as explained below. Moreover, when the numerical integration does not provide the exact result, the corresponding approximation tends to compensate for the otherwise too large stiffness. Hence, inaccurate numerical integration may provide a more accurate FE solution.

It is emphasized that such inaccurate integration may also involve certain pitfalls in terms of so-called *spurious zero-energy modes*, which we shall consider in the following section. It turns out that the only sure way to avoid such pitfalls is to adopt an integration of a sufficiently high order.

We shall now give a heuristic proof that the stiffness of the body is overestimated by the FE method when exact integration is used. A concise proof may be found, for instance, in Hughes (1987, p. 187). We consider an elastic body, and we note that the FE solution permits only those displacements that can be represented by the adopted shape functions. In general, the exact displacement field will differ from the assumed displacements. Effectively, the assumed displacement field imposes constraints that prevent the structure from deforming the way it wants to. That is, the FE method provides constraints on the displacements and, in effect, the FE approach creates a substitute structure that is stiffer than the real one. Similar arguments hold for heat flow in a body. We conclude that

> Exact integrations provide a structure that is too stiff.

20.4 Reduced integration and spurious zero-energy modes

When the Jacobian matrix $\mathbf{J}$ is constant and when the order of integration is such that it yields the exact result, we speak of *full integration*. Examples are 2×2 and 3×3 integration for the undistorted four-node and eight-node isoparametric elements, respectively. *Reduced integration* means that the order of integration is lower than that of full integration.

Before investigating the implications of reduced integration, we note from (13.10) that the strain energy U of a three-dimensional elastic body is given by

$$U = \int_V W \, dV = \tfrac{1}{2} \int_V \boldsymbol{\varepsilon}^{\mathrm{T}} \mathbf{D} \boldsymbol{\varepsilon} \, dV \tag{20.17}$$

where W is the strain energy per unit volume. With

$$\boldsymbol{\varepsilon} = \mathbf{B}\mathbf{a} \quad \text{and} \quad \mathbf{K} = \int_V \mathbf{B}^{\mathrm{T}} \mathbf{D} \mathbf{B}^{\mathrm{T}} \, dV$$

it follows that the strain energy of the body can be written as

$$U = \tfrac{1}{2} \mathbf{a}^{\mathrm{T}} \mathbf{K} \mathbf{a} \tag{20.18}$$

The same expression can easily be shown to apply also to one- and two-dimensional bodies.

As the stiffness matrix $\mathbf{K}$ is positive semi-definite (cf. (16.30)), it follows that

$$U \geq 0 \tag{20.19}$$

where the equality sign holds only for rigid-body motions; cf. the discussion of (16.30). The rigid-body motions of a rectangular element are illustrated in Figure 20.5, and the corresponding displacement modes are termed *zero-energy modes*, since no elastic energy is created by these modes. It is emphasized that the above observation – that $U = 0$ holds if, and only if, rigid-body motions are considered – is only true when the stiffness matrix is derived by exact integrations. When reduced integration is adopted, it will turn out that other displacement modes than those corresponding to

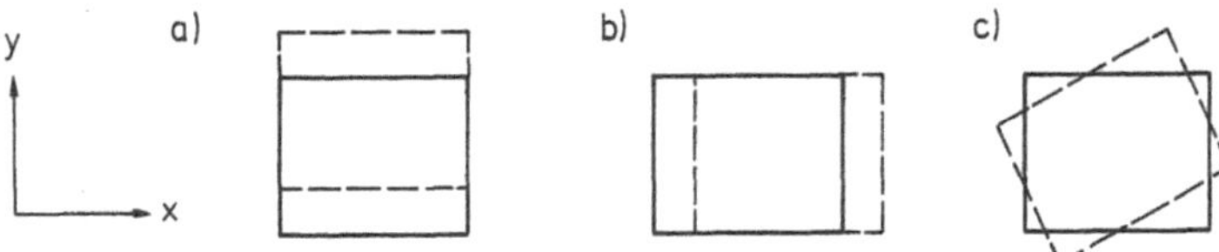

Figure 20.5 Rigid-body motions: (a) translation in the direction of the *y*-axis; (b) translation in the direction of the *x*-axis; (c) rotation

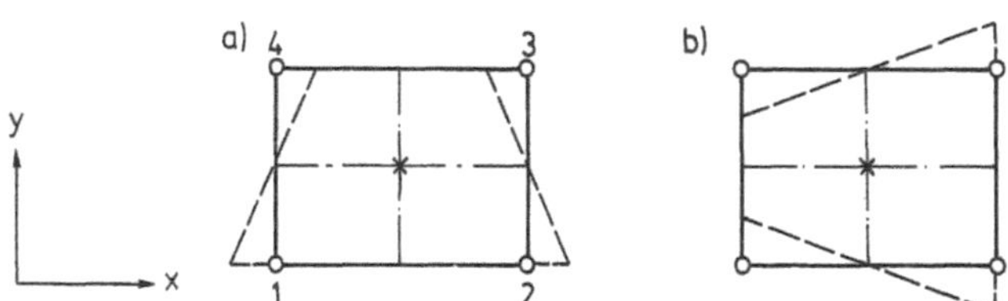

Figure 20.6 Spurious zero-energy modes for four-node isoparametric element with 1×1 point integration

rigid-body motions may create zero strain energy. Such displacement modes are termed *spurious zero-energy modes*.

In order to illustrate such spurious zero-energy modes, we next consider the four-node isoparametric element used in two-dimensional solid mechanics. A 1×1 point integration, i.e. a reduced integration, is adopted; cf. Figure 20.4. In this case, it follows that the element stiffness matrix $\mathbf{K}^e$ takes the form

$$\mathbf{K}^e = \alpha(\mathbf{B}^{eT}\mathbf{D}\mathbf{B}^e t)_{\xi=\eta=0} \tag{20.20}$$

where the coefficient α depends on the weight and the determinant of the Jacobian matrix. Moreover, the position of the Gauss point is given by $\xi = \eta = 0$; cf. Table 20.1 and Figure 20.4. As usual, we have $\varepsilon = \mathbf{B}^e\mathbf{a}^e$, i.e.

$$\mathbf{K}^e\mathbf{a}^e = \alpha(\mathbf{B}^{eT}\mathbf{D}\varepsilon t)_{\xi=\eta=0} \tag{20.21}$$

Consider the four-node isoparametric element when it takes the form of a rectangle; see Figure 20.6. Suppose that the nodal points of this element are displaced so that

$$\mathbf{a}^{eT} = \begin{bmatrix} -C & 0 & C & 0 & -C & 0 & C & 0 \end{bmatrix} \tag{20.22}$$

where C is an arbitrary constant. The displacements given by (20.22) are illustrated in Figure 20.6(a). From (15.5) it follows that

$$u_x = (-N_1^e + N_2^e - N_3^e + N_4^e)C; \quad u_y = 0$$

which, with (19.6), reduce to

$$u_x = -C\xi\eta; \quad u_y = 0 \tag{20.23}$$

In order to derive the strains, the Jacobian matrix $\mathbf{J}$ has to be derived. Referring to Figure 19.14 and (19.60), we have

$$\mathbf{J} = \begin{bmatrix} \frac{1}{2}(x_2 - x_1) & \frac{1}{2}(x_3 - x_2) \\ \frac{1}{2}(y_2 - y_1) & \frac{1}{2}(y_3 - y_2) \end{bmatrix}$$

In the present case, the element boundaries are parallel to the coordinate axis, i.e. $x_3 = x_2$ and $y_2 = y_1$, which leads to

$$\mathbf{J} = \begin{bmatrix} \frac{1}{2}(x_2 - x_1) & 0 \\ 0 & \frac{1}{2}(y_3 - y_1) \end{bmatrix};$$

i.e. $(\mathbf{J}^{\mathrm{T}})^{-1} = \dfrac{2}{(x_2 - x_1)(y_3 - y_1)} \begin{bmatrix} y_3 - y_1 & 0 \\ 0 & x_2 - x_1 \end{bmatrix}$ (20.24)

As in (19.59) we have

$$\begin{bmatrix} \dfrac{\partial u_x}{\partial x} \\ \dfrac{\partial u_x}{\partial y} \end{bmatrix} = (\mathbf{J}^{\mathrm{T}})^{-1} \begin{bmatrix} \dfrac{\partial u_x}{\partial \xi} \\ \dfrac{\partial u_x}{\partial \eta} \end{bmatrix}$$

Use of (20.23) and (20.24) then gives

$$\frac{\partial u_x}{\partial x} = -\frac{2}{x_2 - x_1} C\eta; \quad \frac{\partial u_x}{\partial y} = -\frac{2}{y_3 - y_1} C\xi$$

and as $u_y = 0$, we find that

$$\varepsilon_{xx} = -\frac{2}{x_2 - x_1} C\eta; \quad \varepsilon_{yy} = 0; \quad \gamma_{xy} = -\frac{2}{y_3 - y_1} C\xi \tag{20.25}$$

i.e.

$$\varepsilon_{\xi = \eta = 0} = \mathbf{0} \tag{20.26}$$

This result may, in fact, be concluded directly by investigating Figure 20.6(a). Use of (20.26) in (20.21) yields

$$\mathbf{K}^{\mathrm{e}}\mathbf{a}^{\mathrm{e}} = \mathbf{0}; \quad \text{i.e. } U = \tfrac{1}{2}\mathbf{a}^{\mathrm{eT}}\mathbf{K}^{\mathrm{e}}\mathbf{a}^{\mathrm{e}} = 0 \tag{20.27}$$

It appears that the displacements given by (20.22) and illustrated in Figure 20.6(a) result in a spurious zero-energy mode. Another such mode is illustrated in Figure 20.6(b).

The existence of spurious zero-energy modes should, of course, be avoided in an FE model, and we also observe that these zero-energy modes emerge as a result of reduced integration. Moreover, as the body provides no resistance at all to spurious zero-energy modes, the effect of reduced integration is to soften the stiffness of the FE model. However, we have previously shown that with an exact integration of the stiffness matrix, a model that is too stiff is created, and in this respect reduced integration may be beneficial to the accuracy of the FE solution. In light of these facts, we may draw the following conclusions:

> Reduced integration may result in spurious zero-energy modes that destroy the FE solution. If spurious zero-energy modes are not created, reduced integration may increase the accuracy of the FE solution, since it tends to soften the stiffness of the model.

Having considered reduced integration of one four-node isoparametric element, we shall now turn to the identification of spurious zero-energy modes for an FE mesh of arbitrary elements and with an arbitrary number of Gauss points. It turns out that it is relatively easy to establish an expression for the number of spurious zero-energy modes that *at least* must be present, but it is more difficult to identify the exact number of spurious zero-energy modes. We shall therefore concentrate on the former case. For this purpose we assume that a total of n_{int} integration points is adopted for the *entire* body (this notation is used here to distinguish n_{int} from n, the number of nodal points). With (20.8), the stiffness matrix $\mathbf{K}$ may be written as

$$\mathbf{K} = \sum_{i=1}^{n_{\text{int}}} \alpha_i \mathbf{B}_i^{\mathrm{T}} \mathbf{D}_i \mathbf{B}_i \tag{20.28}$$

where $\mathbf{B}_i$ and $\mathbf{D}_i$ indicate that $\mathbf{B}$ and $\mathbf{D}$ are evaluated at Gauss point i. Moreover, the parameters α_i depend on the Gauss weight and det $\mathbf{J}_i$ which are both positive, i.e.

$$\alpha_i > 0 \tag{20.29}$$

From (20.28), the elastic strain energy U given by (20.18) becomes

$$U = \tfrac{1}{2}\mathbf{a}^{\mathrm{T}}\mathbf{K}\mathbf{a} = \tfrac{1}{2} \sum_{i=1}^{n_{\text{int}}} \alpha_i \mathbf{a}^{\mathrm{T}} \mathbf{B}_i^{\mathrm{T}} \mathbf{D}_i \mathbf{B}_i \mathbf{a} = \tfrac{1}{2} \sum_{i=1}^{n_{\text{int}}} \alpha_i \boldsymbol{\varepsilon}_i^{\mathrm{T}} \mathbf{D}_i \boldsymbol{\varepsilon}_i \tag{20.30}$$

where

$$\boldsymbol{\varepsilon}_i = \mathbf{B}_i \mathbf{a} \tag{20.31}$$

By definition, zero-energy modes exist when the strain energy $U = 0$ for $\mathbf{a} \neq \mathbf{0}$ and we want to identify such modes. As the constitutive matrix $\mathbf{D}$ is positive definite (cf. (13.11)), it follows from (20.29) and (20.30) that $U = 0$ if, and only if, $\boldsymbol{\varepsilon}_i = \mathbf{0}$ for $i = 1, \ldots, n_{\text{int}}$. According to (20.31), this requires a non-trivial solution of the following homogeneous system of equations:

$$\begin{bmatrix} \mathbf{B}_1 \\ \mathbf{B}_2 \\ \vdots \\ \mathbf{B}_{n_{\text{int}}} \end{bmatrix} \mathbf{a} = \mathbf{0} \tag{20.32}$$

In the following let us consider the analysis of two-dimensional elasticity. The results obtained are, however, easily carried over to one- and three-dimensional elasticity as well as to heat flow. For two-dimensional elasticity, and with n denoting the number of nodal points for the entire body, the dimensions of $\mathbf{B}_i$ and $\mathbf{a}$ are given by

$$\mathbf{B}_i = 3 \times 2n; \quad \mathbf{a} = 2n \times 1 \tag{20.33}$$

(cf. (15.13) and (15.14)). It appears that the dimensions of the matrices entering (20.32) are given as shown below:

$$\begin{bmatrix} \mathbf{B}_1 \\ \mathbf{B}_2 \\ \vdots \\ \mathbf{B}_{n_{int}} \end{bmatrix} \mathbf{a} = 0 \tag{20.34}$$

$$(3n_{int} \times 2n)(2n \times 1) = 3n_{int} \times 1$$

That is, in this homogeneous system of equations, we have $2n$ unknowns and $3n_{int}$ equations. It is evident that if $2n > 3n_{int}$, then there are more unknowns than equations and we have at least $2n - 3n_{int}$ non-trivial solutions of $\mathbf{a}$ (cf. (2.65)). Therefore, we have

$$\text{at least} \quad 2n - 3n_{int} \quad \text{zero-energy modes} \tag{20.35}$$

In the case of a four-node isoparametric element with one point integration, we have $n = 4$ and $n_{int} = 1$, i.e. we have at least $2 \times 4 - 3 \times 1 = 5$ zero-energy modes.

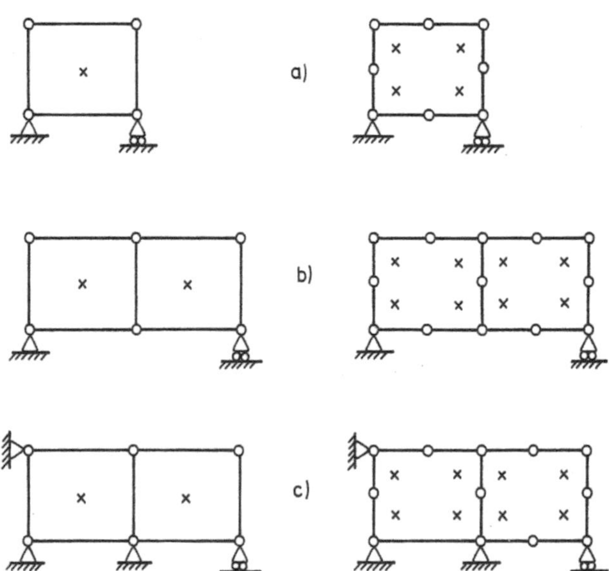

Figure 20.7   Illustration of identification of spurious zero-energy modes

Table 20.2   Spurious modes for the structures in Figure 20.7

	Spurious modes for four-node element	Spurious modes for eight-node element
Structure (a)	2	1
Structure (b)	3	none
Structure (c)	none	none

Table 20.3   Gauss integration techniques used for two-dimensional elements

	Element	Reliable integration order	Reduced integration used in practice (with spurious zero-energy mode(s))
Four-node		2×2	–
Four-node distorted		2×2	–
Eight-node		3×3	2×2
Eight-node distorted		3×3	2×2
Nine-node		3×3	2×2
Nine-node distorted		3×3	2×2

Of these, we have already identified the three modes corresponding to rigid-body motions (cf. Figure 20.5), as well as the two spurious zero-energy modes (cf. Figure 20.6).

In practice, the structure is supported at some nodal points, i.e. some of the components of the **a**-vector are prescribed. Let n_{pre} denote the number of prescribed displacements. Then we conclude from (20.35) that we have

$$\boxed{\text{at least} \quad (2n - n_{\text{pre}}) - 3n_{\text{int}} \quad \text{zero-energy modes}} \tag{20.36}$$

In practice, the number n_{pre} is sufficiently large to prevent rigid-body motions (i.e. $n_{\text{pre}} \geq 3$) and the zero-energy modes given by (20.36) are therefore spurious modes. We recall that (20.36) refers to two-dimensional solid mechanics and that similar expressions may be derived for other situations.

In order to illustrate the use of (20.36), the structures shown in Figure 20.7 are investigated. All these structures are supported so that rigid-body motions are prevented, i.e. (20.36) provides the number of spurious zero-energy modes. As an example, consider the structure modelled by four-node elements with one point integration shown in Figure 20.7(b). We have $n = 6$; $n_{\text{pre}} = 3$; $n_{\text{int}} = 2$; that is, (20.36) gives $(2 \times 6 - 3) - 3 \times 2 = 3$ spurious zero-energy modes. If we proceed in this manner with all the structures shown in Figure 20.7, the results shown in Table 20.2 are obtained.

It appears that expression (20.36) may be used to determine quickly whether spurious zero-energy modes are present, and in this case reduced integration is evidently to be avoided.

20.4.1 *Suggested order of Gauss integration of isoparametric elements*

From the discussion above, we may conclude that full integration is always a reliable technique by which pitfalls are avoided. However, reduced integration may improve the accuracy, provided that spurious zero-energy modes are not present. We also observe that it is advantageous to distort the isoparametric elements as little as possible. Table 20.3, taken from Bathe (1982, p. 286), summarizes some of these general conclusions.

In conclusion, the discussion above suggests that choice of the order of integration is far from being trivial and, in effect, many special techniques have been developed for efficient reduced integration techniques. Such issues are discussed at length by Zienkiewicz and Taylor (1989), Hughes (1987) as well as by Bathe (1982).

References

Adini, A. and Clough, R. W. (1961) 'Analysis of plate bending by the finite element method', Report to National Science Foundation, USA, G.7337.

Barsoum, R. S. (1976) 'On the use of isoparametric finite elements in linear fracture mechanics', *International Journal for Numerical Methods in Engineering*, **10**, 25–37.

Bathe, K.-J. (1982) *Finite Element Procedures in Engineering Analysis*, Prentice Hall: Englewood Cliffs, NJ.

Bear, J. (1979) *Hydraulics of Groundwater*, McGraw-Hill: New York.

Becker, E. B., Carey, G. F. and Oden, J. T. (1981) *Finite Elements. An Introduction*, vol. 1, Prentice Hall: Englewood Cliffs, NJ.

Boresi, A. P., Sidebottom, O. M., Seely, F. B. and Smith, J. O. (1978) *Advanced Mechanics of Materials*, 3rd edn, Wiley: New York.

Carslaw, H. S. and Jaeger, J. C. (1959) *Conduction of Heat in Solids*, 2nd edn, Clarendon Press: Oxford.

Cook, R. D., Malkus, D. S. and Plesha, M. E. (1989) *Concepts and Applications of Finite Element Analysis*, 3rd edn, Wiley: New York.

Crandall, S. H. (1956) *Engineering Analysis. A Survey of Numerical Procedures*, McGraw-Hill: New York.

Crisfield, M. A. (1986) *Finite Elements and Solutions Procedures for Structural Analysis*, vol. 1, *Linear Analysis*, Pineridge Press.

Dahlquist, G. and Björck, A. (1974) *Numerical Analysis*, Prentice Hall: Englewood Cliffs, NJ.

Ergatoudis, J. G., Irons, B. M. and Zienkiewicz, O. C. (1968) 'Curved isoparametric quadrilateral elements for finite element analysis', *International Journal of Solids and Structures*, **4**, 31–42.

Finlayson, B. A. and Scriven, L. E. (1966) 'The method of weighted residuals – a review', *Applied Mechanics Reviews*, **19**, 735–48.

Finlayson, B. A. (1972) *The Method of Weighted Residuals and Variational Principles*, Academic Press: New York.

Fröberg, C.-F. (1965) *Introduction to Numerical Analysis*, Addison-Wesley: Reading, MA.

Fung, Y. C. (1965) *Foundations of Solid Mechanics*, Prentice Hall: Englewood Cliffs, NJ.

Galerkin, B. G. (1915) 'Series solution of some problems of elastic equilibrium of rods and plates' (in Russian), *Vestn. Inzh. Tech.*, **19**, 897–908.

Gallagher, R. H. (1975) *Finite Element Analysis. Fundamentals*, Prentice Hall: Englewood Cliffs, NJ.

Henshell, R. D. and Shaw, K. G. (1975) 'Crack tip elements are unnecessary', *International Journal for Numerical Methods in Engineering*', **9**, 495–507.

Hildebrand, F. B. (1965) *Methods of Applied Mathematics*, 2nd edn, Prentice Hall: Englewood Cliffs, NJ.

Hughes, T. J. R. (1987) *The Finite Element Method. Linear Static and Dynamic Finite Element Analysis*, Prentice Hall: Englewood Cliffs, NJ.

Hughes, W. F. and Brighton, J. A. (1967) *Theory and Problems of Fluid Dynamics*, Schaum.

Irons, B. M. (1966a) 'Engineering application of numerical integration in stiffness method', *Journal of AIAA*, **14**, 2035–7.

Irons, B. M. (1966b) 'Numerical integration applied to finite element methods', *Conference on the Use of Digital Computers in Structural Engineering, University of Newcastle*.

Johnson, C. (1987) *Numerical Solutions of Partial Differential Equations by the Finite Element Method*, Student litteratur: Lund.

Kaplan, W. (1981) *Advanced Mathematics for Engineers*, Addison-Wesley: Reading, MA.

Kirchhoff, G. (1850) 'Über das Gleichgwicht und die Bewegung einer elastichen Scheibe', *Journal für die reine und angewandte Mathematik (Crelle)*, **40**, 51–8.

Kollbrunner, C. F. and Hajdin, N. (1972) *Dünnwangidge stabe*, Springer Verlag: Berlin.

Kreyszig, E. (1979) *Advanced Engineering Analysis*, 4th edn, Wiley: New York.

Lekhnitskii, S. G. (1981) *Theory of Elasticity of an Anisotropic Body*, Mir: Moscow.

Love, A. E. H. (1944) *A Treatise on the Mathematical Theory of Elasticity*, 4th edn, Dover: New York.

Malvern, L. E. (1969) *Introduction to the Mechanics of a Continuous Medium*, Prentice Hall: Englewood Cliffs, NJ.

Melosh, R. J. (1963) 'Basis of derivation of matrices for the direct stiffness method', *Journal of AIAA*, **1**, 1631–7.

Mindlin, R. D. (1951) 'Influence of rotary inertia and shear on flexural motion of isotropic elastic plates', *Journal of Applied Mechanics*, **18**, 31–8.

Morley, L. S. D. (1971) 'The constant bending-moment plate bending element', *Journal of Strain Analysis*, **6**, no. 1.

Owen, D. R. J. and Hinton, E. (1980) *Finite Elements in Plasticity*, Pineridge Press.

Press, W. H., Flannery, B. P., Teukolsky, S. A. and Vetterling, W. T. (1986) *Numerical Recipes*, Cambridge: Cambridge University Press.

Przemieniecki, J. S. (1968) *Theory of Matrix Structural Analysis*, McGraw-Hill: New York.

Reissner, E. (1945) 'The effect of transverse shear deformations on the bending of elastic plates', *Journal of Applied Mechanics*, **12**, A69–77.

Roark, R. J. (1975) *Formulas for Stress and Strain*, 5th edn, McGraw-Hill: New York.

Segerlind, L. J. (1976) *Applied Finite Element Analysis*, Wiley: New York.

Sokolnikoff, I. S. and Redheffer, R. M. (1958) *Mathematics and Physics of Modern Engineering*, McGraw-Hill: New York.

Sokolnikoff, T. S. (1946) *Mathematical Theory of Elasticity*, McGraw-Hill: New York.

Spencer, A. J. M. (1980) *Continuum Mechanics*, Longman: Harlow.

Stasa, F. L. (1985) *Applied Finite Element Analysis for Engineers*, CBS International Editions.

Strang, G. (1980) *Linear Algebra and its Applications*, 2nd edn, Academic Press: New York.

Strang, G. and Fix, G. J. (1973) *An Analysis of the Finite Element Method*, Prentice Hall: Englewood Cliffs, NJ.

Taig, I. C. (1961) 'Structural analysis by the matrix displacement method', English Electric Aviation Report no. So17.

Thelandersson, S. (1984) *Konstruktionsberäkningar med dator*, Studentlitteratur: Lund.

Timoshenko, S. and Goodier, J. N. (1970) *Theory of Elasticity*, 3rd edn, McGraw-Hill: New York.

Timoshenko, S. P. and Gere, J. M. (1972) *Mechanics of Materials*, Van Nostrand: New York.

Timoshenko, S. P. and Woinowsky-Krieger, S. (1959) *Theory of Plates and Shells*, 2nd edn, McGraw-Hill: New York.

Turner, M. J., Clough, R. W., Martin, H. C. and Topp, L. P. (1956) 'Stiffness and deflection analysis of complex structures', *Journal of Aeronautical Sciences*, **23**, 805–23.

Zienkiewicz, O. C. (1970) 'The finite element method: From intuition to generality', *Applied Mechanics Review*, **23**, 249–56.

Zienkiewicz, O. C. (1977) *The Finite Element Method*, 3rd edn, McGraw-Hill: New York.

Zienkiewicz, O. C. (1983) 'The generalized finite element method – State of art and future directions', *Journal of Applied Mechanics*, **50**, 1210–17.

Zienkiewicz, O. C. and Taylor, R. L. (1989) *The Finite Element Method*, 4th edn, vol. 1, McGraw-Hill: New York.

Index